U

Dynamical Systems

Dynamical Systems

Differential equations, maps and chaotic behaviour

D. K. Arrowsmith BSc, PhD
Senior Lecturer
School of Mathematical Sciences
Queen Mary and Westfield College
University of London

and

C. M. Place BSc, PhD
Lecturer
(formerly Department of Mathematics
Westfield College
University of London)

CHAPMAN & HALL
London · Glasgow · New York · Tokyo · Melbourne · Madras

Published by Chapman & Hall, 2-6 Boundary Row, London SE1 8HN

Chapman & Hall, 2-6 Boundary Row, London SE1 8HN, UK

Blackie Academic & Professional, Wester Cleddens Road, Bishopbriggs, Glasgow G64 2NZ, UK

Chapman & Hall GmbH, Pappelallee 3, 69469 Weinheim, Germany

Chapman & Hall USA, 115 Fifth Avenue, New York, NY 10003, USA

Chapman & Hall Japan, ITP - Japan, Kyowa Building, 3F, 2-2-1 Hirakawacho, Chiyoda-ku, Tokyo 102, Japan

Chapman & Hall Australia, 102 Dodds Street, South Melbourne, Victoria 3205, Australia

Chapman & Hall India, R. Seshadri, 32 Second Main Road, CIT East, Madras 600 035, India

First edition 1992
Reprinted 1995

Typeset in 10/12pt Times by Thomson Press(India) Ltd, New Dehli
Printed in Great Britain by T.J. Press, Padstow, Cornwall

ISBN 0 412 39070 1 (HB) 0 412 39080 9 (PB)

A Catalogue record for this book is available from the British Library

Library of Congress Cataloging-in-Publication Data available

∞ Printed on acid-free text paper, manufactured in accordance with ANSI/NISO Z39.48-1992 and ANSI/NISO Z39.48-1984 (Permanence of Paper).

Contents

Preface ix

1 Introduction 1
1.1 Preliminary ideas 1
1.1.1 Existence and uniqueness 1
1.1.2 Geometrical representation 3
1.2 Autonomous equations 6
1.2.1 Solution curves and the phase portrait 6
1.2.2 Phase portraits and dynamics 11
1.3 Autonomous systems in the plane 12
1.4 Construction of phase portraits in the plane 17
1.4.1 Use of calculus 17
1.4.2 Isoclines 20
1.5 Flows and evolution 23
Exercises 27

2 Linear systems 35
2.1 Linear changes of variable 35
2.2 Similarity types for 2 × 2 real matrices 38
2.3 Phase portraits for canonical systems in the plane 43
2.3.1 Simple canonical systems 43
2.3.2 Non-simple canonical systems 46
2.4 Classification of simple linear phase portraits in the plane 48
2.4.1 Phase portrait of a simple linear system 48
2.4.2 Types of canonical system and qualitative equivalence 50
2.4.3 Classification of linear systems 52
2.5 The evolution operator 52
2.6 Affine systems 55
2.7 Linear systems of dimension greater than two 57
2.7.1 Three-dimensional systems 57

2.7.2 Four-dimensional systems 61
2.7.3 n-Dimensional systems 62
Exercises 63

3 Non-linear systems in the plane 71
3.1 Local and global behaviour 71
3.2 Linearization at a fixed point 74
3.3 The linearization theorem 77
3.4 Non-simple fixed points 81
3.5 Stability of fixed points 84
3.6 Ordinary points and global behaviour 93
3.6.1 Ordinary points 93
3.6.2 Global phase portraits 95
3.7 First integrals 96
3.8 Limit points and limit cycles 101
3.9 Poincaré–Bendixson theory 105
Exercises 110

4 Flows on non-planar phase spaces 120
4.1 Fixed points 120
4.1.1 Hyperbolic fixed points 120
4.1.2 Non-hyperbolic fixed points 125
4.2 Closed orbits 129
4.2.1 Poincaré maps and hyperbolic closed orbits 129
4.2.2 Topological classification of hyperbolic closed orbits 132
4.2.3 Periodic orbits and quasi-periodic motion 136
4.3 Attracting sets and attractors 138
4.3.1 Trapping regions for Poincaré maps 140
4.3.2 Saddle points in attracting sets 143
4.4 Further integrals 147
4.4.1 Hamilton's equations 148
4.4.2 Poincaré maps of Hamiltonian flows 152
Exercises 155

5 Applications I: planar phase spaces 162
5.1 Linear models 162
5.1.1 A mechanical oscillator 162
5.1.2 Electrical circuits 167
5.1.3 Economics 170
5.1.4 Coupled oscillators 172
5.2 Affine models 175
5.2.1 The forced harmonic oscillator 176
5.2.2 Resonance 177
5.3 Non-linear models 179
5.3.1 Competing species 180

5.3.2 Volterra–Lotka equations 183
5.3.3 The Holling–Tanner model 185
5.4 Relaxation oscillations 188
5.4.1 Van der Pol oscillator 188
5.4.2 Jumps and regularization 192
5.5 Piecewise modelling 195
5.5.1 The jump assumption and piecewise models 196
5.5.2 A limit cycle from linear equations 198
Exercises 202

6 Applications II: non-planar phase spaces, families of systems and bifurcations 212
6.1 The Zeeman models of heartbeat and nerve impulse 212
6.2 A model of animal conflict 218
6.3 Families of differential equations and bifurcations 223
6.3.1 Introductory remarks 223
6.3.2 Saddle–node bifurcation 226
6.3.3 Hopf bifurcation 228
6.4 A mathematical model of tumour growth 232
6.4.1 Construction of the model 232
6.4.2 An analysis of the dynamics 233
6.5 Some bifurcations in families of one-dimensional maps 240
6.5.1 The fold bifurcation 240
6.5.2 The flip bifurcation 242
6.5.3 The logistic map 245
6.6 Some bifurcations in families of two-dimensional maps 251
6.6.1 The child on a swing 251
6.6.2 The Duffing equation 254
6.7 Area-preserving maps, homoclinic tangles and strange attractors 259
6.7.1 Introductory remarks 259
6.7.2 Periodic orbits and island chains 261
6.7.3 Chaotic orbits and homoclinic tangles 264
6.7.4 Strange attracting sets 267
6.8 Symbolic dynamics 271
6.9 New directions 279
6.9.1 Introductory remarks 279
6.9.2 Iterated function schemes 280
6.9.3 Cellular automata 284
Exercises 288

Bibliography 303

Hints to Exercises 306

Index 326

Preface

In recent years there has been unprecedented popular interest in the chaotic behaviour of discrete dynamical systems. The ease with which a modest microcomputer can produce graphics of extraordinary complexity has fired the interest of mathematically-minded people from pupils in schools to postgraduate students. At undergraduate level, there is a need to give a basic account of the computed complexity within a recognized framework of mathematical theory. In producing this replacement for *Ordinary Differential Equations* (*ODE*) we have responded to this need by extending our treatment of the qualitative behaviour of differential equations.

This book is aimed at second and third year undergraduate students who have completed first courses in Calculus of Several Variables and Linear Algebra. Our approach is to use examples to illustrate the significance of the results presented. The text is supported by a mix of manageable and challenging exercises that give readers the opportunity to both consolidate and develop the ideas they encounter. As in *ODE*, we wish to highlight the significance of important theorems, to show how they are used and to stimulate interest in a deeper understanding of them.

We have retained our earlier introduction and discussion of linear systems (Chapters 1 and 2). Our treatment of non-linear differential equations has been extended to include Poincaré maps and phase spaces of dimension greater than two (Chapters 3 and 4). Applications involving planar phase spaces (covered in Chapter 4 of *ODE*) appear in Chapter 5. Problems involving non-planar phase spaces and families of systems are considered in Chapter 6, where elementary bifurcation theory is introduced and its application to chaotic behaviour is examined. Although ordinary differential equations remain the driving force behind the book, a substantial part of the new material concerns discrete dynamical systems and the title *Ordinary Differential Equations* is no longer appropriate. We have therefore chosen a new title for the extended text that clarifies its connection with the broader field of dynamical systems.

In addition to Professors Brown and Eastham, and Drs Knowles and Smith, who read and commented on the manuscript for *ODE*, we would like to thank those readers who kindly drew our attention to some failings of that book. In relation to the new material appearing in this text we must thank Dr A. Lansbury for her help with the Duffing problem and Mrs G. A. Place for her assistance with the continuity of the assembled manuscript. We are also grateful to Cambridge University Press, the *Quarterly Journal of Applied Mathematics* and the American Institute of Physics for allowing us to use diagrams from some of their publications. Once again we must acknowledge the forbearance of our families and one of us (CMP) would like to thank the Brayshay Foundation for its financial support.

1

Introduction

In this chapter we illustrate the qualitative approach to differential equations and introduce some key ideas such as phase portraits and qualitative equivalence.

1.1 PRELIMINARY IDEAS

1.1.1 Existence and uniqueness

Definition 1.1.1

Let $X(t, x)$ be a real-valued function of the real variables t and x, with domain $D \subseteq \mathbb{R}^2$. A function $x(t)$, with t in some open interval $I \subseteq \mathbb{R}$, which satisfies

$$\dot{x}(t) = \frac{\mathrm{d}x}{\mathrm{d}t} = X(t, x(t)) \tag{1.1}$$

is said to be a **solution** of the differential equation (1.1).

A necessary condition for $x(t)$ to be a solution is that $(t, x(t)) \in D$ for each $t \in I$; so that D limits the domain and range of $x(t)$. If $x(t)$, with domain I, is a solution to (1.1) then so is its restriction to any interval $J \subset I$. To prevent any confusion, we will always take I to be the largest interval for which $x(t)$ satisfies (1.1). Solutions with this property are called **maximal** solutions. Thus, unless otherwise stated, we will use the word 'solution' to mean 'maximal solution'. Consider the following examples of (1.1) and their solutions; we give

$$\dot{x} = X(t, x), \qquad D, \qquad x(t), \qquad I$$

in each case (C and C' are real numbers):

1. $\dot{x} = x - t$, $\quad \mathbb{R}^2$, $\quad 1 + t + Ce^t$, $\quad \mathbb{R}$;
2. $\dot{x} = x^2$, $\quad \mathbb{R}^2$, $\quad (C - t)^{-1}$, $\quad (-\infty, C)$
   $\quad 0$, $\quad \mathbb{R}$
   $\quad (C' - t)^{-1}$, $\quad (C', \infty)$;

3. $\dot{x} = -x/t$, $\{(t,x) \mid t \neq 0\}$, C/t, $(-\infty, 0)$
C'/t, $(0, \infty)$;

4. $\dot{x} = 2x^{1/2}$, $\{(t,x) \mid x \geqslant 0\}$, $\begin{cases} 0, & (-\infty, C) \\ (t-C)^2, & [C, \infty) \end{cases}$
0, $\mathbb{R}$;

5. $\dot{x} = 2xt$, $\mathbb{R}^2$, Ce^{t^2}, $\mathbb{R}$;
6. $\dot{x} = -x/\tanh t$, $\{(t,x) \mid t \neq 0\}$, $C/\sinh t$, $(-\infty, 0)$
$C'/\sinh t$, $(0, \infty)$.

The existence of solutions is determined by the properties of X. The following proposition is stated without proof (Petrovski, 1966).

Proposition 1.1.1
If X is continuous in an open domain, $D' \subseteq D$, then given any pair $(t_0, x_0) \in D'$, there exists a solution $x(t)$, $t \in I$, of $\dot{x} = X(t,x)$ such that $t_0 \in I$ and $x(t_0) = x_0$.

For example, consider

$$\dot{x} = 2|x|^{1/2}, \tag{1.2}$$

where $D = \mathbb{R}^2$. Any pair (t_0, x_0) with $x_0 \geqslant 0$ is given by $(t_0, x(t_0))$ when $x(t)$ is the solution

$$x(t) = \begin{cases} 0, & t \in (-\infty, C) \\ (t-C)^2, & t \in [C, \infty) \end{cases} \tag{1.3}$$

and $C = t_0 - \sqrt{x_0}$. A solution can similarly be found for pairs (t_0, x_0) when $x_0 < 0$.

Observe that Proposition 1.1.1 does not exclude the possibility that $x(t_0) = x_0$ for more than one solution $x(t)$. For example, for (1.2) infinitely many solutions $x(t)$ satisfy $x(t_0) = 0$; namely every solution of the form (1.3) for which $C > t_0$ and solution $x(t) \equiv 0$.

The following proposition gives a sufficient condition for each pair in D' to occur in one and only one solution of (1.1).

Proposition 1.1.2
If X and $\partial X/\partial x$ are continuous in an open domain $D' \subseteq D$, then given any $(t_0, x_0) \in D'$ there exists a unique *solution $x(t)$ of $\dot{x} = X(t,x)$ such that $x(t_0) = x_0$.*

Notice that, while $X = 2|x|^{1/2}$ is continuous on $D(= \mathbb{R}^2)$, $\partial X/\partial x (= |x|^{-1/2}$ for $x > 0$ and $-|x|^{-1/2}$ for $x < 0)$ is continuous only on $D' = \{(t,x) \mid x \neq 0\}$; it is undefined for $x = 0$. We have already observed that the pair $(t_0, 0)$, $t_0 \in \mathbb{R}$ occurs in infinitely many solutions of $\dot{x} = 2|x|^{1/2}$.

On the other hand, $X(t,x) = x - t$ and $\partial X/\partial x = 1$ are continuous throughout the domain $D = \mathbb{R}^2$. Any (t_0, x_0) occurs in one and only one solution of $\dot{x} = x - t$; namely

$$x(t) = 1 + t + Ce^t \tag{1.4}$$

when $C = (x_0 - t_0 - 1)e^{-t_0}$.

Weaker sufficient conditions for existence and uniqueness do exist (Petrovski, 1966). However, Propositions 1.1.1 and 1.1.2 illustrate the kind of properties required for $X(t, x)$.

1.1.2 Geometrical representation

A solution $x(t)$ of $\dot{x} = X(t, x)$ is represented geometrically by the graph of $x(t)$. This graph defines a **solution curve** in the t, x-plane.

If X is continuous in D, then Proposition 1.1.1 implies that the solution curves fill the region D of the t, x-plane. This follows because each point in D must lie on at least one solution curve. The solutions of the differential equation are, therefore, represented by a **family of solution curves** in D (as illustrated in Figs 1.1–1.8).

If both X and $\partial X/\partial x$ are continuous in D then Proposition 1.1.2 implies that there is a unique solution curve passing through every point of D (as shown in Figs 1.1–1.6).

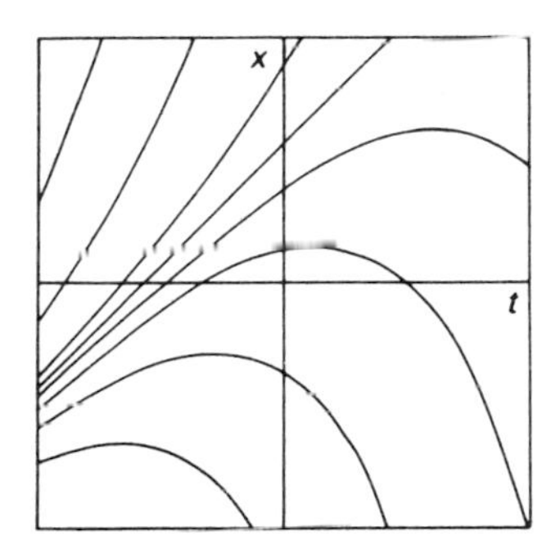

Fig. 1.1. $\dot{x} = x - t$.

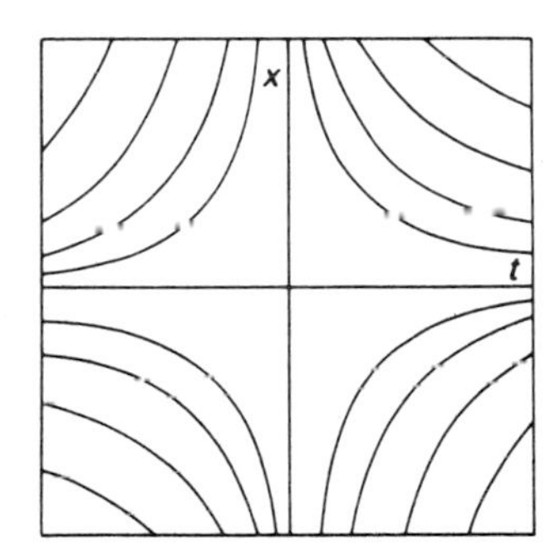

Fig. 1.2. $\dot{x} = -x/t, t \neq 0$.

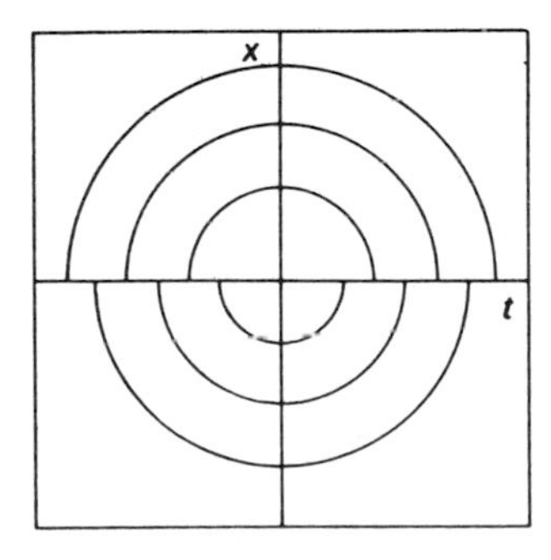

Fig. 1.3. $\dot{x} = -t/x$.

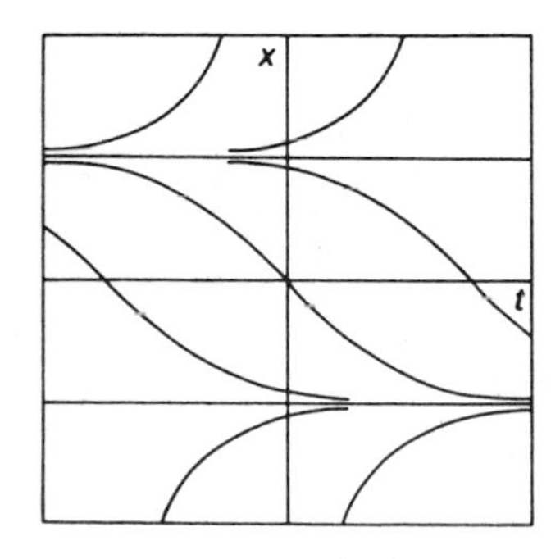

Fig. 1.4. $\dot{x} = \frac{1}{2}(x^2 - 1)$.

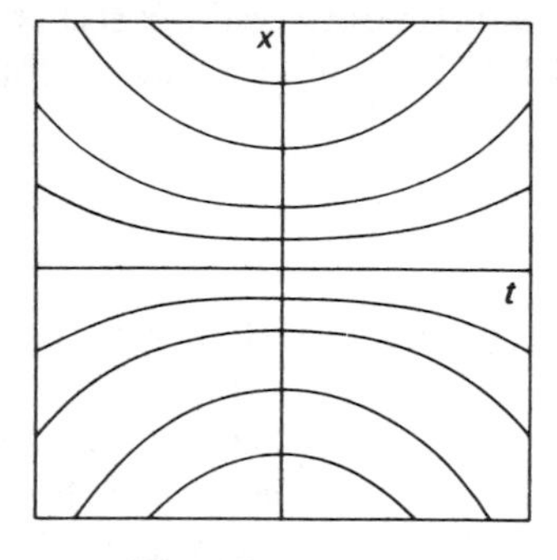

Fig. 1.5. $\dot{x} = 2xt$.

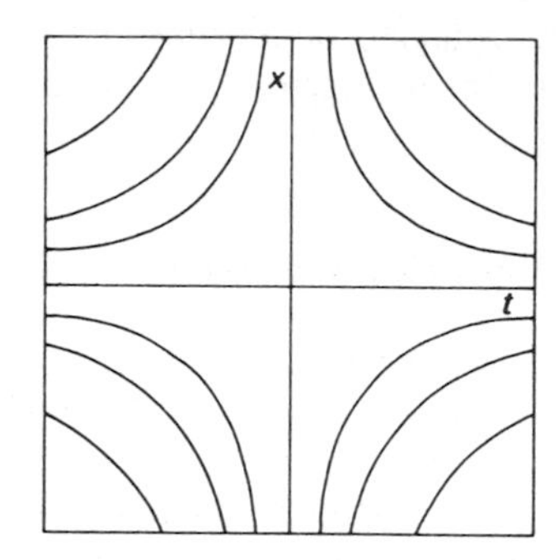

Fig. 1.6. $\dot{x} = -x/\tanh t, t \neq 0$.

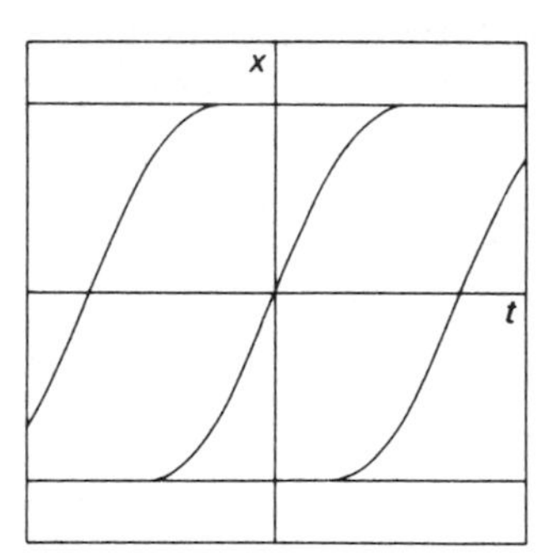

Fig. 1.7. $\dot{x} = \sqrt{(1 - x^2)}, |x| \leqslant 1$.

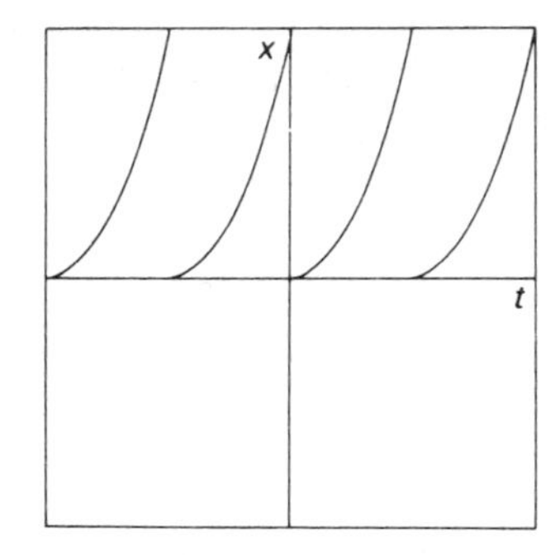

Fig. 1.8. $\dot{x} = 2x^{1/2}, x \geqslant 0$.

Observe that the families of solution curves in Figs 1.2 and 1.6 bear a marked resemblance to one another. Every solution curve in one figure has a counterpart in the other; they are similar in shape, have the same asymptotes, etc., but they are not identical curves. The relationship between these two families of solution curves is an example of what we call **qualitative equivalence** (also described in sections 1.3, 2.4 and 3.3). We say that the **qualitative behaviour** of the solution curves in Fig. 1.2 is the same as those in Fig. 1.6.

Accurate plots of the solution curves are not always necessary to obtain their qualitative behaviour; a sketch is often sufficient. We can sometimes obtain a sketch of the family of solution curves directly from the differential equation.

Example 1.1.1

Sketch the solution curves of the differential equation

$$\dot{x} = t + t/x \tag{1.5}$$

in the region D of the t, x-plane where $x \neq 0$.

Solution

We make the following observations.

1. The differential equation gives the slope of the solution curves at all points of the region D. Thus, in particular, the solution curves cross the curve $t + t/x = k$, a constant, with slope k. This curve is called the **isocline** of slope k. The set of isoclines, obtained by taking different real values for k, is family of hyperbolae

$$x = \frac{t}{k - t}, \tag{1.6}$$

with asymptotes $x = -1$ and $t = k$. A selection of these isoclines is shown in Fig. 1.9.

2. The sign of $\ddot{x}$ determines where in D the solution curves are concave and convex. If $\ddot{x} > 0$ (< 0) then $\dot{x}$ is increasing (decreasing) with t and the solution curve is said to be **convex (concave)**. The region D can therefore be divided into subsets on which the solution curves are either concave or convex separated by boundaries where $\ddot{x} = 0$. For (1.5) we find

$$\ddot{x} = x^{-3}(x + 1)(x - t)(x + t) \tag{1.7}$$

and D can be split up into regions $P(\ddot{x} > 0)$ and $N(\ddot{x} < 0)$ as shown in Fig. 1.10.

3. The isoclines are symmetrically placed relative to $t = 0$ and so there must also be symmetry of the solution curves. The function $X(t, x) = t + t/x$ satisfies $X(-t, x) = -X(t, x)$ and thus if $x(t)$ is a solution to $\dot{x} = X(t, x)$ then so is $x(-t)$ (cf. Exercise 1.5).

These three observations allow us to produce a sketch of the solution curves for $\dot{x} = t + t/x$ as in Fig. 1.11. Notice that both $X(t, x) = t + t/x$ and

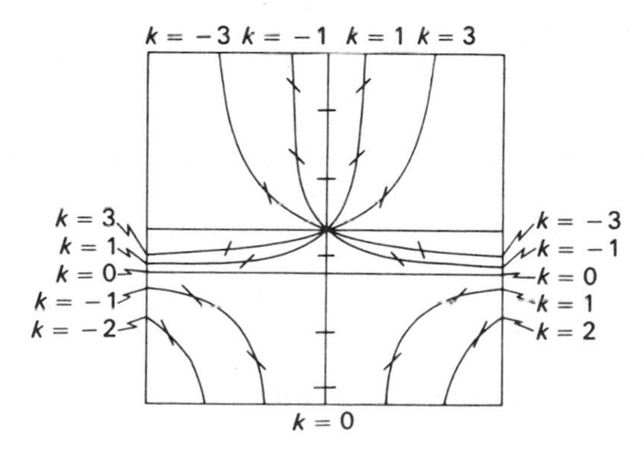

Fig. 1.9. Selected isoclines for the equation $\dot{x} = t + t/x$. The short line segments on the isoclines have slope k and indicate how the solution curves cross them.

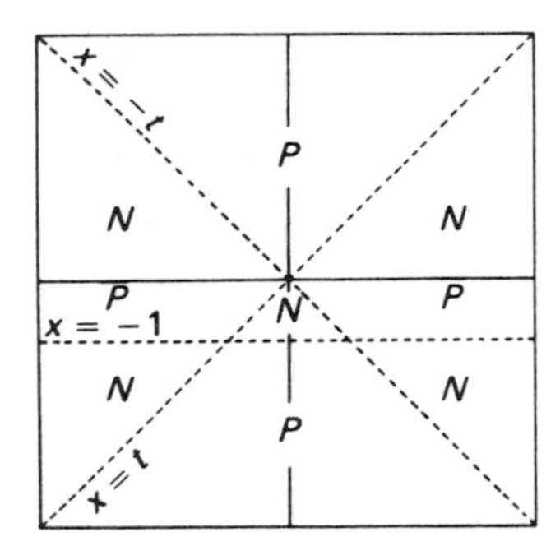

Fig. 1.10. Regions of convexity (P) and concavity (N) for solutions of $\dot{x} = t + t/x$.

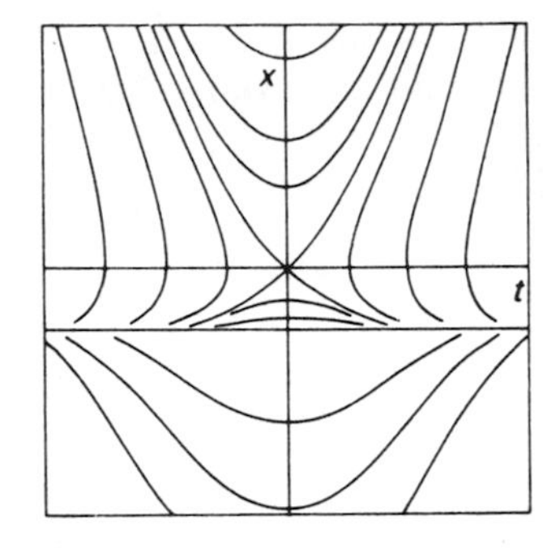

Fig. 1.11. The solution curves of the differential equation $\dot{x} = t + t/x$ in the t, x-plane.

$\partial X/\partial x = -t/x^2$ are continuous on $D = \{(t, x) | x \neq 0\}$, so there is a unique solution curve passing through each point of D. □

It is possible to find the solutions of

$$\dot{x} = t + t/x \tag{1.8}$$

by separation of the variable (defined in Exercise 1.2). We obtain the equation

$$x - \ln|x + 1| = \tfrac{1}{2}t^2 + C, \tag{1.9}$$

C a constant, for the family of solution curves as well as the solution $x(t) \equiv -1$. However, to sketch the solution curves from (1.9) is less straightforward than to use (1.8) itself.

The above discussion has introduced two important ideas:

1. that two different differential equations can have solutions that exhibit the same qualitative behaviour; and
2. that the qualitative behaviour of solutions is determined by $X(t, x)$.

We will now put these two ideas together and illustrate the qualitative approach to differential equations for the special case of equations of the form $\dot{x} = X(x)$. We shall see that such equations can be classified into qualitatively equivalent types.

1.2 AUTONOMOUS EQUATIONS

1.2.1 Solution curves and the phase portrait

A differential equation of the form

$$\dot{x} = X(x), \qquad x \in S \subseteq \mathbb{R}, \qquad (D = \mathbb{R} \times S) \tag{1.10}$$

is said to be **autonomous**, because $\dot{x}$ is determined by x alone and so the solutions are, as it were, self-governing.

The solutions of autonomous equations have the following important property. If $\xi(t)$ is a solution of (1.10) with domain I and range $\xi(I)$ then $\eta(t) = \xi(t + C)$, for any real C, is also a solution with the same range, but with domain $\{t \mid t + C \in I\}$. This follows because

$$\dot{\eta}(t) = \dot{\xi}(t + C) = X(\xi(t + C)) = X(\eta(t)). \tag{1.11}$$

The solution curve $x = \xi(t)$ is obtained by translating the solution curve $x = \eta(t)$ by the amount C in the positive t-direction.

Furthermore if there exists a unique solution curve passing through each point of strip $D' = \mathbb{R} \times \xi\ (I)$ then all solution curves on D' are translations of $x = \xi(t)$. The domain D is therefore divided into strips where the solution curves are all obtained by shifting a single curve in the t-direction (as shown in Figs 1.12–1.15). For example.

$$\dot{x} = x \tag{1.12}$$

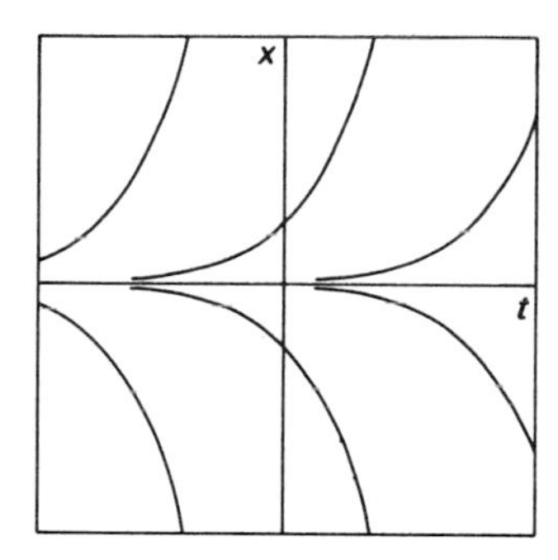

Fig. 1.12. $\dot{x} = x$: strips D' consist of the half-planes $x < 0$ and $x > 0$.

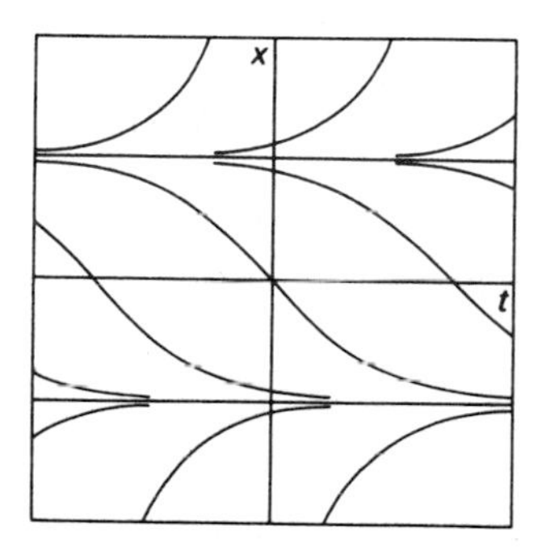

Fig. 1.13. $\dot{x} = \frac{1}{2}(x^2 - 1)$: strips $D' = \mathbb{R} \times \xi(I)$ with $\xi(I) = (-\infty, -1)$, $(-1, 1)$ and $(1, \infty)$.

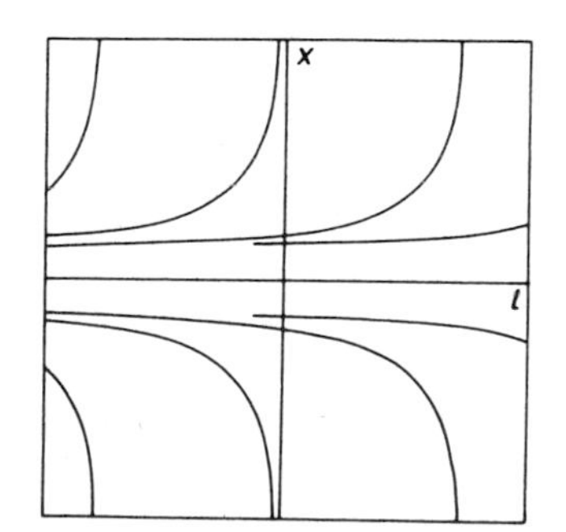

Fig. 1.14. Solution curves for $\dot{x} = x^3$.

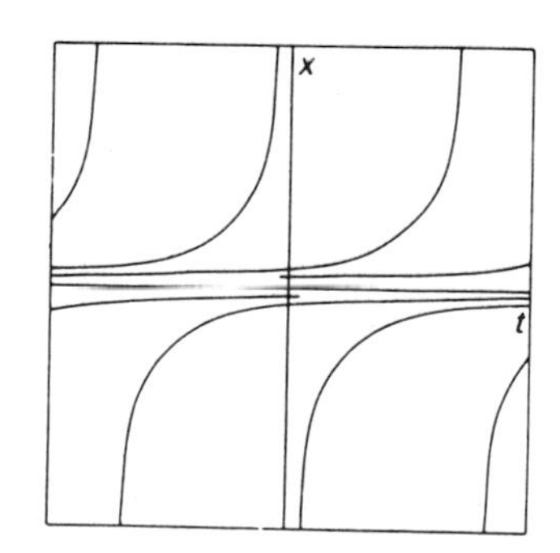

Fig. 1.15. Solution curves for $\dot{x} = x^2$.

has solutions:

$$\xi(t) = e^t, \qquad I = \mathbb{R}, \qquad \xi(I) = (0, \infty) \tag{1.13}$$

$$\xi(t) \equiv 0, \qquad I = \mathbb{R}, \qquad \xi(I) = \{0\} \tag{1.14}$$

$$\xi(t) = -e^t, \qquad I = \mathbb{R}, \qquad \xi(I) = (-\infty, 0). \tag{1.15}$$

All the solution curves in the strip D' defined by $x \in (0, \infty)$, $t \in \mathbb{R}$ are translations of e^t. Similarly, those in $D' = \{(t, x) | x \in (-\infty, 0),\ t \in \mathbb{R}\}$ are translations of $-e^t$.

For families of solution curves related by translations in t, the qualitative behaviour of the family of solutions is determined by that of any individual member. The qualitative behaviour of such a sample curve is determined by $X(x)$. When $X(x) \neq 0$, then the solution is either increasing or decreasing; when $X(c) = 0$ there is a solution $x(t) \equiv c$.

This information can be represented on the x-line rather than the t, x-plane. If $X(x) \neq 0$ for $x \in (a, b)$ then the interval is labelled with an arrow showing the sense in which x is changing. When $X(c) = 0$, the solution $x(t) \equiv c$ is represented by the point $x = c$. These solutions are called **fixed points** of the equation because x remains at c for all t. This geometrical representation of the qualitative behaviour of $\dot{x} = X(x)$ is called its **phase portrait**. Some examples of phase portraits are shown, in relation to X, in Figs 1.16–1.19. The corresponding families of solution curves are given in Figs 1.12–1.15.

If x is not stationary it must either be increasing or decreasing. Thus for a given finite number of fixed points there can only be a finite number of 'different' phase portraits. By 'different', we mean with distinct assignments of where x is increasing or decreasing. For example, consider a single fixed point $x = c$ (Fig. 1.20). For $x < c$, $X(x)$ must be either positive or negative and similarly for $x > c$. Hence, one of the four phase portraits shown must occur. This means that the qualitative behaviour of any autonomous differential equation with one fixed point must correspond to one of the phase

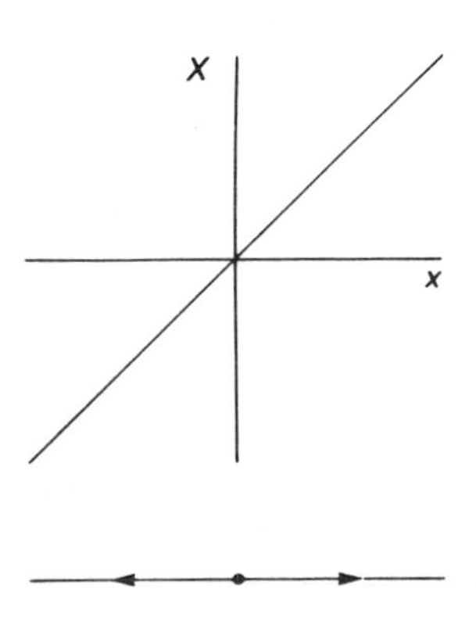

Fig. 1.16. $\dot{x} = x$, $x = 0$ is a fixed point.

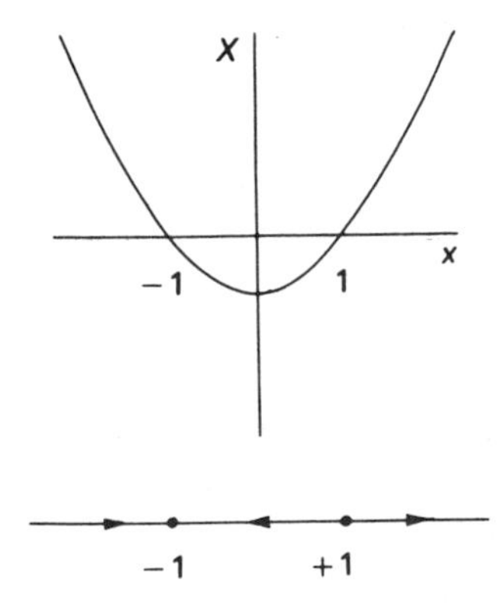

Fig. 1.17. $\dot{x} = \frac{1}{2}(x^2 - 1)$, $x = \pm 1$ are fixed points.

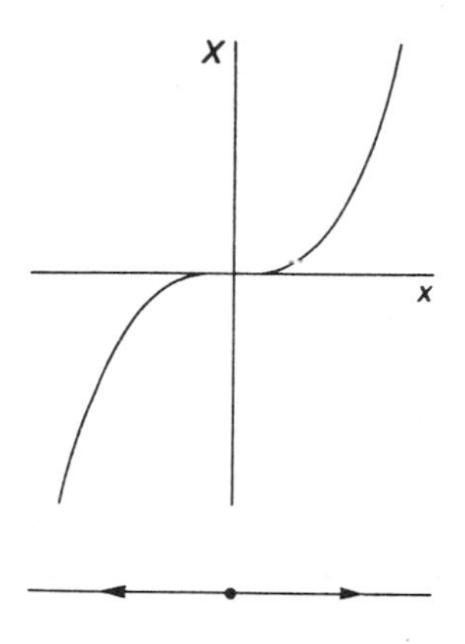

Fig. 1.18. $\dot{x} = x^3$, $x = 0$ is a fixed point.

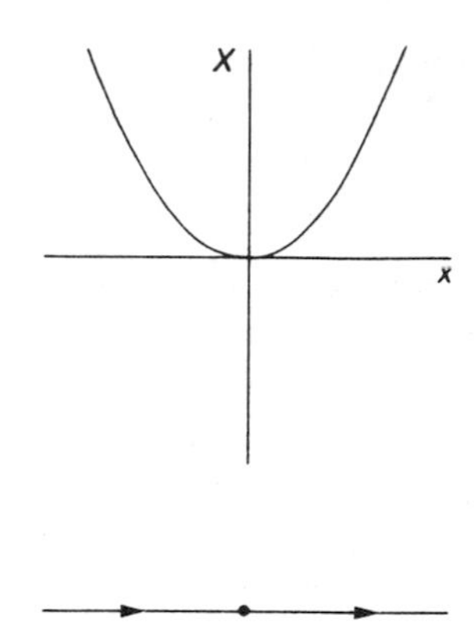

Fig. 1.19. $\dot{x} = x^2$, $x = 0$ is a fixed point.

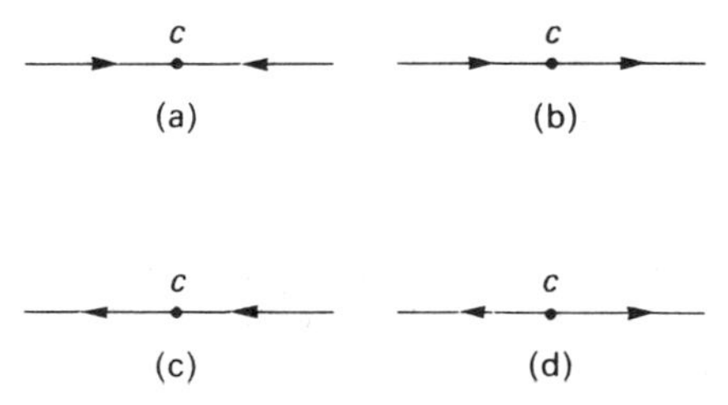

Fig. 1.20. The four possible phase portraits associated with a single fixed point. The fixed point is described as an **attractor** in (a), a **shunt** in (b) and (c) and a **repellor** in (d).

portraits in Fig. 1.20 for some value of c. For example, $\dot{x} = x$, $\dot{x} = x^3$, $\dot{x} = x - a$, $\dot{x} = (x - a)^3$, $\dot{x} = \sinh x$, $\dot{x} = \sinh(x - a)$ all correspond to Fig. 1.20(d) for $c = 0$ or a. Of course, two different equations, each having one fixed point, that correspond to the same phase portrait in Fig. 1.20 have the same qualitative behaviour. We say that two such differential equations are **qualitatively equivalent**.

Now observe that the argument leading to Fig. 1.20 holds equally well if the fixed point at $x = c$ is one of many in a phase portrait. In other words, the qualitative behaviour of x in the neighbourhood of any fixed point must be one of those illustrated in Fig. 1.20(a)–(d). We say that this behaviour determines the **nature of the fixed point** and use the terminology defined in the caption to Fig. 1.20 to describe this.

This is an important step because it implies that the phase portrait of any autonomous equation is determined completely by the nature of its fixed points. We can make the following definition.

Definition 1.2.1
Two differential equations of the form $\dot{x} = X(x)$ are **qualitatively equivalent** if they have the same number of fixed points, of the same nature, arranged in the same order along the phase line.

For example, $\dot{x} = (x+2)(x+1)$ is qualitatively equivalent to $\dot{x} = \frac{1}{2}(x^2 - 1)$. Both have two fixed points, one attractor and one repellor, with the attractor occurring at the smaller value of x. The equation $\dot{x} = -(x+2)(x+1)$ is not qualitatively equivalent to $\dot{x} = \frac{1}{2}(x^2 - 1)$ because the attractor and repellor occur in the reverse order. If we think of the equation $\dot{x} = 1$ as having a repellor at $x = -\infty$ and an attractor at $x = +\infty$, then $\dot{x} = 1$ and $\dot{x} = -1$ are not qualitatively equivalent because the latter has an attractor at $x = -\infty$ and a repellor at $x = +\infty$.

Example 1.2.1
Arrange the following differential equations in qualitatively equivalent groups:

(1) $\dot{x} = \cosh x$; (2) $\dot{x} = (x-a)^2$; (3) $\dot{x} = \sin x$;
(4) $\dot{x} = \cos x - 1$; (5) $\dot{x} = \cosh x - 1$; (6) $\dot{x} = \sin 2x$;
(7) $\dot{x} = e^x$; (8) $\dot{x} = \sinh^2(x-b)$.

Solution
Equations (1) and (7) have no fixed points; in both cases $X(x) > 0$ for all x. They both have the phase portrait shown in Fig. 1.21(a) and are therefore qualitatively equivalent.

Equation (2) has a single shunt at $x = a$ with $X(x) \geqslant 0$ for all x. Another equation with a single fixed point is (5); this has the same kind of shunt at $x = 0$. Equation (8) also has a single shunt, at $x = b$; again $X(x) \geqslant 0$ for all x. These equations form a second qualitatively equivalent group with the phase portrait shown in Fig. 1.21(b).

The remaining equations, (3), (4) and (6), all have infinitely many fixed points: (3) at $x = n\pi$, (4) at $x = 2n\pi$ and (6) at $x = n\pi/2$. However, equation

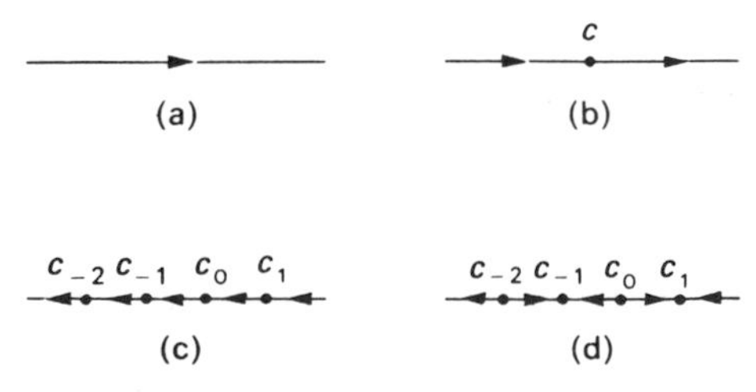

Fig. 1.21. Phase portraits for qualitatively equivalent groups in Example 1.2.1. In (b) $c = a$ for (2), 0 for (5) and b for (8). In (c) $c_n = 2n\pi$ for (4) and in (d) $c_n = n\pi$ for (3) and $n\pi/2$ for (6).

(4) has $X(x) \leqslant 0$ for all x, so that every fixed point is a shunt, whereas (3) and (6) have alternating attractors and repellors (shown in Fig. 1.21(c) and (d)). Thus only (3) and (6) are qualitatively equivalent. □

Example 1.2.1 draws attention to the fact that 'qualitative equivalence' is an equivalence relation on the set of all autonomous differential equations. We can consequently divide this set into disjoint classes according to their qualitative behaviour. However, if we only demand uniqueness of solutions then there are infinitely many distinct qualitative classes. This follows because arbitrarily many fixed points can occur.

If we place other limitations on X, there may only be a finite number of classes. For example, suppose we require X to be **linear**, i.e. $X(x) = ax$, a real. For any $a \neq 0$, there is only a single fixed point at $x = 0$. If $a > 0$, this is a repellor and if $a < 0$ it is an attractor. For the special case $a = 0$ every point of the phase portrait is a fixed point. Thus, $\{\dot{x} = X(x) | X(x) = ax\}$ can be divided into three classes according to qualitative behaviour.

For non-linear X, each isolated fixed point can only be one of the four possibilities shown in Fig. 1.20. Thus, although there are infinitely many distinct phase portraits, they contain, at most, four distinct types of fixed point. This limitation arises because $\dot{x} = X(x)$ is a differential equation for the single real variable x. There is consequently a one-dimensional phase portrait with x either increasing or decreasing at a non-stationary point.

1.2.2 Phase portraits and dynamics

In applications the differential equation $\dot{x} = X(x)$ models the time dependence of a property, x, of some physical system. We say that the **state** of the system is specified by x. For example, the equation

$$\dot{p} = ap; \qquad p, a > 0 \tag{1.16}$$

models the growth of the population, p, of an isolated species. Within this model, the state of the species at time t is given by the number of individuals, $p(t)$, living at that time. Another example is Newton's law of cooling. The temperature, T, of a body cooling in a draught of temperature τ is given by

$$\dot{T} = -a(T - \tau), \qquad a > 0. \tag{1.17}$$

Here the state of the body is taken to be determined by its temperature.

We can represent the state $x(t_0)$ of a model at any time t_0 by a point on the phase line of $\dot{x} = X(x)$. As time increases, the state changes and the phase point representing it moves along the line with velocity $\dot{x} = X(x)$. Thus, the dynamics of the physical system are represented by the motion of a phase point on the phase line.

The phase portrait records only the direction of the velocity of the phase point and therefore represents the dynamics in a qualitative way. Such

Fig. 1.22. Phase portraits for (a) $\dot{p} = ap$ and (b) $\dot{p} = p(a - bp)$, $p_e = a/b$. In both cases, we are interested only in the behaviour for non-negative populations ($p \geqslant 0$). The equation in (b) is known as the **logistic law** of population growth.

qualitative information can be helpful when constructing models. For example, consider the model (1.16) of an isolated population. Observe that $\dot{p} > 0$, for all $p > 0$, the phase portrait, in Fig. 1.22(a), shows that the population increases indefinitely. This feature is clearly unrealistic; the environment in which the species live must have limits and could not support an ever-increasing population.

Let us suppose that the environment can support a population p_e. Then how could (1.16) be modified to take account of this? Obviously, the indefinite increase of p should be interrupted. One possibility is to introduce an attractor at p_e as shown in Fig. 1.22(b). This means that populations greater than p_e decline, while populations less than p_e increase. Finally, equilibrium is reached at $p = p_e$. The fixed point at $p = p_e$, as well as $p = 0$, requires a non-linear $X(p)$ in (1.16). The form

$$\dot{p} = p(a - bp) \tag{1.18}$$

has the advantage of reducing to (1.16) when $b = 0$; otherwise $p_e = a/b$. The population p_e is known as the **carrying capacity** of the environment.

Of course, models of physical systems frequently involve more than a single state variable. If we are to be able to use qualitative ideas in modelling these systems, then we must examine autonomous equations involving more than one variable.

1.3 AUTONOMOUS SYSTEMS IN THE PLANE

Consider the differential equation

$$\dot{\mathbf{x}} \equiv \frac{\mathrm{d}\mathbf{x}}{\mathrm{d}t} = \mathbf{X}(\mathbf{x}) \tag{1.19}$$

where $\mathbf{x} = (x_1, x_2)$ is a vector in $\mathbb{R}^2$. This equation is equivalent to the system of two coupled equations

$$\dot{x}_1 = X_1(x_1, x_2), \qquad \dot{x}_2 = X_2(x_1, x_2), \tag{1.20}$$

where $\mathbf{X}(\mathbf{x}) = (X_1(x_1, x_2),\ X_2(x_1, x_2))$, because $\dot{\mathbf{x}} = (\dot{x}_1, \dot{x}_2)$. A solution to (1.19) consists of a pair of functions $(x_1(t), x_2(t))$, $t \in I \subseteq \mathbb{R}$, which satisfy (1.20). In general, both $x_1(t)$ and $x_2(t)$ involve an arbitrary constant so that there is a two-parameter family of solutions.

The qualitative behaviour of this family is determined by how x_1 and x_2 behave as t increases. Instead of simply indicating whether x is increasing or decreasing on the phase line, we must indicate how $\mathbf{x}$ behaves in the **phase plane**. The phase portrait is therefore a two-dimensional figure and the qualitative behaviour is represented by a family of curves, directed with increasing t, known as **trajectories** or **orbits**.

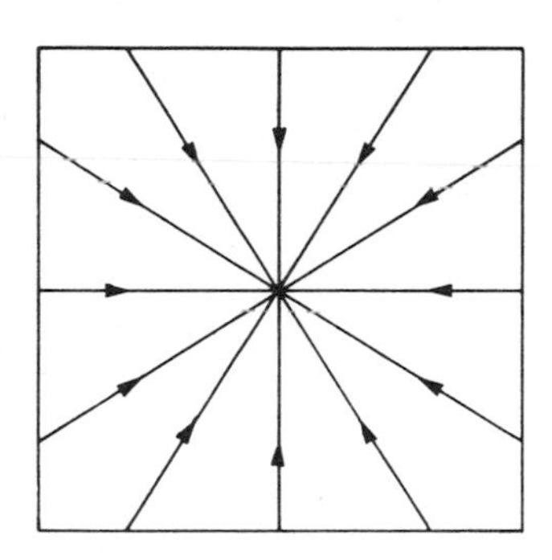

Fig. 1.23. $\dot{x}_1 = -x_1, \dot{x}_2 = -x_2$.

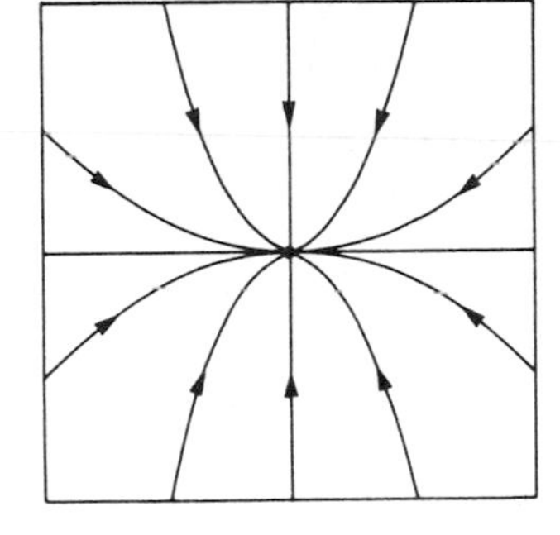

Fig. 1.24. $\dot{x}_1 = -x_1, \dot{x}_2 = -2x_2$.

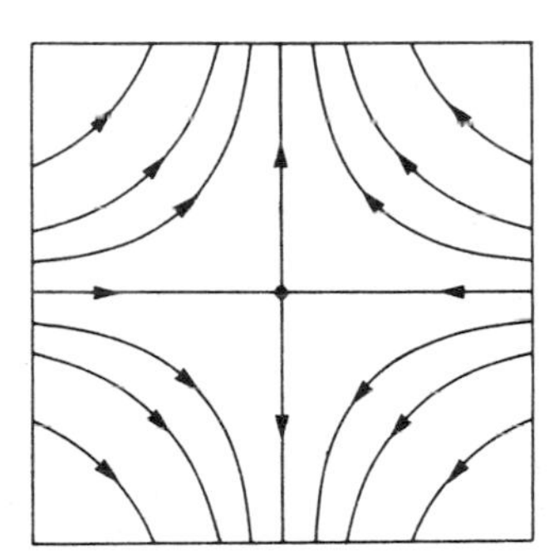

Fig. 1.25. $\dot{x}_1 = -x_1, \dot{x}_2 = x_2$.

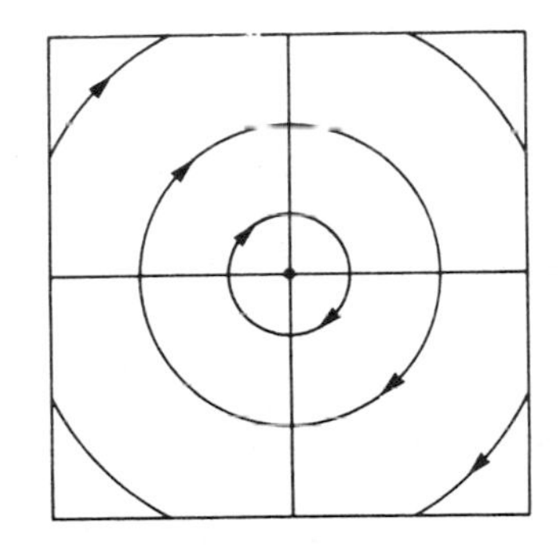

Fig. 1.26. $\dot{x}_1 = x_2, \dot{x}_2 = -x_1$.

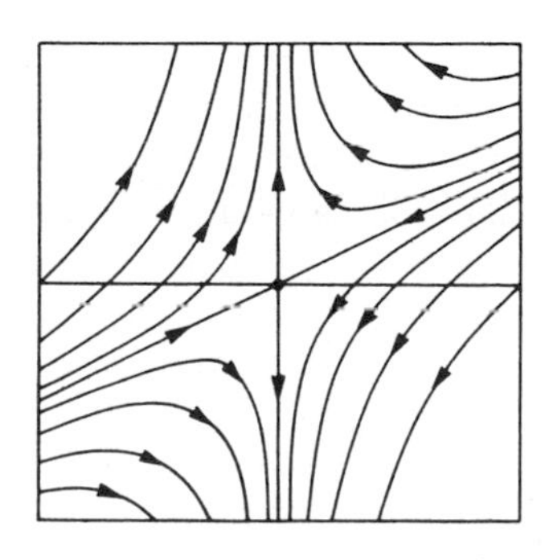

Fig. 1.27. $\dot{x}_1 = -x_1, \dot{x}_2 = -x_1 + x_2$.

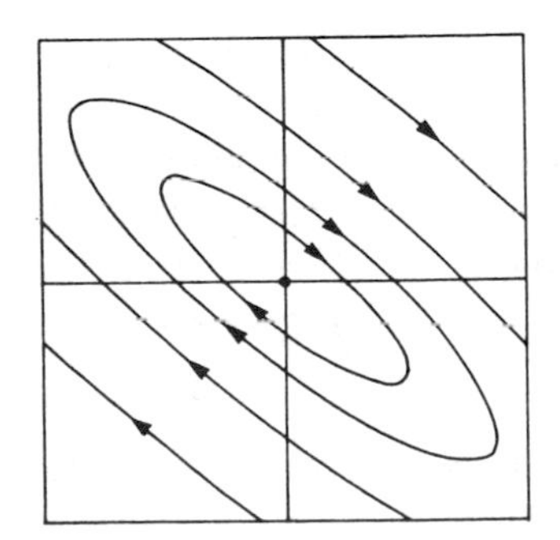

Fig. 1.28. $\dot{x}_1 = 3x_1 + 4x_2$,
$\dot{x}_2 = -3x_1 - 3x_2$.

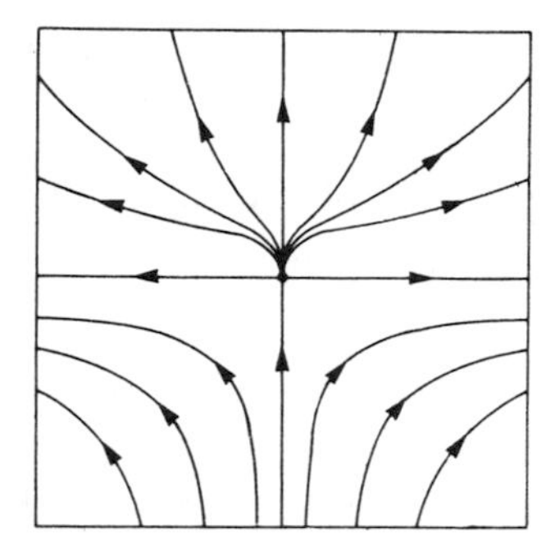

Fig. 1.29. $\dot{x}_1 = x_1$, $\dot{x}_2 = x_2^2$.

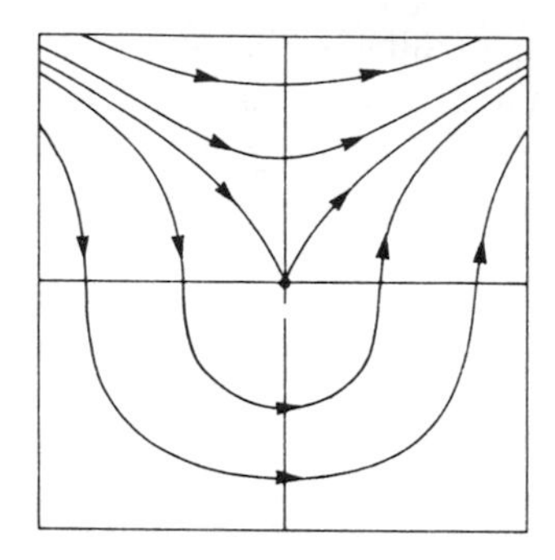

Fig. 1.30. $\dot{x}_1 = x_2^2$, $\dot{x}_2 = x_1$.

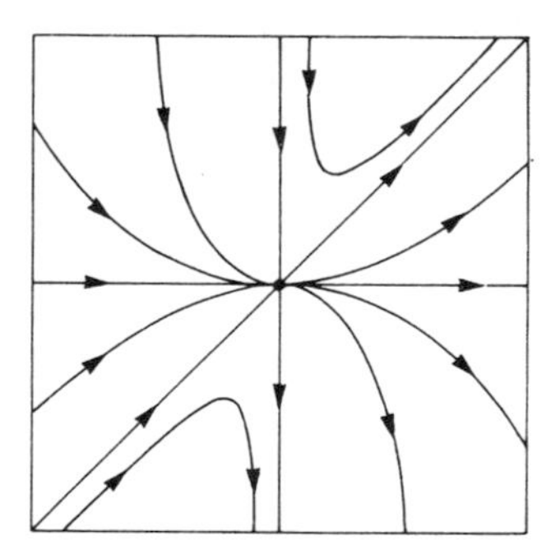

Fig. 1.31. $\dot{x}_1 = x_1^2$, $\dot{x}_2 = x_2(2x_1 - x_2)$.

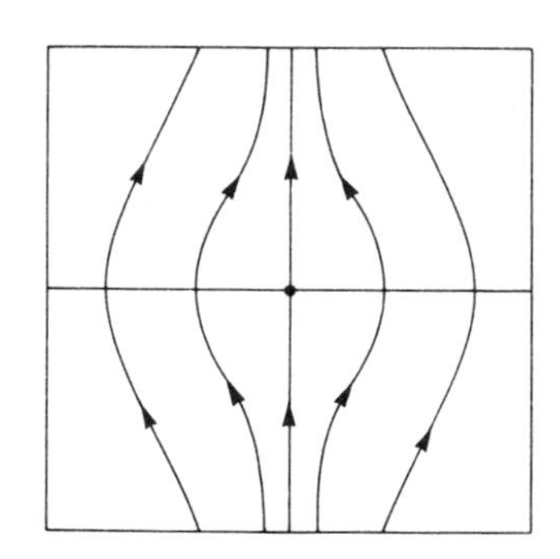

Fig. 1.32. $\dot{x}_1 = -x_1 x_2$, $\dot{x}_2 = x_1^2 + x_2^2$.

To examine qualitative behaviour in the plane, we begin (as in Section 1.2) by looking at any fixed points of (1.19). These are solutions of the form $\mathbf{x}(t) \equiv \mathbf{c} = (c_1, c_2)$ which arise when

$$X_1(c_1, c_2) = 0 \text{ and } X_2(c_1, c_2) = 0. \tag{1.21}$$

The corresponding trajectory is the point (c_1, c_2) in the phase plane. In section 1.2, the 'nature' of the fixed points determined the phase portrait. Let us consider some examples of isolated fixed points in the plane, with a view to determining their nature. Figures 1.23–1.32 show a small selection of the possibilities.

Consider Fig. 1.23; the system

$$\dot{x}_1 = -x_1, \qquad \dot{x}_2 = -x_2 \tag{1.22}$$

has a fixed point at $(0, 0)$ and solutions

$$x_1(t) = C_1 \mathrm{e}^{-t}, \qquad x_2(t) = C_2 \mathrm{e}^{-t} \tag{1.23}$$

where C_1, C_2 are real constants. Clearly, every member of the family (1.23) satisfies

$$x_2(t) = Kx_1(t), \tag{1.24}$$

where $K = C_2/C_1$, for every t. Thus every member is associated with a radial straight line in the x_1x_2-plane. Equation (1.23) shows that, for any choice of C_1 and C_2, $|x_1(t)|$ and $|x_2(t)|$ decrease as t increases and approach zero as $t \to \infty$. This is indicated by the direction of the arrow on the trajectory; i.e. if $(x_1(t), x_2(t))$ represents the value of $\mathbf{x}$ at t then this **phase point** will move along the radial straight line towards the origin as t increases. The directed straight line is sufficient to describe this qualitative behaviour. In Fig. 1.24,

$$x_2 = Kx_1^2,$$

which alters the shape of the trajectories. However, they are still all directed towards the fixed point at the origin.

Figure 1.25 shows another possibility: here as t increases, $|x_1(t)|$ decreases but $|x_2(t)|$ increases. In fact, $\dot{x}_1 = -x_1$, $\dot{x}_2 = x_2$ has solutions

$$x_1(t) = C_1 e^{-t}, \qquad x_2(t) = C_2 e^{t}, \tag{1.25}$$

C_1, C_2 real; so that

$$x_2 = Kx_1^{-1}, \tag{1.26}$$

with $K = C_1C_2$. In this case, only two special trajectories approach the fixed point at (0, 0), the remainder all turn away sooner or later and $|x_2| \to \infty$ as $|x_1| \to 0$. This qualitative behaviour is obviously quite different from that in Figs 1.23 and 1.24.

In Fig. 1.26 the trajectories close on themselves so that the same set of points in the phase plane recur time and time again as t increases. In section 1.4, we show that the system

$$\dot{x}_1 = x_2, \qquad \dot{x}_2 = -x_1 \tag{1.27}$$

has solutions

$$x_1(t) = C_1 \cos(-t + C_2), \qquad x_2(t) = C_1 \sin(-t + C_2). \tag{1.28}$$

It follows that

$$x_1^2 + x_2^2 = C_1^2 \tag{1.29}$$

and the trajectories are a family of concentric circles centred on the fixed point at (0, 0). This obviously corresponds to yet another kind of qualitative behaviour. The fact that $x_1(t)$ and $x_2(t)$ are periodic with the same period is reflected in the closed trajectories.

These examples show that qualitatively different solutions, $(x_1(t), x_2(t))$, lead to trajectories with different geometrical properties. The problem of recognizing different types of fixed point becomes one of recognizing 'distinct' geometrical configurations of trajectories. As in section 1.2, we must decide what we mean by 'distinct' and there is an element of choice in the criteria that we set.

For example, in Figs 1.23 and 1.24 all the trajectories are directed towards the origin. It would be reasonable to argue that this is the dominant qualitative feature and that the differences in shape of the trajectories are unimportant. We would then say that the nature of the fixed point at (0, 0) was the same in both cases. Of course, its nature would be completely changed if we replaced $\dot{x}_1$ by $-\dot{x}_1$ and $\dot{x}_2$ by $-\dot{x}_2$. Under these circumstances all trajectories would be directed away from the origin (see Exercise 1.22) corresponding to quite different qualitative behaviour of the solutions.

Let us compare Figs 1.25 and 1.27. Are the fixed points of the same nature? In both cases $|x_1(t)|$ tends to zero while $|x_2(t)|$ becomes infinite and only two special trajectories approach the fixed point itself. Yes, we would argue, they are the same. If the orientation of the trajectories is reversed in these examples is the nature of the fixed point changed as in our previous example? Orientation reversal would mean that the roles x_1 and x_2 were interchanged. However, the features which distinguish Figs 1.25 and 1.27 from the remaining ten diagrams still persist and we conclude that the nature of the fixed point does not change. Similarly, we would say that Figs 1.26, 1.28 and their counterparts with orientation reversed all had the same kind of fixed point at the origin.

The intuitive approach used in discussing the above examples is sufficient for us to recognize that seven distinct types of fixed point are illustrated in Figs 1.23–1.32. In fact, there are infinitely many qualitatively different planar phase portraits containing a single fixed point. Some examples of phase portraits with more than one isolated fixed point are shown in Figs 1.33–1.36. As can be seen, it is not difficult to produce complicated families of trajectories.

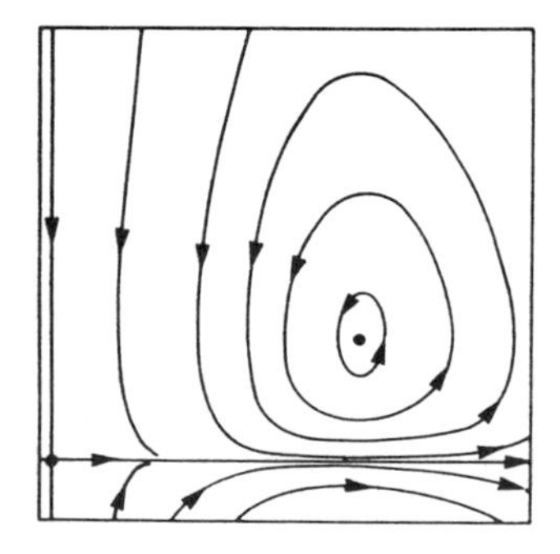

Fig. 1.33. Phase portrait for $\dot{x}_1 = x_1(a - bx_2)$, $\dot{x}_2 = -x_2(c - dx_1)$ with $a, b, c, d > 0$. There are fixed points at (0, 0) and $(c/d, a/b)$ in the phase plane.

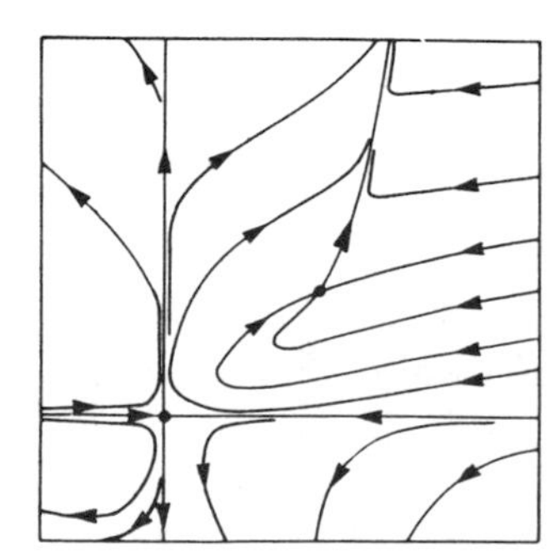

Fig. 1.34. Phase portrait for $\dot{x}_1 = -x_1[1 - 3x_2^{2/3}(1 - x_1)/(1 + x_1)]$, $\dot{x}_2 = x_2 - 3x_1x_2^{2/3}/(1 + x_1)$; $x_1 > -1$. Fixed points occur at (0, 0) and $(\frac{1}{2}, 1)$.

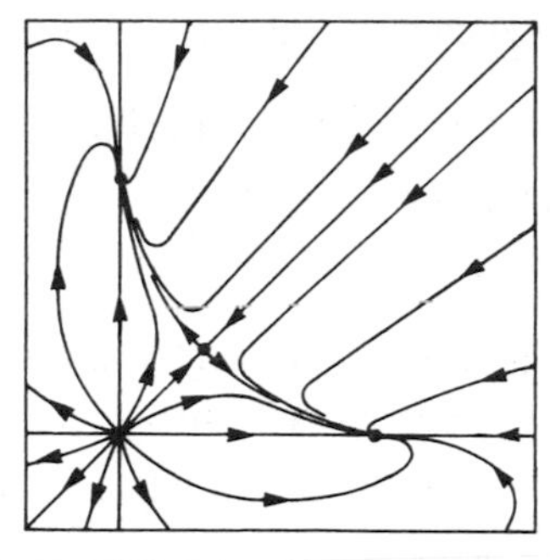

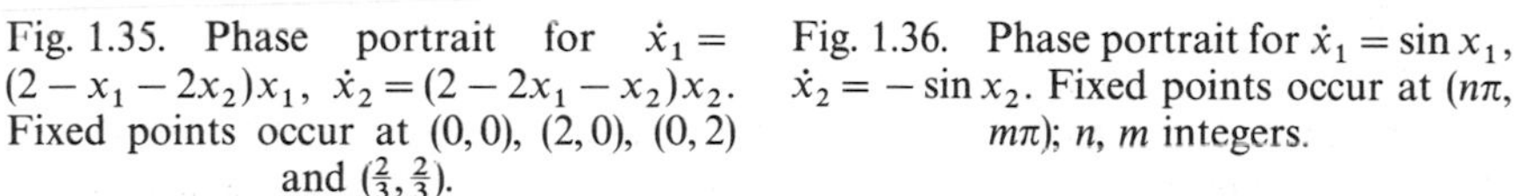

Fig. 1.35. Phase portrait for $\dot{x}_1 = (2 - x_1 - 2x_2)x_1$, $\dot{x}_2 = (2 - 2x_1 - x_2)x_2$. Fixed points occur at (0, 0), (2, 0), (0, 2) and $(\frac{2}{3}, \frac{2}{3})$.

Fig. 1.36. Phase portrait for $\dot{x}_1 = \sin x_1$, $\dot{x}_2 = -\sin x_2$. Fixed points occur at $(n\pi, m\pi)$; n, m integers.

1.4 CONSTRUCTION OF PHASE PORTRAITS IN THE PLANE

Methods of obtaining information about the trajectories associated with systems like

$$\dot{x}_1 = X_1(x_1, x_2); \qquad \dot{x}_2 = X_2(x_1, x_2), \qquad (x_1, x_2) \in S \subseteq \mathbb{R}^2,$$

are straightforward extensions of the ideas used to obtain solutions in section 1.1.

1.4.1 Use of calculus

In section 1.3, we were able to obtain the trajectories shown in Figs 1.23–1.25 by solving the system equations separately. This was possible because $\dot{x}_1$ depended only on x_1 and $\dot{x}_2$ only on x_2. A system in which each equation contains one, and only one, variable is said to be **decoupled**. This is not usually the case and if calculus is to be used in this way, new variables must be found which bring about this isolation of the variables.

Example 1.4.1
Consider the system

$$\dot{x} = x_2, \qquad \dot{x}_2 = -x_1 \tag{1.30}$$

whose phase portrait appears in Fig. 1.26. Use plane polar coordinates (r, θ) such that

$$x_1 = r\cos\theta, \qquad x_2 = r\sin\theta \tag{1.31}$$

to re-express (1.30) and hence obtain x_1 and x_2 as functions of t.

Solution
We establish differential equations for r and θ by observing that

$$r^2 = x_1^2 + x_2^2 \quad \text{and} \quad \tan\theta = x_2/x_1, \quad x_1 \neq 0, \tag{1.32}$$

and differentiating with respect to t. We find

$$2r\dot{r} = 2x_1\dot{x}_1 + 2x_2\dot{x}_2 \quad \text{and} \quad \sec^2\theta\dot{\theta} = x_1^{-2}(\dot{x}_2x_1 - \dot{x}_1x_2). \tag{1.33}$$

On substituting for $\dot{x}_1$ and $\dot{x}_2$ from (1.30) we conclude that

$$\dot{r} = 0 \quad \text{and} \quad \dot{\theta} = -1. \tag{1.34}$$

These equations can be solved separately to obtain $r(t) \equiv C_1$ and $\theta(t) = -t + C_2$, where C_1 and C_2 are real constants. Finally, we have

$$x_1(t) = C_1\cos(-t + C_2) \quad \text{and} \quad x_2(t) = C_1\sin(-t + C_2) \tag{1.35}$$

from (1.31). □

Sometimes solutions can be found when only one of the variables is isolated in one of the equations. Such systems are said to be **partially decoupled**.

Example 1.4.2
Find solutions to the system

$$\dot{x}_1 = x_1, \qquad \dot{x}_2 = x_1 + x_2 \tag{1.36}$$

and hence construct its phase portrait.

Solution
The system (1.36) can be solved without introducing new variables. The first equation has solutions

$$x_1(t) = C_1\mathrm{e}^t, \qquad C_1 \text{ real.} \tag{1.37}$$

Substitution in the second gives

$$\dot{x}_2 = x_2 + C_1\mathrm{e}^t \tag{1.38}$$

which in turn has solutions

$$x_2(t) = \mathrm{e}^t(C_1t + C_2) \tag{1.39}$$

(cf. Exercise 1.1).

To construct the phase portrait examine (1.37) and (1.39) and note the following.

1. For $C_1 = K > 0$, $x_1(t)$ strictly increases through all positive values as t increases through $(-\infty, \infty)$.
2. For $C_1 = K > 0$, $x_2(t) \to 0$ as $t \to -\infty$; $x_2(t) < 0$ for $C_1t + C_2 < 0$; $x_2(t) = 0$ for $C_1t + C_2 = 0$; and $x_2(t) \to \infty$ as $t \to \infty$.
3. For $C_1 = 0$, we obtain the solution

$$x_1(t) \equiv 0, \qquad x_2(t) = C_2\mathrm{e}^t. \tag{1.40}$$

4. For $C_1 = -K < 0$, both $x_1(t)$ and $x_2(t)$ assume precisely minus one times the values on a trajectory in 1 and 2 above.

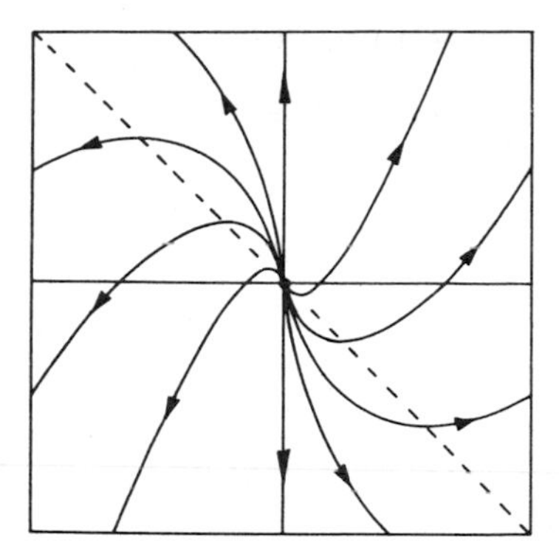

Fig. 1.37. Phase portrait for the system $\dot{x}_1 = x_1$, $\dot{x}_2 = x_1 + x_2$. The dashed line is $x_2 = -x_1$ where $\dot{x}_2 = 0$. The origin is a fixed point.

Notice also that the turning point in x_2 implied in 2, is given more precisely in (1.36) where $\dot{x}_2 = 0$ when $x_2 = -x_1$. Furthermore, the symmetry described in 4 is also apparent in (1.36) because this system is invariant under the transformation $x_1 \to -x_1$, $x_2 \to -x_2$ (Exercise 1.26). Finally, we sketch the phase portrait in Fig. 1.37. □

In Section 1.3, we sometimes found it advantageous to eliminate t between $x_1(t)$ and $x_2(t)$ and obtain a non-parametric form for the trajectories. These equations can often be found directly by solving

$$\frac{dx_2}{dx_1} = \frac{\dot{x}_2}{\dot{x}_1} = \frac{X_2(x_1, x_2)}{X_1(x_1, x_2)}. \tag{1.41}$$

Example 1.4.3
Consider the system

$$\dot{x}_1 = x_2, \qquad \dot{x}_2 - x_1. \tag{1.42}$$

Use dx_2/dx_1 to determine the nature of the trajectories and hence construct the phase portrait for the system.

Solution
For the system (1.42) we have

$$\frac{dx_2}{dx_1} = \frac{x_1}{x_2}, \qquad x_2 \neq 0, \tag{1.43}$$

which has solutions satisfying

$$x_1^2 - x_2^2 = C, \tag{1.44}$$

C real. This family of hyperbolae is easily sketched and their orientation as

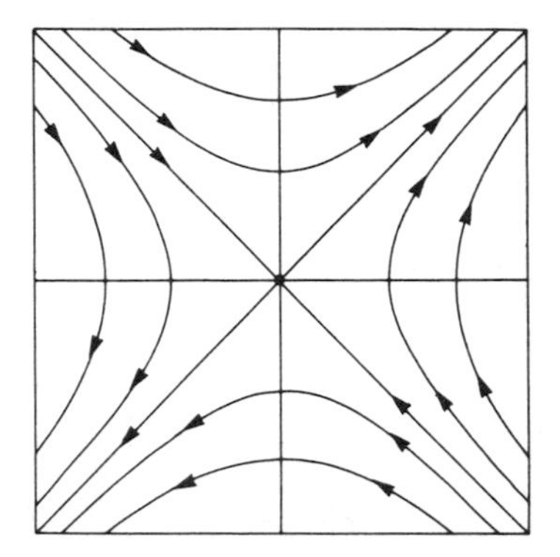

Fig. 1.38. Phase portrait for $\dot{x}_1 = x_2$, $\dot{x}_2 = x_1$. The origin is a fixed point and the trajectories are hyperbolae.

trajectories is given by (1.42). For example, note that both $x_1, x_2 > 0$ implies that both $x_1(t)$ and $x_2(t)$ increase with t. This provides directions for all the trajectories in $x_2 > -x_1$. Similarly, $x_1, x_2 < 0$ means $\dot{x}_1, \dot{x}_2 < 0$ and gives orientation to the trajectories in $x_2 < -x_1$. Thus the phase portrait is as shown in Fig. 1.38. □

1.4.2 Isoclines

As in section 1.1, it may be necessary to construct a phase portrait when calculus fails to give tractable solutions. This can be done by extending the method of isoclines to the plane. The vector-valued function or **vector field** $\mathbf{X}$: $S \to \mathbb{R}^2$ now gives $\dot{\mathbf{x}}$ at each point of the plane where $\mathbf{X}$ is defined. For qualitative purposes it is usually sufficient to record the direction of $\mathbf{X}(\mathbf{x})$. This is constant on the isoclines of (1.41). The zeros and divergences of $\mathrm{d}x_2/\mathrm{d}x_1$ are of particular interest and such lines will be referred to as the '$\dot{x}_2 = 0$' or '$\dot{x}_1 = 0$'-isoclines respectively.

If a unique solution $\mathbf{x}(t)$ of

$$\dot{\mathbf{x}} = \mathbf{X}(\mathbf{x}), \qquad \mathbf{x} \in S \text{ and with } \mathbf{x}(t_0) = \mathbf{x}_0 \tag{1.45}$$

exists for any $\mathbf{x}_0 \in S$ and $t_0 \in \mathbb{R}$, then each point of S lies on one, and only one, trajectory. All of the examples presented thus far in sections 1.3 and 1.4 have been of this kind. Furthermore, when non-uniqueness occurs, it is frequently confined to lines in the domain S. It is only in the neighbourhood of such lines that the behaviour of the trajectories is not immediately apparent from $\mathbf{X}(\mathbf{x})$.

Example 1.4.4
Obtain solutions to the system

$$\dot{x}_1 = 3x_1^{2/3}, \qquad \dot{x}_2 = 1 \tag{1.46}$$

and show that any point $(0, c)$, c real, lies on more than one solution. Interpret this result on the phase portrait for (1.46).

Solution

The system (1.46) is already decoupled. Calculus gives

$$x_1(t) = (t + C_1)^3 \quad \text{and} \quad x_2(t) = t + C_2 \tag{1.47}$$

as functions which satisfy the system equations. However, we know that $x_1(t) \equiv 0$ is a solution of $\dot{x} = 3x_1^{2/3}$ and so there are solutions of the form

$$x_1(t) = 0, \qquad x_2(t) = t + C_2. \tag{1.48}$$

If we let $C_2 = 0$ and $C_1 = -c$ in (1.47) then $(x_1(c), x_2(c)) = (0, c)$. Equally, if $C_2 = 0$ in (1.48) then $(x_1(c), x_2(c)) = (0, c)$, also. The solutions in (1.47) and (1.48) are distinct so $(0, c)$ lies on more than one solution.

The phase portrait obtained by representing (1.47) and (1.48) as trajectories is shown in Fig. 1.39. Notice the trajectories derived from (1.47) all touch the x_2-axis. This is itself a trajectory, oriented as shown, corresponding to (1.48). It is no longer clear what we should mean by a single trajectory because

$$x_1(t) = \begin{cases} (t-a)^3, & t < a \\ 0, & a \leqslant t \leqslant b, \\ (t-b)^3, & t > b \end{cases} \qquad x_2 = t + C \tag{1.49}$$

also satisfies (1.46) □

In the remainder of this book we will be concerned only with systems that have unique solutions. However, (cf. Section 1.1.1) a useful sufficient condition for existence and uniqueness of solutions to $\dot{\mathbf{x}} = \mathbf{X}(\mathbf{x})$ is that $\mathbf{X}$ be continuously differentiable.

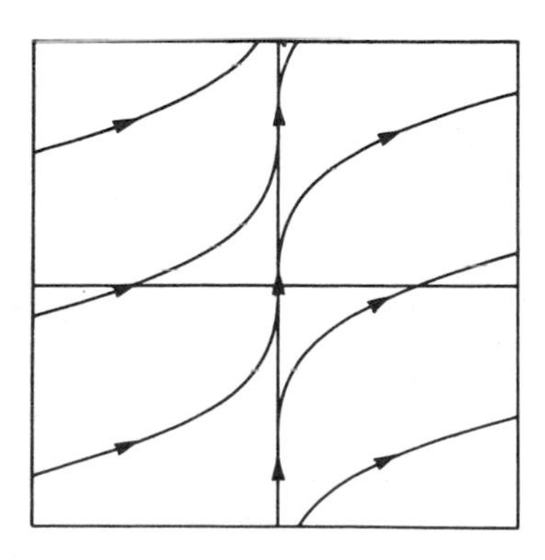

Fig. 1.39. Phase portrait for $\dot{x}_1 = 3x_1^{2/3}$, $\dot{x}_2 = 1$. There is no fixed point for this system and trajectories touch the x_2-axis.

Example 1.4.5

Sketch the phase portrait for the system

$$\dot{x}_1 = x_1^2, \qquad \dot{x}_2 = x_2(2x_1 - x_2) \tag{1.50}$$

without using calculus to determine the trajectories.

Solution

Examine (1.50) and note the following features.

1. There is a single fixed point at the origin of the x_1x_2-plane.
2. The vector field $\mathbf{X}(\mathbf{x}) = (x_1^2, x_2(2x_1 - x_2))$, corresponding to (1.50), satisfies $\mathbf{X}(\mathbf{x}) = \mathbf{X}(-\mathbf{x})$. This means that the shape of the trajectories is invariant under the transformation $x_1 \to -x_1$ and $x_2 \to -x_2$ (as shown in Exercise 1.26). We will, therefore, focus attention on the half-plane $x_1 \geqslant 0$.
3. The $\dot{x}_1 = 0$ isocline coincides with the x_2-axis and on it $\dot{x}_2 = -x_2^2 < 0$ for $x_2 \neq 0$. Hence, there is a trajectory coincident with the positive x_2-axis directed towards the origin and one coincident with the negative x_2-axis directed away from the origin.

   Consider

$$\frac{dx_2}{dx_1} = \frac{x_2(2x_1 - x_2)}{x_1^2}, \qquad x_1 \neq 0. \tag{1.51}$$

The isocline of slope C is given by

$$x_2(2x_1 - x_2) = x_1^2 - (x_1 - x_2)^2 = Cx_1^2$$

or

$$x_2 = x_1(1 \pm (1 - C)^{1/2}), \qquad C \leqslant 1. \tag{1.52}$$

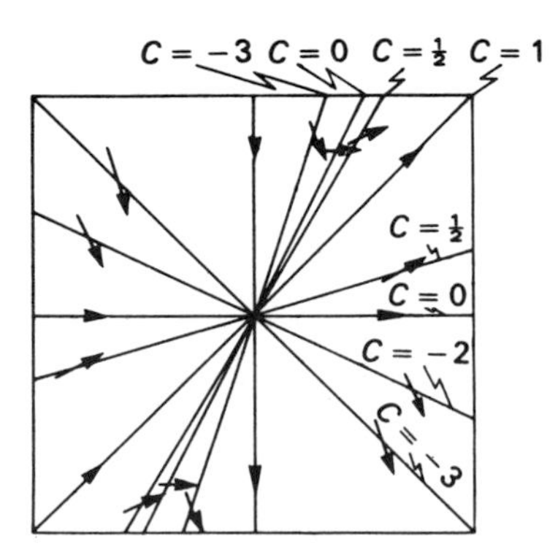

Fig. 1.40. Summary of information about $\dot{x}_1 = x_1^2$, $\dot{x}_2 = x_2(2x_1 - x_2)$. Note $C = 1$ on $x_2 = x_1$ means trajectories coincide with this line for $x_1 \neq 0$ as shown.

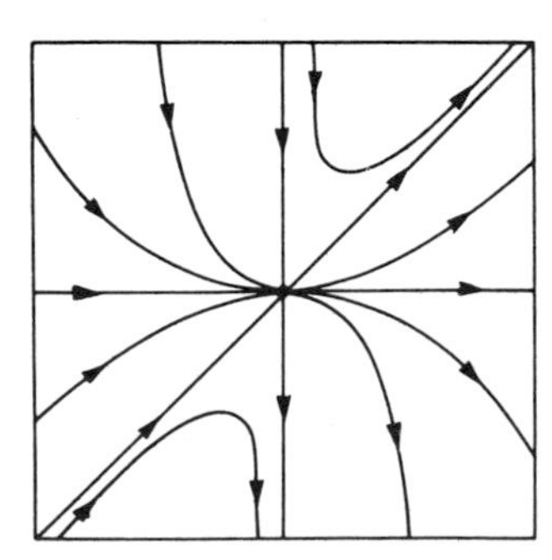

Fig. 1.41. Sketch of phase portrait for system (1.50). Notice curvature of trajectories between lines $x_2 = 0$ and $x_2 = x_1$ follows because C is increasing as shown in Fig. 1.40.

This equation allows us to obtain the slope of the trajectories on a selection of isoclines. For example, the slope C is given as:

1. $C = 0$ on the lines $x_2 = 0$ and $x_2 = 2x_1$;
2. $C = 1$ on $x_2 = x_1$;
3. $C = \frac{1}{2}$ on $x_2 = (1 \pm 1/\sqrt{2})x_1$;
4. $C = -3$ on $x_2 = -x_1$ and $x_2 = 3x_1$;
5. $C = -2$ on $x_2 = (1 \pm \sqrt{3})x_1$.

The orientation of the trajectories is fixed by recognizing that $\dot{x}_1 > 0$ for $x_1 \neq 0$.

This information allows us to construct Fig. 1.40 and we can, using uniqueness, sketch the phase portrait shown in Fig. 1.41. □

1.5 FLOWS AND EVOLUTION

The dynamic interpretation of the differential equation

$$\dot{x} = X(x), \qquad x \in S \subseteq \mathbb{R} \tag{1.53}$$

as the velocity of a point on the phase line (cf. section 1.2.2), leads to a new way of looking at both the differential equation and its solutions. Equation (1.53) can be thought of as defining a **flow** of phase points along the phase line. The function X gives the velocity of the flow at each value of $x \in S$.

The solution, $x(t)$, of (1.53) which satisfies $x(t_0) = x_0$ gives the past ($t < t_0$) and future ($t > t_0$) positions, or **evolution**, of the phase point which is at x_0 when $t = t_0$. This idea can be formalized by introducing a function $\phi_t: S \to S$ referred to either as the **flow** or the **evolution operator**.

The term 'evolution operator' is usually used in applications where ϕ_t describes the time development of the state of a real physical system. The word 'flow' is more frequently used when discussing the dynamics as a whole rather than the evolution of a particular point.

The function ϕ_t maps any $x_0 \in S$ onto the point $\phi_t(x_0)$ obtained by evolving for time t along a solution curve of (1.53) through x_0. Clearly, both existence and uniqueness of solutions are required for ϕ_t to be well defined. The point $\phi_t(x_0)$ is equal to $x(t + t_0)$ for any solution, $x(t)$, of (1.53) which satisfies $x(t_0) = x_0$ for some t_0. This property is illustrated in Fig. 1.42. As can be seen, it arises because the solutions of autonomous equations are related by translations in t. Thus the solution to $\dot{x} = X(x)$ with $x(t_0) = x_0$ is

$$x(t) = \phi_{t - t_0}(x_0). \tag{1.54}$$

A simple example of (1.54) is provided by the linear equation $\dot{x} = ax$. In this case

$$x(t) = \exp(a(t - t_0))x_0 \tag{1.55}$$

so that $\phi_{t - t_0}$ is simply multiplication by $\exp(a(t - t_0))$. However, it is only for linear equations that the evolution operator takes this simple form. In

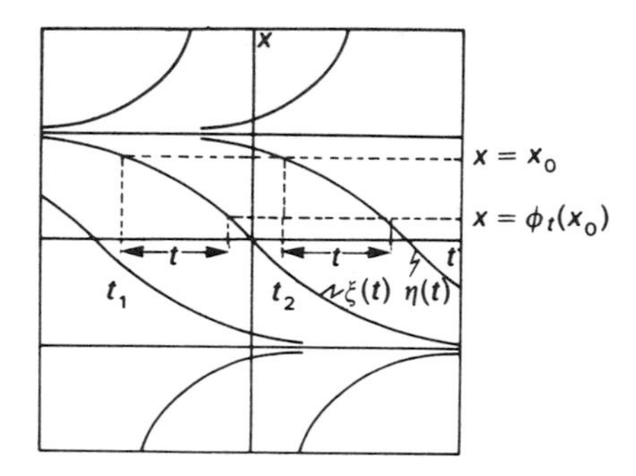

Fig. 1.42: Solution curves for $\dot{x} = \frac{1}{2}(x^2 - 1)$. Observe that the solutions $\xi(t)$ and $\eta(t)$ satisfy $\xi(t_1) = \eta(t_2) = x_0$, and $\xi(t + t_1) = \eta(t + t_2) = \phi_t(x_0)$, $t \in \mathbb{R}$.

fact, knowing ϕ_t is equivalent to having solved (1.53) and so finding it is of comparable difficulty.

The flow ϕ_t has simple properties which follow directly from its definition. Uniqueness conditions ensure that

$$\phi_{s+t}(x) = \phi_s(\phi_t(x)); \qquad s, t \in \mathbb{R} \tag{1.56}$$

providing both sides exist. In particular,

$$\phi_t(\phi_{-t}(x)) = \phi_{-t}(\phi_t(x)) = \phi_0(x) = x \tag{1.57}$$

and so

$$\phi_t^{-1} = \phi_{-t}. \tag{1.58}$$

For the flow in (1.55) the equalities (1.56) and (1.58) follow immediately from the properties of the exponential function. However, these relations are not always so apparent.

Example 1.5.1
Find the evolution operator ϕ_t for the equation

$$\dot{x} = x - x^2. \tag{1.59}$$

Verify (1.56) for this example.

Solution
The solutions to (1.59) satisfy

$$\int^x \frac{du}{u - u^2} = \ln\left|\frac{x}{x-1}\right| = t + C \tag{1.60}$$

for $x \neq 0$ or 1. This relation can be rearranged to give

$$\frac{x}{x-1} = K e^t$$

with $K = \pm e^C$, and thus

$$x(t) = Ke^t/(Ke^t - 1). \tag{1.61}$$

If we let $x = x_0$ at $t = 0$, then (1.61) implies $K = x_0/(x_0 - 1)$ and so

$$x(t) = \phi_t(x_0) = x_0 e^t/(x_0 e^t - x_0 + 1) \tag{1.62}$$

for t real and $x_0 \neq 0$ or 1. The points $x = 0$ and 1 were excluded in (1.60) because the integral is not defined for intervals including them. However, we see from (1.59) that they are the fixed points of the equation, which means that

$$\phi_t(0) = 0 \quad \text{and} \quad \phi_t(1) = 1. \tag{1.63}$$

for all $t \in \mathbb{R}$. The form for ϕ_t as given in (1.62) has precisely these properties and we can take

$$\phi_t(x) = xe^t/(xe^t - x + 1) \tag{1.64}$$

for all real x and t.

To check the basic property (1.56) of the evolution operator, observe

$$\phi_s(\phi_t(x)) = \phi_s(x_1) \tag{1.65}$$

where $x_1 = \phi_t(x)$. Thus

$$\phi_s(\phi_t(x)) = x_1 e^s/(x_1 e^s - x_1 + 1) \tag{1.66}$$

with

$$x_1 = xe^t/(xe^t - x + 1). \tag{1.67}$$

Substituting (1.67) into (1.66) gives

$$\phi_s(\phi_t(x)) = xe^{s+t}/(xe^{s+t} - x + 1) = \phi_{s+t}(x). \tag{1.68}$$ □

In section 1.1, we saw that solutions of $\dot{x} = X(x)$ may not be defined for all real t. The following example illustrates how this is reflected in the domain of definition of $\phi_t(x)$.

Example 1.5.2

Find the evolution operator ϕ_t for the equation

$$\dot{x} = x^2 \tag{1.69}$$

and give the intervals of t on which it is defined for each real x.

Solution

Solutions to (1.69) satisfy

$$\int^x u^{-2}\,du = -x^{-1} = t + C \tag{1.70}$$

where C is a constant, in any interval which does not contain $x = 0$. If $x = x_0$

when $t = 0$, then $C = -x_0^{-1}$ and we obtain

$$x(t) = x_0/(1 - x_0 t), \qquad t \neq x_0^{-1}. \tag{1.71}$$

In terms of the evolution operator, (1.71) means that

$$\phi_t(x) = x/(1 - xt) \tag{1.72}$$

for any non-zero x. As in Example 1.5.1, $\phi_t(x)$ given in (1.72) is also valid at $x = 0$; i.e.

$$\phi_t(0) = 0 \text{ for all real } t, \tag{1.73}$$

as required by the fixed point at $x = 0$ in (1.69). Thus (1.72) is valid for all real x. However, $\phi_t(x)$ is not defined for all t; consider, for example, $t = x^{-1}$ in (1.72). In fact, the interval in t for which $\phi_t(x)$ is defined is determined by x as follows.

1. $x > 0$: the entire evolution of x is given by (1.72) for $-\infty < t < x^{-1}$.
2. $x = 0$: (1.73) shows that $\phi_t(0)$ is defined for $-\infty < t < \infty$.
3. $x < 0$: (1.72) again describes the evolution of x but with $x^{-1} < t < \infty$.

In case 1, $\phi_t(x)$ increases from arbitrarily small, positive values at large negative t through x at $t = 0$, and tends to infinity as $t \to x^{-1}$. Similarly, in case 3 $\phi_t(x)$ takes arbitrarily large, negative values for t close to x^{-1}; as t increases, x increases strictly and approaches zero as $t \to \infty$ (cf. Fig. 1.15). □

The evolution operator plays an analogous role in the plane. The autonomous nature of (1.19) again ensures that solutions are related by translations in t and ϕ_t maps $\mathbf{x} \in \mathbb{R}^2$ to the point obtained by evolving for time t from $\mathbf{x}$ according to $\dot{\mathbf{x}} = \mathbf{X}(\mathbf{x})$, i.e. ϕ_t: $\mathbb{R}^2 \to \mathbb{R}^2$. In these terms the orbit or trajectory passing through $\mathbf{x}$ is simply $\{\phi_t(\mathbf{x})$: t real$\}$ oriented by increasing t. As we shall see in Chapters 3, 4 and 6, this notation can be a more useful description of the solutions.

In section 2.5 we discuss a general method for obtaining the evolution operator for linear systems in the plane as a 2×2 matrix. For example, the solutions

$$x_1(t) = C_1 e^{-t}, \qquad x_2(t) = C_2 e^{-2t} \tag{1.74}$$

for $\dot{x}_1 = -x_1$, $\dot{x}_2 = -2x_2$ can be written as

$$\begin{bmatrix} x_1(t) \\ x_2(t) \end{bmatrix} = \begin{bmatrix} e^{-t} & 0 \\ 0 & e^{-2t} \end{bmatrix} \begin{bmatrix} C_1 \\ C_2 \end{bmatrix}. \tag{1.75}$$

It follows that

$$\phi_t = \begin{bmatrix} e^{-t} & 0 \\ 0 & e^{-2t} \end{bmatrix} \tag{1.76}$$

is the evolution operator for (1.74).

The dynamic interpretation of a differential equation can also be extended to the two-dimensional problem defined by (1.19). The vector field $\mathbf{X}(\mathbf{x})$ defines a flow velocity at each point $\mathbf{x}$ of the plane. In the manner of Winnie the Pooh, we can imagine dropping a 'Pooh point' onto the plane and watching it move according to the flow. The path followed by such a phase point can be represented by an oriented curve on the plane corresponding to a trajectory in the phase portrait of (1.19). At each point $\mathbf{x}$ of its path the phase point has velocity $\mathbf{X}(\mathbf{x})$ and this velocity must be tangential to the path at $\mathbf{x}$. It follows that $\mathbf{X}(\mathbf{x})$ is directed along the tangent to the trajectory through $\mathbf{x}$ at each point $\mathbf{x}$ in the plane. This fundamental result provides a vectorial interpretation of the method of isoclines.

EXERCISES

Section 1.1

1.1 Show that a differential equation of the form

$$\dot{x} + p(t)x = q(t)$$

can be written as

$$\frac{\mathrm{d}}{\mathrm{d}t}(x\mathrm{e}^{P(t)}) = q(t)\mathrm{e}^{P(t)},$$

where $P(t) = \int^t p(s)\mathrm{d}s$. The function $\mathrm{e}^{P(t)}$ is called the **integrating factor** for the equation.

Use this observation to solve the following equations:
(a) $\dot{x} = x - t$; (b) $\dot{x} = x + \mathrm{e}^t$;
(c) $\dot{x} = -x\cot t + 5\mathrm{e}^{\cos t}$, $t \neq n\pi$, n integer.

1.2 A differential equation of the form

$$\dot{x} = f(x)g(t)$$

is said to be **separable**, because the solution passing through the point (t_0, x_0) of the t, x-plane satisfies

$$\int_{x_0}^{x} \frac{\mathrm{d}u}{f(u)} = \int_{t_0}^{t} g(s)\,\mathrm{d}s,$$

provided these integrals exist. The variables x and t are separated in this relation. Use this result to find solutions to:
(a) $\dot{x} = xt$; (b) $\dot{x} = -x/t$, $t \neq 0$;
(c) $\dot{x} = -t/x$, $x \neq 0$; (d) $\dot{x} = -x/\tanh t$, $t \neq 0$.

1.3 A differential equation of the form

$$M(t,x) + N(t,x)\dot{x} = 0$$

is said to be **exact** if there is a function $F(t,x)$ with continuous second partial derivatives such that $\partial F/\partial t \equiv M(t,x)$ and $\partial F/\partial x \equiv N(t,x)$. Show that a necessary condition for such a function to exist is that

$$\frac{\partial M}{\partial x} = \frac{\partial N}{\partial t}$$

and that any solution to the differential equation satisfies

$$F(t,x) = \text{constant}.$$

Show that

$$\frac{x}{t} + [\ln(xt) + 1]\dot{x} = 0, \qquad t, x > 0$$

is an exact differential equation. Find $F(t,x)$ and plot several solution curves.

1.4 Use the calculus to find solutions for the following differential equations:
(a) $\dot{x} = x^2$; (b) $\dot{x} = \frac{1}{2}x^3$;
(c) $\dot{x} = \frac{1}{2}(x^2 - 1)$; (d) $\dot{x} = 3x^{2/3}$;
(e) $\dot{x} = \sqrt{(1 - x^2)}$, $|x| \leqslant 1$; (f) $\dot{x} = 2x^{1/2}$, $x \geqslant 0$.
Equations (e) and (f) are defined on the **closed** domains $D = \{(t,x) \mid |x| \leqslant 1\}$ and $D = \{(t,x) \mid x \geqslant 0\}$, resepectively. Proposition 1.1.1 ensures existence of solutions on the **open** domains $D' = \{(t,x) \mid |x| < 1\}$ and $D' = \{t,x) \mid x > 0\}$ for these examples. Do the solutions you have found exist on the boundary of D?

1.5 Show how to construct infinitely many solutions, satisfying the initial condition $x(0) = 0$, for the differential equations given in Exercise 1.4 (d) and (f). Can the same be done for the equation in Exercise 1.4 (e) subject to the condition (a) $x(0) = 1$, (b) $x(0) = -1$? Explain your answer.

1.6 Suppose the differential equation $\dot{x} = X(t,x)$ has the property $X(t,x) = X(-t,-x)$. Prove that if $x = \xi(t)$ is a solution then so is $x = -\xi(-t)$. Find similar results on the symmetry of solutions when: (a) $X(t,x) = -X(-t,x)$; (b) $X(t,x) = -X(t,-x)$. Which of these symmetries appear in Figs 1.1–1.8?

1.7 A differential equation of the form

$$\dot{x} = h(t,x) \tag{1}$$

is said to be **homogeneous** if $h(t,x)$ satisfies $h(\alpha t, \alpha x) \equiv h(t,x)$ for all non-zero real α. Show that the isoclines of such an equation are always straight lines through the origin of the t, x-plane.
Use this result to sketch the solution curves of

$$\dot{x} = e^{x/t}, \qquad t \neq 0. \tag{2}$$

Show that the change of variable $x = ut$ allows (1) to be written as a separable equation (Exercise 1.2) for u when $t \neq 0$. Does this result help to obtain the family of solution curves for (2)?

1.8 Sketch the family of solutions of the differential equation

$$\dot{x} = ax - bx^2, \qquad x > 0, \qquad a \text{ and } b > 0.$$

Obtain the sketch directly from the differential equation itself. Prove that $\dot{x}$ is increasing for $0 < x < a/2b$ and decreasing for $a/2b < x < a/b$. How does this result influence your sketches? How does $\dot{x}$ behave for $a/b < x < \infty$?

1.9 Show that the substitution $y = x^{-1}$, $x \neq 0$, allows

$$\dot{x} = ax - bx^2$$

to be written as a differential equation for y with the form described in Exercise 1.1. Solve this equation and show that

$$x(t) = ax_0/\{bx_0 + (a - bx_0)\exp(-a(t - t_0))\},$$

where $x(t_0) = x_0$. Verify that the sketches obtained in Exercise 1.8 agree with this result. Can you identify any new qualitative features of the solutions which are not apparent from the original differential equation?

1.10 Sketch the solution curves of the differential equations:
(a) $\dot{x} = x^2 - t^2 - 1$; (b) $\dot{x} = t - t/x$, $x \neq 0$;
(c) $\dot{x} = (2t + x)/(t - 2x)$, $t \neq 2x$; (d) $\dot{x} = x^2 + t^2$;
by using isoclines and the regions of convexity and concavity.

1.11 Obtain isoclines and sketch the family of solutions for the following differential equations without finding x as a function of t.
(a) $\dot{x} = x + t$; (b) $\dot{x} = x^3 - x$;
(c) $\dot{x} = xt^2$; (d) $\dot{x} = x \ln x$, $x > 0$;
(e) $\dot{x} = \sinh x$; (f) $\dot{x} = t(x + 1)/(t^2 + 1)$.
What geometrical feature do the isoclines of (b), (d) and (e) have in common? Finally, verify your results using calculus.

Section 1.2

1.12 Find the fixed points of the following autonomous differential equations:
(a) $\dot{x} = x + 1$; (b) $\dot{x} = x - x^3$; (c) $\dot{x} = \sinh(x^2)$;
(d) $\dot{x} = x^4 - x^3 - 2x^2$; (e) $\dot{x} = x^2 + 1$.
Determine the nature (attractor, repellor or shunt) of each fixed point and hence construct the phase portrait of each equation.

1.13 Which differential equations in the following list have the same phase portrait?

(a) $\dot{x} = \sinh x$; (b) $\dot{x} = ax$, $a > 0$; (c) $\dot{x} = \begin{cases} x \ln|x|, & x \neq 0 \\ 0 & x = 0; \end{cases}$

(d) $\dot{x} = \sin x$; (e) $\dot{x} = x^3 - x$; (f) $\dot{x} = \tanh x$.
Explain, in your own words, the significance of two differential equations having the same phase portrait.

1.14 Consider the parameter-dependent differential equation

$$\dot{x} = (x - \lambda)(x^2 - \lambda), \qquad \lambda \text{ real.}$$

Find all possible phase portraits that could occur for this equation together with the intervals of λ in which they occur.

1.15 How many distinct qualitative types of phase portrait can occur on the phase line for a differential equation with three fixed points? What is the formula for the number of distinct phase portraits in the general case with n fixed points?

1.16 Show that the phase portrait of

$$\dot{x} = (a - x)(b - x)$$

is qualitatively the same as that of

$$\dot{y} = y(y - c)$$

for all real a, b, c; $a \neq b$, $c \neq 0$. Show, however, that a transformation, $y = kx + l$, which takes the first equation into the second, exists if and only if $c = b - a$ or $a - b$.

1.17 Consider the differential equation

$$\dot{x} = x^3 + ax - b.$$

Show that there is a curve C in the a, b-plane which separates this plane into two regions A and B such that: if $(a, b) \in A$ the phase portrait consists of a single repellor and if $(a, b) \in B$ it has two repellors separated by an attractor. Let $a < 0$ be fixed; describe the change in configuration of the fixed points as b varies from $-\infty$ to ∞.

1.18 A substance γ is formed in a chemical reaction between substances α and β. In the reaction each gram of γ is produced by the combination of p grams of α and $q = 1 - p$ grams of β. The rate of formation of γ at any instant of time t, is equal to the product of the masses of α and β that remain uncombined at that instant. If a grams of α and b grams of β are brought together at $t = 0$, show that the differential equation governing the mass, $x(t)$, of γ present at time $t > 0$ is

$$\dot{x} = (a - px)(b - qx).$$

Assume $a/p > b/q$ and construct the phase portrait for this equation. What is the maximum amount of γ that can possibly be produced in this experiment?

Section 1.3

1.19 Find the fixed points of the following systems of differential equations in the plane:

(a) $\dot{x}_1 = x_1(a - bx_2)$
$\dot{x}_2 = -x_2(c - dx_1)$
$a, b, c, d > 0$;

(b) $\dot{x}_1 = x_2$
$\dot{x}_2 = -\sin x_1$;

(c) $\dot{x}_1 = x_2$
$\dot{x}_2 = x_2(1 - x_1^2) - x_1$;

(d) $\dot{x}_1 = x_1(2 - x_1 - 2x_2)$
$\dot{x}_2 = x_2(2 - 2x_1 - x_2)$;

(e) $\dot{x}_1 = \sin x_1$
$\dot{x}_2 = \cos x_2$.

1.20 Sketch the following parametrized families of curves in the plane:
(a) $(x_1, x_2) = (a\cos t, a\sin t)$; (b) $(x_1, x_2) = (a\cos t, 2a\sin t)$;
(c) $(x_1, x_2) = (ae^t, be^{-2t})$; (d) $(x_1, x_2) = (ae^t + be^{-t}, ae^t - be^{-t})$;
(e) $(x_1, x_2) = (ae^t + be^{2t}, be^{2t})$;
where $a, b \in \mathbb{R}$. Find the systems of differential equations in the plane for which these curves form the phase portrait.

1.21 Use the sketches obtained in Exercise 1.20, to arrange the families of curves (a)–(e) in groups with the same type of fixed point at the origin of the x_1x_2-plane.

1.22 Consider the phase portrait of the system

$$\dot{x}_1 = X_1(x_1, x_2), \qquad \dot{x}_2 = X_2(x_1, x_2). \tag{1}$$

Show that the system

$$\dot{y}_1 = -X_1(y_1, y_2), \qquad \dot{y}_2 = -X_2(y_1, y_2)$$

has trajectories with the same shape but with the reverse orientation to those of (1). Verify your result by obtaining solutions for:
(a) $\dot{x}_1 = x_1$
$\dot{x}_2 = x_2$;
(b) $\dot{x}_1 = x_1$
$\dot{x}_2 = 2x_2$
and comparing with Figs 1.23 and 1.24.

Section 1.4

1.23 Use the change of variable

$$x_1 = y_1 + y_2, \qquad x_2 = y_1 - y_2$$

to decouple the pair of differential equations

$$\dot{x}_1 = x_2, \qquad \dot{x}_2 = x_1.$$

Hence construct the phase portrait for the system.

1.24 Show that a solution of the system of differential equations

$$\dot{x}_1 = -2x_1, \qquad \dot{x}_2 = x_1 - 2x_2$$

satisfying $x_1(0) = 1, x_2(0) = 2$ is given by

$$x_1(t) = e^{-2t}, \qquad x_2(t) = e^{-2t}(t + 2).$$

Prove that this solution is unique.

1.25 Consider the system of non-linear first-order differential equations

$$\dot{x}_1 = -x_2 + x_1(1 - x_1^2 - x_2^2), \qquad \dot{x}_2 = x_1 + x_2(1 - x_1^2 - x_2^2). \tag{1}$$

Use the change of variables $x_1 = r\cos\theta, x_2 = r\sin\theta$ to show that (1) is equivalent to the system

$$\dot{r} = r(1 - r^2), \qquad \dot{\theta} = 1. \tag{2}$$

Solve (2) with initial conditions $r(0) = r_0$ and $\theta(0) = \theta_0$ to obtain

$$r(t) = r_0/[r_0^2 + (1 - r_0^2)e^{-2t}]^{1/2}, \qquad \theta(t) = t + \theta_0.$$

1.26 (a) Suppose the autonomous system

$$\dot{\mathbf{x}} = \mathbf{X}(\mathbf{x}) \tag{1}$$

is invariant under the transformation $\mathbf{x} \to -\mathbf{x}$. Show that if $\boldsymbol{\xi}(t)$ satisfies (1) then so does $\boldsymbol{\eta}(t) = -\boldsymbol{\xi}(t)$.

(b) Suppose, instead, that $\mathbf{X}$ satisfies $\mathbf{X}(\mathbf{x}) = \mathbf{X}(-\mathbf{x})$. Show, in this case, that if $\boldsymbol{\xi}(t)$ is a solution to (1) then $\boldsymbol{\eta}(t) = -\boldsymbol{\xi}(-t)$ is also a solution.

Illustrate the relations obtained in (a) and (b) by examining typical trajectories $\boldsymbol{\xi}(t)$ and $\boldsymbol{\eta}(t)$ for:

(i) $\dot{x}_1 = x_1$, $\dot{x}_2 = x_1 + x_2$; (ii) $\dot{x}_1 = x_1^2$, $\dot{x}_2 = x_2^4$.

1.27 (a) Consider the non-autonomous equation

$$\dot{x} = X(t, x) \tag{1}$$

discussed in section 1.1. Use the substitution $x_1 = t$, $x_2 = x$ to show that (1) is equivalent to the autonomous system

$$\dot{x}_1 = 1, \qquad \dot{x}_2 = X(x_1, x_2).$$

(b) Use the substitution $x_1 = x$, $x_2 = \dot{x}$ to show that any second-order equation $\ddot{x} = F(x, \dot{x})$ is equivalent to the autonomous system

$$\dot{x}_1 = x_2, \qquad \dot{x}_2 = F(x_1, x_2).$$

Use these observations to convert the following equations into equivalent first order *autonomous* systems:

(i) $\dot{x} = x - t$; (ii) $\dot{x} = xt$; (iii) $\ddot{x} + \sin x = 0$;
(iv) $\ddot{x} + 2a\dot{x} + bx = 0$; (v) $\ddot{x} + f(x)\dot{x} + g(x) = 0$.

Show that (b) is not a unique procedure by verifying that $\dot{x}_1 = x_2 - \int^{x_1} f(u)\,du$, $\dot{x}_2 = -g(x_1)$ gives alternative systems for (iv) and (v).

1.28 Reduce the following sets of equations to a system of autonomous first-order equations in an appropriate number of variables:

(a) $\ddot{x} + x = 1$, $\ddot{y} + \dot{y} + y = 0$; (b) $\dot{x} + t = x$, $\dot{y} + y^3 = t$;
(c) $\ddot{x} + tx + 1 = 0$, $\ddot{y} + t^2\dot{x}^2 + x + y = 0$.

1.29 Use the method of isoclines to sketch the phase portraits of the following systems:

(a) $\dot{x}_1 = x_1$, $\dot{x}_2 = x_1 - x_2$; (b) $\dot{x}_1 = x_1 x_2$, $\dot{x}_2 = x_2^2$.
(c) $\dot{x}_1 = \ln x_1$; $\dot{x}_2 = x_2$, $x_1 > 0$.

1.30 The following systems of differential equations in the plane all have a single fixed point. Find the fixed point and use the method of isoclines to determine which systems do *not* have closed orbits in their phase portraits.

(a) $\dot{x}_1 = x_2 - 1$, $\dot{x}_2 = -(x_1 - 2)$; (b) $\dot{x}_1 = x_1 + x_2$, $\dot{x}_2 = x_1$; (c) $\dot{x}_1 = x_1 - x_2$, $\dot{x}_2 = x_1 + x_2$.

Confirm your results using calculus to find the solutions of $dx_2/dx_1 = \dot{x}_2/\dot{x}_1$.

1.31 Consider the differential equation

$$\ddot{x} + \dot{x}^2 + x = 0. \tag{1}$$

Use Exercise 1.27 (b) to convert it to first-order form. Show that the isocline of slope k

$$\mathscr{I}_k = \{(x_1, x_2) | \dot{x}_2 = k\dot{x}_1\}$$

is a parabola with vertex $(k^2/4, -k/2)$ and that these vertices themselves lie on a parabola $x_2^2 = x_1$. Use the substitution $x_2^2 = w$ to solve for x_2 as a function of x_1 and hence sketch the phase portrait of (1).

Section 1.5

1.32 Show that the evolution operator of the differential equation $\dot{x} = x - x^3$ for $x > 1$ is given by

$$\phi_t(x) = xe^t/\sqrt{(x^2e^{2t} - x^2 + 1)}.$$

Check that $\phi_s(\phi_t(x)) = \phi_{s+t}(x)$.

1.33 Find the evolution operators of the differential equations:
(a) $\dot{x} = \tanh x$; (b) $\dot{x} = x \ln x$, $x > 0$.

1.34 Find the evolution operators of the following systems:
(a) $\dot{x}_1 = x_1$, $\dot{x}_2 = x_1 - x_2$; (b) $\dot{x}_1 = x_1x_2$, $\dot{x}_2 = x_2^2$.
Give the intervals of time for which each operator is defined and verify that $\phi_{t+s}(\mathbf{x}) = \phi_t(\phi_s(\mathbf{x}))$ provided t, s and $t + s$ belong to the same interval of definition.

1.35 Let ϕ_t describe a flow in the plane. Draw phase portraits with at most two fixed points that satisfy each of the constraints given below:
(a) $\lim_{t\to\infty} \phi_t(\mathbf{x}) = \mathbf{0}$ for all points $\mathbf{x}$ of the plane;
(b) there is a point $\mathbf{x}_0$ such that $\lim_{t\to\infty} \phi_t(\mathbf{x}_0) = \lim_{t\to-\infty} \phi_t(\mathbf{x}_0) = \mathbf{0}$;
(c) $\lim_{t\to-\infty} \phi_t(\mathbf{x}) = \mathbf{0}$ for all points $\mathbf{x}$ of the plane;
(d) there is a trajectory through $\mathbf{x}_0$ such that $\lim_{t\to\infty} \phi_t(\mathbf{x}_0) = \mathbf{0}$ and a trajectory through $\mathbf{x}'_0$ such that $\lim_{t\to-\infty} \phi_t(\mathbf{x}'_0) = \mathbf{0}$.

1.36 Show that the polar form of the non-linear system

$$\dot{x}_1 = -x_2 + x_1[1 - (x_1^2 + x_2^2)^{\frac{1}{2}}], \qquad \dot{x}_2 = x_1 + x_2[1 - (x_1^2 + x_2^2)^{\frac{1}{2}}]$$

is given by

$$\dot{r} = r(1 - r), \qquad \dot{\theta} = 1.$$

Solve this equation to obtain the flow of (1) in the polar form

$$\phi_t(r,\theta) = (r/[r-(r-1)\mathrm{e}^{-t}], \theta + t).$$

Can the phase portrait of (1) be sketched more easily from the polar differential equation than from the expression for $\phi_t(r,\theta)$?

1.37 Let $\mathbf{x} = \mathbf{u}(t)$ define a parametrized curve in the plane. Draw diagrams showing how $[\mathbf{u}(t+h) - \mathbf{u}(t)]$ varies as $h > 0$ decreases towards 0. What do these illustrations indicate about the direction of $\dot{\mathbf{u}}(t)$? If $\mathbf{u}(t)$ represents the position of a point in the plane at time t, how is $\dot{\mathbf{u}}(t)$ to be interpreted?

Consider the flow defined by the matrix equation

$$\phi_t(\mathbf{x}) = \begin{bmatrix} \cos\beta t & -\sin\beta t \\ \sin\beta t & \cos\beta t \end{bmatrix} \begin{bmatrix} x_1 \\ x_2 \end{bmatrix},$$

where $\mathbf{x} = (x_1, x_2)$. Find the vector field $\mathbf{X}(\mathbf{x})$ appearing in the autonomous differential equation describing this flow and show that it is tangent to a trajectory of the system at every point of the plane.

1.38 Given the non-autonomous differential equation $\dot{x} = xt$, x, $t \geqslant 0$, prove that the evolution of the point x_0 at $t = t_0$ does not depend only on x_0 and $t - t_0$ as in the autonomous case by showing that

$$\phi(t, t_0, x_0) = x_0 \exp\{(t-t_0)(t+t_0)/2\}$$

2

Linear systems

A system $\dot{\mathbf{x}} = \mathbf{X}(\mathbf{x})$, where $\mathbf{x}$ is a vector in $\mathbb{R}^n$, is called a **linear system** of dimension n, if $\mathbf{X}$: $\mathbb{R}^n \rightarrow \mathbb{R}^n$ is a linear mapping. We will show that only a finite number of qualitatively different phase portraits can arise for linear systems. To do this we will first consider how such a system is affected by a linear change of variables.

2.1 LINEAR CHANGES OF VARIABLE

If the mapping $\mathbf{X}$: $\mathbb{R}^n \rightarrow \mathbb{R}^n$, where $\mathbb{R}^n = \{(x_1, \ldots, x_n)\} | x_i \in \mathbb{R},\ i = 1, \ldots, n\}$, is linear, then it can be written in the matrix form

$$\boldsymbol{X}(\boldsymbol{x}) = \begin{bmatrix} X_1(x_1, \ldots, x_n) \\ \vdots \\ X_n(x_1, \ldots, x_n) \end{bmatrix} = \begin{bmatrix} a_{11} & \cdots & a_{1n} \\ \vdots & & \vdots \\ a_{n1} & \cdots & a_{nn} \end{bmatrix} \begin{bmatrix} x_1 \\ \vdots \\ x_n \end{bmatrix}. \tag{2.1}$$

Correspondingly, $\dot{\mathbf{x}} = \mathbf{X}(\mathbf{x})$ becomes

$$\dot{\boldsymbol{x}} = \boldsymbol{X}(\boldsymbol{x}) = \boldsymbol{A}\boldsymbol{x}, \tag{2.2}$$

where $\boldsymbol{A}$ is the **coefficient matrix**. Each component X_i $(i = 1, \ldots, n)$ of $\dot{\mathbf{x}}$ is a linear function of the variables $x_1, \ldots, x_n$. These variables are, of course, simply the coordinates of $\mathbf{x} = (x_1, \ldots, x_n)$ relative to the natural basis in $\mathbb{R}^n$ (i.e. $\{\mathbf{e}_i\}_{i=1}^n$, where $\mathbf{e}_i = (0, \ldots, 0, 1, 0, \ldots, 0)$ with 1 in the ith position). Thus

$$\mathbf{x} = \sum_{i=1}^{n} x_i \mathbf{e}_i \quad \text{and} \quad \mathbf{X}(\mathbf{x}) = \sum_{i=1}^{n} X_i(\mathbf{x}) \mathbf{e}_i. \tag{2.3}$$

In order to make a change of variables we must express each $x_i (i = 1, \ldots, n)$ as a function of the new variables. We consider the effect on (2.2), of making a linear change of variable,

$$x_i = \sum_{j=1}^{n} m_{ij} y_j \ (i = 1, \ldots, n) \quad \text{or} \quad \boldsymbol{x} = \boldsymbol{M}\boldsymbol{y} \tag{2.4}$$

where m_{ij} is a real constant for all i and j. Of course, there must be a unique set of new variables $(y_1, \ldots, y_n)$ corresponding to a given set of old ones $(x_1, \ldots, x_n)$, and vice versa. This means that (2.4) must be a bijection and therefore $\boldsymbol{M}$ is a non-singular matrix. It follows that the columns $\boldsymbol{m}_i$, $i = 1, \ldots, n$, of $\boldsymbol{M}$ are linearly independent. Equation (2.4) implies

$$\boldsymbol{x} = \sum_{i=1}^{n} y_i \boldsymbol{m}_i, \tag{2.5}$$

and we recognize $y_1, \ldots, y_n$ as the coordinates of $\mathbf{x} \in \mathbb{R}^n$ relative to the new basis $\{\mathbf{m}_i\}_{i=1}^n$.

It is easy to express (2.2) in terms of the new variables; we find

$$\dot{\boldsymbol{x}} = \boldsymbol{M}\dot{\boldsymbol{y}} = \boldsymbol{A}\boldsymbol{M}\boldsymbol{y} \tag{2.6}$$

so that

$$\dot{\boldsymbol{y}} = \boldsymbol{B}\boldsymbol{y}, \tag{2.7}$$

with

$$\boldsymbol{B} = \boldsymbol{M}^{-1}\boldsymbol{A}\boldsymbol{M}. \tag{2.8}$$

Thus the coefficient matrix $\boldsymbol{B}$ of the transformed system is **similar** to $\boldsymbol{A}$.

Example 2.1.1

The system

$$\dot{x}_1 = x_2, \qquad \dot{x}_2 = x_1 \tag{2.9}$$

is transformed into the decoupled system

$$\dot{y}_1 = y_1, \qquad \dot{y}_2 = -y_2 \tag{2.10}$$

by the change of variable

$$x_1 = y_1 + y_2; \qquad x_2 = y_1 - y_2. \tag{2.11}$$

Use this result to illustrate (2.8).

Solution

Expressing these systems of equations in matrix form we obtain

$$\begin{bmatrix} \dot{x}_1 \\ \dot{x}_2 \end{bmatrix} = \begin{bmatrix} 0 & 1 \\ 1 & 0 \end{bmatrix} \begin{bmatrix} x_1 \\ x_2 \end{bmatrix}, \qquad \begin{bmatrix} x_1 \\ x_2 \end{bmatrix} = \begin{bmatrix} 1 & 1 \\ 1 & -1 \end{bmatrix} \begin{bmatrix} y_1 \\ y_2 \end{bmatrix}$$

and

$$\begin{bmatrix} \dot{y}_1 \\ \dot{y}_2 \end{bmatrix} = \begin{bmatrix} 1 & 0 \\ 0 & -1 \end{bmatrix} \begin{bmatrix} y_1 \\ y_2 \end{bmatrix}.$$

Thus

$$\boldsymbol{A} = \begin{bmatrix} 0 & 1 \\ 1 & 0 \end{bmatrix}, \qquad \boldsymbol{M} = \begin{bmatrix} 1 & 1 \\ 1 & -1 \end{bmatrix} \text{ and } \boldsymbol{B} = \begin{bmatrix} 1 & 0 \\ 0 & -1 \end{bmatrix}.$$

Observe that $\boldsymbol{AM} = \boldsymbol{MB}$ so that (2.8) is satisfied. □

Example 2.1.2
Find the matrix representation of the linear system

$$\dot{x}_1 = x_1 + 2x_2, \qquad \dot{x}_2 = 2x_2 \tag{2.12}$$

under the change of variables

$$x_1 = y_1 + 2y_2, \qquad x_2 = y_2. \tag{2.13}$$

What is the basis for which y_1, y_2 are the corresponding coordinates?

Solution
The change of variables (2.13) can be written in the form

$$\begin{bmatrix} x_1 \\ x_2 \end{bmatrix} = \begin{bmatrix} 1 & 2 \\ 0 & 1 \end{bmatrix} \begin{bmatrix} y_1 \\ y_2 \end{bmatrix} \tag{2.14}$$

so that

$$\boldsymbol{M} = \begin{bmatrix} 1 & 2 \\ 0 & 1 \end{bmatrix}. \tag{2.15}$$

The system (2.12) has the matrix form

$$\dot{\boldsymbol{x}} = \begin{bmatrix} 1 & 2 \\ 0 & 2 \end{bmatrix} \boldsymbol{x} = \boldsymbol{A}\boldsymbol{x}. \tag{2.16}$$

Equations (2.7) and (2.8) give

$$\boldsymbol{B} = \boldsymbol{M}^{-1}\boldsymbol{A}\boldsymbol{M} = \begin{bmatrix} 1 & 0 \\ 0 & 2 \end{bmatrix} \quad \text{and} \quad \dot{\boldsymbol{y}} = \boldsymbol{B}\boldsymbol{y} \tag{2.17}$$

as the required matrix representation.

The basis with coordinates y_1, y_2 is given by the columns of $\boldsymbol{M}$ in (2.15), that is $\{(1,0),(2,1)\}$. □

Similarity is an equivalence relation on the set of $n \times n$ real matrices (as shown in Exercise 2.2) and it follows that this set can be disjointly decomposed into equivalence or **similarity classes**. For any two matrices $\boldsymbol{A}$ and $\boldsymbol{B}$ in the same similarity class, the solutions of the systems $\dot{\boldsymbol{x}} = \boldsymbol{A}\boldsymbol{x}$ and $\dot{\boldsymbol{y}} = \boldsymbol{B}\boldsymbol{y}$ are related by $\boldsymbol{x} = \boldsymbol{M}\boldsymbol{y}$ if $\boldsymbol{M}^{-1}\boldsymbol{A}\boldsymbol{M} = \boldsymbol{B}$. Thus, if one such system can be solved, solutions can be obtained for each member of the class.

In section 1.4.1, we saw that decoupled or partially decoupled systems could be solved easily and we are led to consider whether each similarity class contains at least one correspondingly simple (e.g. diagonal or triangular) member. The answer to this algebraic problem is known and we will illustrate it for $n = 2$ in the next section.

2.2 SIMILARITY TYPES FOR 2 × 2 REAL MATRICES

For each positive integer n, there are infinitely many similarity classes of $n \times n$ real matrices. These similarity classes can be grouped into just finitely many types. In the following proposition we give these types for $n = 2$.

Proposition 2.2.1
Let $\boldsymbol{A}$ be a real 2×2 matrix, then there is a real, non-singular matrix $\boldsymbol{M}$ such that $\boldsymbol{J} = \boldsymbol{M}^{-1}\boldsymbol{A}\boldsymbol{M}$ is one of the types:

$$\text{(a)} \begin{bmatrix} \lambda_1 & 0 \\ 0 & \lambda_2 \end{bmatrix}, \quad \lambda_1 > \lambda_2; \qquad \text{(b)} \begin{bmatrix} \lambda_0 & 0 \\ 0 & \lambda_0 \end{bmatrix};$$

$$\text{(c)} \begin{bmatrix} \lambda_0 & 1 \\ 0 & \lambda_0 \end{bmatrix}; \qquad \text{(d)} \begin{bmatrix} \alpha & -\beta \\ \beta & \alpha \end{bmatrix}, \quad \beta > 0 \tag{2.18}$$

where $\lambda_0, \lambda_1, \lambda_2, \alpha, \beta$ are real numbers.

The matrix $\boldsymbol{J}$ is said to be the **Jordan form** of $\boldsymbol{A}$. The eigenvalues of the matrix $\boldsymbol{A}$ (and $\boldsymbol{J}$) are the values of λ for which

$$p_A(\lambda) = \lambda^2 - \operatorname{tr}(\boldsymbol{A})\lambda + \det(\boldsymbol{A}) = 0. \tag{2.19}$$

Here $\operatorname{tr}(\boldsymbol{A}) = a_{11} + a_{22}$ is the trace of $\boldsymbol{A}$ and $\det(\boldsymbol{A}) = a_{11}a_{22} - a_{12}a_{21}$ is its determinant. Thus the eigenvalues of $\boldsymbol{A}$ are

$$\lambda_1 = \tfrac{1}{2}(\operatorname{tr}(\boldsymbol{A}) + \sqrt{\Delta}) \quad \text{and} \quad \lambda_2 = \tfrac{1}{2}(\operatorname{tr}(\boldsymbol{A}) - \sqrt{\Delta}). \tag{2.20}$$

with

$$\Delta = (\operatorname{tr}(\boldsymbol{A}))^2 - 4\det(\boldsymbol{A}). \tag{2.21}$$

It is the nature of the eigenvalues: real distinct ($\Delta > 0$), real equal ($\Delta = 0$), and complex ($\Delta < 0$), that determine the type of the Jordan form $\boldsymbol{J}$ of $\boldsymbol{A}$.

(*a*) *Real distinct eigenvalues* ($\Delta > 0$)

The eigenvectors $\boldsymbol{u}_1$, $\boldsymbol{u}_2$ of $\boldsymbol{A}$ are given by

$$\boldsymbol{A}\boldsymbol{u}_i = \lambda_i \boldsymbol{u}_i, \qquad (i = 1, 2) \tag{2.22}$$

with $\lambda_1, \lambda_2, \lambda_1 > \lambda_2$, the distinct eigenvalues. Let

$$\boldsymbol{M} = [\boldsymbol{u}_1 \mid \boldsymbol{u}_2] \tag{2.23}$$

be the matrix with the eigenvectors $\boldsymbol{u}_1$, $\boldsymbol{u}_2$ as columns. Then

$$\boldsymbol{A}\boldsymbol{M} = [\boldsymbol{A}\boldsymbol{u}_1 \mid \boldsymbol{A}\boldsymbol{u}_2] = [\lambda_1 \boldsymbol{u}_1 \mid \lambda_2 \boldsymbol{u}_2] = \boldsymbol{M}\boldsymbol{J} \tag{2.24}$$

where

$$\boldsymbol{J} = \begin{bmatrix} \lambda_1 & 0 \\ 0 & \lambda_2 \end{bmatrix}.$$

For distinct eigenvalues the eigenvectors $\boldsymbol{u}_1$ and $\boldsymbol{u}_2$ are linearly independent and therefore $\boldsymbol{M}$ is non-singular. Thus

$$\boldsymbol{M}^{-1}\boldsymbol{A}\boldsymbol{M}=\boldsymbol{J}=\begin{bmatrix}\lambda_1 & 0\\ 0 & \lambda_2\end{bmatrix}. \tag{2.25}$$

(*b*) *Equal eigenvalues* ($\Delta=0$)

Equation (2.20) gives $\lambda_1=\lambda_2=\frac{1}{2}\mathrm{tr}(A)=\lambda_0$, and we must consider the following possibilities.

(i) $\boldsymbol{A}$ *is diagonal*

$$\boldsymbol{A}=\begin{bmatrix}\lambda_0 & 0\\ 0 & \lambda_0\end{bmatrix}=\lambda_0\boldsymbol{I} \tag{2.26}$$

which is (2.18(b)). In this case for any non-singular matrix $\boldsymbol{M}$, $\boldsymbol{M}^{-1}\boldsymbol{A}\boldsymbol{M}=\boldsymbol{A}$. Therefore the matrix $\boldsymbol{A}$ is only similar to itself and hence is in a similarity class of its own.

(ii) $\boldsymbol{A}$ *is not diagonal*

In this case $\lambda_1=\lambda_2=\lambda_0$, rank $(\boldsymbol{A}-\lambda_0\boldsymbol{I})=1$ and there are not two linearly independent eigenvectors. Let $\boldsymbol{u}_0$ be an eigenvector of $\boldsymbol{A}$. If we take $\boldsymbol{m}_1=\boldsymbol{u}_0$ and choose $\boldsymbol{m}_2$ so that $\boldsymbol{M}=[\boldsymbol{u}_0 \mid \boldsymbol{m}_2]$ is a nonsingular maxtrix,

$$\boldsymbol{A}\boldsymbol{M}=[\lambda_0\boldsymbol{u}_0 \mid \boldsymbol{A}\boldsymbol{m}_2]=M[\lambda_0\boldsymbol{e}_1 \mid \boldsymbol{M}^{-1}\boldsymbol{A}\boldsymbol{m}_2], \tag{2.27}$$

where $\boldsymbol{e}_1$ is the first column of $\boldsymbol{I}$.

The matrices $\boldsymbol{A}$ and $\boldsymbol{M}^{-1}\boldsymbol{A}\boldsymbol{M}$ have the same eigenvalues, and so

$$\boldsymbol{M}^{-1}\boldsymbol{A}\boldsymbol{M}=\begin{bmatrix}\lambda_0 & C\\ 0 & \lambda_0\end{bmatrix} \tag{2.28}$$

for some non-zero C. However, the simple modification of $\boldsymbol{M}$ to

$$\boldsymbol{M}_1=\boldsymbol{M}\begin{bmatrix}1 & 0\\ 0 & C^{-1}\end{bmatrix} \tag{2.29}$$

results in

$$\boldsymbol{M}_1^{-1}\boldsymbol{A}\boldsymbol{M}_1=\begin{bmatrix}\lambda_0 & 1\\ 0 & \lambda_0\end{bmatrix}$$

which is (2.18(c)).

(*c*) *Complex eigenvalues* ($\Delta<0$)

We can write $\lambda_1=\alpha+\mathrm{i}\beta$ and $\lambda_2=\alpha-\mathrm{i}\beta$, where $\alpha=\frac{1}{2}\mathrm{tr}(A)$ and $\beta=+\frac{1}{2}\sqrt{(-\Delta)}$. We wish to show that there is a non-singular matrix $\boldsymbol{M}=[\boldsymbol{m}_1 \mid \boldsymbol{m}_2]$, such that $M^{-1}AM$ is given by (2.18(d)), or equivalently that

$$AM = M\begin{bmatrix} \alpha & -\beta \\ \beta & \alpha \end{bmatrix}. \tag{2.30}$$

Partitioning M into its columns, we obtain

$$[Am_1 \mid Am_2] = [\alpha m_1 + \beta m_2 \mid -\beta m_1 + \alpha m_2]$$

or

$$[(A - \alpha I)m_1 - \beta I m_2 \mid \beta I m_1 + (A - \alpha I)m_2] = [0 \mid 0]. \tag{2.31}$$

This matrix equation can be written as four homogeneous linear equations for the unknown elements of M, i.e.

$$\left[\begin{array}{c|c} A - \alpha I & -\beta I \\ \hline \beta I & A - \alpha I \end{array}\right]\left[\begin{array}{c} m_1 \\ \hline m_2 \end{array}\right] = \left[\begin{array}{c} 0 \\ \hline 0 \end{array}\right]. \tag{2.32}$$

To solve these equations, let P be the coefficient matrix in (2.32) and let

$$Q = \left[\begin{array}{c|c} A - \alpha I & \beta I \\ \hline -\beta I & A - \alpha I \end{array}\right]. \tag{2.33}$$

Now observe that

$$PQ = \left[\begin{array}{c|c} p_A(A) & 0 \\ \hline 0 & p_A(A) \end{array}\right] \tag{2.34}$$

where $p_A(\lambda) = \lambda^2 - 2\alpha\lambda + (\alpha^2 + \beta^2)$ is the characteristic polynomial of A. The Cayley–Hamilton theorem (Hartley and Hawkes, 1970) states that $p_A(A) = 0$ (some examples are given in Exercise 2.7) and we conclude that

$$PQ = 0. \tag{2.35}$$

the 4×4 null matrix. Thus, the columns of Q must be solutions to (2.32). The first column of Q gives

$$M = \begin{bmatrix} a_{11} - \alpha & -\beta \\ a_{21} & 0 \end{bmatrix}. \tag{2.36}$$

Note that the discriminant Δ of $p_A(\lambda) = 0$ can be written

$$\Delta = (a_{11} - a_{22})^2 + 4a_{12}a_{21}. \tag{2.37}$$

If $\Delta < 0$ then $a_{12}a_{21} \neq 0$ and hence $a_{21} \neq 0$. Further, $\beta = +\frac{1}{2}\sqrt{(-\Delta)} \neq 0$ and we conclude that det $(M) = \beta a_{21} \neq 0$. Thus, (2.36) provides a non-singular matrix M such that $M^{-1}AM$ is given by (2.18(d)).

Any given 2×2 real matrix A falls into one, and only one, of the cases set out above (as shown in Exercise 2.6) and Proposition 2.2.1 is justified. Let us consider some simple illustrations of this result.

Example 2.2.1
Find the Jordan forms of each of the following matrices:

$$\text{(a)} \begin{bmatrix} 1 & 2 \\ 1 & 1 \end{bmatrix}; \quad \text{(b)} \begin{bmatrix} 2 & 1 \\ -2 & 4 \end{bmatrix}; \quad \text{(c)} \begin{bmatrix} 3 & -1 \\ 1 & 1 \end{bmatrix}.$$

Solution
The pairs of eigenvalues of the matrices (a)–(c) are $(\lambda_1, \lambda_2) = (1+\sqrt{2}, 1-\sqrt{2})$, $(3+\mathrm{i}, 3-\mathrm{i})$ and $(2, 2)$, respectively. Thus, the Jordan forms are

$$\begin{aligned} &\text{(a)} \begin{bmatrix} \lambda_1 & 0 \\ 0 & \lambda_2 \end{bmatrix} = \begin{bmatrix} 1+\sqrt{2} & 0 \\ 0 & 1-\sqrt{2} \end{bmatrix} \\ &\text{(b)} \begin{bmatrix} \alpha & -\beta \\ \beta & \alpha \end{bmatrix} = \begin{bmatrix} 3 & -1 \\ 1 & 3 \end{bmatrix} \\ &\text{(c)} \begin{bmatrix} \lambda_0 & 1 \\ 0 & \lambda_0 \end{bmatrix} = \begin{bmatrix} 2 & 1 \\ 0 & 2 \end{bmatrix} \end{aligned} \tag{2.38}$$

respectively. Notice (2.18(c)) is chosen for (2.38(c)) above because the original matrix is non-diagonal. □

Example 2.2.2
Find a matrix $\boldsymbol{M}$ which converts each of the matrices in Example 2.2.1 into their appropriate Jordan forms.

Solution
Matrix (a) has real distinct eigenvalues and so the columns of $\boldsymbol{M}$ can be taken as the two corresponding eigenvectors. The eigenvector $\boldsymbol{u}_1 = \begin{bmatrix} u_{11} \\ u_{21} \end{bmatrix}$ satisfies $[\boldsymbol{A} - (1+\sqrt{2})\boldsymbol{I}] = \boldsymbol{0}$ which implies only that $u_{11}\sqrt{2} = 2u_{21}$. We can take $\boldsymbol{u}_1 = \begin{bmatrix} \sqrt{2} \\ 1 \end{bmatrix}$. Similarly, $\boldsymbol{u}_2$ corresponding to $\lambda_2 = 1-\sqrt{2}$ can be taken as $\begin{bmatrix} -\sqrt{2} \\ 1 \end{bmatrix}$. Thus $\boldsymbol{M} = \begin{bmatrix} \sqrt{2} & -\sqrt{2} \\ 1 & 1 \end{bmatrix}$ and

$$\boldsymbol{M}^{-1} \begin{bmatrix} 1 & 2 \\ 1 & 1 \end{bmatrix} \boldsymbol{M} = \begin{bmatrix} 1+\sqrt{2} & 0 \\ 0 & 1-\sqrt{2} \end{bmatrix}.$$

Matrix (b) has complex conjugate eigenvalues $3+\mathrm{i}$, $3-\mathrm{i}$ and equation (2.36) gives an explicit form for $\boldsymbol{M}$. In this case, we find

$$\boldsymbol{M} = \begin{bmatrix} -1 & -1 \\ -2 & 0 \end{bmatrix}$$

and verify that

$$\begin{bmatrix} 2 & 1 \\ -2 & 4 \end{bmatrix} M = M \begin{bmatrix} 3 & -1 \\ 1 & 3 \end{bmatrix}.$$

Finally, matrix (c) has equal eigenvalues and a single eigenvector $\begin{bmatrix} 1 \\ 1 \end{bmatrix}$ obtained by solving

$$\left\{ \begin{bmatrix} 3 & -1 \\ 1 & 1 \end{bmatrix} - 2 \begin{bmatrix} 1 & 0 \\ 0 & 1 \end{bmatrix} \right\} \begin{bmatrix} u_{10} \\ u_{20} \end{bmatrix} = \begin{bmatrix} 0 \\ 0 \end{bmatrix}.$$

This gives the first column of $\boldsymbol{M}$. Provided the second column is chosen to make $\boldsymbol{M}$ non-singular, matrix (c) can be reduced to upper triangular form. For example, let

$$M = \begin{bmatrix} 1 & -1 \\ 1 & 0 \end{bmatrix};$$

this is a simple choice which makes $\det(\boldsymbol{M}) = 1$, then

$$M^{-1} \begin{bmatrix} 3 & -1 \\ 1 & 1 \end{bmatrix} M = \begin{bmatrix} 2 & -1 \\ 0 & 2 \end{bmatrix}.$$

This is not the Jordan form given in (2.38), however, if we replace $\boldsymbol{M}$ by $\boldsymbol{M}_1$ as suggested in (2.29), i.e.

$$M_1 = M \begin{bmatrix} 1 & 0 \\ 0 & -1 \end{bmatrix} = \begin{bmatrix} 1 & 1 \\ 1 & 0 \end{bmatrix},$$

The result $J = \begin{bmatrix} 2 & 1 \\ 0 & 2 \end{bmatrix}$ is achieved. □

We will discuss the Jordan forms of matrices with dimension greater than two in section 2.7, but first let us examine the significance of the above results for linear systems in the plane. We can summarize the situation as follows.

1. Every two-dimensional linear system

$$\dot{x} = Ax \tag{2.39}$$

can be transformed into an equivalent **canonical system**

$$\dot{y} = Jy, \tag{2.40}$$

where $\boldsymbol{J} = \boldsymbol{M}^{-1}\boldsymbol{A}\boldsymbol{M}$ is the Jordan form of $\boldsymbol{A}$ and

$$x = My. \tag{2.41}$$

2. The Jordan matrix $\boldsymbol{J}$ must belong to one of the four types given in the Proposition 2.2.1.

We can now find solutions to any system like (2.39) by solving the corresponding canonical system (2.40) for $\boldsymbol{y}$ and using (2.41) to find $\boldsymbol{x}$.

2.3 PHASE PORTRAITS FOR CANONICAL SYSTEMS IN THE PLANE

A linear system $\dot{\boldsymbol{x}} = \boldsymbol{A}\boldsymbol{x}$ is said to be **simple** if the matrix $\boldsymbol{A}$ is non-singular, (i.e. $\det(\boldsymbol{A}) \neq 0$ and $\boldsymbol{A}$ has non-zero eigenvalues). The only solution to

$$\boldsymbol{A}\boldsymbol{x} = \boldsymbol{0} \tag{2.42}$$

is then $\boldsymbol{x} = \boldsymbol{0}$ and the system has a single isolated fixed point at the origin of the phase plane. The canonical system corresponding to a simple linear system is also simple because the eigenvalues of $\boldsymbol{A}$ and $\boldsymbol{J}$ are the same.

2.3.1 Simple canonical systems

(a) Real, distinct eigenvalues

In this case $\boldsymbol{J}$ is given by (2.18(a)) with λ_1 and λ_2 non-zero. Thus

$$\dot{y}_1 = \lambda_1 y_1, \qquad \dot{y}_2 = \lambda_2 y_2 \tag{2.43}$$

and hencc

$$y_1(t) = C_1 e^{\lambda_1 t}, \qquad y_2(t) = C_2 e^{\lambda_2 t}, \tag{2.44}$$

where C_1, C_2 are real constants.

If λ_1 and λ_2 have thc same sign phase portraits like those in Fig. 2.1 arise. The fixed point at the origin of the $y_1 y_2$-plane is said to be a **node**. When all the trajectories are oriented towards (away from) the origin the node is said to be **stable** (**unstable**). The shape of the trajectories is determined by

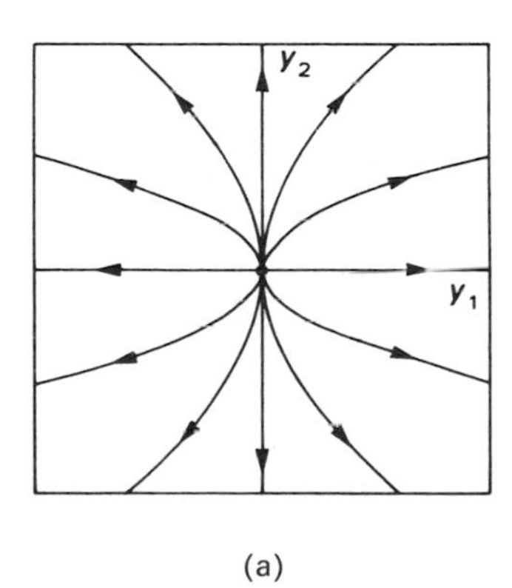

(a)

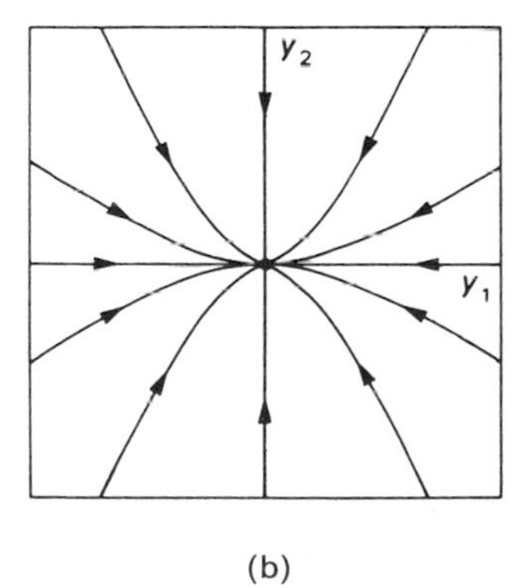

(b)

Fig. 2.1. Real distinct eigenvalues of the same sign give rise to nodes: (a) unstable ($\lambda_1 > \lambda_2 > 0$); (b) stable ($\lambda_2 < \lambda_1 < 0$).

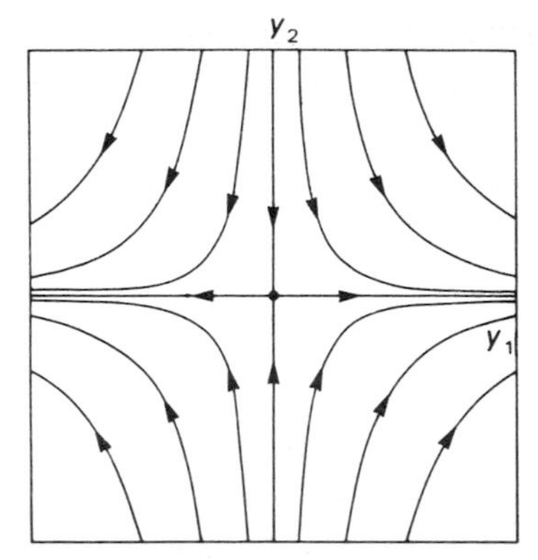

Fig. 2.2. Real eigenvalues of opposite sign ($\lambda_2 < 0 < \lambda_1$) give rise to saddle points.

the ratio $\gamma = \lambda_2/\lambda_1$. Notice that (2.43) and (2.44) give

$$\frac{dy_2}{dy_1} = K y_1^{(\gamma - 1)}, \tag{2.45}$$

where $K = \gamma C_2/C_1^{\gamma}$. Therefore, as $y_1 \to 0$

$$\frac{dy_2}{dy_1} \to \begin{cases} 0, & \text{if } \gamma > 1 \\ \infty, & \text{if } \gamma < 1. \end{cases} \tag{2.46}$$

If λ_1 and λ_2 have opposite signs, the phase portrait in Fig. 2.2 occurs. The coordinate axes (excluding the origin) are the unions of special trajectories, called **separatrices**. These are the only trajectories that are radial straight lines. A particular coordinate axis contains a pair of separatrices (remember the origin is a trajectory in its own right) which are directed towards (away from) the origin if the corresponding eigenvalue is negative (positive). The remaining trajectories have the separatrices as asymptotes; first approaching the fixed point as t increases from $-\infty$, passing through a point of closest approach and finally moving away again. In this case the origin is said to be a **saddle point**.

(b) Equal eigenvalues

If $\boldsymbol{J}$ is diagonal, the canonical system has solutions given by (2.44) with $\lambda_1 = \lambda_2 = \lambda_0 \neq 0$. Thus (2.18(b)) corresponds to a special node, called a **star** (stable if $\lambda_0 < 0$; unstable if $\lambda_0 > 0$), in which the non-trivial trajectories are all radial straight lines (as shown in Fig. 2.3).

If $\boldsymbol{J}$ is not diagonal (i.e. (2.18(c)) then we must consider

$$\dot{y}_1 = \lambda_0 y_1 + y_2, \qquad \dot{y}_2 = \lambda_0 y_2; \qquad \lambda_0 \neq 0. \tag{2.47}$$

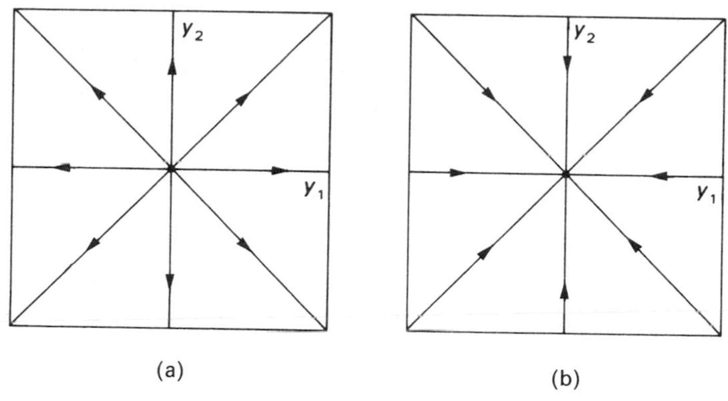

Fig. 2.3. Equal eigenvalues ($\lambda_1 = \lambda_2 = \lambda_0$) give rise to star nodes: (a) unstable; (b) stable; when A is diagonal.

This system has solutions

$$y_1(t) = (C_1 + tC_2)e^{\lambda_0 t}; \qquad y_2(t) = C_2 e^{\lambda_0 t} \tag{2.48}$$

and phase portraits like those in Fig. 2.4 are obtained (cf. Example 1.4.2).

The origin is said to be an **improper node** (stable $\lambda_0 < 0$; unstable $\lambda_0 > 0$). The line on which the trajectories change direction is the locus of extreme values for y_1. This is given by the $\dot{y}_1 = 0$ isocline, namely

$$y_2 = -\lambda_0 y_1. \tag{2.49}$$

(c) Complex eigenvalues

The Jordan matrix is given by (2.18(d)) so the canonical system is

$$\dot{y}_1 = \alpha y_1 - \beta y_2; \qquad \dot{y}_2 = \beta y_1 + \alpha y_2. \tag{2.50}$$

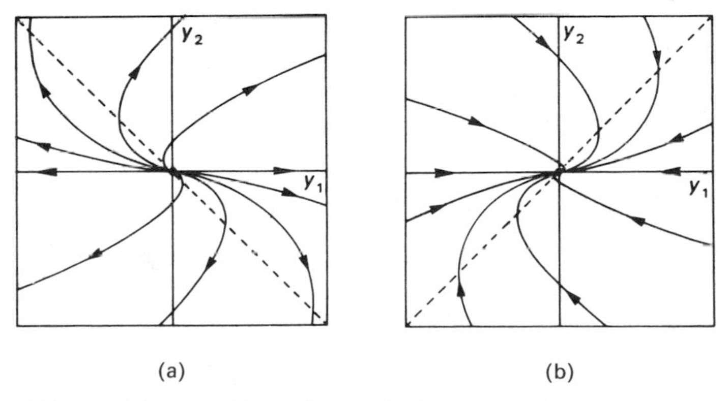

Fig. 2.4. When A is not diagonal, equal eigenvalues indicate that the origin is an improper node: (a) unstable ($\lambda_0 > 0$); (b) stable ($\lambda_0 < 0$).

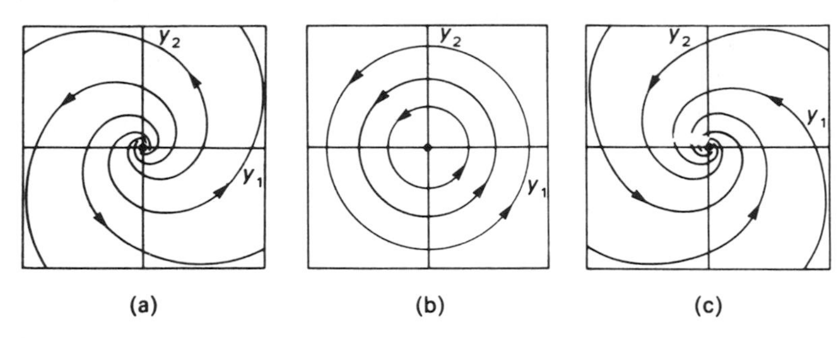

Fig. 2.5. Complex eigenvalues give rise to (a) unstable foci ($\alpha > 0$), (b) centres ($\alpha = 0$) and (c) stable foci ($\alpha < 0$).

This kind of system can be solved by introducing plane polar coordinates such that $y_1 = r\cos\theta$, $y_2 = r\sin\theta$. We obtain, as in Example 1.4.1,

$$\dot{r} = \alpha r, \qquad \dot{\theta} = \beta \tag{2.51}$$

with solutions

$$r(t) = r_0 e^{\alpha t}, \qquad \theta(t) = \beta t + \theta_0. \tag{2.52}$$

Typical phase portraits are shown in Fig. 2.5.

If $\alpha \neq 0$ the origin is said to be a **focus** (stable if $\alpha < 0$; unstable if $\alpha > 0$). The phase portrait is often said to consist of an **attracting** ($\alpha < 0$) or **repelling** ($\alpha > 0$) **spiral**. The parameter $\beta > 0$ determines the angular speed of description of the spiral.

When $\alpha = 0$, the origin is said to be a **centre** and the phase portrait consists of a continuum of concentric circles. This is the only non-trivial way in which recurrent or **periodic** behaviour occurs in planar linear systems. Every point (excluding the origin) in the phase plane recurs at intervals of $T = 2\pi/\beta$. The coordinates are periodic in t with this period, i.e.,

$$y_1 = r_0\cos(\beta t + \theta_0), \qquad y_2 = r_0\sin(\beta t + \theta_0). \tag{2.53}$$

The other (trivial) form of recurrence that occurs in simple linear systems is the fixed point. In a sense, the fixed point is the ultimate in recurrence; it recurs instantaneously with period zero.

2.3.2 Non-simple canonical systems

A linear system $\dot{\boldsymbol{x}} = \boldsymbol{A}\boldsymbol{x}$ is non-simple if $\boldsymbol{A}$ is singular (i.e. $\det(\boldsymbol{A}) = 0$ and at least one of the eigenvalues of $\boldsymbol{A}$ is zero). It follows that there are non-trivial solutions to $\boldsymbol{A}\boldsymbol{x} = 0$ and the system has fixed points other than $\boldsymbol{x} = 0$. For linear systems in the plane, there are only two possibilities: either the rank of $\boldsymbol{A}$ is one; or $\boldsymbol{A}$ is null. In the first case there is a line of fixed points passing through the origin; in the second, every point in the plane is a fixed point. Of course, the rank of $\boldsymbol{J}$ is equal to the rank of $\boldsymbol{A}$, so that the canonical systems

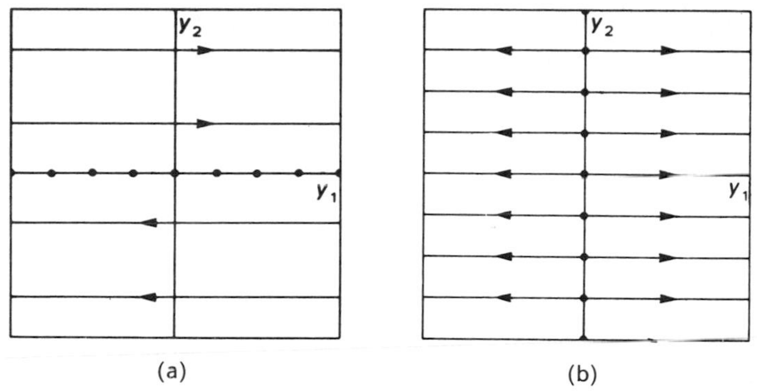

Fig. 2.6. (a)

$$\begin{bmatrix}\dot{y}_1\\ \dot{y}_2\end{bmatrix}=\begin{bmatrix}0 & 1\\ 0 & 0\end{bmatrix}\begin{bmatrix}y_1\\ y_2\end{bmatrix}, \quad \text{i.e. } \lambda_0=0.$$

Every point on the y_1-axis is a fixed point; cf. the $\lambda_0 \to 0$ limit of Fig. 2.4(a).
(b)

$$\begin{bmatrix}\dot{y}_1\\ \dot{y}_2\end{bmatrix}=\begin{bmatrix}1 & 0\\ 0 & 0\end{bmatrix}\begin{bmatrix}y_1\\ y_2\end{bmatrix}, \quad \text{i.e. } \lambda_1=1,\ \lambda_2=0.$$

Every point on the y_2-axis is a fixed point; cf. the $\lambda_2 \to 0$ limit of Fig. 2.1(a).

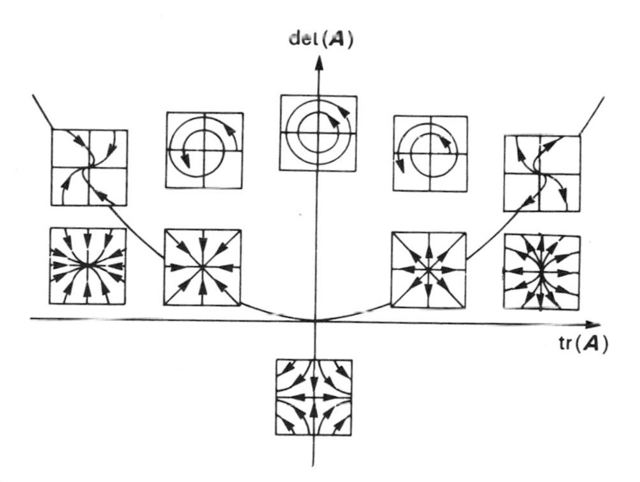

Fig. 2.7. Summary of how the phase portraits of the systems $\dot{\boldsymbol{x}}=\boldsymbol{A}\boldsymbol{x}$ depend on the trace and determinant of $\boldsymbol{A}$.

exhibit corresponding non-simple behaviour. Figure 2.6 shows two examples when $\boldsymbol{J}$ has rank one.

We shall, in the remainder of this chapter, focus attention on simple linear systems. These systems play a key role in understanding the nature of the fixed points in non-linear systems (described in section 3.3).

The results obtained in this section are summarized in Fig. 2.7, where each type of phase portrait is associated with a set of points in the $\text{tr}(\boldsymbol{A})$–$\det(\boldsymbol{A})$ plane. Each point of this plane corresponds to a particular pair of eigenvalues for $\boldsymbol{A}$ and therefore to a particular canonical system.

2.4 CLASSIFICATION OF SIMPLE LINEAR PHASE PORTRAITS IN THE PLANE

2.4.1 Phase portrait of a simple linear system

The phase portrait of any linear system $\dot{\boldsymbol{x}} = \boldsymbol{A}\boldsymbol{x}$ can be obtained from that of its canonical form $\dot{\boldsymbol{y}} = \boldsymbol{J}\boldsymbol{y}$ by applying the transformation $\boldsymbol{x} = \boldsymbol{M}\boldsymbol{y}$. Let us consider how the phase portrait is changed under this transformation. For example, the canonical system

$$\dot{y}_1 = y_1; \qquad \dot{y}_2 = -y_2 \tag{2.54}$$

has the phase portrait shown in Fig. 2.8(a). Figure 2.8(b)–(f) shows the phase portraits for some linear systems that have (2.54) as their canonical system.

As noted in (2.5) the variables y_1 and y_2 are the coordinates of $\boldsymbol{x}$ relative to the basis $\{\boldsymbol{m}_1, \boldsymbol{m}_2\}$ obtained from the columns of $\boldsymbol{M}$. Thus the y_1- and y_2-axes are represented by straight lines, through the origin of the x_1x_2-plane, in the directions of $\boldsymbol{m}_1$ and $\boldsymbol{m}_2$ respectively.

The directions defined by $\boldsymbol{m}_1$ and $\boldsymbol{m}_2$ are called the **principal directions** at the origin. The mapping is a bijection so every point of the y_1y_2-plane is uniquely represented in the x_1x_2-plane and vice versa. Furthermore, $\boldsymbol{x}$ is a continuous function of $\boldsymbol{y}$, so that trajectories map onto trajectories. Orientation of the trajectories is also preserved. In particular, the directions of the separatrices in the x_1x_2-plane are given by the eigenvectors of $\boldsymbol{A}$.

These properties of the linear transformation $\boldsymbol{x} = \boldsymbol{M}\boldsymbol{y}$ are sufficient to ensure that, although the phase portrait is distorted, the origin is still a saddle point. The phase portraits in Fig. 2.8(a)–(f) are typical of all systems $\dot{\boldsymbol{x}} = \boldsymbol{A}\boldsymbol{x}$ for which $\boldsymbol{A}$ has eigenvalues $+1$ and -1; they all have saddle points at the origin. In other words, linear transformations preserve the qualitative behaviour of the solutions.

Example 2.4.1

Sketch the phase portrait of the system

$$\dot{y}_1 = 2y_1, \qquad \dot{y}_2 = y_2, \tag{2.55}$$

and the corresponding phase portraits in the x_1x_2-plane where

$$x_1 = y_1 + y_2, \qquad x_2 = y_1 + 2y_2, \tag{2.56}$$

and

$$x_1 = y_1, \qquad x_2 = -y_1 + y_2 \tag{2.57}$$

respectively.

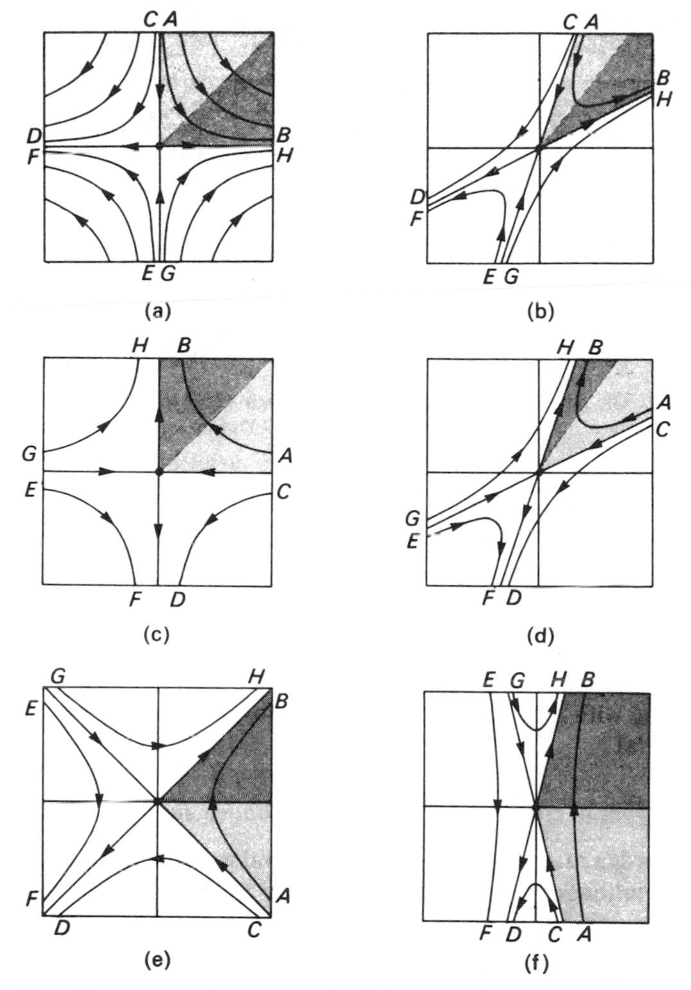

Fig. 2.8. The effect on the phase portrait of the canonical system $\dot{y}_1 = y_1$, $\dot{y}_2 = -y_2$ (shown in (a)) of the linear transformation $\boldsymbol{x} = \boldsymbol{M}\boldsymbol{y}$ with:

(b) $\boldsymbol{M} = \begin{bmatrix} 2 & 1 \\ 1 & 3 \end{bmatrix}$; (c) $\boldsymbol{M} = \begin{bmatrix} 0 & 1 \\ 1 & 0 \end{bmatrix}$;

(d) $\boldsymbol{M} = \begin{bmatrix} 1 & 2 \\ 3 & 1 \end{bmatrix}$; (e) $\boldsymbol{M} = \begin{bmatrix} 1 & 1 \\ 1 & -1 \end{bmatrix}$;

(f) $\boldsymbol{M} = \begin{bmatrix} 1 & 1 \\ 4 & -4 \end{bmatrix}$.

The pairs AB, CD, EF and GH label corresponding segments of trajectories in the phase portraits.

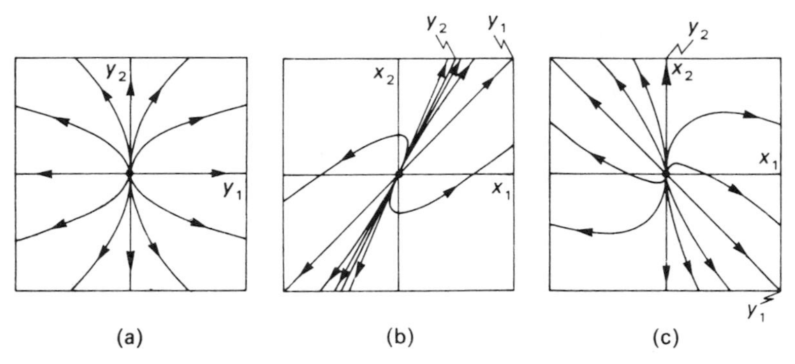

Fig. 2.9. The phase portraits of the canonical system (2.55) in the y_1, y_2-plane [(a)] and in the x_1, x_2-plane defined by (2.56) and (2.57) [(b) and (c) respectively].

Solution

The phase portrait of system (2.55) is sketched in Fig. 2.9(a).

Observe that the change of variables given by (2.56) and (2.57) can be written in the matrix form $\boldsymbol{x} = \boldsymbol{M}\boldsymbol{y}$ with

$$\boldsymbol{M} = \begin{bmatrix} 1 & 1 \\ 1 & 2 \end{bmatrix} \quad \text{and} \quad \begin{bmatrix} 1 & 0 \\ -1 & 1 \end{bmatrix}$$

respectively. The basis for which y_1 and y_2 are coordinates is $\{(1,1), (1,2)\}$ for (2.56) and $\{(1,-1), (0,1)\}$ for (2.57). The y_1 and y_2-coordinate directions are shown in the $x_1 x_2$-plane in Fig. 2.9(b) and (c) along with the correspondingly transformed phase portrait from Fig. 2.9(a). □

2.4.2 Types of canonical system and qualitative equivalence

Two Jordan matrices with different eigenvalues are not similar and so their solutions cannot be related by a linear transformation. However, the fixed point at the origin of every simple canonical system is one of a small number of possibilities, namely a node, a focus, a saddle, etc. How are these phase portraits related?

Consider, for example, the canonical systems

$$\dot{y}_1 = \alpha y_1 - \beta y_2, \qquad \dot{y}_2 = \beta y_1 + \alpha y_2, \tag{2.58}$$

$\alpha, \beta > 0$ and

$$\dot{z}_1 = z_1 - z_2, \qquad \dot{z}_2 = z_1 + z_2. \tag{2.59}$$

Their phase portraits are both unstable foci (as illustrated in Fig. 2.10).

The trajectories are given by

$$r = r_0 e^{\alpha t}, \qquad \theta = \beta t + \theta_0 \tag{2.60}$$

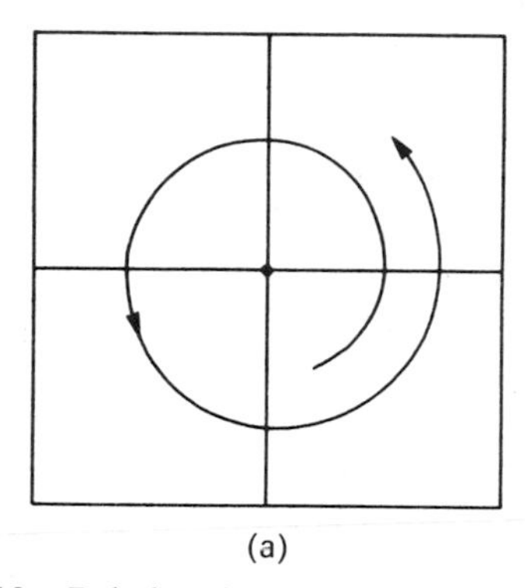

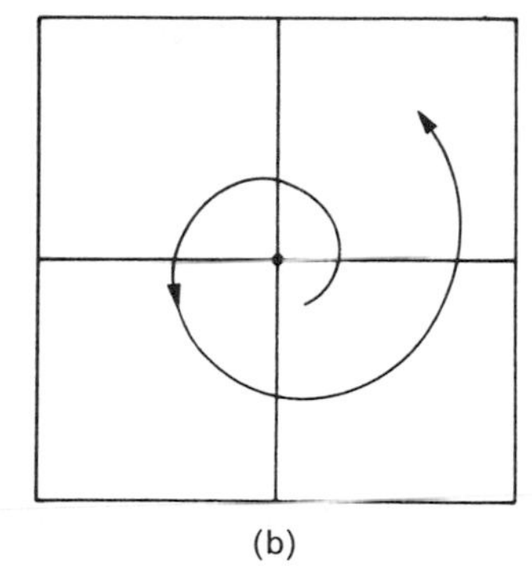

Fig. 2.10. Relation between phase portraits of canonical systems: (a) $y_1 y_2$-plane: trajectory through (r_0, θ_0) of the system (2.60); (b) $z_1 z_2$-plane: trajectory through (R_0, ϕ_0) of the system (2.61).

and

$$R = R_0 e^t, \qquad \phi = t + \phi_0, \tag{2.61}$$

respectively, where (r, θ) and (R, ϕ) are polar coordinates in the $y_1 y_2$- and $z_1 z_2$-planes. The trajectory passing through (r_0, θ_0), $r_0 \neq 0$, in the y_1, y_2-plane can be mapped onto the trajectory passing through (R_0, ϕ_0) in the $z_1 z_2$-plane by the transformation

$$R = R_0 \left(\frac{r}{r_0} \right)^{1/\alpha}, \qquad \phi = \frac{\theta - \theta_0}{\beta} + \phi_0. \tag{2.62}$$

This transformation allows the phase portrait in Fig. 2.10(a) to be mapped—trajectory by trajectory—onto that in Fig. 2.10(b). The transformation in (2.62) is a bijection; it is continuous; it preserves the orientation of the trajectories. However, it does *not* map the $y_1 y_2$-plane onto the $z_1 z_2$-plane in a *linear* way, unless $\alpha = \beta = 1$.

In sections 2.4.1 and 2.4.2 we have examined more closely the relationship between systems with the same qualitative behaviour (see also Exercise 2.18). We are led to make the intuitive ideas of section 1.3 more precise in the following definition.

Definition 2.4.1
Two systems of first-order differential equations are said to be **qualitatively equivalent** if there is a continuous bijection which maps the phase portrait of one onto that of the other in such a way as to preserve the orientation of the trajectories.

Note that, for consistency with Definition 1.2.1, the map relating the two phase portraits must also be order-preserving in the one-dimensional case.

2.4.3 Classification of linear systems

In this chapter, we have illustrated the fact that every simple linear system in the plane is qualitatively equivalent to one of the systems whose phase portraits are shown in Fig. 2.11. The ten phase portraits shown are representative of the **algebraic type** of the linear system.

Qualitative equivalence goes further than this. It can be shown that all stable (unstable) nodes, improper nodes and foci are also related in the manner described in Definition 2.4.1. This means that the algebraic types can be further grouped into **qualitative** (or **topological**) **types** as indicated in Fig. 2.11. In these terms, simple linear systems only exhibit four kinds of qualitative behaviour: stable, centre, saddle and unstable.

2.5 THE EVOLUTION OPERATOR

In Section 1.5 we suggested that the trajectories of linear systems in the plane could be described by an evolution matrix. We will now consider this idea in more detail.

For any real $n \times n$ matrix $\boldsymbol{P}$ define the **exponential matrix** $\mathrm{e}^{\boldsymbol{P}}$ by

$$\mathrm{e}^{\boldsymbol{P}} = \exp(\boldsymbol{P}) = \sum_{k=0}^{\infty} \frac{\boldsymbol{P}^k}{k!} \tag{2.63}$$

with $\boldsymbol{P}^0 = \boldsymbol{I}_n$, the $n \times n$ unit matrix. Let $\boldsymbol{Q}$ be a real $n \times n$ matrix such that $\boldsymbol{PQ} = \boldsymbol{QP}$ then it follows from (2.63) that

$$\mathrm{e}^{\boldsymbol{P}+\boldsymbol{Q}} = \mathrm{e}^{\boldsymbol{P}}\mathrm{e}^{\boldsymbol{Q}}, \tag{2.64}$$

(proved in Exercise 2.25). This result allows us to conclude that

$$\mathrm{e}^{\boldsymbol{P}}\mathrm{e}^{-\boldsymbol{P}} = \mathrm{e}^{-\boldsymbol{P}}\mathrm{e}^{\boldsymbol{P}} = \mathrm{e}^{\boldsymbol{O}} = \boldsymbol{I}_n, \tag{2.65}$$

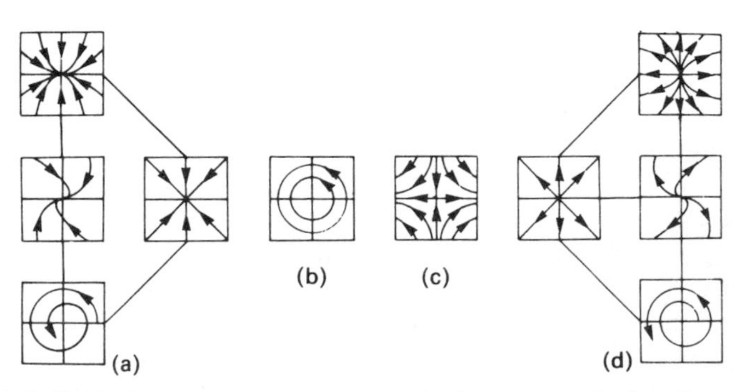

Fig. 2.11. The distinct qualitative types of phase portrait for linear systems: (a) stable; (b) centre; (c) saddle; (d) unstable; in relation to the algebraic types obtained in Section 2.3.

so that

$$(e^{P})^{-1} = e^{-P}. \tag{2.66}$$

If we let $P = At$ and differentiate (2.63) with respect to t, we find

$$\frac{d}{dt}(e^{At}) = Ae^{At} = e^{At}A. \tag{2.67}$$

This means that

$$\frac{d}{dt}(e^{-At}x) = e^{-At}\dot{x} - e^{-At}Ax = e^{-At}(\dot{x} - Ax) \equiv 0. \tag{2.68}$$

If we are given $x(t_0) = x_0$, then integrating (2.68) from t_0, to t, gives

$$e^{-At}x(t) = e^{-At_0}x_0 \tag{2.69}$$

and hence

$$x(t) = e^{A(t-t_0)}x_0 = \phi_{t-t_0}(x_0), \tag{2.70}$$

by (2.66). Thus, matrix multiplication by $e^{A(t-t_0)}$ gives the evolution of the phase point which is at x_0 when $t = t_0$. It follows that e^{At} is the evolution operator for the linear system $\dot{x} = Ax$.

Of course, in order to make use of (2.70) we must know $e^{A(t-t_0)}$. For simple canonical systems in the plane we can write down e^{Jt} from the solutions given in section 2.3.1. We find:

$$e^{Jt} = \begin{bmatrix} e^{\lambda_1 t} & 0 \\ 0 & e^{\lambda_2 t} \end{bmatrix} \quad \text{for } J = \begin{bmatrix} \lambda_1 & 0 \\ 0 & \lambda_2 \end{bmatrix} \tag{2.71}$$

(including $\lambda_1 = \lambda_2 = \lambda_0$) from (2.44);

$$e^{Jt} = e^{\lambda_0 t}\begin{bmatrix} 1 & t \\ 0 & 1 \end{bmatrix} \quad \text{for } J = \begin{bmatrix} \lambda_0 & 1 \\ 0 & \lambda_0 \end{bmatrix} \tag{2.72}$$

from (2.48); and

$$e^{Jt} = e^{\alpha t}\begin{bmatrix} \cos\beta t & -\sin\beta t \\ \sin\beta t & \cos\beta t \end{bmatrix} \quad \text{for } J = \begin{bmatrix} \alpha & -\beta \\ \beta & \alpha \end{bmatrix} \tag{2.73}$$

from (2.52). These results can also be obtained directly from (2.63) (as shown in Exercise 2.26).

Equations (2.71)–(2.73) can be used to obtain e^{At} for any non-singular 2×2 martrix A by exploiting the relation between A and its Jordan form J. Recall,

$$x(t) = My(t) = Me^{Jt}y(0) = (Me^{Jt}M^{-1})x(0), \tag{2.74}$$

so that

$$e^{At} = Me^{Jt}M^{-1}. \tag{2.75}$$

While this gives us an understandable way of obtaining e^{At} it is not always convenient in practice. An alternative way is to adapt a general method of Sylvester (Barnett, 1975) for finding functions of matrices. We consider the method for $n = 2$.

Let the 2×2 real matrix $\boldsymbol{A}$ have two distinct real or complex eigenvalues λ_1, λ_2. Define the (possibly complex) matrices $\boldsymbol{Q}_1, \boldsymbol{Q}_2$ by

$$\boldsymbol{Q}_1 = \frac{\boldsymbol{A} - \lambda_2 \boldsymbol{I}}{\lambda_1 - \lambda_2}, \qquad \boldsymbol{Q}_2 = \frac{\boldsymbol{A} - \lambda_1 \boldsymbol{I}}{\lambda_2 - \lambda_1} \tag{2.76}$$

and observe that $\boldsymbol{A} = \lambda_1 \boldsymbol{Q}_1 + \lambda_2 \boldsymbol{Q}_2$. It can be shown (Exercise 2.27) that $\boldsymbol{Q}_1 \boldsymbol{Q}_2 = \boldsymbol{Q}_2 \boldsymbol{Q}_1 = \boldsymbol{0}$, $\boldsymbol{Q}_1^2 = \boldsymbol{Q}_1$ and $\boldsymbol{Q}_2^2 = \boldsymbol{Q}_2$ and therefore

$$\boldsymbol{A}^k = (\lambda_1 \boldsymbol{Q}_1 + \lambda_2 \boldsymbol{Q}_2)^k = \lambda_1^k \boldsymbol{Q}_1 + \lambda_2^k \boldsymbol{Q}_2 \tag{2.77}$$

for any positive integer k. Thus

$$\begin{aligned} \mathrm{e}^{\boldsymbol{A}t} &= \sum_{k=0}^{\infty} \frac{\boldsymbol{A}^k t^k}{k!} = \sum_{k=0}^{\infty} \left(\frac{\lambda_1^k t^k}{k!} \boldsymbol{Q}_1 + \frac{\lambda_2^k t^k}{k!} \boldsymbol{Q}_2 \right) \\ &= \mathrm{e}^{\lambda_1 t} \boldsymbol{Q}_1 + \mathrm{e}^{\lambda_2 t} \boldsymbol{Q}_2. \end{aligned} \tag{2.78}$$

When the eigenvalues λ_1, λ_2 of $\boldsymbol{A}$ are equal, define $\boldsymbol{Q} = \boldsymbol{A} - \lambda_0 \boldsymbol{I}$, where $\lambda_1 = \lambda_2 = \lambda_0$. In this case $\boldsymbol{Q}^2 = \boldsymbol{0}$ and

$$\boldsymbol{A}^k = (\lambda_0 \boldsymbol{I} + \boldsymbol{Q})^k = \lambda_0^k \boldsymbol{I} + k \lambda_0^{k-1} \boldsymbol{Q} \tag{2.79}$$

because $\boldsymbol{Q}^k = \boldsymbol{0}$ for $k \geqslant 2$. Thus

$$\begin{aligned} \mathrm{e}^{\boldsymbol{A}t} &= \sum_{k=0}^{\infty} \left(\frac{\lambda_0^k}{k!} \boldsymbol{I} + \frac{k \lambda_0^{k-1}}{k!} \boldsymbol{Q} \right) t^k \\ &= \mathrm{e}^{\lambda_0 t} (\boldsymbol{I} + t(\boldsymbol{A} - \lambda_0 \boldsymbol{I})). \end{aligned} \tag{2.80}$$

Example 2.5.1

Find $\mathrm{e}^{\boldsymbol{A}t}$ where

$$\boldsymbol{A} = \begin{bmatrix} 1 & 1 \\ -1 & 3 \end{bmatrix} \tag{2.81}$$

by (a) the method of reducing to canonical form, and (b) the use of formula (2.78) or (2.80).

Solution

(a) The eigenvalues $\boldsymbol{A}$ are both $\lambda_0 = 2$. An eigenvector $\boldsymbol{u}_0$ is given by $\boldsymbol{u}_0 = \begin{bmatrix} 1 \\ 1 \end{bmatrix}$. Let

$$\boldsymbol{M} = \begin{bmatrix} 1 & 1 \\ 1 & -0 \end{bmatrix}, \tag{2.82}$$

then

$$M^{-1}AM = \begin{bmatrix} 2 & 1 \\ 0 & 2 \end{bmatrix} = J; \tag{2.83}$$

the required Jordan matrix. The exponential matrix

$$e^{Jt} = e^{2t}\begin{bmatrix} 1 & t \\ 0 & 1 \end{bmatrix}$$

by (2.72), and (2.75) gives

$$\begin{aligned} e^{At} = Me^{Jt}M^{-1} &= e^{2t}\begin{bmatrix} 1 & -1 \\ 1 & 0 \end{bmatrix}\begin{bmatrix} 1 & t \\ 0 & 1 \end{bmatrix}\begin{bmatrix} 0 & 1 \\ -1 & 1 \end{bmatrix} \\ &= e^{2t}\begin{bmatrix} 1-t & t \\ -t & 1+t \end{bmatrix}. \end{aligned} \tag{2.84}$$

(b) The eigenvalues of A have been calculated in (a). Substituting for A and $\lambda_0 = 2$ in (2.80), we obtain

$$\begin{aligned} e^{At} &= e^{2t}\left\{I + t\left\{\begin{bmatrix} 1 & 1 \\ -1 & 3 \end{bmatrix} - 2\begin{bmatrix} 1 & 0 \\ 0 & 1 \end{bmatrix}\right\}\right\} \\ &= e^{2t}\begin{bmatrix} 1-t & t \\ -t & 1+t \end{bmatrix}. \end{aligned} \tag{2.85}$$

□

2.6 AFFINE SYSTEMS

In this section we consider systems where the vector field **X** is of the form

$$X(x) = Ax + h \tag{2.86}$$

where $h \in \mathbb{R}^n$. The mapping **X** is said to be **affine**. In the strict algebraic sense $Ax + h$ is a non-linear function of x but it is clearly closely related to linear functions. In fact, systems $\dot{x} = X(x)$ with X given by (2.86) are frequently called **non-homogeneous linear** systems as opposed to the **homogeneous linear** system $\dot{x} = Ax$.

The change of variable

$$x = y - x_0, \tag{2.87}$$

where x_0 is a solution of $Ax_0 = h$, if one exists, allows (2.86) to be written as

$$Y(y) = X(y - x_0) = A(y - x_0) + h = Ay. \tag{2.88}$$

In such cases, the change of variable (2.87) allows the affine system

$$\dot{x} = Ax + h \tag{2.89}$$

to be written as the linear system

$$\dot{y} = Ay. \tag{2.90}$$

However, what if $\boldsymbol{h}$ is not in the image of $\boldsymbol{A}$ or if $\boldsymbol{h}$ in (2.89) is replaced by a time-dependent vector $\boldsymbol{h}(t)$? Non-autonomous systems of the latter form frequently occur in practical problems such as those described in section 5.2. The evolution operator provides a solution to both of these difficulties.

The procedure is a matrix equivalent of the 'integrating factor method' used in Exercise 1.1. Multiply (2.89) throughout by e^{-At} and rearrange as

$$\frac{\mathrm{d}}{\mathrm{d}t}(\mathrm{e}^{-At}\boldsymbol{x}) = \mathrm{e}^{-At}\boldsymbol{h}(t). \tag{2.91}$$

Here the evolution operator acts as a 'matrix integrating factor'. Suppose we are given $\boldsymbol{x}(t_0) = \boldsymbol{x}_0$, then integrating (2.91) from t_0 to t gives

$$\mathrm{e}^{-At}\boldsymbol{x}(t) - \mathrm{e}^{-At_0}\boldsymbol{x}_0 = \int_{t_0}^{t} \mathrm{e}^{-As}\boldsymbol{h}(s)\mathrm{d}s.$$

Since $\mathrm{e}^{At}\mathrm{e}^{-At} = \boldsymbol{I}$ we can solve for $\boldsymbol{x}(t)$; thus

$$\boldsymbol{x}(t) = \boldsymbol{x}_{\mathrm{CF}}(t) + \boldsymbol{x}_{\mathrm{PI}}(t), \tag{2.92}$$

where

$$\boldsymbol{x}_{\mathrm{CF}}(t) = \mathrm{e}^{A(t-t_0)}\boldsymbol{x}_0 \tag{2.93}$$

and

$$\boldsymbol{x}_{\mathrm{PI}}(t) = \mathrm{e}^{At}\int_{t_0}^{t} \mathrm{e}^{-As}\boldsymbol{h}(s)\mathrm{d}s. \tag{2.94}$$

Equation (2.92) gives the general solution to the affine system (2.89), whether or not $\boldsymbol{h}$ depends on t.

The first term $\boldsymbol{x}_{\mathrm{CF}}(t)$ is the solution of the linear system $\dot{\boldsymbol{x}} = \boldsymbol{A}\boldsymbol{x}$ which satisfies the initial condition $\boldsymbol{x}_{\mathrm{CF}}(t_0) = \boldsymbol{x}_0$. It corresponds to the **complementary function** of traditional treatments of second-order, non-homogeneous linear equations. The second term $\boldsymbol{x}_{\mathrm{PI}}(t)$, gives a general expression for the '**particular integral**' of such treatments. It is easy to verify that $\boldsymbol{x}_{\mathrm{PI}}(t)$ satisfies (2.89) and that $\boldsymbol{x}_{\mathrm{PI}}(0) = \boldsymbol{0}$, so that $\boldsymbol{x}(t)$ in (2.92) satisfies both (2.89) and $\boldsymbol{x}(t_0) = \boldsymbol{x}_0$.

Example 2.6.1

Consider the second-order equation

$$\ddot{x} + 2\dot{x} + 2x = u(t). \tag{2.95}$$

Write this equation a first-order system and obtain a general expression for the solution which satisfies $x(0) = \dot{x}(0) = 0$.

Solution

Let $x_1 = x$ and $x_2 = \dot{x}$, then (2.95) becomes

$$\dot{x}_1 = x_2, \qquad \dot{x}_2 = -2x_1 - 2x_2 + u(t). \tag{2.96}$$

In matrix notation, (2.96) takes the form $\dot{\boldsymbol{x}} = \boldsymbol{A}\boldsymbol{x} + \boldsymbol{h}(t)$ with

$$\boldsymbol{A} = \begin{bmatrix} 0 & 1 \\ -2 & -2 \end{bmatrix} \quad \text{and} \quad \boldsymbol{h}(t) = \begin{bmatrix} 0 \\ u(t) \end{bmatrix}. \tag{2.97}$$

The matrix $\boldsymbol{A}$ has eigenvalues $\lambda_1 = -1 + \mathrm{i}$, $\lambda_2 = -1 - \mathrm{i}$ and (2.78) gives

$$\mathrm{e}^{At} = \mathrm{e}^{-t}\left\{\frac{\mathrm{e}^{\mathrm{i}t}}{2\mathrm{i}}\begin{bmatrix} 1+\mathrm{i} & 1 \\ -2 & -1+\mathrm{i} \end{bmatrix} - \frac{\mathrm{e}^{-\mathrm{i}t}}{2\mathrm{i}}\begin{bmatrix} 1-\mathrm{i} & 1 \\ -2 & -1-\mathrm{i} \end{bmatrix}\right\}$$

$$\mathrm{e}^{At} = \mathrm{e}^{-t}\begin{bmatrix} \cos t + \sin t & \sin t \\ -2\sin t & \cos t - \sin t \end{bmatrix}. \tag{2.98}$$

The initial condition given corresponds to $\boldsymbol{x}(0) = \boldsymbol{0}$, so $\boldsymbol{x}_{\mathrm{CF}}(t) \equiv \boldsymbol{0}$ in (2.92) and $\boldsymbol{x}(t) = \boldsymbol{x}_{\mathrm{PI}}(t)$, which is

$$\mathrm{e}^{-t}\begin{bmatrix} \cos t + \sin t & \sin t \\ -2\sin t & \cos t - \sin t \end{bmatrix}\int_0^t \mathrm{e}^s u(s)\begin{bmatrix} -\sin s \\ \cos s + \sin s \end{bmatrix}\mathrm{d}s. \tag{2.99}$$

Hence

$$x_1(t) = x(t) = \mathrm{e}^{-t}\left\{\sin t\int_0^t \mathrm{e}^s u(s)\cos s\,\mathrm{d}s - \cos t\int_0^t \mathrm{e}^s u(s)\sin s\,\mathrm{d}s\right\}. \tag{2.100}$$

□

2.7 LINEAR SYSTEMS OF DIMENSION GREATER THAN TWO

2.7.1 Three-dimensional systems

The real Jordan forms for 3×3 matrices are

$$\begin{bmatrix} \lambda_0 & 0 & 0 \\ 0 & \lambda_1 & 0 \\ 0 & 0 & \lambda_2 \end{bmatrix}, \quad \begin{bmatrix} \alpha & -\beta & 0 \\ \beta & \alpha & 0 \\ 0 & 0 & \lambda_1 \end{bmatrix},$$

$$\begin{bmatrix} \lambda_0 & 1 & 0 \\ 0 & \lambda_0 & 0 \\ 0 & 0 & \lambda_1 \end{bmatrix} \quad \text{and} \quad \begin{bmatrix} \lambda_0 & 1 & 0 \\ 0 & \lambda_0 & 1 \\ 0 & 0 & \lambda_0 \end{bmatrix}, \tag{2.101}$$

where α, β, λ_0, λ_1, λ_2 are real (Hirsch and Smale, 1974). Observe that all, except the last form, in (2.101) can be partitioned in diagonal blocks of dimension 1 and 2. For example, consider

$$\dot{\boldsymbol{y}} = \boldsymbol{J}\boldsymbol{y} \quad \text{with} \quad \boldsymbol{J} = \left[\begin{array}{cc|c} \alpha & -\beta & 0 \\ \beta & \alpha & 0 \\ \hline 0 & 0 & \lambda_1 \end{array}\right] \tag{2.102}$$

where $\lambda_1 < \alpha < 0$, $\beta > 0$. The diagonal block structure ensures that the y_3-equation is decoupled from the other two. In fact, (2.102) is equivalent to the two systems

$$\begin{bmatrix} \dot{y}_1 \\ \dot{y}_2 \end{bmatrix} = \begin{bmatrix} \alpha & -\beta \\ \beta & \alpha \end{bmatrix} \begin{bmatrix} y_1 \\ y_2 \end{bmatrix}, \tag{2.103}$$

$$\dot{y}_3 = \lambda_1 y_3. \tag{2.104}$$

This decoupling allows us to construct the qualitative features of the three-dimensional phase portrait.

Equation (2.103) shows that the projection of the trajectories onto a plane of constant y_3 is a stable ($\alpha < 0$) focus, the attracting spiral being described in an anticlockwise sense ($\beta > 0$). Since $\lambda_1 < 0$, $|y_3(t)|$ decreases exponentially with increasing t. A typical trajectory of (2.102) passing through (a, b, c) at $t = 0$, with all three coordinates non-zero, is shown in Fig. 2.12. The spiral in the constant y_3 projection is drawn out into a kind of helix with decreasing radius and pitch.

Returning to (2.101), the Jordan form that is not amenable to block decomposition corresponds to a system $\dot{\boldsymbol{y}} = \boldsymbol{J}\boldsymbol{y}$ that can be solved by finding $\mathrm{e}^{\boldsymbol{J}t}$. The evolution operator can be obtained by using the expansion (2.63); observe

$$\boldsymbol{J} = \begin{bmatrix} \lambda_0 & 1 & 0 \\ 0 & \lambda_0 & 1 \\ 0 & 0 & \lambda_0 \end{bmatrix} = \begin{bmatrix} \lambda_0 & 0 & 0 \\ 0 & \lambda_0 & 0 \\ 0 & 0 & \lambda_0 \end{bmatrix} + \begin{bmatrix} 0 & 1 & 0 \\ 0 & 0 & 1 \\ 0 & 0 & 0 \end{bmatrix}$$

$$\boldsymbol{J} = \boldsymbol{D} + \boldsymbol{C}. \tag{2.105}$$

As in Exercise 2.25, $\boldsymbol{DC} = \boldsymbol{CD}$ and (2.64) reduces the problem to one of finding $\mathrm{e}^{\boldsymbol{C}t}$. In this case, $\boldsymbol{C}^3 = \boldsymbol{0}$, $\mathrm{e}^{\boldsymbol{C}t} = \boldsymbol{I} + t\boldsymbol{C} + \frac{1}{2}t^2\boldsymbol{C}^2$ and

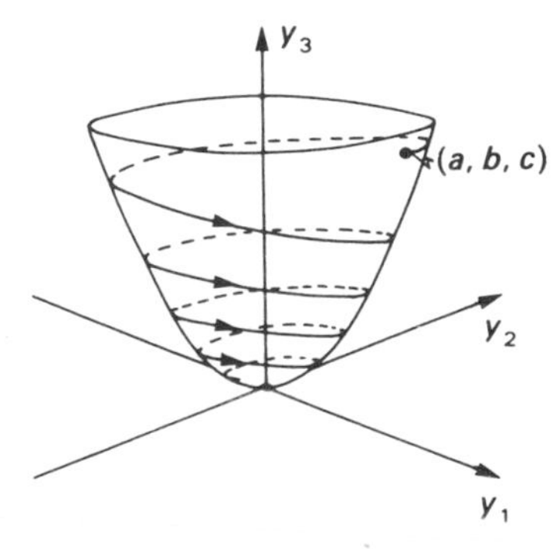

Fig. 2.12. The trajectory of (2.102) passing through (a, b, c). Each trajectory lies in a surface of the form $y_3 = K(y_1^2 + y_2^2)^{\lambda_1/2\alpha}$.

$$e^{Jt} = e^{\lambda_0 t} \begin{bmatrix} 1 & t & \frac{1}{2}t^2 \\ 0 & 1 & t \\ 0 & 0 & 1 \end{bmatrix}. \tag{2.106}$$

Example 2.7.1

Sketch the trajectory which satisfies $\boldsymbol{x}(0) = (0, b, c)$; $b, c > 0$, for the system

$$\dot{\boldsymbol{x}} = \boldsymbol{J}\boldsymbol{x}, \tag{2.107}$$

where $\boldsymbol{J}$ is given by (2.105) with $\lambda_0 > 0$.

Solution

Observe that the projection of (2.107) onto the x_2x_3-plane, i.e.

$$\begin{bmatrix} \dot{x}_2 \\ \dot{x}_3 \end{bmatrix} = \begin{bmatrix} \lambda_0 & 1 \\ 0 & \lambda_0 \end{bmatrix} \begin{bmatrix} x_2 \\ x_3 \end{bmatrix} \tag{2.108}$$

is independent of x_1. This means that each trajectory of (2.107) can be projected onto one of the trajectories of (2.108) in the x_2x_3-plane. The system (2.108) has an unstable ($\lambda_0 > 0$) improper node at $x_2 = x_3 = 0$ and so its trajectories are like those in Fig. 2.13.

Let ϕ_t denote the evolution operator for (2.107); then it follows that the trajectory $\{\phi_t(0, b, c) \mid t \in \mathbb{R}\}$ lies on the surface S generated by translating the trajectory,

$$\begin{bmatrix} x_2(t) \\ x_3(t) \end{bmatrix} = e^{\lambda_0 t} \begin{bmatrix} 1 & t \\ 0 & 1 \end{bmatrix} \begin{bmatrix} b \\ c \end{bmatrix}, \tag{2.109}$$

of (2.108), in the x_1-direction (as illustrated in Fig. 2.14). Notice that (2.109)

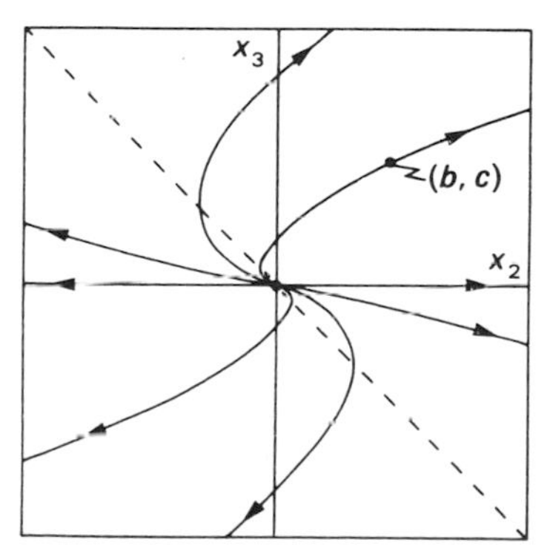

Fig. 2.13. Projections of the trajectories of system (2.107) onto the x_2x_3-plane are given by the phase portrait of (2.109), which has an unstable improper node at the origin. The line $x_3 = -\lambda_0 x_2$, where $\dot{x}_2 = 0$, is shown dashed.

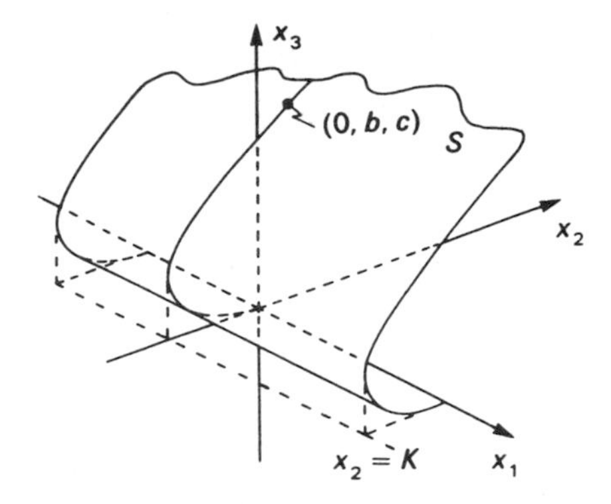

Fig. 2.14. The surface S containing the trajectory $\{\phi_t(0,b,c)|t\in\mathbb{R}\}$ of (2.107), obtained by translating the curve given by (2.109) in the x_1-direction.

is the x_2x_3-projection of

$$\phi_t(0,b,c) = e^{\lambda_0 t}\begin{bmatrix}1 & t & \frac{1}{2}t^2\\ 0 & 1 & t\\ 0 & 0 & 1\end{bmatrix}\begin{bmatrix}0\\ b\\ c\end{bmatrix} \tag{2.110}$$

(cf. e^{Jt} in (2.106)).

Now consider the x_1x_2-projection of the required trajectory. The system equations (2.107) give $\dot{x}_1 = 0$ for $x_2 = -\lambda_0 x_1$ and $\dot{x}_2 = 0$ for $x_3 = -\lambda_0 x_2$. The latter plane intersects S in its fold line (see Fig. 2.13). The x_1x_2-projection

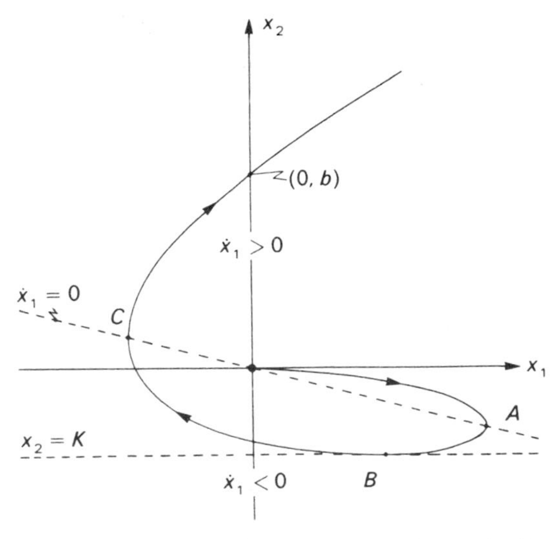

Fig. 2.15. Projection of $\{\phi_t(0,b,c)|t\in\mathbb{R}\}$ onto x_1,x_2-plane. A, B and C label points of the projection where $\dot{x}_1 = 0$ or $\dot{x}_2 = 0$. The turning point at $x_2 = K$ coincides with the fold in S shown in Fig. 2.14.

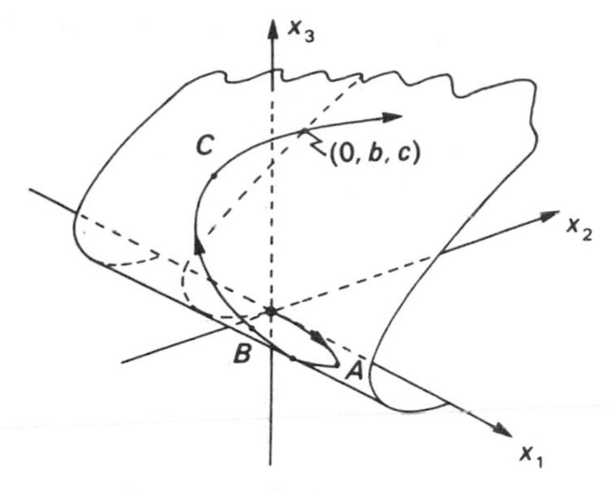

Fig. 2.16. The trajectory $\{\phi_t(0,b,c)|t\in\mathbb{R}\}$ of (2.107). The extreme points at A, B and C are those shown in Fig. 2.15. The curve (2.109) in the x_2,x_3-plane is shown dashed.

of the fold is the line $x_2 = K$, where K is the value of x_2 for which $\dot{x}_2 = 0$ in (2.109). We can then sketch the x_1x_2-projection shown in Fig. 2.15. Finally we show the required trajectory on S in Fig. 2.16. □

2.7.2 Four-dimensional systems

The Jordan forms $\boldsymbol{J}$ of 4×4 real matrices can be grouped as follows:

1. $$\left[\begin{array}{c|c} \boldsymbol{B} & \boldsymbol{0} \\ \hline \boldsymbol{0} & \boldsymbol{C} \end{array}\right]; \tag{2.111}$$

2. $$\left[\begin{array}{ccc|c} \lambda_0 & 1 & 0 & 0 \\ 0 & \lambda_0 & 1 & 0 \\ 0 & 0 & \lambda_0 & 0 \\ \hline 0 & 0 & 0 & \lambda_1 \end{array}\right]; \tag{2.112}$$

3. $$\begin{bmatrix} \lambda_0 & 1 & 0 & 0 \\ 0 & \lambda_0 & 1 & 0 \\ 0 & 0 & \lambda_0 & 1 \\ 0 & 0 & 0 & \lambda_0 \end{bmatrix}; \tag{2.113}$$

   and

4. $$\begin{bmatrix} \alpha & -\beta & 1 & 0 \\ \beta & \alpha & 0 & 1 \\ 0 & 0 & \alpha & -\beta \\ 0 & 0 & \beta & \alpha \end{bmatrix}; \tag{2.114}$$

In 1, the submatrices $\boldsymbol{B}$ and $\boldsymbol{C}$ are 2×2 Jordan matrices and $\boldsymbol{0}$ is the 2×2 null matrix. In 2 and 3 λ_0, λ_1 are real numbers.

The matrices in both 1 and 2 allow $\dot{\boldsymbol{y}} = \boldsymbol{J}\boldsymbol{y}$ to be split into subsystems of lower dimension. We have for 1

$$\begin{bmatrix} \dot{x}_1 \\ \dot{x}_2 \end{bmatrix} = \boldsymbol{B}\begin{bmatrix} x_1 \\ x_2 \end{bmatrix} \quad \text{and} \quad \begin{bmatrix} \dot{x}_3 \\ \dot{x}_4 \end{bmatrix} = \boldsymbol{C}\begin{bmatrix} x_3 \\ x_4 \end{bmatrix} \tag{2.115}$$

and for 2

$$\begin{bmatrix} \dot{x}_1 \\ \dot{x}_2 \\ \dot{x}_3 \end{bmatrix} = \begin{bmatrix} \lambda_0 & 1 & 0 \\ 0 & \lambda_0 & 1 \\ 0 & 0 & \lambda_0 \end{bmatrix}\begin{bmatrix} x_1 \\ x_2 \\ x_3 \end{bmatrix} \quad \text{and} \quad \dot{x}_4 = \lambda_1 x_4. \tag{2.116}$$

The solution curves of all these subsystems have already been discussed.

There are two new systems to consider. The trajectories for the system with coefficient matrix of type 3 can be found by extending the method used to obtain (2.106). We find

$$\boldsymbol{x}(t) = \phi_t(\boldsymbol{x}_0) = \mathrm{e}^{\lambda_0 t}\begin{bmatrix} 1 & t & \frac{1}{2}t^2 & \frac{1}{6}t^3 \\ 0 & 1 & t & \frac{1}{2}t^2 \\ 0 & 0 & 1 & t \\ 0 & 0 & 0 & 1 \end{bmatrix}\boldsymbol{x}_0. \tag{2.117}$$

The remaining system, with coefficient matrix of type 4, can be solved by substituting the known solutions for x_3 and x_4 into the differential equations for x_1 and x_2. Then resulting equations are affine and the techniques of section 2.6 can be used to obtain

$$\boldsymbol{x}(t) = \phi_t(\boldsymbol{x}_0) = \mathrm{e}^{\alpha t}\begin{bmatrix} \cos\beta t & -\sin\beta t & t\cos\beta t & -t\sin\beta t \\ \sin\beta t & \cos\beta t & t\sin\beta t & t\cos\beta t \\ 0 & 0 & \cos\beta t & -\sin\beta t \\ 0 & 0 & \sin\beta t & \cos\beta t \end{bmatrix}\boldsymbol{x}_0. \tag{2.118}$$

Thus it is possible to find solutions for all the canonical systems when $n = 4$.

2.7.3 *n*-Dimensional systems

The pattern of block decomposition which we have illustrated for the Jordan forms of 3×3 and 4×4 real matrices extends to arbitrary n. This means that the corresponding canonical systems $\dot{\boldsymbol{y}} = \boldsymbol{J}\boldsymbol{y}$, with $\boldsymbol{J} = \boldsymbol{M}^{-1}\boldsymbol{A}\boldsymbol{M}$, decouple into subsystems. These subsystems may have:

dimension 1

$$\dot{y}_i = \lambda_i y_i, \qquad \lambda_i \text{ real;}$$

or dimension 2

$$\begin{bmatrix} \dot{y}_i \\ \dot{y}_{i+1} \end{bmatrix} = \begin{bmatrix} \alpha & -\beta \\ \beta & \alpha \end{bmatrix}\begin{bmatrix} y_i \\ y_{i+1} \end{bmatrix}, \qquad \beta > 0,\ \alpha \text{ real;}$$

or dimension j, with $2 \leqslant j \leqslant n$

$$\begin{bmatrix} \dot{y}_i \\ \cdot \\ \cdot \\ \cdot \\ \cdot \\ \cdot \\ \dot{y}_{i+j-1} \end{bmatrix} = \begin{bmatrix} \lambda & 1 & 0 & \cdot & \cdot & \cdot & 0 \\ 0 & \lambda & 1 & \cdot & \cdot & \cdot & \cdot \\ \cdot & \cdot & \cdot & \cdot & \cdot & \cdot & \cdot \\ \cdot & 0 & \cdot & \cdot & \cdot & \cdot & \cdot \\ \cdot & \cdot & \cdot & \cdot & \cdot & \cdot & \cdot \\ \cdot & \cdot & \cdot & \cdot & \cdot & \lambda & 1 \\ 0 & 0 & \cdot & \cdot & \cdot & 0 & \lambda \end{bmatrix} \begin{bmatrix} y_i \\ \cdot \\ \cdot \\ \cdot \\ \cdot \\ \cdot \\ y_{i+j-1} \end{bmatrix}, \qquad \lambda \text{ real.}$$

or even dimension j, $4 \leqslant j \leqslant n$,

$$\begin{bmatrix} \dot{y}_i \\ \cdot \\ \cdot \\ \cdot \\ \cdot \\ \cdot \\ \cdot \\ \cdot \\ \dot{y}_{i+j-1} \end{bmatrix} = \begin{bmatrix} \alpha & -\beta & 1 & 0 & \cdot & \cdot & 0 & 0 \\ \beta & \alpha & 0 & 1 & \cdot & \cdot & 0 & 0 \\ 0 & 0 & \alpha & -\beta & \cdot & \cdot & \cdot & \cdot \\ 0 & 0 & \beta & \alpha & \cdot & \cdot & \cdot & \cdot \\ \cdot & \cdot & \cdot & \cdot & \cdot & \cdot & \cdot & \cdot \\ \cdot & \cdot & \cdot & \cdot & \cdot & \cdot & 1 & 0 \\ \cdot & \cdot & \cdot & \cdot & \cdot & \cdot & 0 & 1 \\ 0 & 0 & \cdot & \cdot & 0 & 0 & \alpha & -\beta \\ 0 & 0 & \cdot & \cdot & 0 & 0 & \beta & \alpha \end{bmatrix} \begin{bmatrix} \dot{y}_i \\ \cdot \\ \cdot \\ \cdot \\ \cdot \\ \cdot \\ \cdot \\ \cdot \\ y_{i+j-1} \end{bmatrix},$$

with $\beta > 0$, α real. However, as we have seen for $n = 3$ and 4, solutions can be found for all of these subsystems and we conclude any canonical system can be solved. Of course, this means that the solutions of all linear systems $\dot{\boldsymbol{x}} = \boldsymbol{A}\boldsymbol{x}$ can also be found by using $\boldsymbol{x} = \boldsymbol{M}\boldsymbol{y}$.

EXERCISES

Sections 2.1 and 2.2

2.1 Indicate the effect of each of the following linear transformations of the x_1x_2-plane by shading the image of the square $S = \{(x_1, x_2) \mid 0 \leqslant x_1, x_2 \leqslant 1\}$:

$$\text{(a)} \begin{bmatrix} 1 & 0 \\ 0 & 2 \end{bmatrix}; \quad \text{(b)} \begin{bmatrix} -1 & 2 \\ -2 & -1 \end{bmatrix}; \quad \text{(c)} \begin{bmatrix} -7 & 3 \\ -8 & 3 \end{bmatrix}; \quad \text{(d)} \begin{bmatrix} 1 & -2 \\ 1 & 1 \end{bmatrix}.$$

Use the images of the square to help to sketch the images of the following sets:
(a) the circle $x_1^2 + x_2^2 = 1$; (b) the curve $x_1x_2 = 1$, $x_1, x_2 > 0$.

2.2 Prove that the relation $\sim$ on pairs $\boldsymbol{A}$, $\boldsymbol{B}$ of real $n \times n$ matrices given by

$$\boldsymbol{A} \sim \boldsymbol{B} \Leftrightarrow \boldsymbol{B} = \boldsymbol{M}^{-1}\boldsymbol{A}\boldsymbol{M},$$

where $\boldsymbol{M}$ is a non-singular matrix, is an equivalence relation. Furthermore, show that all the matrices in any given equivalence class have the same eigenvalues. Group the following systems under this equivalence on their coefficient matrices:

(a) $\dot{\boldsymbol{x}} = \begin{bmatrix} 2 & 0 \\ 0 & 2 \end{bmatrix} \boldsymbol{x}$; (b) $\dot{\boldsymbol{x}} = \begin{bmatrix} 2 & 1 \\ 0 & 2 \end{bmatrix} \boldsymbol{x}$;

(c) $\dot{\boldsymbol{x}} = \begin{bmatrix} 7 & -2 \\ 15 & -6 \end{bmatrix} \boldsymbol{x}$; (d) $\dot{\boldsymbol{x}} = \begin{bmatrix} -4 & 2 \\ -4 & 5 \end{bmatrix} \boldsymbol{x}$;

(e) $\dot{\boldsymbol{x}} = \begin{bmatrix} 2 & 2 \\ 3 & 1 \end{bmatrix} \boldsymbol{x}$; (f) $\dot{\boldsymbol{x}} = \begin{bmatrix} 2 & 2 \\ 0 & 2 \end{bmatrix} \boldsymbol{x}$.

2.3 Show that each of the following systems can be transformed into a canonical system by the given change of variable:
(a) $\dot{x}_1 = 4x_1 + x_2$, $\dot{x}_2 = -x_1 + 2x_2$,
$y_1 = 2x_1 + x_2$, $y_2 = x_1 + x_2$;
(b) $\dot{x}_1 = 12x_1 + 4x_2$, $\dot{x}_2 = -26x_1 - 8x_2$,
$y_1 = 2x_1 + x_2$, $y_2 = 3x_1 + x_2$;
(c) $\dot{x}_1 = 10x_1 + 2x_2$, $\dot{x}_2 = -28x_1 - 5x_2$,
$y_1 = 7x_1 + 2x_2$, $y_2 = 4x_1 + x_2$.
Write down the coefficient matrices $\boldsymbol{A}$ of each of the above systems, the Jordan forms $\boldsymbol{J}$ and the change of variable matrices $\boldsymbol{M}$. Check that $\boldsymbol{J} = \boldsymbol{M}^{-1}\boldsymbol{A}\boldsymbol{M}$ is satisfied.

2.4 For each matrix $\boldsymbol{A}$ given below

$$\begin{bmatrix} 0 & 1 \\ -2 & 3 \end{bmatrix}, \quad \begin{bmatrix} 41 & -29 \\ 58 & -41 \end{bmatrix}, \quad \begin{bmatrix} 9 & 4 \\ -9 & -3 \end{bmatrix}$$

find its Jordan form $\boldsymbol{J}$ and a matrix $\boldsymbol{M}$ that satisfies

$$\boldsymbol{J} = \boldsymbol{M}^{-1}\boldsymbol{A}\boldsymbol{M}.$$

2.5 What is the effect on the system

$$\begin{aligned} \dot{x}_1 &= -7x_1 - 4x_2 - 6x_3 \\ \dot{x}_2 &= -3x_1 - 2x_2 - 3x_3 \\ \dot{x}_3 &= 3x_1 + 2x_2 + 2x_3 \end{aligned}$$

of the change of variables $y_1 = x_1 + 2x_3$, $y_2 = x_2$, $y_3 = x_1 + x_3$? How does the introduction of these new variables help to investigate the solution curves of the system?

2.6 Prove that no two different 2×2 Jordan matrices are similar. Hence prove that there is one and only one Jordan matrix similar to any given 2×2 real matrix.

2.7 Find the characteristic equations of the matrices

$$\begin{bmatrix} 2 & 1 \\ 0 & 2 \end{bmatrix}, \quad \begin{bmatrix} 3 & 1 \\ -2 & 7 \end{bmatrix}, \quad \begin{bmatrix} 8 & -4 \\ 6 & -3 \end{bmatrix},$$

and illustrate the Cayley–Hamilton theorem for each matrix.

Section 2.3

2.8 Find the solution curves $\boldsymbol{x}(t)$ satisfying $\dot{\boldsymbol{x}} = \boldsymbol{A}\boldsymbol{x}$ subject to $\boldsymbol{x}(0) = \boldsymbol{x}_0$, where the matrix $\boldsymbol{A}$ is taken to be each of the matrices in Exercise 2.4.

2.9 Sketch phase portraits for the linear system

$$\begin{bmatrix} \dot{x}_1 \\ \dot{x}_2 \end{bmatrix} = \boldsymbol{A} \begin{bmatrix} x_1 \\ x_2 \end{bmatrix}$$

when $\boldsymbol{A}$ is given by:

(a) $\begin{bmatrix} 1 & 0 \\ 0 & 2 \end{bmatrix}$; (b) $\begin{bmatrix} -1 & 0 \\ 0 & 2 \end{bmatrix}$; (c) $\begin{bmatrix} 3 & 1 \\ -1 & 3 \end{bmatrix}$;

(d) $\begin{bmatrix} 2 & 1 \\ 0 & 2 \end{bmatrix}$; (e) $\begin{bmatrix} 1 & 0 \\ 0 & \frac{1}{2} \end{bmatrix}$; (f) $\begin{bmatrix} -3 & 0 \\ 0 & -3 \end{bmatrix}$;

(g) $\begin{bmatrix} 0 & 2 \\ -2 & 0 \end{bmatrix}$; (h) $\begin{bmatrix} 3 & 0 \\ 0 & -1 \end{bmatrix}$; (i) $\begin{bmatrix} 3 & 0 \\ 0 & 0 \end{bmatrix}$;

(j) $\begin{bmatrix} 0 & -2 \\ 2 & 0 \end{bmatrix}$.

2.10 Indicate the effect of the linear transformation

$$\begin{bmatrix} y_1 \\ y_2 \end{bmatrix} = \begin{bmatrix} 2 & 1 \\ 1 & 1 \end{bmatrix} \begin{bmatrix} x_1 \\ x_2 \end{bmatrix}$$

on each of the systems in Exercise 2.9 by sketching each phase portrait in the $y_1 y_2$-plane.

2.11 Find the 2×2 matrix $\boldsymbol{A}$ such that the system

$$\dot{\boldsymbol{x}} = \boldsymbol{A}\boldsymbol{x}$$

has a solution curve

$$\boldsymbol{x}(t) = \begin{bmatrix} e^{-t}(\cos t + 2\sin t) \\ e^{-t}\cos t \end{bmatrix}.$$

2.12 Let

$$\dot{\boldsymbol{x}} = \begin{bmatrix} -1 & -1 \\ 2 & -4 \end{bmatrix} \boldsymbol{x}$$

and $y = x_1 + 3x_2$. For this situation the system is said to be 'observable' if by knowing $y(t)$ the solution curve $\boldsymbol{x}(t)$ can be derived. Prove that the system is observable when $y(t) = 4e^{-2t}$ by finding $\boldsymbol{x}(0)$.

Section 2.4

2.13 Locate each of the following linear systems in the tr–det plane and hence state their phase portrait type:

(a) $\dot{x}_1 = 2x_1 + x_2$, $\quad \dot{x}_2 = x_1 + 2x_2$;
(b) $\dot{x}_1 = 2x_1 + x_2$, $\quad \dot{x}_2 = x_1 - 3x_2$;
(c) $\dot{x}_1 = x_1 - 4x_2$, $\quad \dot{x}_2 = 2x_1 - x_2$;
(d) $\dot{x}_1 = 2x_2$, $\quad \dot{x}_2 = -3x_1 - x_2$,
(e) $\dot{x}_1 = -x_1 + 8x_2$, $\quad \dot{x}_2 = -2x_1 + 7x_2$.

2.14 Express the system

$$\dot{x}_1 = -7x_1 + 3x_2, \; \dot{x}_2 = -8x_1 + 3x_2$$

in coordinates y_1, y_2 relative to the basis $\{(1,2), (3,4)\}$. Hence sketch the phase portrait of the system in the x_1x_2-plane.

2.15 Let $\boldsymbol{x}(t)$ be a trajectory of the linear system $\dot{\boldsymbol{x}} = \boldsymbol{A}\boldsymbol{x}$ and suppose $\lim_{t\to\infty} \boldsymbol{x}(t) = \boldsymbol{0}$. Use the continuity of the linear transformation $\boldsymbol{y} = \boldsymbol{N}\boldsymbol{x}$, where $\boldsymbol{N}$ is a non-singular matrix, to show that the trajectory $\boldsymbol{y}(t) = \boldsymbol{N}\boldsymbol{x}(t)$ of the system

$$\dot{\boldsymbol{y}} = \boldsymbol{N}\boldsymbol{A}\boldsymbol{N}^{-1}\boldsymbol{y}$$

also satisfies the property

$$\lim_{t\to\infty} \boldsymbol{y}(t) = \boldsymbol{0}.$$

Hence prove that if $\dot{\boldsymbol{x}} = \boldsymbol{A}\boldsymbol{x}$ has a fixed point which is a stable node, improper node or focus then so has the system $\dot{\boldsymbol{y}} = \boldsymbol{N}\boldsymbol{A}\boldsymbol{N}^{-1}\boldsymbol{y}$.

2.16 Find a corresponding result to that of Exercise 2.15 when a trajectory $\boldsymbol{x}(t)$ of the system $\dot{\boldsymbol{x}} = \boldsymbol{A}\boldsymbol{x}$ satisfies $\lim_{t\to-\infty} \boldsymbol{x}(t) = \boldsymbol{0}$. Use both results to prove that if the system $\dot{\boldsymbol{x}} = \boldsymbol{A}\boldsymbol{x}$ has a saddle point at the origin then so does the system $\dot{\boldsymbol{y}} = \boldsymbol{N}\boldsymbol{A}\boldsymbol{N}^{-1}\boldsymbol{y}$.

2.17 The transformation $y = -x$ is an example of an **order reversing** continuous bijection that maps the real line onto itself. Make this change of variable in each of the following one-dimensional systems:

(a) $\dot{x} = 1$; $\quad$ (b) $\dot{x} = x^2$;
(c) $\dot{x} = x(1 - x)$; $\quad$ (d) $\dot{x} = f(x)$, $\; f$ even.

Sketch phase portraits for each system before and after transformation and confirm that the members of each pair are qualitatively distinct in the sense of Definition 1.2.1 which only admits order-preserving transformations.

Why could the admission of order-reversing transformations have mathematical merit and yet be inappropriate to applications?

2.18 Show that there is a linear mapping of the family of curves $x_2 = Cx_1^{\mu}$ onto the family of curves $y_2 = C'y_1^{\mu'}$ (μ, μ' constant) if and only if $\mu = \mu'$ or $\mu = 1/\mu'$. Hence show that there is a linear mapping of the trajectories of

$$\dot{x}_1 = \lambda_1 x_1, \qquad \dot{x}_2 = \lambda_2 x_2, \qquad \lambda_1\lambda_2 \neq 0,$$

onto the phase portrait of the system

$$\dot{y}_1 = \nu_1 y_1, \qquad \dot{y}_2 = \nu_2 y_2$$

if and only if $\lambda_1/\lambda_2 = \nu_1/\nu_2$ or $\lambda_1/\lambda_2 = \nu_2/\nu_1$.

2.19 Show that there is a real positive number k such that the continuous bijection of the plane

$$y_1 = x_1$$

$$y_2 = \begin{cases} x_2^k, & x_2 \geqslant 0 \\ -(-x_2)^k, & x_2 < 0 \end{cases}$$

maps the trajectories of the node $\dot{x}_1 = \lambda_1 x_1$, $\dot{x}_2 = \lambda_2 x_2$ onto the trajectories of the star node $\dot{y}_1 = \varepsilon y_1, \dot{y}_2 = \varepsilon y_2, \varepsilon = +1$ or -1, preserving orientation.

2.20 Show that the results developed in Exercise 2.19 can be used to find a mapping which maps trajectories of any saddle point onto the trajectories of the saddle $\dot{x}_1 = x_1$, $\dot{x}_2 = -x_2$.

2.21 Sketch the phase portrait of the system

$$\dot{x}_1 = 2x_1 + x_2, \qquad \dot{x}_2 = 3x_2$$

and the phase portraits obtained by

(a) reflection in the x_1-axis;
(b) a half turn in the $x_1 x_2$-plane;
(c) an anticlockwise rotation of $\pi/2$;
(d) interchanging the axes x_1 and x_2.

State transforming matrices for each of the cases above and find the transformed system equations. Check that each of the phase portraits you have obtained corresponds to the appropriate transformed system.

Section 2.5

2.22 Calculate e^{At} for the following matrices A:

(a) $\begin{bmatrix} 2 & 0 \\ 0 & 3 \end{bmatrix}$, (b) $\begin{bmatrix} 1 & 2 \\ 0 & 2 \end{bmatrix}$, (c) $\begin{bmatrix} 2 & 4 \\ 3 & 3 \end{bmatrix}$,

(d) $\begin{bmatrix} -2 & 2 \\ -4 & 2 \end{bmatrix}$, (e) $\begin{bmatrix} -4 & 1 \\ -1 & -2 \end{bmatrix}$.

2.23 Calculate e^{At} when A is equal to:

(a) $\begin{bmatrix} 2 & -7 \\ 3 & -8 \end{bmatrix}$; (b) $\begin{bmatrix} 2 & 4 \\ -2 & 6 \end{bmatrix}$.

Use the results to calculate e^{At} when A equals:

(c) $\begin{bmatrix} 2 & -7 & 0 \\ 3 & -8 & 0 \\ 0 & 0 & 1 \end{bmatrix}$; (d) $\begin{bmatrix} 2 & -7 & 0 & 0 \\ 3 & -8 & 0 & 0 \\ 0 & 0 & 2 & 4 \\ 0 & 0 & -2 & 6 \end{bmatrix}$.

2.24 Re-order the variables of the system

$$\dot{x} = \begin{bmatrix} 2 & 0 & 1 & 0 \\ 0 & -5 & 0 & 9 \\ 0 & 0 & 2 & 0 \\ 0 & -4 & 0 & 8 \end{bmatrix} x$$

so that the coefficient matrix takes the same form as that in Exercise 2.23(d). Hence, or otherwise, find the trajectory passing through $(2, 1, 0, 1)$ at $t = 1$.

2.25 Prove that if P and Q are $n \times n$ commuting matrices then $P^r Q^s = Q^s P^r$ for r, s non-negative integers.
Hence show that

$$\sum_{k=0}^{\infty} \frac{(P+Q)^k}{k!} = \left(\sum_{k=0}^{\infty} \frac{P^k}{k!} \right) \left(\sum_{k=0}^{\infty} \frac{Q^k}{k!} \right)$$

and deduce

$$e^{P+Q} = e^P e^Q.$$

2.26 Use (2.63) to evaluate e^{At} when:

(a) $A = \begin{bmatrix} \lambda_0 & 1 \\ 0 & \lambda_0 \end{bmatrix}$ by writing $A = \lambda_0 I + C$ where $C = \begin{bmatrix} 0 & 1 \\ 0 & 0 \end{bmatrix}$;

(b) $A = \begin{bmatrix} \alpha & -\beta \\ \beta & \alpha \end{bmatrix}$ by writing $A = \alpha I + \beta D$, where

$$D = \begin{bmatrix} 0 & -1 \\ 1 & 0 \end{bmatrix}.$$

2.27 Write down the characteristic equation of a 2×2 matrix A. Use the Cayley–Hamilton theorem to obtain an expression for A^2 in terms of A and I. Use this result to prove that if λ_1, λ_2 ($\lambda_1 \neq \lambda_2$) are the eigenvalues of A, then

$$\left\{ \frac{A - \lambda_1 I}{\lambda_2 - \lambda_1} \right\} \left\{ \frac{A - \lambda_2 I}{\lambda_1 - \lambda_2} \right\} = 0,$$

and

$$\left\{ \frac{A - \lambda_1 I}{\lambda_2 - \lambda_1} \right\}^2 = \left\{ \frac{A - \lambda_1 I}{\lambda_2 - \lambda_1} \right\}$$

(cf. section 2.5).

2.28 Let $\boldsymbol{A}$ be a 3×3 real matrix with eigenvalues $\lambda_1 = \lambda_2 = \lambda_3 \ (= \lambda_0)$. Show that

$$e^{At} = e^{\lambda_0 t}(\boldsymbol{I} + t\boldsymbol{Q} + \tfrac{1}{2}t^2\boldsymbol{Q}^2)$$

where $\boldsymbol{Q} = \boldsymbol{A} - \lambda_0 \boldsymbol{I}$. Generalize this result to $n \times n$ matrices.

Section 2.6

2.29 Find, where possible, a change of variables which converts each affine system into an equivalent linear system.
(a) $\dot{x}_1 = x_1 + x_2 + 2, \quad \dot{x}_2 = x_1 + 2x_2 + 3;$
(b) $\dot{x}_1 = x_2 + 1, \quad \dot{x}_2 = 3;$
(c) $\dot{x}_1 = 2x_1 - 3x_2 + 1, \quad \dot{x}_2 = 6x_1 - 9x_2;$
(d) $\dot{x}_1 = 2x_1 - x_2 + 1, \quad \dot{x}_2 = 6x_1 + 3x_2;$
(e) $\dot{x}_1 = 2x_1 + x_2 + 1, \quad \dot{x}_2 = x_1 + x_3, \ \dot{x}_3 = x_2 + x_3 + 2;$
(f) $\dot{x}_1 = x_1 + x_2 + x_3 + 1, \quad \dot{x}_2 = -x_2, \ \dot{x}_3 = x_1 + x_3 + 1.$
For those affine systems equivalent to a linear system, state their algebraic type.

2.30 Find the solution curve $(x_1(t), x_2(t))$ which satisfies

$$\dot{x}_1 = x_1 + x_2 + 1, \qquad \dot{x}_2 = x_1 + x_2$$

subject to the initial condition $x_1(0) = a$, $x_2(0) = b$.

2.31 When the non-singular change of variables $\boldsymbol{x} = \boldsymbol{M}\boldsymbol{y}$, $\boldsymbol{x}, \boldsymbol{y} \in \mathbb{R}^n$, is applied to the affine system $\dot{\boldsymbol{x}} = \boldsymbol{A}\boldsymbol{x} + \boldsymbol{h}$, what is the transformed system? Hence show, when $n = 2$, that every affine system can be changed to an affine system with a Jordan coefficient matrix. In the case when $\boldsymbol{A}$ has two real distinct eigenvalues show that the system $\dot{\boldsymbol{x}} = \boldsymbol{A}\boldsymbol{x} + \boldsymbol{h}(t)$ can be decoupled.

Is it possible to change an affine system to a linear system by a linear transformation?

2.32 Sketch phase portraits of the affine systems
(a) $\dot{x}_1 = 2x_2 + 1, \ \dot{x}_2 = -x_1 + 1$ (b) $\dot{x}_1 = x_1 + 2, \ \dot{x}_2 = 3.$

Section 2.7

2.33 Show that the system

$$\begin{aligned} \dot{x}_1 &= -3x_1 + 10x_2 \\ \dot{x}_2 &= -2x_1 + 5x_2 \\ \dot{x}_3 &= -2x_1 + 2x_2 + 3x_3 \end{aligned}$$

can be transformed into

$$\begin{aligned} \dot{y}_1 &= ay_1 - by_2 \\ \dot{y}_2 &= by_1 + ay_2 \\ \dot{y}_3 &= cy_3 \end{aligned}$$

where a, b and c are constants. What are their values? Find a matrix $\boldsymbol{M}$ of the form

$$\begin{bmatrix} m_{11} & m_{12} & 0 \\ 0 & m_{22} & 0 \\ 0 & m_{32} & m_{33} \end{bmatrix}$$

such that

$$\boldsymbol{x} = \boldsymbol{M}\boldsymbol{y}.$$

What form does the phase portrait take when projected onto the y_1y_2-plane and the y_3-axis respectively? Make a sketch of some trajectories of the transformed system.

2.34 Find a solution curve of the system

$$\begin{aligned} \dot{x}_1 &= x_1 + 2x_2 + x_3 \\ \dot{x}_2 &= x_2 + 2x_3 \\ \dot{x}_3 &= 2x_3 \end{aligned}$$

which satisfies $x_1(0) = x_2(0) = 0$ and $x_3(0) = 1$.

2.35 Find the form of the solution curves of the systems $\dot{\boldsymbol{x}} = \boldsymbol{A}\boldsymbol{x}$, $\boldsymbol{x} \in \mathbb{R}^4$, where $\boldsymbol{A}$ equals:

$$\text{(a)} \begin{bmatrix} \lambda & 1 & 0 & 0 \\ 0 & \lambda & 1 & 0 \\ 0 & 0 & \lambda & 1 \\ 0 & 0 & 0 & \lambda \end{bmatrix}; \quad \text{(b)} \begin{bmatrix} \alpha & -\beta & 0 & 0 \\ \beta & \alpha & 0 & 0 \\ 0 & 0 & \lambda & 1 \\ 0 & 0 & 0 & \lambda \end{bmatrix};$$

$$\text{(c)} \begin{bmatrix} \alpha & -\beta & 0 & 0 \\ \beta & \alpha & 0 & 0 \\ 0 & 0 & \gamma & -\delta \\ 0 & 0 & \delta & \gamma \end{bmatrix}; \quad \text{(d)} \begin{bmatrix} \alpha & -\beta & 1 & 0 \\ \beta & \alpha & 0 & 1 \\ 0 & 0 & \alpha & -\beta \\ 0 & 0 & \beta & \alpha \end{bmatrix}.$$

2.36 Consider the six-dimensional system

$$\begin{bmatrix} \dot{x}_1 \\ \dot{x}_2 \\ \dot{x}_3 \\ \dot{x}_4 \\ \dot{x}_5 \\ \dot{x}_6 \end{bmatrix} = \begin{bmatrix} 2 & 0 & 0 & 0 & 0 & 0 \\ 0 & 0 & -1 & 0 & 0 & 0 \\ 0 & 1 & 0 & 0 & 0 & 0 \\ 0 & 0 & 0 & 1 & 0 & 0 \\ 0 & 0 & 0 & 0 & 1 & 1 \\ 0 & 0 & 0 & 0 & 0 & 1 \end{bmatrix} \begin{bmatrix} x_1 \\ x_2 \\ x_3 \\ x_4 \\ x_5 \\ x_6 \end{bmatrix}$$

Divide the system into subsystems and describe the phase portrait behaviour of each subsystem.

3

Non-linear systems in the plane

In this chapter we consider the phase portraits of systems $\dot{\mathbf{x}} = \mathbf{X}(\mathbf{x})$, $\mathbf{x} \in S \subseteq \mathbb{R}^2$, where $\mathbf{X}$ is a continuously differentiable, non-linear function. In contrast to sections 1.2 and 2.3 these phase portraits are not always determined by the nature of the fixed points of the system.

3.1 LOCAL AND GLOBAL BEHAVIOUR

Definition 3.1.1
A **neighbourhood**, N, of a point $\mathbf{x}_0 \in \mathbb{R}^2$ is a subset of $\mathbb{R}^2$ containing a disc $\{\mathbf{x} \mid |\mathbf{x} - \mathbf{x}_0| < r\}$ for some $r > 0$.

Definition 3.1.2
The part of the phase portrait of a system that occurs in a neighbourhood N of $\mathbf{x}_0$ is called the **restriction** of the phase portrait to N.

These definitions are illustrated in Fig. 3.1 using a simple linear system. When analysing non-linear systems, we often consider a restriction of the complete or **global** phase portrait to a neighbourhood of $\mathbf{x}_0$ that is as small as we please (to be described in section 3.3). Such a restriction will be referred to as the **local phase portrait** at $\mathbf{x}_0$.

Consider the restriction of a simple linear system to a neighbourhood N of the origin. There is a neighbourhood $N' \subseteq N$ such that the restriction of this phase portrait to N' is qualitatively equivalent to the global phase portrait of the simple linear system itself. That is, there is a continuous bijection between N' and $\mathbb{R}^2$ which maps the phase portrait restricted to N' onto the complete phase portrait. This result is illustrated for a centre in Fig. 3.2. The neighbourhood N is taken to be $\{(x_1, x_2) \mid a < x_1 < b,\ c < x_2 < d;\ a, c < 0;\ b, d > 0\}$. Let N' be the set of (x_1, x_2) lying inside the critical trajectory T

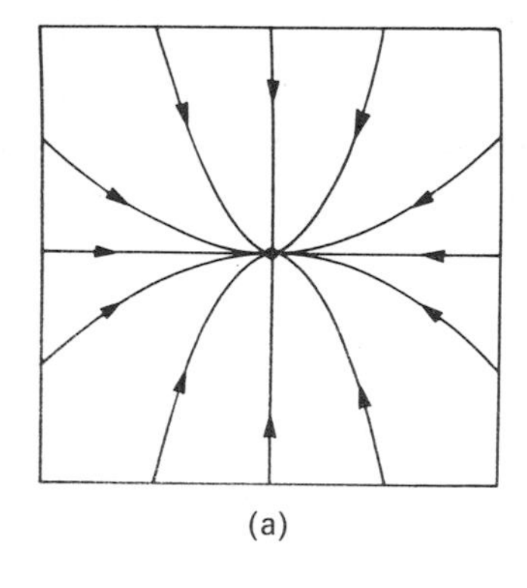
(a)

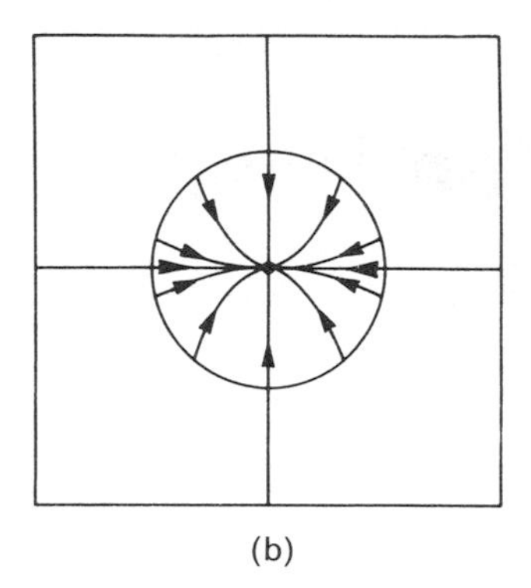
(b)

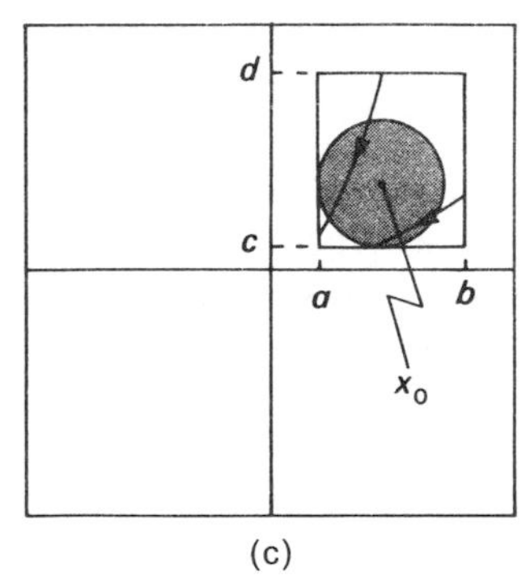

(c)

Fig. 3.1. (a) Phase portrait for $\dot{x}_1 = -x_1, \dot{x}_2 = -2x_2$. (b) Restriction of (a) to $N = \{x \,||x| < a\}$, $a > 0$, of the fixed point at (0, 0). (c) Restriction of (a) to $N = \{x | a < x_1 < b, \quad c < x_2 < d\} \quad a, b, c, d > 0$, of point $\mathbf{x}_0$ where $\dot{\mathbf{x}} \neq \mathbf{0}$. A disc radius $r > 0$ centred on $\mathbf{x}_0$ is shown shaded.

shown in Fig. 3.2(b). Every trajectory in the global phase portrait (Fig. 3.2(a)) has a counterpart in its restriction to N' and vice versa. If we consider the restriction of the stable node shown in Fig. 3.1(b), then again the result holds but in this case with $N' = N$.

This qualitative equivalence of the phase portrait and its restrictions is what we really mean by saying that the phase portrait of a simple linear system is determined by the 'nature' of its fixed point. In other words, the local phase portrait at the origin is qualitatively equivalent to the global phase portrait of the system.

Non-linear systems can have more than one fixed point and we can often obtain the local phase portraits at all of them. However, as Fig. 3.3 shows, local phase portraits do not always determine the global phase portrait. The figure shows three qualitatively different global phase portraits, each containing three fixed points. The local phase portraits at the fixed points are the same in all three diagrams. These phase portraits are realized by the non-linear system

$$\begin{aligned} \dot{x}_1 &= -\alpha x_2 + x_1(1 - x_1^2 - x_2^2) - x_2(x_1^2 + x_2^2) \\ \dot{x}_2 &= \alpha x_1 + x_2(1 - x_1^2 - x_2^2) + x_1(x_1^2 + x_2^2) + \beta \end{aligned} \tag{3.1}$$

for certain choices of the pair (α, β).

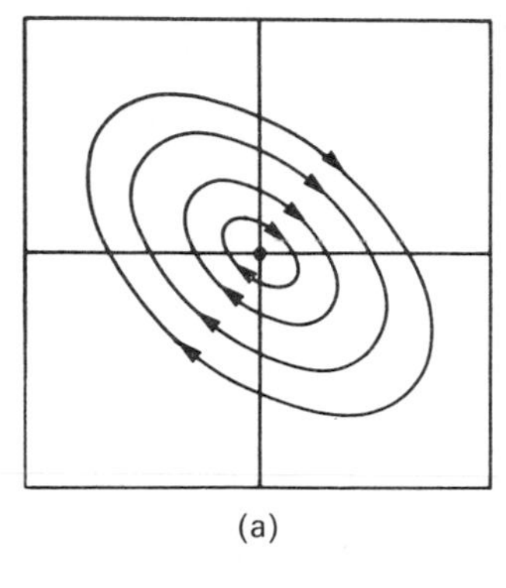

(a)

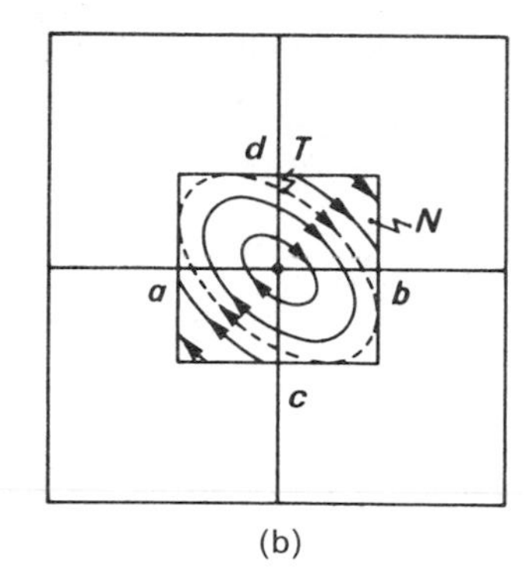

(b)

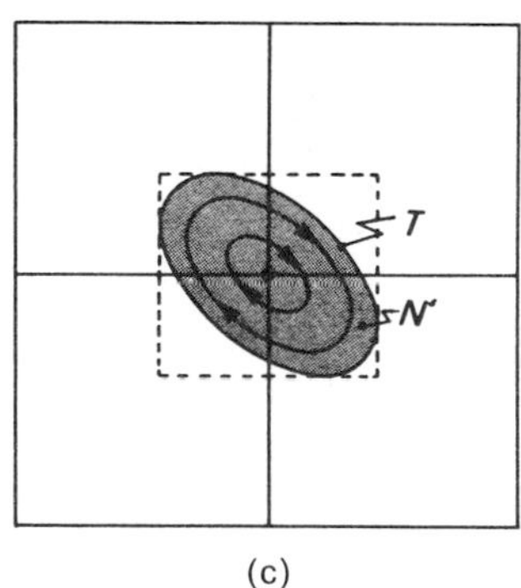

(c)

Fig. 3.2. (a) Phase portrait for $\dot{x}_1 = 3x_1 + 4x_2$, $\dot{x}_2 = -3x_1 - 3x_2$. (b) Restriction of (a) to $N = \{(x_1, x_2) | a < x_1 < b, c < x_2 < d;\ a, c < 0;\ b, d > 0\}$. The critical trajectory T is shown dashed. (c) Restriction to neighbourhood N' (shaded) $= \{\mathbf{x} | \mathbf{x} \text{ inside } T\}$. This restriction is qualitatively equivalent to (a).

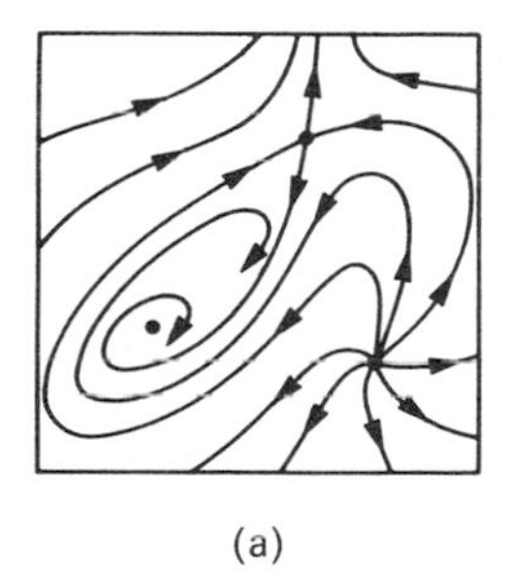

(a)

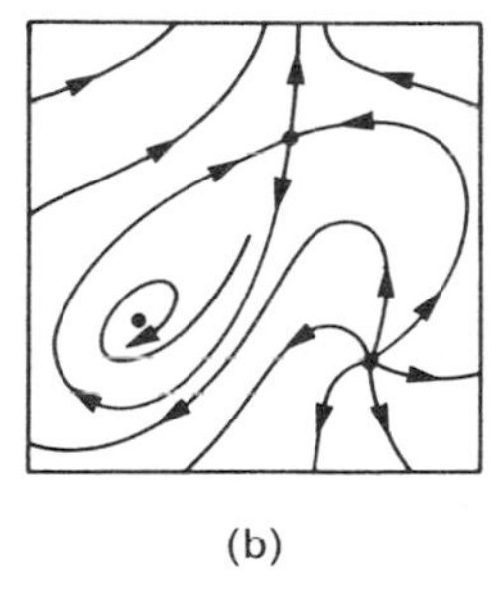

(b)

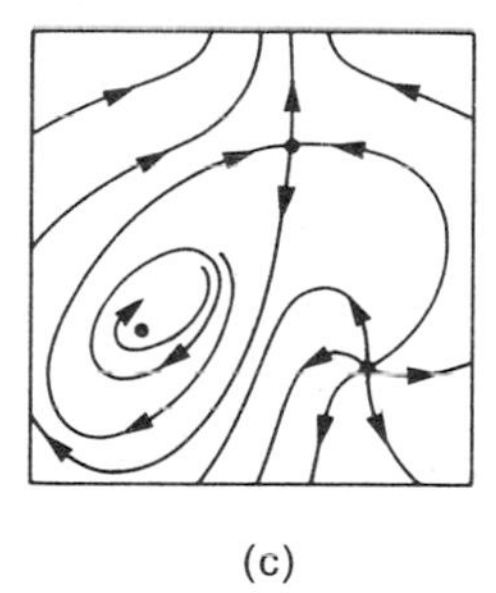

(c)

Fig. 3.3. Qualitatively different global phase portraits consistent with given local behaviour at three fixed points.

Figure 3.3(c) also illustrates another global feature of non linear phase portraits which is not revealed by a study of fixed points. The isolated closed orbit around one of the fixed points is called a limit cycle. The detection of limit cycles requires a global approach (described in sections 3.8 and 3.9).

The treatment of non-linear systems, therefore, involves techniques relating to both local and global behaviour. The former are discussed in sections 3.2–3.6.1 inclusive, while the latter are dealt with in sections 3.6.2–3.9.

3.2 LINEARIZATION AT A FIXED POINT

We begin by examining non-linear systems with a fixed point at the origin of their phase plane.

Definition 3.2.1
Suppose the system $\dot{\mathbf{y}} = \mathbf{Y}(\mathbf{y})$ can be written in the form

$$\begin{aligned}\dot{y}_1 &= ay_1 + by_2 + g_1(y_1, y_2)\\ \dot{y}_2 &= cy_1 + dy_2 + g_2(y_1, y_2),\end{aligned} \tag{3.2}$$

where $[g_i(y_1, y_2)/r] \to 0$ as $r = (y_1^2 + y_2^2)^{1/2} \to 0$. The linear system

$$\dot{y}_1 = ay_1 + by_2, \qquad \dot{y}_2 = cy_1 + dy_2 \tag{3.3}$$

is said to be the **linearization** (or **linearized system**) of (3.2) at the origin. The components of the linear vector field in (3.3) are said to form the **linear part** of **Y**.

Example 3.2.1
Find the linearizations of the following systems at the origin:

1. $\dot{y}_1 = y_1 + y_1^2 + y_1 y_2^2, \qquad \dot{y}_2 = y_2 + y_2^{3/2};$
2. $\dot{y}_1 = y_1^3, \qquad \dot{y}_2 = y_2 + y_2 \sin y_1;$
3. $\dot{y}_1 = y_1^2 \mathrm{e}^{y_2}, \qquad \dot{y}_2 = y_2(\mathrm{e}^{y_1} - 1).$

Solution
For each system, we tabulate the real numbers a, b, c, d and the functions g_1 and g_2 below. The functions $g_i (i = 1, 2)$ are given in polar coordinates so that the requirement

$$\lim_{r \to 0} [g_i(y_1, y_2)/r] = 0, \qquad i = 1, 2, \tag{3.4}$$

can be checked.

System	a	b	c	d	g_1	g_2
1.	1	0	0	1	$r^2(\cos^2\theta + r\cos\theta\sin^2\theta)$	$r^{3/2}\sin^{3/2}\theta$
2.	0	0	0	1	$r^3\cos^3\theta$	$r\sin\theta\sin(r\cos\theta)$
3.	0	0	0	0	$r^2\cos^2\theta\, \mathrm{e}^{r\sin\theta}$	$r\sin\theta(r\cos\theta +$ $+ \frac{1}{2}r^2\cos^2\theta + \cdots)$

Hence, the linearizations are:

1. $\dot{y}_1 = y_1, \qquad \dot{y}_2 = y_2;$
2. $\dot{y}_1 = 0, \qquad \dot{y}_2 = y_2;$
3. $\dot{y}_1 = 0, \qquad \dot{y}_2 = 0.$

□

Definition 3.2.1 can be applied at fixed points which do not occur at the origin by introducing **local coordinates**. Suppose (ξ, η) is a fixed point of the non-linear system $\dot{\mathbf{x}} = \mathbf{X}(\mathbf{x})$, $\mathbf{x} = (x_1, x_2)$. The variables

$$y_1 = x_1 - \xi, \qquad y_2 = x_2 - \eta \tag{3.5}$$

are a set of cartesian coordinates for the phase plane with their origin at $(x_1, x_2) = (\xi, \eta)$. They are said to be local coordinates at (ξ, η). It follows that

$$\dot{y}_i = \dot{x}_i = X_i(y_1 + \xi, y_2 + \eta), \qquad i = 1, 2, \tag{3.6}$$

where X_1, X_2 are the component functions of $\mathbf{X}$. If we define

$$Y_i(y_1, y_2) = X_i(y_1 + \xi, y_2 + \eta), \tag{3.7}$$

(3.6) becomes

$$\dot{y}_i = Y_i(y_1, y_2), \qquad i = 1,2 \quad \text{or} \quad \dot{\mathbf{y}} = \mathbf{Y}(\mathbf{y}). \tag{3.8}$$

The system in (3.8) has the fixed point of interest at the origin of its phase plane and Definition 3.2.1 can be applied.

Example 3.2.2

Show that the system

$$\dot{x}_1 = \mathrm{e}^{x_1 + x_2} - x_2, \qquad \dot{x}_2 = -x_1 + x_1 x_2 \tag{3.9}$$

has only one fixed point. Find the linearization of (3.9) at this point.

Solution

The fixed points of the system satisfy

$$\mathrm{e}^{x_1 + x_2} - x_2 = 0 \tag{3.10}$$

and

$$x_1(x_2 - 1) = 0. \tag{3.11}$$

Equation (3.11) is satisfied only by $x_1 = 0$ or $x_2 = 1$. If $x_1 = 0$, (3.10) becomes $\mathrm{e}^{x_2} = x_2$ which has no real solution ($\mathrm{e}^{x_2} > x_2$, for all real x_2) and there is no fixed point with $x_1 = 0$. If $x_2 = 1$, (3.10) gives $\mathrm{e}^{x_1 + 1} = 1$ which has only one real solution, $x_1 = -1$. Thus $(x_1, x_2) = (-1, 1)$ is the only fixed point of (3.9).

To find the linearized system at $(-1, 1)$, introduce local coordinates $y_1 = x_1 + 1$ and $y_2 = x_2 - 1$. We find

$$\dot{y}_1 = \mathrm{e}^{y_1 + y_2} - y_2 - 1, \qquad \dot{y}_2 = -y_2 + y_1 y_2. \tag{3.12}$$

This can be written in the form (3.2), by using the power series expansion of $\mathrm{e}^{y_1 + y_2}$:

$$\dot{y}_1 = y_1 + \left\{ \frac{(y_1 + y_2)^2}{2!} + \frac{(y_1 + y_2)^3}{3!} + \cdots \right\}$$

$$\dot{y}_2 = -y_2 + y_1 y_2. \tag{3.13}$$

Finally, we recognize the linearization as

$$\dot{y}_1 = y_1, \qquad \dot{y}_2 = -y_2. \tag{3.14}$$ □

Example 3.2.2 suggests a systematic way of obtaining linearizations by utilizing Taylor expansions. If the component functions $X_i(x_1, x_2)(i = 1, 2)$ are continuously differentiable in some neighbourhood of the point (ξ, η) then for each i

$$X_i(x_1, x_2) = X_i(\xi, \eta) + (x_1 - \xi)\frac{\partial X_i}{\partial x_1}(\xi, \eta)$$
$$+ (x_2 - \eta)\frac{\partial X_i}{\partial x_2}(\xi, \eta) + R_i(x_1, x_2). \tag{3.15}$$

The remainder functions $R_i(x_1, x_2)$ satisfy

$$\lim_{r \to 0}[R_i(x_1, x_2)/r] = 0, \tag{3.16}$$

where $r = \{(x_1 - \xi)^2 + (x_2 - \eta)^2\}^{1/2}$. If (ξ, η) is a fixed point of $\dot{\mathbf{x}} = \mathbf{X}(\mathbf{x})$, then $X_i(\xi, \eta) = 0\,(i = 1, 2)$ and on introducing local coordinates (3.5), we obtain

$$\dot{y}_1 = y_1\frac{\partial X_1}{\partial x_1}(\xi, \eta) + y_2\frac{\partial X_1}{\partial x_2}(\xi, \eta) + R_1(y_1 + \xi, y_2 + \eta)$$

$$\dot{y}_2 = y_1\frac{\partial X_2}{\partial x_1}(\xi, \eta) + y_2\frac{\partial X_2}{\partial x_2}(\xi, \eta) + R_2(y_1 + \xi, y_2 + \eta). \tag{3.17}$$

Equation (3.16) ensures that (3.17) is in the form (3.2) with $g_i(y_1, y_2) \equiv R_i(y_1 + \xi, y_2 + \eta)$ $(i = 1, 2)$ and the linearization at (ξ, η) is given by

$$a = \frac{\partial X_1}{\partial x_1}, \qquad b = \frac{\partial X_1}{\partial x_2}, \qquad c = \frac{\partial X_2}{\partial x_1}, \qquad d = \frac{\partial X_2}{\partial x_2}, \tag{3.18}$$

all evaluated at (ξ, η). Thus in matrix form the linearization is $\dot{\boldsymbol{y}} = \boldsymbol{A}\boldsymbol{y}$, where

$$\boldsymbol{A} = \left.\begin{bmatrix} \dfrac{\partial X_1}{\partial x_1} & \dfrac{\partial X_1}{\partial x_2} \\ \dfrac{\partial X_2}{\partial x_1} & \dfrac{\partial X_2}{\partial x_2} \end{bmatrix}\right|_{(x_2, x_2) = (\xi, \eta)} \tag{3.19}$$

Example 3.2.3
Obtain the linearization of (3.9) by using the Taylor expansion of X_1 and X_2 about $(-1, 1)$.

Solution
Recall

$$X_1(x_1, x_2) = \mathrm{e}^{x_1 + x_2} - x_2, \qquad X_2(x_1, x_2) = -x_1 + x_1 x_2.$$

The matrix A in (3.19) is

$$A = \begin{bmatrix} e^{x_1+x_2} & e^{x_1+x_2} - 1 \\ -1 + x_2 & x_1 \end{bmatrix}\Bigg|_{(x_1,x_2)=(-1,1)} = \begin{bmatrix} 1 & 0 \\ 0 & -1 \end{bmatrix}.$$

Therefore the linearization at $(-1, 1)$ is

$$\dot{y}_1 = y_1, \qquad \dot{y}_2 = -y_2;$$

in agreement with (3.14). □

3.3 THE LINEARIZATION THEOREM

This theorem relates the phase portrait of a non-linear system in the neighbourhood of a fixed point to that of its linearization.

Definition 3.3.1
A fixed point at the origin of a non-linear system $\dot{\mathbf{y}} = \mathbf{Y}(\mathbf{y})$, $\mathbf{y} \in S \subseteq \mathbb{R}^2$, is said to be **simple** if its linearized system is simple.

This definition extends the idea of simplicity (introduced in section 2.3) to the fixed points of non-linear systems. It can be used when the fixed point of interest is not at the origin by introducing local coordinates as in section 3.2.

Theorem 3.3.1 (Linearization theorem)
Let the non-linear system

$$\dot{\mathbf{y}} = \mathbf{Y}(\mathbf{y}) \tag{3.20}$$

have a simple fixed point at $\mathbf{y} = \mathbf{0}$. *Then, in a neighbourhood of the origin the phase portraits of the system and its linearization are qualitatively equivalent provided the linearized system is not a centre.*

This important theorem is not proved in this book. The interested reader may consult Hartman (1964) if a proof is desired.

The linearization theorem is the basis of one of the main methods—'linear stability analysis'—of investigating non-linear systems. However, it is easy to confuse the theorem and the techniques so that the significance of the theorem is not appreciated.

For example, one might mistakenly argue that the definition of the linearized system implies that, in a sufficiently small neighbourhood of the origin, the linear part of the vector field $\mathbf{Y}$ quantitatively approximates $\mathbf{Y}$ itself; and so their qualitative behaviour will obviously be the same. This is not a valid argument as the case when the linearization is a centre (excluded in the theorem) clearly shows.

Example 3.3.1
Show that the two systems

$$\dot{x}_1 = -x_2 + x_1(x_1^2 + x_2^2), \qquad \dot{x}_2 = x_1 + x_2(x_1^2 + x_2^2) \tag{3.21}$$

and

$$\dot{x}_1 = -x_2 - x_1(x_1^2 + x_2^2), \qquad \dot{x}_2 = x_1 - x_2(x_1^2 + x_2^2) \tag{3.22}$$

both have the same linearized systems at the origin, but that their phase portraits are qualitatively different.

Solution
Both systems (3.21) and (3.22) are already in the form (3.2), since both

$$\lim_{r \to 0}[x_1(x_1^2 + x_2^2)/r] \quad \text{and} \quad \lim_{r \to 0}[x_2(x_1^2 + x_2^2)/r]$$

are zero. Thus both systems have the linearization

$$\dot{x}_1 = -x_2, \qquad \dot{x}_2 = x_1 \tag{3.23}$$

which has a centre at $x_1 = x_2 = 0$. However, in polar coordinates (3.21) becomes

$$\dot{r} = r^3, \qquad \dot{\theta} = +1 \tag{3.24}$$

while (3.22) gives

$$\dot{r} = -r^3, \qquad \dot{\theta} = +1. \tag{3.25}$$

Equation (3.24) shows that $\dot{r} > 0$ for all $r > 0$ and so the trajectories of (3.21) spiral outwards as t increases. On the other hand, (3.25) has $\dot{r} < 0$ for all $r > 0$ and the trajectories of (3.22) spiral inwards (as shown in Fig. 3.4). Thus (3.21) shows unstable behaviour while (3.22) is stable, i.e. they are qualitatively different. However, sufficiently close to the origin the vector fields of both (3.21) and (3.22) are quantitatively approximated, to whatever accuracy we choose, by the linear vector field in (3.23). □

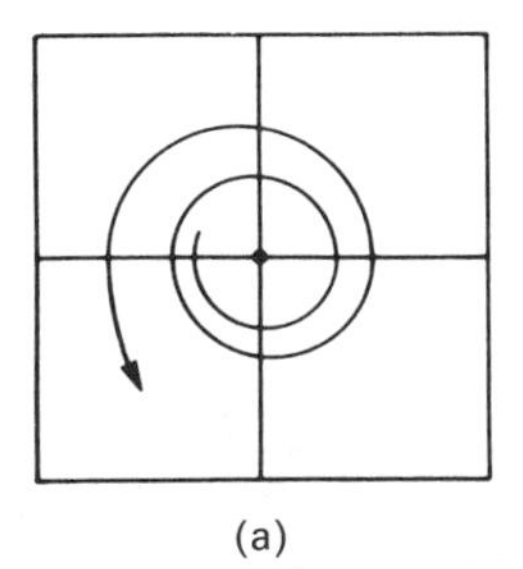

(a)

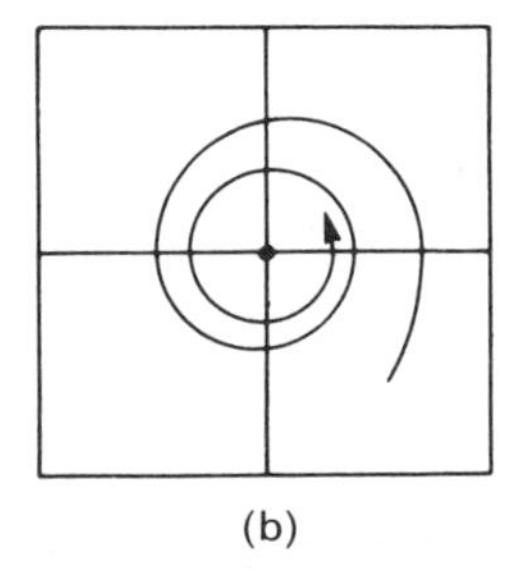

(b)

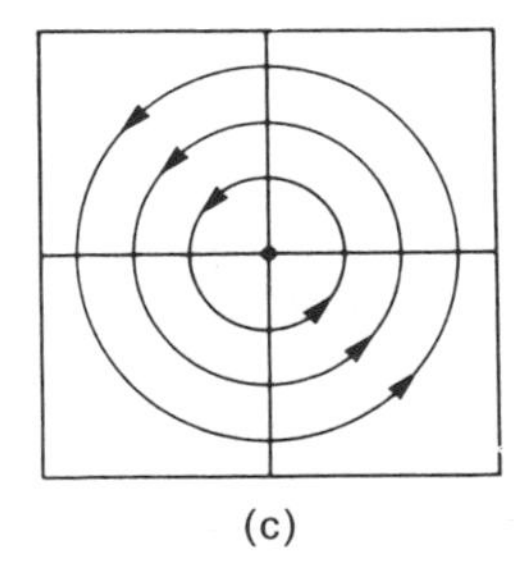

(c)

Fig. 3.4. Phase portraits for systems (a) (3.21); (b) (3.22) and (c) their linearization (3.23).

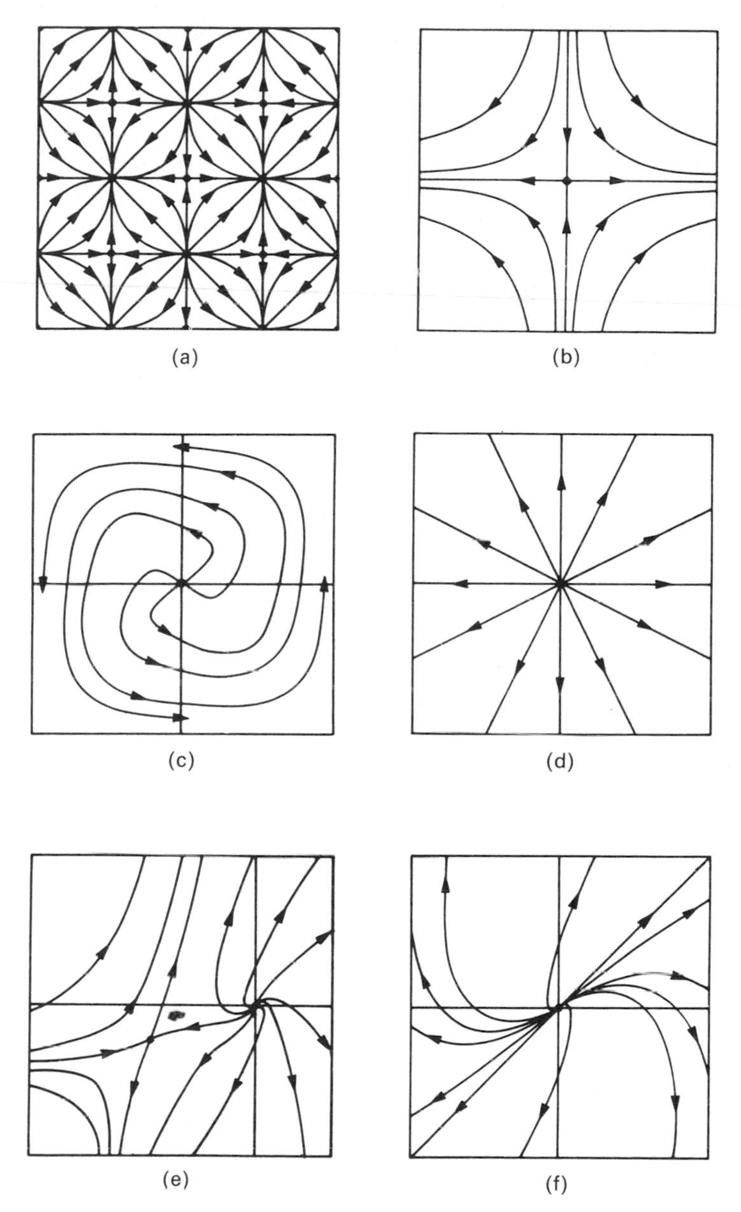

Fig. 3.5. Non-linear phase portraits and their linearizations at the origin: (a) non-linear system $\dot{x}_1 = \sin x_1$, $\dot{x}_2 = -\sin x_2$; (b) linearized system for (a) $\dot{x}_1 = x_1$, $\dot{x}_2 = -x_2$, a saddle point; (c) non-linear system $\dot{x}_1 = x_1 - x_2^3$, $\dot{x}_2 = x_2 + x_1^3$; (d) linearized system for (c) $\dot{x}_1 = x_1$, $\dot{x}_2 = x_2$, an unstable star node; (e) non-linear system $\dot{x}_1 = \frac{1}{2}(x_1 + x_2) + x_1^2$, $\dot{x}_2 = \frac{1}{2}(3x_2 - x_1)$; (f) linearized system for (e) $\dot{x}_1 = \frac{1}{2}(x_1 + x_2)$, $\dot{x}_2 = \frac{1}{2}(3x_2 - x_1)$. All these fixed points are hyperbolic.

Example 3.3.1 shows that the quantitative proximity of **Y** and its linear part does not guarantee qualitative equivalence of the non-linear system and its linearization. The content of the linearization theorem is that the centre is the *only* exception. That is, provided the eigenvalues of the linearized system have non-zero real part, the phase portraits of the non-linear system and its linearization are qualitatively equivalent in a neighbourhood of the fixed point. Such fixed points are said to be **hyperbolic**. Some examples are shown in Fig. 3.5.

Observe that the qualitative equivalence is not confined to the ultimate classification of stable, saddle, unstable of Fig. 2.11. In fact, within the stable or unstable categories, nodes, improper nodes and foci in the linearization are preserved in the phase portrait of the non-linear system. We can therefore refer to the non-linear fixed points as nodes, foci, etc. when their linearizations are of that type.

This general property of hyperbolic fixed points arises from the special character of the continuous bijection involved in the qualitative equivalence. The mapping must reflect the quantitative agreement between **Y** and its linear part near the fixed point. On small enough neighbourhoods it must, in some sense, be close to the identity. This property of the continuous bijection allows us to deduce some more information from the linearized system.

A **separatrix** is a trajectory which approaches, or leaves, a fixed point tangent to a fixed radial direction. It follows that the tangents to the separatrices of the linearized system at the fixed point are also the tangents to the non-linear separatrices.

Example 3.3.2

Use the linearization theorem to determine the local phase portrait of the system

$$\dot{x}_1 = x_1 + 4x_2 + e^{x_1} - 1, \qquad \dot{x}_2 = -x_2 - x_2 e^{x_1} \tag{3.26}$$

at the origin.

Solution

The component functions of the vector field in (3.26) are twice differentiable and so (3.19) is applicable. We obtain

$$\boldsymbol{A} = \left.\begin{bmatrix} \dfrac{\partial}{\partial x_1}(x_1 + 4x_2 + e^{x_1} - 1) & \dfrac{\partial}{\partial x_2}(x_1 + 4x_2 + e^{x_1} - 1) \\ \dfrac{\partial}{\partial x_1}(-x_2 - x_2 e^{x_1}) & \dfrac{\partial}{\partial x_2}(-x_2 - x_2 e^{x_1}) \end{bmatrix}\right|_{(x_1,x_2)=(0,0)} \tag{3.27}$$

Hence the linearization is

$$\dot{x}_1 = 2x_1 + 4x_2, \qquad \dot{x}_2 = -2x_2, \tag{3.28}$$

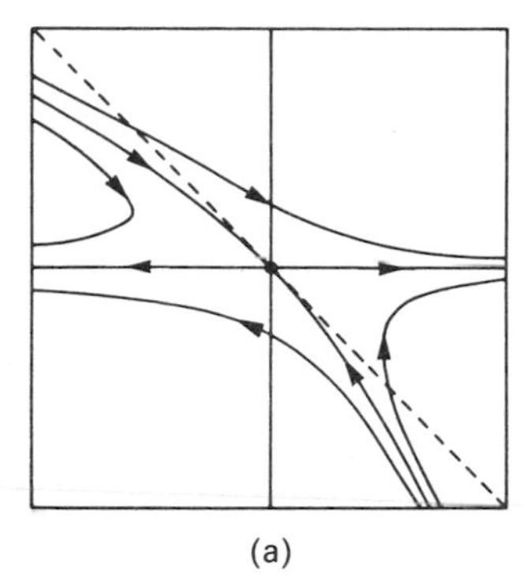

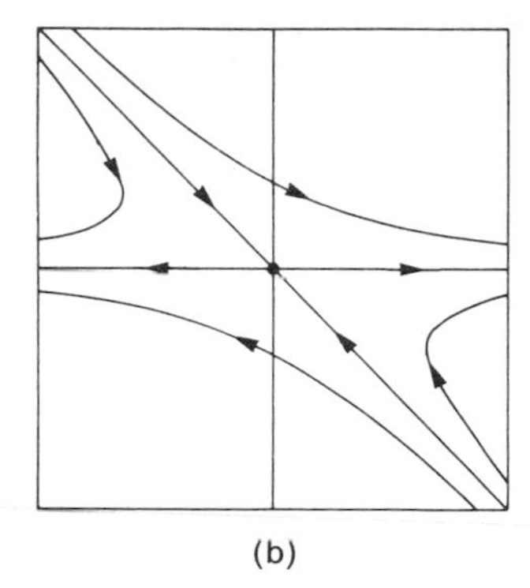

Fig. 3.6. Phase portraits for (a) the non-linear system (3.26) and (b) its linearized system (3.28). Notice that the x_1-axis is the unstable separatrix for both systems because both (3.26) and (3.28) imply $\dot{x}_2 = 0$ when $x_2 = 0$. The stable separatrices of (3.26) and (3.28), however, only become tangential at the fixed point.

which is a simple linear system with a saddle point at the origin ($\det(\boldsymbol{A}) < 0$).

The directions of the separatrices are given by the eigenvectors $(1, 0)$ and $(1, -1)$ of $\boldsymbol{A}$. The corresponding eigenvalues are $+2$ (unstable) and -2 (stable), respectively. Therefore, the unstable separatrices lie on the line $x_2 = 0$ while the stable ones lie on $x_2 = -x_1$ (as shown in Fig. 3.6(b)).

For (3.26), there is a neighbourhood of the origin in which the non-linear separatrices are as shown in Fig. 3.6(a). This follows because, on the line $x_2 = -x_1$, $\mathrm{d}x_2/\mathrm{d}x_1 \gtrless -1$ when $x_1 \lessgtr 0$. □

Further examples of the tangential relation between separatrices in non-linear systems and their linearizations appear in Fig. 3.5. The star node in items (c) and (d) of this figure is particularly interesting because every trajectory of both linear and non-linear systems is a separatrix. Thus every non-linear trajectory ultimately becomes tangential to its radial counterpart in the linearized system.

Our examples show that in analysing non-linear systems it is the directions of the straight line separatrices in the linearization that are important. They provide us with the direction of the non-linear separatrices at the fixed point. These directions are referred to as the **principal directions** at the fixed point and they are usually obtained (as in Example 3.3.2) from the linear transformation relating the linearized system to its canonical system.

3.4 NON-SIMPLE FIXED POINTS

A fixed point of a non-linear system is said to be **non-simple** if the corresponding linearized system is non-simple. Recall that such linear systems contain a straight line, or possibly a whole plane, of fixed points. The non-

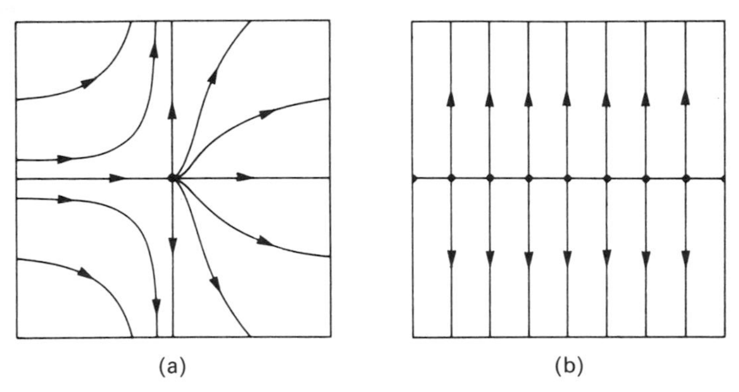

Fig. 3.7. Phase portraits for (a) the system $\dot{x}_1 = x_1^2$, $\dot{x}_2 = x_2$ and (b) its linearization at $(0, 0)$; $\dot{x}_1 = 0$, $\dot{x}_2 = x_2$, where the x_1-axis is a line of fixed points.

linear terms in g_1 and g_2 can drastically modify this behaviour as illustrated in Fig. 3.7 for example.

The nature of the local phase portrait is now determined by non-linear terms. Therefore, in contrast to the simple fixed points of section 3.3, there are infinitely many different types of local phase portrait. Some examples of what can occur for just low-order polynomials in x_1 and x_2 are shown in Figs 3.8–3.10. The linearizations of all of the systems shown in these figures have at least a line of fixed points as their phase portraits. These diagrams illustrate comparatively straightforward non-linear vector fields.

Lines of fixed points can also occur in non-linear systems; they are not necessarily straight and always consist of non-simple fixed points. Consider the system

$$\dot{x}_1 = x_1 - x_2^2, \qquad \dot{x}_2 = x_2(x_1 - x_2^2), \tag{3.29}$$

for example. The fixed points of (3.29) lie on the parabola $x_2^2 = x_1$. A typical

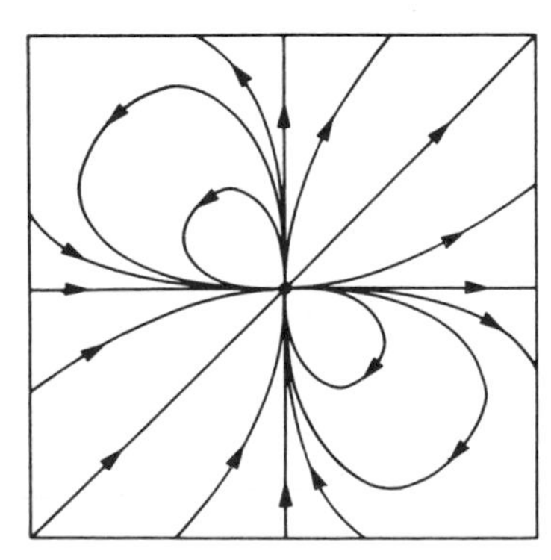

Fig. 3.8. $\dot{x}_1 = x_1(x_1 + 2x_2)$, $\dot{x}_2 = x_2(2x_1 + x_2)$; linearization $\dot{x}_1 = 0$, $\dot{x}_2 = 0$.

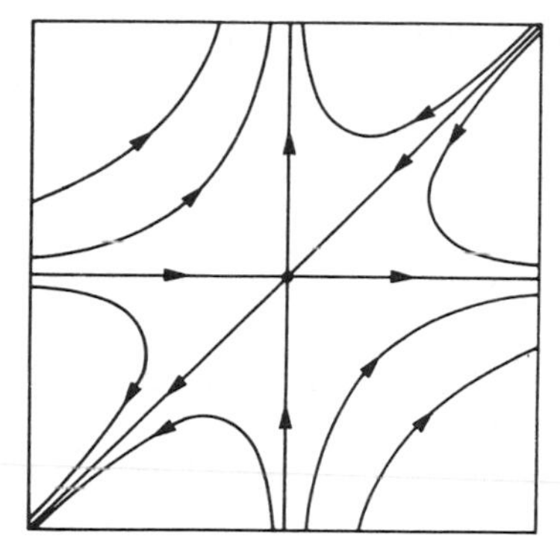

Fig. 3.9. $\dot{x}_1 = x_1(x_1 - 2x_2)$, $\dot{x}_2 = -x_2(2x_1 - x_2)$; linearization $\dot{x}_1 = 0$, $\dot{x}_2 = 0$.

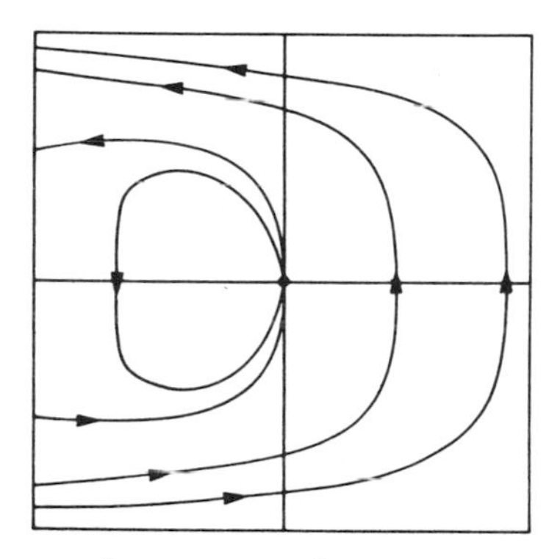

Fig. 3.10. $\dot{x}_1 = -x_2^5$, $\dot{x}_2 = x_1 + x_2^2$; linearization $\dot{x}_1 = 0$, $\dot{x}_2 = x_1$.

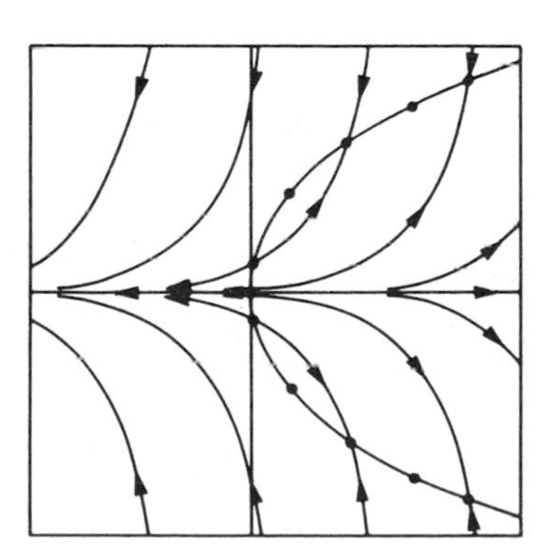

Fig. 3.11. The phase portrait for the system $\dot{x}_1 = x_1 - x_2^2$, $\dot{x}_2 = x_2(x_1 - x_2^2)$ with non-simple fixed points on the parabola $x_2^2 = x_1$.

fixed point (k^2, k), k real, has linearization $\dot{\boldsymbol{y}} = \boldsymbol{A}\boldsymbol{y}$, where

$$\boldsymbol{A} = \begin{bmatrix} 1 & -2k \\ k & -2k^2 \end{bmatrix}. \tag{3.30}$$

Clearly, $\det(\boldsymbol{A}) = 0$ for all k and so the fixed points are all non-simple. The phase portrait for (3.29) is shown in Fig. 3.11.

In view of the above observations, it is not surprising that there is no detailed classification of non-simple fixed points. However, the following definitions of stability (which apply to both simple and non-simple fixed points) do provide a coarse classification of qualitative behaviour.

3.5 STABILITY OF FIXED POINTS

It can be shown that the local phase portrait in the neighbourhood of any fixed point falls into one, and only one, of three stability types: asymptotically stable, neutrally stable or unstable. The following definition of a stable fixed point plays a central role in distinguishing these stability types.

Definition 3.5.1
A fixed point $\mathbf{x}_0$ of the system $\dot{\mathbf{x}} = \mathbf{X}(\mathbf{x})$ is said to be **stable** if, for every neighbourhood N of $\mathbf{x}_0$, there is a smaller neighbourhood $N' \subseteq N$ of $\mathbf{x}_0$ such that every trajectory which passes through N' remains in N as t increases.

This characterization of a stable fixed point is associated with the Russian mathematician Liapunov; indeed, it is often referred to as 'stability in the sense of Liapunov'.

Definition 3.5.2
A fixed point $\mathbf{x}_0$ of the system $\dot{\mathbf{x}} = \mathbf{X}(\mathbf{x})$ is said to be **asymptotically stable** if it is stable and there is a neighbourhood N of $\mathbf{x}_0$ such that every trajectory passing through N approaches $\mathbf{x}_0$ as t tends to infinity.

We have encountered asymptotic stability in connection with linear systems whose coefficient matrix $\boldsymbol{A}$ satisfies $\mathrm{tr}(\boldsymbol{A}) < 0$, $\det(\boldsymbol{A}) > 0$ (illustrated in Fig. 2.7). For the canonical systems of this type, the trajectory of every point outside the disc $D_r = \{(y_1, y_2) | (y_1^2 + y_2^2)^{1/2} < r\}$ enters the disc once and subsequently remains within it. Let N be any neighbourhood of the origin and $D_{r'}$ be a disc, centred on the origin, that is wholly contained in N. Then Definition 3.5.1 is satisfied with $N' = D_{r'}$ and the origin is a stable fixed point. What is more, the trajectories of all points within $D_{r'}$ approach the origin as t tends to infinity and therefore the origin is an asymptotically stable fixed point by Definition 3.5.2.

We can widen this result to linear systems of the above kind that are not in canonical form by using the non-singular matrix, $\boldsymbol{M}$, relating the system and its canonical form (described in section 2.4). The neighbourhoods that satisfy Definitions 3.5.1 and 3.5.2 for the canonical system provide corresponding neighbourhoods for the non-canonical system when mapped into the non-canonical phase plane. This is an example of the more general result that qualitatively equivalent fixed points must have the same stability type. Simple, non-linear fixed points can be tackled in a similar manner by making use of the local qualitative equivalence that is given by the linearization theorem. For example, the non-linear system

$$\dot{x}_1 = -x_1 + x_2 - x_1^3, \qquad \dot{x}_2 = -x_1 - x_2 + x_2^2 \tag{3.31}$$

has an asymptotically stable fixed point at the origin. This follows because: the linearized system

$$\dot{x}_1 = -x_1 + x_2, \qquad \dot{x}_2 = -x_1 - x_2 \tag{3.32}$$

has eigenvalues $-1 \pm \mathrm{i}$, so that the origin is a stable focus; and the existence of neighbourhoods of the kind required by Definitions 3.5.1 and 3.5.2 in the phase plane of (3.31) can be demonstrated by using the continuous bijection that relates the local phase portrait of (3.31) to that of (3.32) at the origin.

It is tempting to think that a fixed point must be asymptotically stable if the trajectories of the system tend to it as $t \to \infty$. However, this is not the case. The non-simple system

$$\dot{x}_1 = \frac{x_1^2(x_2 - x_1) + x_2^5}{(x_1^2 + x_2^2)(1 + (x_1^2 + x_2^2)^2)},$$
$$\dot{x}_2 = \frac{x_2^2(x_2 - 2x_1)}{(x_1^2 + x_2^2)(1 + (x_1^2 + x_2^2)^2)}, \tag{3.33}$$

shows that such fixed points may not be stable in the sense of Definition 3.5.1 (as shown in Exercise 3.12) and, consequently, they fail to be asymptotically stable according to Definition 3.5.2.

Every asymptotically stable fixed point is stable by Definition 3.5.2; however, the converse is not true.

Example 3.5.1

Show that the system

$$\dot{x}_1 = x_2, \qquad \dot{x}_2 = -x_1^3 \tag{3.34}$$

is stable at the origin but not asymptotically stable.

Solution

The fixed point at the origin of (3.34) is non-simple (the linearized system is $\dot{x}_1 = x_2$, $\dot{x}_2 = 0$) so that the linearization theorem does not provide a local phase portrait. However, the shape of the trajectories is given by

$$\frac{\mathrm{d}x_2}{\mathrm{d}x_1} = -\frac{x_1^3}{x_2} \tag{3.35}$$

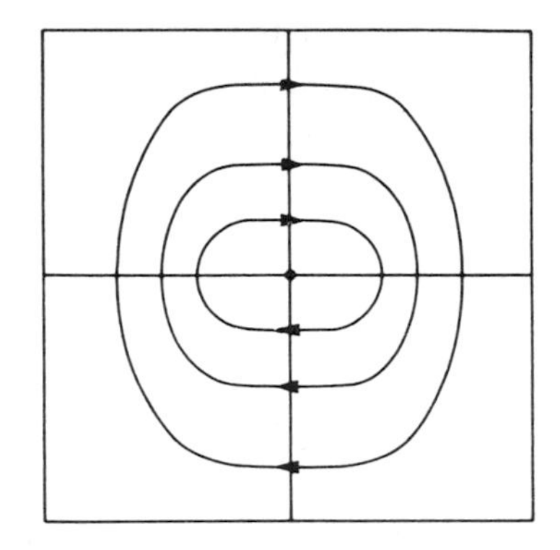

Fig. 3.12. Phase portrait for the system (3.34). Trajectories satisfy $\frac{1}{2}x_1^4 + x_2^2 = C$. Orientation given by $\dot{x}_1 > 0$ for $x_2 > 0$.

which has solutions satisfying

$$\tfrac{1}{2}x_1^4 + x_2^2 = C, \tag{3.36}$$

where C is a real constant. The phase portrait is shown in Fig. 3.12.

None of the trajectories approach the origin as $t \to \infty$, so the fixed point is not asymptotically stable. However, as Fig. 3.13 shows, for every disc N centred on the origin, there is a smaller disc N' such that every trajectory passing through N' remains in N. Thus the origin is stable. □

Definition 3.5.3

A fixed point of the system $\dot{\mathbf{x}} = \mathbf{X}(\mathbf{x})$ which is stable but not asymptotically stable is said to be **neutrally stable**.

There are many examples of neutrally stable fixed points similar to Example 3.5.1. For instance, the non-trivial fixed point of the Volterra–Lotka equations

$$\dot{x}_1 = x_1(a - bx_2), \qquad \dot{x}_2 = -x_2(c - dx_1), \tag{3.37}$$

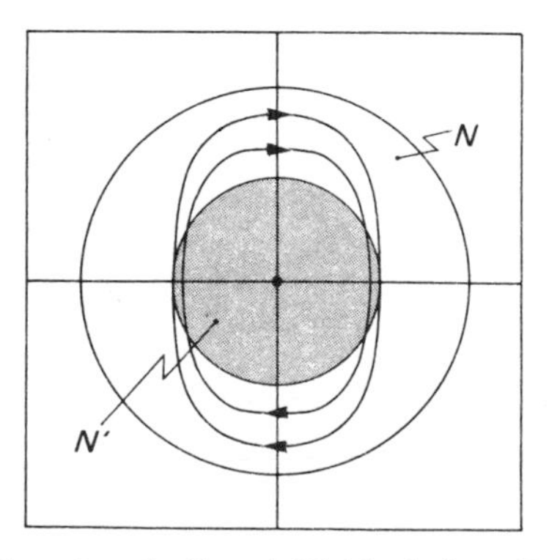

Fig. 3.13. Typical neighbourhoods N and N' (shaded) or Definition 3.5.1. Observe all trajectories passing through N' remain in N.

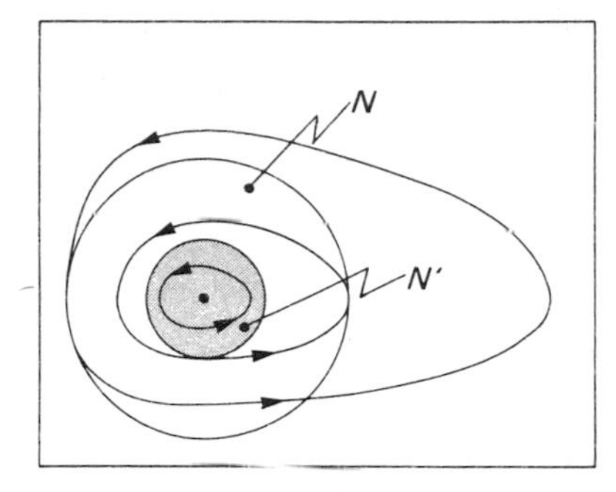

Fig. 3.14. Typical neighbourhoods N and N' for the Volterra–Lotka equations showing neutral stability.

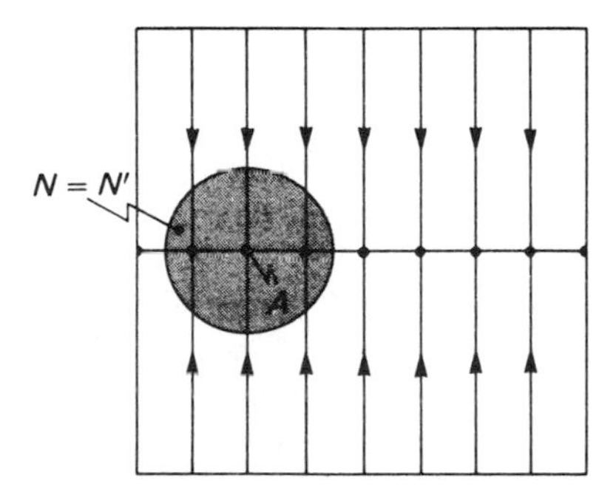

Fig. 3.15. Neutral stability of the system $\dot{x}_1 = 0$, $\dot{x}_2 = -x_2$ at A follows with $N = N'$.

$a, b, c, d > 0$, is neutrally stable. The phase portrait for these equations was given in Fig. 1.33. Neutral stability of the fixed point at $(c/d, a/b)$ follows from the existence of neighbourhoods N and N' satisfying Definition 3.5.1 as indicated in Fig. 3.14. Clearly, the fixed point is not asymptotically stable.

Another example is the non-simple linear fixed point shown in Fig. 3.15. The particular fixed point A is not asymptotically stable because there are trajectories passing through N of Fig. 3.15 which do not approach A as $t \to \infty$. However, with $N' = N$, every trajectory passing through N' remains in N; so A is stable.

Definition 3.5.4
A fixed point of the system $\dot{\mathbf{x}} = \mathbf{X}(\mathbf{x})$ which is not stable is said to be **unstable**.

This means that there is a neighbourhood N of the fixed point such that for every neighbourhood $N' \subseteq N$ there is at least one trajectory which passes through N' and does not remain in N. For example, the saddle point is unstable because there is a separatrix, containing points arbitrarily close to the origin, which escapes to infinity with increasing t.

The examples discussed above do not provide a practical procedure for determining the stability type of any given fixed point. An approach that can work for both simple and non-simple fixed points is to find a **Liapunov function** for the system.

For example, suppose we wish to investigate the nature of the fixed point at the origin of the system

$$\dot{x}_1 = -x_1^3, \qquad \dot{x}_2 = -x_2^3. \tag{3.38}$$

The linearization theorem is of no use here as the linearized system is clearly non-simple. However, we can show that the origin is asymptotically stable by examining how the function $V(x_1, x_2) = x_1^2 + x_2^2$ changes along the trajectories of (3.38).

Let $\mathbf{x}(t) = (x_1(t), x_2(t))$ be any solution curve of system (3.38). Then

$$\dot{V}(\mathbf{x}(t)) = \frac{\partial V}{\partial x_1}\dot{x}_1 + \frac{\partial V}{\partial x_2}\dot{x}_2 = -2(x_1^4 + x_2^4). \tag{3.39}$$

Therefore, $\dot{V}(\mathbf{x}(t))$ is negative at all points other than the origin of the x_1, x_2-plane and so $V(\mathbf{x}(t))$ decreases as t increases. This means that the phase point $\mathbf{x}(t)$ moves towards the origin with increasing t. In fact, $\dot{V}(\mathbf{x}(t)) < 0$ for $\mathbf{x}(t) \neq \mathbf{0}$ implies that $V(\mathbf{x}(t)) \to 0$ as $t \to \infty$ and hence $\mathbf{x}(t) \to \mathbf{0}$ as $t \to \infty$. Thus, the origin is an asymptotically stable fixed point of system (3.38).

The above example is a simple illustration of the use of a Liapunov function. To develop this idea we will need the following definitions.

Definition 3.5.5

A real-valued function $V: N \subseteq \mathbb{R}^2 \to \mathbb{R}$, where N is a neighbourhood of $\mathbf{0} \in \mathbb{R}^2$, is said to be **positive (negative) definite** in N if $V(\mathbf{x}) > 0\,(V(\mathbf{x}) < 0)$ for $\mathbf{x} \in N \setminus \{\mathbf{0}\}$ and $V(\mathbf{0}) = 0$.

Definition 3.5.6

A real-valued function $V: N \subseteq \mathbb{R}^2 \to \mathbb{R}$, where N is a neighbourhood of $\mathbf{0} \in \mathbb{R}^2$, is said to be **positive (negative) semi-definite** in N if $V(\mathbf{x}) \geqslant 0\,(V(\mathbf{x}) \leqslant 0)$ for $\mathbf{x} \in N \setminus \{\mathbf{0}\}$ and $V(\mathbf{0}) = 0$.

Definition 3.5.7

The derivative of $V: N \subseteq \mathbb{R}^2 \to \mathbb{R}$ along a parameterized curve given by $\mathbf{x}(t) = (x_1(t), x_2(t))$ is defined by

$$\dot{V}(\mathbf{x}(t)) = \frac{\partial V(\mathbf{x}(t))}{\partial x_1}\dot{x}_1(t) + \frac{\partial V(\mathbf{x}(t))}{\partial x_2}\dot{x}_2(t). \tag{3.40}$$

The function $V(x_1, x_2) = x_1^2 + x_2^2$ used in the introductory example is positive definite on $\mathbb{R}^2$. This function is typical of the positive definite functions used in this section. Any continuously differentiable, positive definite function V has a continuum of closed level curves (defined by $V(x_1, x_2) = \text{constant}$)

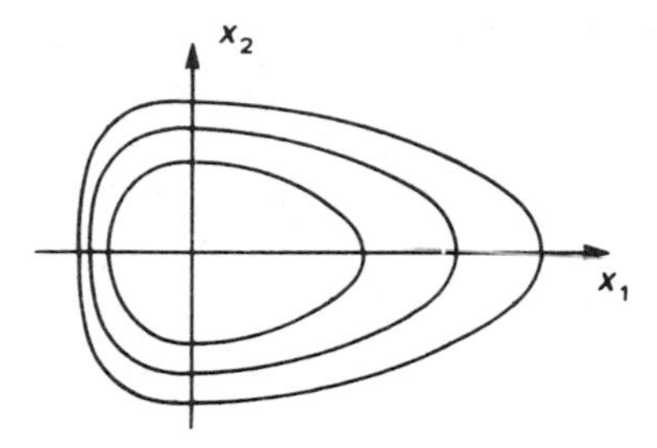

Fig. 3.16. The level curves $V(x_1, x_2) = C$ of the positive definite function $V(x_1, x_2) = x_1 - \log(1 + x_1) + x_2^2$ for $C = 0.5$, 1.0, 1.5.

around the origin. Of course, such curves are not necessarily circular (as shown in Fig. 3.16). However, provided $\dot{V}$ is negative on a trajectory, then that trajectory must still move towards the origin, because V is decreasing. Observe that for any system $\dot{\mathbf{x}} = \mathbf{X}(\mathbf{x})$,

$$\dot{V}(\mathbf{x}(t)) = \left(\frac{\partial V}{\partial x_1} X_1 + \frac{\partial V}{\partial x_2} X_2 \right) \tag{3.41}$$

is a function of x_1 and x_2 only and, for this reason, it is often denoted by $\dot{V}(\mathbf{x})$.

Theorem 3.5.1 (***Liapunov stability theorem***)
Suppose the system $\dot{\mathbf{x}} = \mathbf{X}(\mathbf{x})$, $\mathbf{x} \in S \subseteq \mathbb{R}^2$ has a fixed point at the origin. If there exists a real-valued function V in a neighbourhood N of the origin such that:

1. *the partial derivatives $\partial V/\partial x_1, \partial V/\partial x_2$ exist and are continuous;*
2. *V is positive definite;*
3. *$\dot{V}$ is negative semi-definite;*

then the origin is a **stable** *fixed point of the system.*
If 3 is replaced by the stronger condition
3′. $\dot{V}$ is negative definite,
then the origin is an **asymptotically stable** *fixed point.*

Proof
Properties 1 and 2 imply that the level curves of V form a continuum of closed curves around the origin. Thus, there is a positive k such that $N_1 = \{\mathbf{x} \mid V(\mathbf{x}) < k\}$ is a neighbourhood of the origin contained in N. If $\mathbf{x}_0 \in N_1 \setminus \{\mathbf{0}\}$, then $\dot{V}(\phi_t(\mathbf{x}_0)) \leqslant 0$, for all $t \geqslant 0$, by 3 and $V(\phi_t(\mathbf{x}_0))$ is a non-increasing function of t. Therefore, $V(\phi_t(\mathbf{x}_0)) < k$, for all $t \geqslant 0$, and so $\phi_t(\mathbf{x}_0) \in N_1$ for all $t \geqslant 0$. Consequently, by Definition 3.5.1 the fixed point is stable.

For case 3′ we obtain the asymptotic stability of the origin by the following argument. $V(\phi_t(\mathbf{x}_0))$ is a strictly decreasing function of t and $V(\phi_{t_2}(\mathbf{x}_0)) - V(\phi_{t_1}(\mathbf{x}_0)) < k$ for all $t_2 > t_1 \geqslant 0$. The mean value theorem then gives the existence of a sequence $\{\tau_i\}_{i=1}^{\infty}$, tending to infinity, such that $\dot{V}(\phi_{\tau_i}(\mathbf{x}_0)) \to 0$ as $\tau_i \to \infty$. This, in turn, implies that $\phi_{\tau_i}(\mathbf{x}_0) \to \mathbf{0}$ as $\tau_i \to \infty$ because $\dot{V}$ is negative definite. Now, $V(\phi_t(\mathbf{x}_0)) < V(\phi_{\tau_i}(\mathbf{x}_0))$ for all $t > \tau_i$ because $V(\phi_t(\mathbf{x}_0))$ is decreasing. However, V is positive definite and therefore $\{\phi_t(\mathbf{x}_0) | t > \tau_i\}$ lies inside the level curve of V containing $\phi_{\tau_i}(\mathbf{x}_0)$. This is true for every τ_i, so that '$\phi_{\tau_i}(\mathbf{x}_0) \to \mathbf{0}$ as $\tau_i \to \infty$' implies '$\phi_t(\mathbf{x}_0) \to \mathbf{0}$ as $t \to \infty$'. Moreover, the above argument is valid for all $\mathbf{x}_0$ in N_1 and therefore $\mathbf{x} = \mathbf{0}$ is an asymptotically stable fixed point. □

Definition 3.5.8
A function V satisfying hypotheses 1, 2 and 3 of Theorem 3.5.1 is called a **weak Liapunov function**. If 3 is replaced by 3′ then V is a **strong Liapunov function**.

Example 3.5.2
Prove that the function

$$V(y_1, y_2) = y_1^2 + y_1^2 y_2^2 + y_2^4, \qquad (y_1, y_2) \in \mathbb{R}^2 \tag{3.42}$$

is a strong Liapunov function for the system

$$\begin{aligned} \dot{x}_1 &= 1 - 3x_1 + 3x_1^2 + 2x_2^2 - x_1^3 - 2x_1 x_2^2 \\ \dot{x}_2 &= x_2 - 2x_1 x_2 + x_1^2 x_2 - x_2^3, \end{aligned} \tag{3.43}$$

at the fixed point (1, 0).

Solution
On introducing local coordinates y_1, y_2 at (1, 0), (3.43) becomes

$$\dot{y}_1 = -y_1^3 - 2y_1 y_2^2, \qquad \dot{y}_2 = y_1^2 y_2 - y_2^3. \tag{3.44}$$

The function V in (3.42) is positive definite and

$$\begin{aligned} \dot{V}(y_1, y_2) &= \frac{\partial V}{\partial y_1}\dot{y}_1 + \frac{\partial V}{\partial y_2}\dot{y}_2 \\ &= (2y_1 + 2y_1 y_2^2)(-y_1^3 - 2y_1 y_2^2) \\ &\quad + (2y_1^2 y_2 + 4y_2^3)(y_1^2 y_2 - y_2^3) \\ &= -2y_1^4 - 4y_1^2 y_2^2 - 2y_1^2 y_2^4 - 4y_2^6 \end{aligned}$$

is negative definite. Therefore, V is a strong Liapunov function for (3.43). □

Example 3.5.3
Investigate the stability of the second-order equation

$$\ddot{x} + \dot{x}^3 + x = 0 \tag{3.45}$$

at the origin of its phase plane.

Solution
If $x_1 = x$ and $x_2 = \dot{x}$, then

$$\dot{x}_1 = x_2, \qquad \dot{x}_2 = -x_1 - x_2^3 \tag{3.46}$$

is the first-order system of (3.45). The derivative of the function $V(x_1, x_2) = x_1^2 + x_2^2$ along the trajectories of (3.46) is

$$\dot{V}(x_1, x_2) = -2x_2^4$$

and so $\dot{V}$ is only negative semi-definite. Hence by Theorem 3.5.1 the origin is a stable fixed point of system (3.46). □

In fact, asymptotic stability can be deduced for some systems having a weak Liapunov function similar to that in Example 3.5.3. Observe that $\dot{V}(\mathbf{x})$ only fails to be negative away from the origin on the line $x_2 = 0$. On this line the components of the vector field given by (3.46) are $\dot{x}_1 = 0, \dot{x}_2 = -x_1$. Thus, all trajectories (except the origin) cross the line $x_2 = 0$ and $\dot{V}$ is only momentarily zero. At all other points in the plane it is negative. Under these circumstances the following theorem gives asymptotic stability.

Theorem 3.5.2

If there exists a weak Liapunov function V for the system $\dot{\mathbf{x}} = \mathbf{X}(\mathbf{x})$ in a neighbourhood of an isolated fixed point at the origin, then providing $\dot{V}(\mathbf{x})$ does not vanish identically on any trajectory, other than the fixed point itself, the origin is asymptotically stable.

Example 3.5.4

Show that all trajectories of the system

$$\dot{x}_1 = x_2, \qquad \dot{x}_2 = -x_1 - (1 - x_1^2)x_2 \tag{3.47}$$

passing through points (x_1, x_2), with $x_1^2 + x_2^2 < 1$, tend to the origin with increasing t.

Solution

The function $V(x_1, x_2) = x_1^2 + x_2^2$ is a weak Liapunov function in the region $x_1^2 + x_2^2 < 1$ ($\dot{V}(x_1, x_2) = -2x_2^2(1 - x_1^2)$). The function $\dot{V}$ vanishes only on the lines $x_2 = 0$ and $x_1 = \pm 1$. However, there are no trajectories of (3.47) which lie on these lines because on $x_2 = 0, \dot{x}_2 = -x_1 \neq 0$ and on $x_1 = \pm 1$, $\dot{x}_1 = x_2 \neq 0$. Therefore, by Theorem 3.5.2, the origin is asymptotically stable. Moreover the arguments used to prove Theorem 3.5.1 show that any trajectory $\phi_t(\mathbf{x}_0), |\mathbf{x}_0| < 1$, has the property $\lim_{t \to \infty} \phi_t(\mathbf{x}_0) = \mathbf{0}$. □

The fact that the origin is an asymptotically stable fixed point of system (3.47) can be deduced by using the linearization theorem. However, the above solution provides an explicit 'domain of stability' or 'basin of attraction' $N = \{(x_1, x_2) | x_1^2 + x_2^2 < 1\}$. All trajectories through points of N approach the origin as t increases. The linearization theorem gives the existence of a domain of stability but no indication of its size.

Theorem 3.5.3

Suppose the system $\dot{\mathbf{x}} = \mathbf{X}(\mathbf{x})$ has a fixed point at the origin. If a real-valued, continuous function V exists such that:

1. *the domain of V contains $N = \{\mathbf{x} \,|\, |\mathbf{x}| \leqslant r\}$ for some $r > 0$;*
2. *there are points arbitrarily close to the origin at which V is positive;*
3. *$\dot{V}$ is positive definite; and*
4. *$V(\mathbf{0}) = 0$,*

then the origin is unstable.

Proof

We show that for every point $\mathbf{x}_0$ in N, with $V(\mathbf{x}_0) > 0$, the trajectory $\phi_t(\mathbf{x}_0)$ leaves N for sufficiently large positive t. By hypothesis, such points can be chosen arbitrarily close to the origin and therefore the origin is unstable.

Given r_1, such that $0 < r_1 < r$, there is a point $\mathbf{x}_0 \neq \mathbf{0}$, with $|\mathbf{x}_0| < r_1$ and $V(\mathbf{x}_0) > 0$. The function $\dot{V}$ is positive definite in N and so $V(\phi_t(\mathbf{x}_0))$ is an increasing function of t. Therefore, the trajectory $\phi_t(\mathbf{x}_0)$ does not approach the origin as t increases. Hence $\dot{V}(\phi_t(\mathbf{x}_0))$ will be bounded away from zero, i.e. there exists a positive K such that $\dot{V}(\phi_t(\mathbf{x}_0)) \geqslant K$ for all positive t. If we assume the trajectory $\phi_t(\mathbf{x}_0)$ remains in N, then

$$V(\phi_t(\mathbf{x}_0)) - V(\mathbf{x}_0) \geqslant Kt, \tag{3.48}$$

for all positive t and $V(\phi_t(\mathbf{x}_0))$ becomes arbitrarily large in N. This contradicts the hypothesis that V is a continuous function defined on the closed and bounded set N. Thus the trajectory $\phi_t(\mathbf{x}_0)$ must leave N as t increases. □

Example 3.5.5

Show that the system

$$\dot{x}_1 = x_1^2, \qquad \dot{x}_2 = 2x_2^2 - x_1 x_2 \tag{3.49}$$

is unstable at the origin by using the function

$$V(x_1, x_2) = \alpha x_1^3 + \beta x_1^2 x_2 + \gamma x_1 x_2^2 + \delta x_2^3, \tag{3.50}$$

for a suitable choice of constants $\alpha, \beta, \gamma, \delta$.

Solution

The derivative of V along the trajectories of system (3.49) is

$$\begin{aligned}\dot{V}(x_1, x_2) = {} & 3\alpha x_1^4 + \beta x_1^3 x_2 \\ & + (2\beta - \gamma) x_1^2 x_2^2 + (4\gamma - 3\delta) x_1 x_2^3 \\ & + 6\delta x_2^4.\end{aligned} \tag{3.51}$$

Observe that if we choose $\alpha = \frac{1}{3}, \beta = 4, \gamma = 2, \delta = \frac{4}{3}$ then the various terms of $\dot{V}$ can be grouped together to form

$$\begin{aligned}\dot{V}(x_1, x_2) &= x_1^4 + 4x_1^3 x_2 + 6x_1^2 x_2^2 + 4x_1 x_2^3 + 8x_2^4 \\ &= (x_1 + x_2)^4 + 7x_2^4\end{aligned} \tag{3.52}$$

which is clearly positive definite. The function V is given by

$$V(x, x_2) = \tfrac{1}{3}x_1^3 + 4x_1^2 x_2 + 2x_1 x_2^2 + \tfrac{4}{3}x_2^3. \tag{3.53}$$

This function has the property that $V(x_1, x_2) = \frac{1}{3}x_1^3$ when $x_2 = 0$, and so points arbitrarily close to the origin on the x_1-axis can be found for which V is positive. It follows that the origin is an unstable fixed point by Theorem 3.5.3. □

3.6 ORDINARY POINTS AND GLOBAL BEHAVIOUR

3.6.1 Ordinary points

Any point in the phase plane of the system $\dot{\mathbf{x}} = \mathbf{X}(\mathbf{x})$ which is not a fixed point is said to be an **ordinary point**. Thus, if $\mathbf{x}_0$ is an ordinary point then $\mathbf{X}(\mathbf{x}_0) \neq \mathbf{0}$ and, by the continuity of $\mathbf{X}$, there is a neighbourhood of $\mathbf{x}_0$ containing only ordinary points. This means that the local phase portrait at an ordinary point has no fixed points. There is an important result concerning the qualitative equivalence of such local phase portraits—the **flow box theorem** (Hirsch and Smale, 1974).

Consider the local phase portraits at a typical ordinary point $\mathbf{x}_0$ shown in Figs 3.17–3.20. In each case, a special neighbourhood of $\mathbf{x}_0$, called a flow box, is shown. The trajectories of the system enter at one end and flow out through the other; no trajectories leave through the sides. For each phase portrait shown, we can find new coordinates for the plane such that the local phase portrait in the flow box looks like the one shown in Fig. 3.17. For example, in Fig. 3.18 we take polar coordinates (r, θ). In the r, θ-plane the circles ($r = \text{constant}$) become straight lines parallel to the $\theta = 0$ axis and the radial lines ($\theta = \text{constant}$) become straight lines parallel to the $r = 0$ axis. Thus, the phase portrait in the flow box in Fig. 3.18 is, in the r, θ-plane, the same as Fig. 3.17.

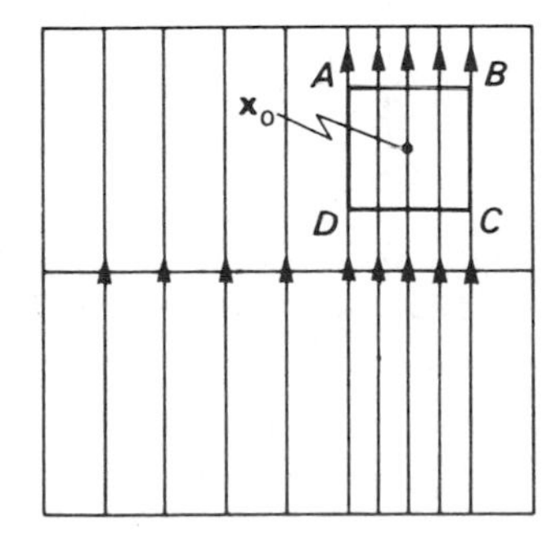

Fig. 3.17. System $\dot{x}_1 = 0$, $\dot{x}_2 = 1$ with typical flow box.

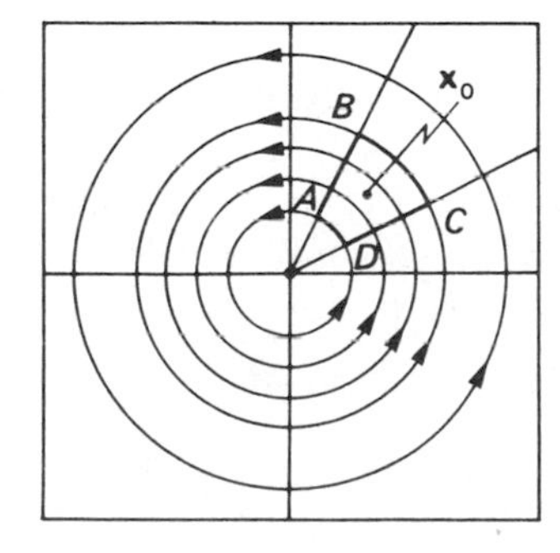

Fig. 3.18. System $\dot{x}_1 = -x_2$, $\dot{x}_2 = x_1$. In polar coordinates this gives $\dot{r} = 0$, $\dot{\theta} = 1$.

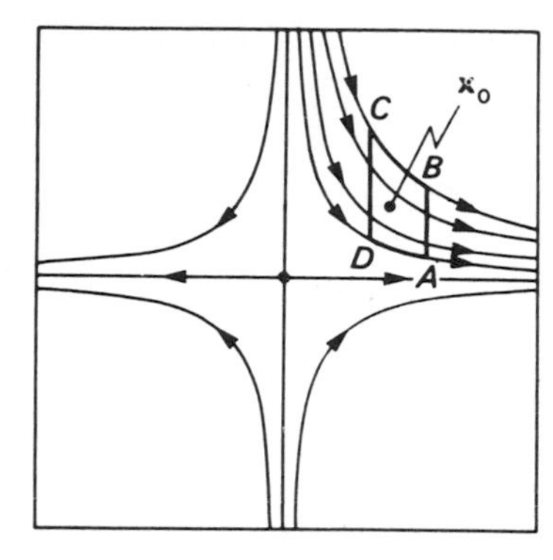

Fig. 3.19. System $\dot{x}_1 = x_1$, $\dot{x}_2 = -x_2$. The variables $y_1 = x_1 x_2$ and $y_2 = \ln x_1$, $x_1 > 0$, satisfy $\dot{y}_1 = 0$, $\dot{y}_2 = 1$.

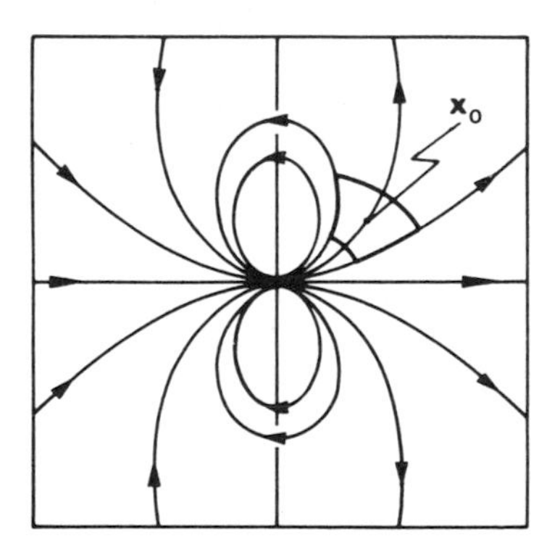

Fig. 3.20. The flow box theorem guarantees the existence of coordinates which transform the local phase portrait at $\mathbf{x}_0$ into the form shown in Fig. 3.17.

For Fig. 3.19, the trajectories in the neighbourhood of $\mathbf{x}_0$ lie on hyperbolae $x_1 x_2 = K > 0$. If we introduce variables $y_1 = x_1 x_2$ and $y_2 = \ln x_1$ then the flow box is bounded by the coordinate lines $y_1 = \text{constant}$ and $y_2 = \text{constant}$ and in the $y_1 y_2$-plane the local phase portrait again looks like Fig. 3.17.

Theorem 3.6.1 (***Flow box theorem***)
In a sufficiently small neighbourhood of an ordinary point $\mathbf{x}_0$ *of the system* $\dot{\mathbf{x}} = \mathbf{X}(\mathbf{x})$ *there is a differentiable change of coordinates* $\mathbf{y} = \mathbf{y}(\mathbf{x})$ *such that* $\dot{\mathbf{y}} = (0, 1)$.

The flow box theorem guarantees the existence of new coordinates with the above property, at least in some neighbourhood of any ordinary point of any system. Thus, local phase portraits at ordinary points are all qualitatively equivalent.

3.6.2 Global phase portraits

The linearization and flow box theorems provide local phase portraits at most simple fixed points and all ordinary points. However, this information does not always determine the complete phase portrait of a system.

Example 3.6.1
Find and classify the fixed points of the system

$$\dot{x}_1 = 2x_1 - x_1^2, \qquad \dot{x}_2 = -x_2 + x_1 x_2. \tag{3.54}$$

Discuss possible phase portraits for the system.

Solution
The system has fixed points at $A = (0, 0)$ and $B = (2, 0)$. The linearized systems are:

$$\dot{x}_1 = 2x_1, \qquad \dot{x}_2 = -x_2 \quad \text{at } A; \tag{3.55}$$

and

$$\dot{y}_1 = -2y_1, \qquad \dot{y}_2 = y_2 \quad \text{at } B. \tag{3.56}$$

The linearization theorem implies that (3.54) has saddle points at A and B. Furthermore, the non-linear separatrices of these saddle points are tangent to the principal directions at A and B. For (3.55) and (3.56) the principal directions coincide with the local coordinate axes.

This information is insufficient to determine the qualitative type of the global phase portrait. For example, Fig. 3.21 shows two phase portraits that are consistent with the local behaviour. The phase portraits shown are not qualitatively equivalent because the two saddle points have a common separatrix or **saddle connection** in Fig. 3.21(a) whereas in Fig. 3.21(b) they do not. This is a qualitative difference; there is no continuous bijection that relates the two phase portraits.

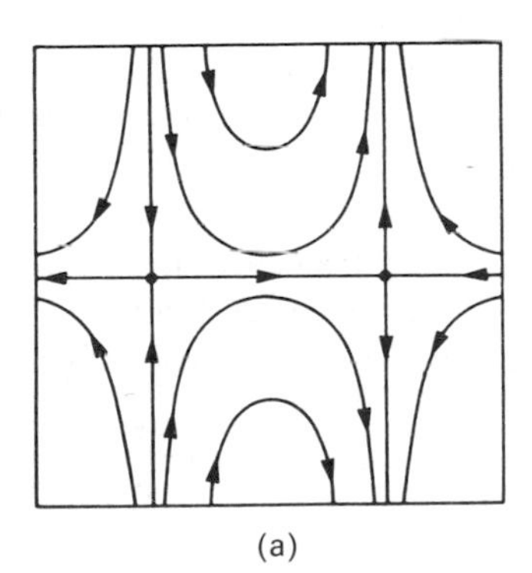

(a)

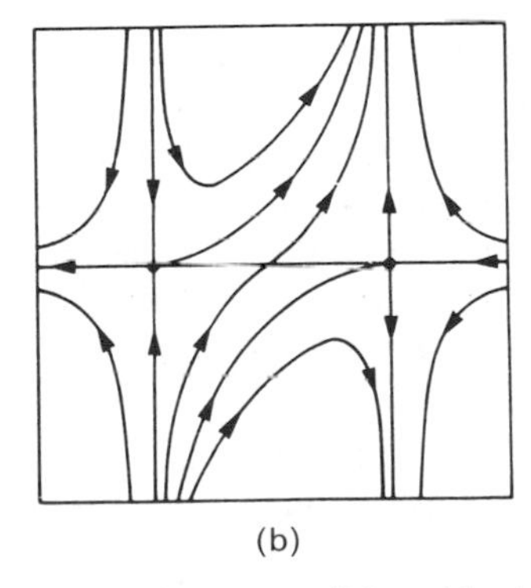

(b)

Fig. 3.21. Two qualitatively different phase portraits compatible with the local phase portraits obtained from the linearization theorem.

Returning to (3.54), observe that $\dot{x}_1 \equiv 0$ on the lines $x_1 = 0$ and $x_1 = 2$ so that trajectories coincide with these lines. Furthermore, $\dot{x}_2 \equiv 0$ when $x_2 = 0$. These observations indicate that Fig. 3.21 (a) gives the correct qualitative type of phase portrait. □

3.7 FIRST INTEGRALS

Definition 3.7.1
A continuously differentiable function $f: D(\subseteq \mathbb{R}^2) \rightarrow \mathbb{R}$ is said to be a **first integral** of the system $\dot{\mathbf{x}} = \mathbf{X}(\mathbf{x})$, $\mathbf{x} \in S \subseteq \mathbb{R}^2$ on the region $D \subseteq S$ if $f(\mathbf{x}(t))$ is constant for any solution $\mathbf{x}(t)$ of the system.

When a first integral exists it is not unique. Clearly if $f(\mathbf{x})$ is a first integral then so is $f(\mathbf{x}) + C$ or $Cf(\mathbf{x})$, C real. The constant C in the former is often chosen to provide a convenient value of the first integral at $\mathbf{x} = \mathbf{0}$. The trivial first integral which is identically constant on D, is always excluded from our considerations.

The fact that f is a first integral for $\dot{\mathbf{x}} = \mathbf{X}(\mathbf{x})$ can be expressed in terms of its (continuous) partial derivatives $f_{x_1} \equiv \partial f / \partial x_1$, $f_{x_2} \equiv \partial f / \partial x_2$. Since f is constant on any solution $\mathbf{x}(t) = (x_1(t), x_2(t))$,

$$\frac{\mathrm{d}}{\mathrm{d}t} f(\mathbf{x}(t)) = 0 = \dot{x}_1(t) f_{x_1}(\mathbf{x}(t)) + \dot{x}_2(t) f_{x_2}(\mathbf{x}(t)) \tag{3.57}$$

$$= X_1(\mathbf{x}(t))\, f_{x_1}(\mathbf{x}(t)) + X_2(\mathbf{x}(t))\, f_{x_2}(\mathbf{x}(t)) \tag{3.58}$$

$$= \lim_{h \rightarrow 0} \left\{ \frac{f(\mathbf{x} + h\mathbf{X}(\mathbf{x})) - f(\mathbf{x})}{h} \right\} \Bigg|_{\mathbf{x} = \mathbf{x}(t)}. \tag{3.59}$$

Equation (3.59) holds at every point of D, so that the directional derivative of f along the vector field $\mathbf{X}$, is identically zero on D.

A first integral is useful because of the connection between its level curves and the trajectories of the system. Consider the level curve $L_C = \{\mathbf{x} \mid f(\mathbf{x}) = C\}$. Let $\mathbf{x}_0 \in L_C$ and let $\xi(t)$ be the trajectory passing through the point $\mathbf{x}_0$ in the phase plane. Since f is a first integral, $f(\xi(t))$ is constant and $f(\xi(t)) = f(\mathbf{x}_0) = C$. Therefore, the trajectory passing through $\mathbf{x}_0$ lies in L_C.

When f is a first integral, it is constant on every trajectory in D. Thus every trajectory is part of some level curve of f. Hence each level curve is a union of trajectories. Uniqueness of the solutions to $\dot{\mathbf{x}} = \mathbf{X}(\mathbf{x})$ ensures that this union is disjoint.

The first integral is so named because it is usually obtained from a single integration of the differential equation

$$\frac{\mathrm{d}x_2}{\mathrm{d}x_1} = \frac{X_2(x_1, x_2)}{X_1(x_1, x_2)}, \qquad (x_1, x_2) \in D' \subseteq S. \tag{3.60}$$

If the solutions of this equation satisfy

$$f(x_1, x_2) = C, \tag{3.61}$$

where $f: D' \to \mathbb{R}$, then f is a first integral of $\dot{\mathbf{x}} = \mathbf{X}(\mathbf{x})$ on D'. This follows by differentiating (3.61),

$$\frac{df}{dx_1} \equiv 0 = f_{x_1} + \frac{dx_2}{dx_1} f_{x_2},$$

substituting from (3.60) and multiplying by $X_1(x_1, x_2)$ to obtain (3.58). Of course, X_1 cannot vanish on D' or (3.60) would not define dx_2/dx_1. However, zeros of X_1 present no such problem in (3.58). Thus, if $f(\mathbf{x})$ is continuously differentiable on a larger set $D \supset D'$ and (3.58) is satisfied there, then f is a first integral of $\dot{\mathbf{x}} = \mathbf{X}(\mathbf{x})$ on D.

Definition 3.7.2
A system that has a first integral on the whole of the plane (i.e. $D = \mathbb{R}^2$) is said to be **conservative**.

Example 3.7.1
Show that the system

$$\dot{x}_1 = -x_2, \qquad \dot{x}_2 = x_1 \tag{3.62}$$

is conservative, while the system

$$\dot{x}_1 = x_1, \qquad \dot{x}_2 = x_2 \tag{3.63}$$

is not.

Solution
The differential equation (3.60) gives

$$\frac{dx_2}{dx_1} = -\frac{x_1}{x_2}, \qquad x_2 \neq 0 \tag{3.64}$$

which has solutions satisfying

$$x_1^2 + x_2^2 = C, \qquad x_2 \neq 0, \tag{3.65}$$

where C is a positive constant. However, (3.58) with

$$f(\mathbf{x}) = x_1^2 + x_2^2 \tag{3.66}$$

is satisfied for all $(x_1, x_2) \in \mathbb{R}^2$. Thus (3.66) is a first integral of the system (3.62) on the whole plane and so (3.62) is a conservative system.

Now consider (3.63); the differential equation

$$\frac{dx_2}{dx_1} = \frac{x_2}{x_1}, \qquad x_1 \neq 0 \tag{3.67}$$

has solutions

$$x_2 = Cx_1 \tag{3.68}$$

with C real. In this case (3.58) is satisfied by

$$f(\mathbf{x}) = x_2/x_1, \qquad x_1 \neq 0 \tag{3.69}$$

so that D' is $\mathbb{R}^2$ less the x_2-axis.

There is no way in which the domain D' of f can be enlarged upon. The only continuous function which is:

1. defined on the whole x_1x_2-plane;
2. constant on every trajectory of (3.63) (i.e. on every radial line in the plane and at the origin itself);

is identically constant. This follows because on any radial line we can easily find a sequence of points $\{\mathbf{x}_i\}_{i=1}^{\infty}$ such that $\lim_{i\to\infty} \mathbf{x}_i = \mathbf{0}$. Thus, by continuity, $f(\mathbf{x}_i) = f(\mathbf{0})$ for all i and f takes the same value at all points of all radial lines. In other words, 1 and 2 are only satisfied by a function which is constant throughout $\mathbb{R}^2$. Thus, there is no first integral on $\mathbb{R}^2$ and (3.63) is not a conservative system. □

Conservative systems play an important part in mechanical problems. The equations of motion of such systems can be constructed from their **Hamiltonian**. For example, a particle moving in one dimension, with position coordinate x, momentum p and Hamiltonian $H(x, p)$ has equations of motion

$$\dot{x} = \frac{\partial H(x,p)}{\partial p}, \qquad \dot{p} = -\frac{\partial H(x,p)}{\partial x}. \tag{3.70}$$

Here $H(x, p)$ is a first integral for (3.70) because

$$\dot{x}\frac{\partial H}{\partial x} + \dot{p}\frac{\partial H}{\partial p} = \frac{\mathrm{d}}{\mathrm{d}t}H \equiv 0$$

(cf. (3.58)) and H remains constant along the trajectories. In other words, H is a conserved quantity or constant of the motion.

Example 3.7.2

Find the Hamiltonian H of the system

$$\dot{x} = p, \qquad \dot{p} = -x + x^3 \tag{3.71}$$

and sketch its phase portrait.

Solution

The differential equation

$$\frac{\mathrm{d}p}{\mathrm{d}x} = \frac{-x + x^3}{p}, \qquad p \neq 0 \tag{3.72}$$

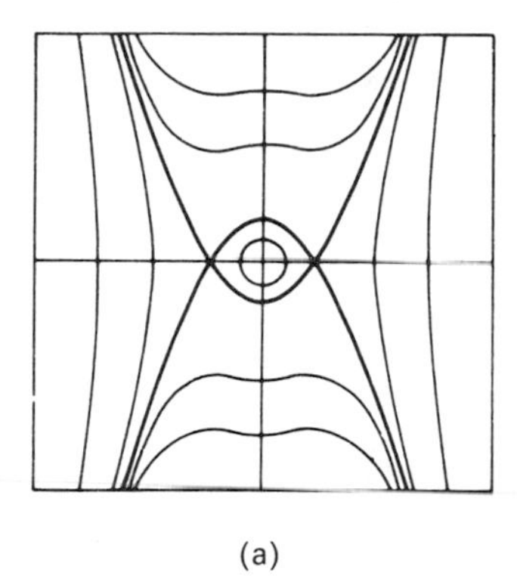

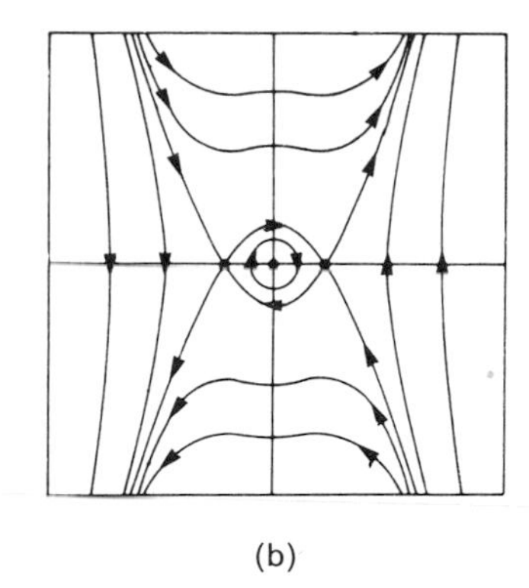

Fig. 3.22. (a) Level curves of $H(x, p) = x^2 - \frac{1}{2}x^4 + p^2 \cdot L_{1/2}$ is marked by a heavy line. (b) Trajectories of (3.71): note saddle connections. The orientations are determined by noting that $\dot{x} > 0$ for $p > 0$ and $\dot{x} < 0$ for $p < 0$.

has solutions on the plane less the x-axis which satisfy

$$x^2 - \tfrac{1}{2}x^4 + p^2 = C, \tag{3.73}$$

where C is a constant. Equation (3.58) shows that this is a first integral on the whole of the plane and $H(x, p) = x^2 - \frac{1}{2}x^4 + p^2$.

The level curves of the first integral are unions of trajectories of (3.71), so we can obtain the global phase portrait for the system by sketching the level curves of $H(x, p)$. These curves are shown in Fig. 3.22(a) and the phase portrait in Fig. 3.22(b). □

We can show that a given level curve is a union of several trajectories by considering $L_{1/2} = \{\mathbf{x} \mid x^2 - \frac{1}{2}x^4 + p^2 = \frac{1}{2}\}$ for $H(x, p)$. This set of points is shown as a heavy line in Fig. 3.22(a). In Fig. 3.22(b), it is made up from eight trajectories.

Notice also that the linearization of (3.71) at the origin is a centre, so that the linearization theorem could not provide a local phase portrait there. In fact, first integrals are one of the main ways of detecting centres in non-linear systems.

Example 3.7.3
Show that the fixed point of the system

$$\dot{x}_1 = x_1 - x_1 x_2, \qquad \dot{x}_2 = -x_2 + x_1 x_2 \tag{3.74}$$

at $(1, 1)$ is a centre.

Solution
The differential equation

$$\frac{dx_2}{dx_1} = \frac{-x_2 + x_1 x_2}{x_1 - x_1 x_2}, \qquad x_1 \neq x_1 x_2, \tag{3.75}$$

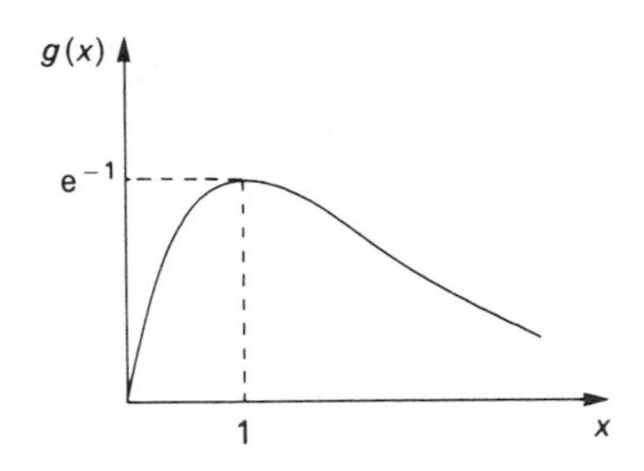

Fig. 3.23. Plot xe^{-x} versus x for $x \geqslant 0$.

is separable, with solutions satisfying

$$g(x_1)g(x_2) = K, \tag{3.76}$$

where $g(x) = xe^{-x}$ and K is a positive constant. The function g is shown in Fig. 3.23 for $x \geqslant 0$; it has a single maximum at $x = 1$ where $g(1) = e^{-1}$. It follows that $(x_1, x_2) = (1, 1)$ is a maximum of the first integral $g(x_1)g(x_2)$. This means that there is a neighbourhood of $(1, 1)$ in which the level curves of $g(x_1)g(x_2)$ are closed. Since, these level curves coincide with trajectories, we conclude that $(1, 1)$ is a centre. □

It is important to realize that first integrals do not give solutions $\mathbf{x}(t)$ for a system; rather they provide the shape of the trajectories.

Example 3.7.4
Show that the systems

$$\dot{x}_1 = x_1, \qquad \dot{x}_2 = -x_2 \tag{3.77}$$

and

$$\dot{x}_1 = x_1(1 - x_2), \qquad \dot{x}_2 = -x_2(1 - x_2) \tag{3.78}$$

have the same first integral and sketch their phase portraits.

Solution
The trajectories of both systems lie on the solutions of

$$\frac{dx_2}{dx_1} = -\frac{x_2}{x_1}, \qquad x_1 \neq 0, \tag{3.79}$$

and both have

$$f(\mathbf{x}) = x_1 x_2 \tag{3.80}$$

as a first integral on $\mathbb{R}^2$. The level curves of f are rectangular hyperbolae which, for (3.77), can be oriented by noting the direction of $\dot{\mathbf{x}}$ on the coordinate axes. This is simply the linear saddle familiar from section 2.3.

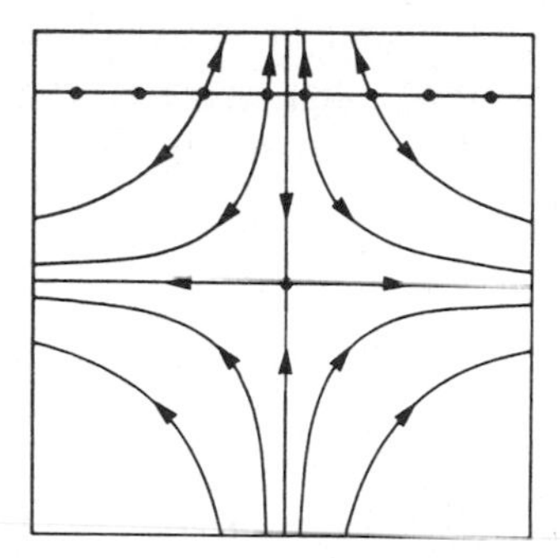

Fig. 3.24. Phase portrait for system (3.78). This system has the same first integral as the linear saddle.

The system (3.78) has fixed points at the origin and everywhere on the line $x_2 = 1$. Furthermore, $\dot{x}_2 > 0$ for $x_2 > 1$ and for $x_2 < 0$ while $\dot{x}_2 < 0$ for $0 < x_2 < 1$. Therefore the phased portrait must be that shown in Fig. 3.24. □

3.8 LIMIT POINTS AND LIMIT CYCLES

Let **x** be a point in the phase portrait of a flow ϕ_t. The α-(ω-) limit set, $L_\alpha(\mathbf{x})$ ($L_\omega(\mathbf{x})$), of **x** contains those points that are approached by the trajectory through **x** as t tends to $-\infty(+\infty)$. These limit points of **x** are defined as follows.

Definition 3.8.1

A point **y** is said to be an α-(ω-) **limit point** of **x** if there exists a sequence of times $\{t_n\}_{n=1}^{\infty}$, tending to minus (plus) infinity as $n \to \infty$, such that $\lim_{n\to\infty} \phi_{t_n}(\mathbf{x}) = \mathbf{y}$.

Consider the phase portrait illustrated in Fig. 3.25. For any ordinary point,

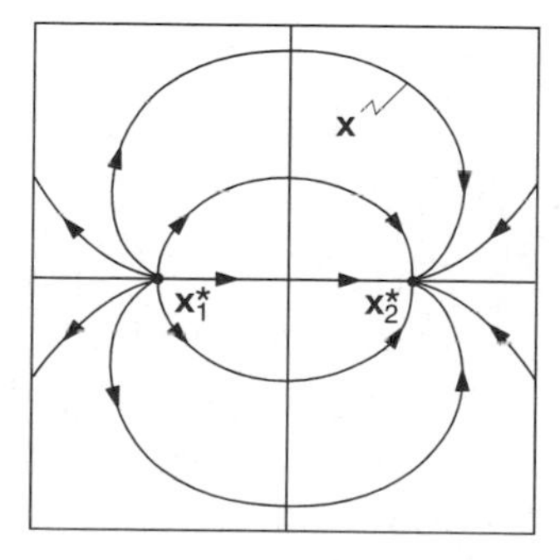

Fig. 3.25. Fixed points are the simplest examples of α- and ω-limit sets. Here $L_\alpha(\mathbf{x}) = \mathbf{x}_1^*$ and $L_\omega(\mathbf{x}) = \mathbf{x}_2^*$ for every ordinary point **x**.

$\mathbf{x}$, $\phi_t(\mathbf{x})$ tends to $\mathbf{x}_1^*$ as $t \to -\infty$ and to $\mathbf{x}_2^*$ as $t \to +\infty$, so that $L_\alpha(\mathbf{x}) = \{\mathbf{x}_1^*\}$ and $L_\omega(\mathbf{x}) = \{\mathbf{x}_2^*\}$. If $\mathbf{x} = \mathbf{x}_1^*$ or $\mathbf{x}_2^*$, then $\phi_t(\mathbf{x}) = \mathbf{x}$ for all t and $L_\alpha(\mathbf{x}) = L_\omega(\mathbf{x}) = \{\mathbf{x}\}$. In both cases, any sequence of times, tending to the appropriate limit, will suffice in Definition 3.8.1.

Example 3.8.1
Let ϕ_t be the flow on $\mathbb{R}^2$ of the differential equation with polar form

$$\dot{r} = ar(1-r), \qquad \dot{\theta} = 1, \tag{3.81}$$

where a is a positive constant. Find $L_\alpha(\mathbf{x})$ and $L_\omega(\mathbf{x})$ for $\mathbf{x} \neq \mathbf{0}$.

Solution
The phase portrait for the system (3.81) is shown in Fig. 3.26. The main features are the unstable focus at the origin and the closed orbit, C, given by $r \equiv 1$.

To find $L_\omega(\mathbf{x})$ for $\mathbf{x} \neq \mathbf{0}$, let $\mathbf{y}$ be any point of C and take $\{t_n\}_{n=1}^\infty$ to be the sequence of $t > 0$ at which the trajectory through $\mathbf{x}$ crosses the radial line through $\mathbf{y}$ (see Fig. 3.26). Clearly, $t_n \to \infty$ as $n \to \infty$ and $\lim_{n\to\infty} \phi_{t_n}(\mathbf{x}) = \mathbf{y}$. In particular, if $\mathbf{x}$ lies in C then $\phi_{t_n}(\mathbf{x}) = \mathbf{y}$ for each n. Thus every point of C is an ω-limit point of $\mathbf{x}$ by Definition 3.8.1 and $L_\omega(\mathbf{x}) = C$ for any $\mathbf{x} \neq \mathbf{0}$.

Let us now turn to $L_\alpha(\mathbf{x})$. If $|\mathbf{x}| \leqslant 1$, then a similar definition of $\{t_n\}_{n=1}^\infty$ with $t < 0$ allows us to show that

$$L_\alpha(\mathbf{x}) = \begin{cases} \{\mathbf{0}\} & \text{if } |x| < 1 \\ C & \text{if } |x| = 1. \end{cases} \tag{3.82}$$

However, if $|\mathbf{x}| > 1$ there is no sequence $\{t_n\}_{n=1}^\infty$, with $t_n \to -\infty$ as $n \to \infty$, such that $\lim_{n\to\infty} \phi_{t_n}(\mathbf{x})$ exists. We conclude therefore that $L_\alpha(\mathbf{x})$ is empty for $|\mathbf{x}| > 1$. □

The closed orbit, C, in Example 3.8.1 is an example of what is called a limit cycle.

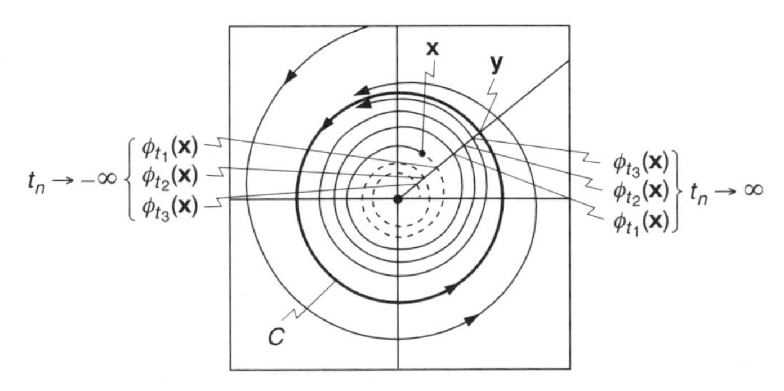

Fig. 3.26. Phase portrait for (3.81) with $a = \frac{1}{5}$. Note that the origin is an unstable focus and the circle $r \equiv 1$ is a closed orbit of period 2π.

Definition 3.8.2

A closed orbit, $\mathscr{C}$, is said to be a **limit cycle** if $\mathscr{C}$ is a subset of $L_\alpha(\mathbf{x})$ or $L_\omega(\mathbf{x})$ for some $\mathbf{x}$ that does not lie in $\mathscr{C}$.

Observe that Definition 3.8.2 does not require that trajectories approach the limit cycle from both sides, as is the case for C in Example 3.8.1. The limit cycle occurring in Example 3.8.1 has the property that the trajectories of all points, $\mathbf{x}$, with $|\mathbf{x}| \neq 0$ or 1 are attracted to it as time increases. It is an example of what is called an **attracting set**. Figure 3.27 shows some limit cycles that are not attracting sets. Thus, while every attracting set is a limit set, not all limit sets are attracting sets. Note also that a closed orbit around a centre is not a limit cycle, because it only contains limit points of points in itself.

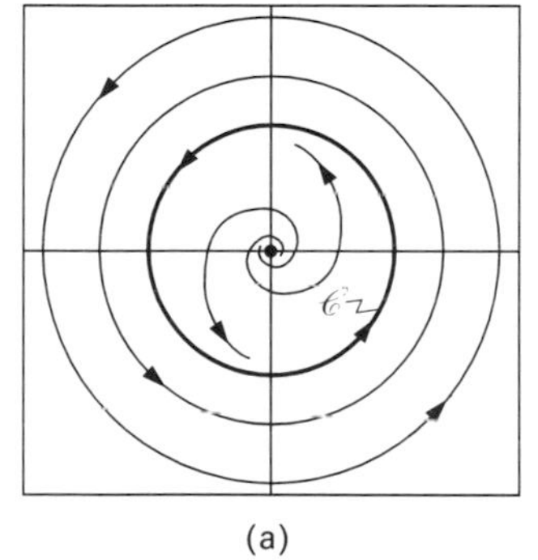

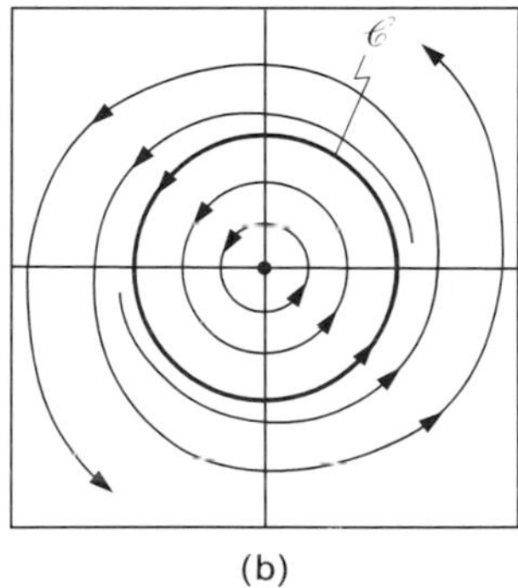

Fig. 3.27. Examples of limit cycles that are not attracting sets. Observe that in: (a) $L_\omega(\mathbf{x}) = \mathscr{C}$ for $|\mathbf{x}| \leqslant 1$; (b) $L_\alpha(\mathbf{x}) = \mathscr{C}$ for $|\mathbf{x}| \geqslant 1$; and $\mathscr{C}$ is therefore a limit cycle in both cases. However, $\mathscr{C}$ is not an attracting set in either case because trajectories do not approach $\mathscr{C}$ from both sides.

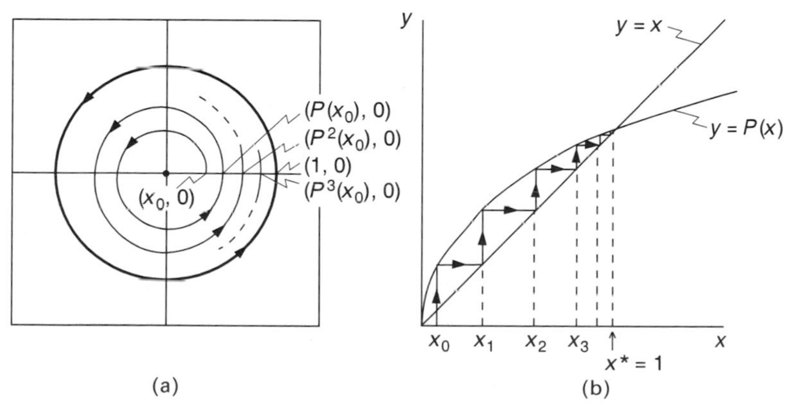

Fig. 3.28. (a) Illustration of the definition of the Poincaré map, P, for the system (3.81). (b) Graphical representation of the fixed point iteration $x_{+1} = P(x_n)$, $n = 0, 1, \ldots$, with attracting fixed point at $x^* = 1$.

The stable nature of the limit cycle in Example 3.8.1 is related to the existence of an attracting fixed point in the corresponding **Poincaré** (or **first return**) **map** of the flow. For example, consider the trajectory passing through the point $(x_0, 0)$ with $x_0 > 0$. Suppose we follow this trajectory and that its first return to the positive x-axis occurs at $x = x_1$. Then we define $x_1 = P(x_0)$ (as shown in Fig. 3.28(a)). This procedure allows us to define the Poincaré map $P(x)$ for every $x \in (0, \infty)$. In particular, $P(1) = 1$, because the trajectory passing through $(1, 0)$ is the closed orbit C. Thus the fixed point $x = x^* = 1$ of P corresponds to the limit cycle C in the flow. What is more, the iteration $x_{n+1} = P(x_n)$, $n = 0, 1, \ldots$, reflects the convergence of the trajectory through $(x_0, 0)$ to the limit cycle C as shown in Fig. 3.28.

In general, the Poincaré map, P, may only be defined on a line segment—or **local section**—transverse to the closed orbit. However, the stability of the limit cycle is still given by the stability of the associated fixed point iteration of P. In particular, x^* is attracting (repelling), and the corresponding limit cycle is **stable (unstable)** if $|\mathrm{d}P/\mathrm{d}x|_{x=x^*}$ is less (greater) than unity. If $|\mathrm{d}P/\mathrm{d}x|_{x=x^*} = 1$ and $\mathrm{d}^2P/\mathrm{d}x^2 \neq 0$ then the iteration will converge on one side of x^* and diverge from the other. In this case the corresponding limit cycles are said to be **semi-stable**.

Example 3.8.2

Find the limit cycles in the following systems and give their types.

1. $\dot{r} = r(r-1)(r-2), \qquad \dot{\theta} = 1;$ (3.83)
2. $\dot{r} = r(r-1)^2, \qquad \dot{\theta} = 1.$ (3.84)

Solution

1. There are closed trajectories given by

$$r(t) \equiv 1, \quad \theta = t \qquad \text{and} \qquad r(t) \equiv 2, \quad \theta = t \tag{3.85}$$

corresponding to fixed points in the Poincaré map defined on any radial line. The stability of these fixed points is given by the sign of $\dot{r}$. Observe

$$\dot{r} \begin{cases} > 0, & 0 < r < 1 \\ < 0, & 1 < r < 2 \\ > 0, & r > 2. \end{cases} \tag{3.86}$$

The system therefore has two circular limit cycles: one stable ($r = 1$) and one unstable ($r = 2$).

2. System (3.84) has a single circular limit cycle of radius one. However, $\dot{r}$ is positive for $0 < r < 1$ and $r > 1$, so the limit cycle is semi-stable. □

Limit cycles are not always circular and are, therefore, not always revealed by simply changing to polar coordinates. For example, consider the 'Van der Pol' equation

$$\ddot{x} - \dot{x}(1 - x^2) + x = 0 \tag{3.87}$$

with its equivalent first-order system

$$\dot{x}_1 = x_2, \qquad \dot{x}_2 = x_2(1 - x_1^2) - x_1. \tag{3.88}$$

In polar coordinates this becomes

$$\begin{aligned} \dot{r} &= r\sin^2\theta(1 - r^2\cos^2\theta) \\ \dot{\theta} &= -1 + \cos\theta\sin\theta(1 - r^2\cos^2\theta). \end{aligned} \tag{3.89}$$

These equations do not give any immediate insight into the nature of the phase portrait which contains a unique attracting limit cycle (described in section 5.4). In fact, the problem of detecting limit cycles in non-linear systems can be a difficult one which we will have to examine more closely.

3.9 POINCARÉ–BENDIXSON THEORY

We have so far encountered limit sets that are fixed points or closed orbits. What other possibilities can occur?

Example 3.9.1

Consider the phase portrait shown in Fig. 3.29. Find $L_\alpha(\mathbf{x})$ and $L_\omega(\mathbf{x})$ for $\mathbf{x}$ lying in the regions A, A', B, B', C, respectively. Comment on the nature of the limit sets you obtain.

Solution

Take straight lines emanating from each of the fixed points P_1 and P_2. Suppose $\mathbf{x}$ lies in each of the regions of interest in turn and examine the intersections of the trajectory through $\mathbf{x}$ with these straight lines. In every case, the intersections provide time sequences, $\{t_n\}_{n=1}^{\infty}$, and limit points, $\mathbf{y}$, satisfying the requirements of Definition 3.8.1. Rotation of each of the straight lines through 2π radians leads to the limit sets required. The results may be

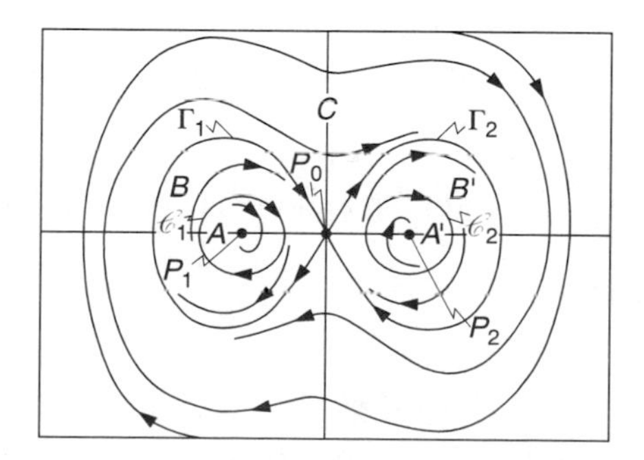

Fig. 3.29. The phase portrait considered in Example 3.9.1. The separatrices at P_0 coincide to form trajectories, Γ_1 and Γ_2, called **saddle-connections**.

summarized as follows.

$$\begin{array}{lll} \mathbf{x}\in A & L_\alpha(\mathbf{x})=\mathscr{C}_1 & L_\omega(\mathbf{x})=P_1 \\ \mathbf{x}\in A' & L_\alpha(\mathbf{x})=\mathscr{C}_2 & L_\omega(\mathbf{x})=P_2 \\ \mathbf{x}\in B & L_\alpha(\mathbf{x})=\mathscr{C}_1 & L_\omega(\mathbf{x})=\Gamma_1\cup P_0 \\ \mathbf{x}\in B' & L_\alpha(\mathbf{x})=\mathscr{C}_2 & L_\omega(\mathbf{x})=\Gamma_2\cup P_0 \\ \mathbf{x}\in C & L_\alpha(\mathbf{x})=\varnothing & L_\omega(\mathbf{x})=\Gamma_1\cup\Gamma_2\cup P_0 \end{array}$$

All the limit sets are closed and bounded and they consist of: (1) fixed points; (2) closed orbits; or (3) unions of fixed points and separatrices. The limit sets of type (3) are closed curves but they are not closed orbits because they are not single trajectories. □

The following theorem, which only holds for planar phase portraits, essentially states that the types of limit set that are found in Example 3.9.1 are the only compact ones that can occur. The reader will recall that a **compact** set in the plane is one that is closed and bounded.

Theorem 3.9.1 (Poincaré–Bendixson)
A non-empty, compact limit set of a phase flow in the plane that does not contain a fixed point is a closed orbit.

This result can be used to prove the existence of an attracting/repelling limit cycle provided we can recognize a bounded region of the phase plane which contains a limit set but does not contain a fixed point. Example 3.8.1 illustrates one scenario in which this is possible.

Refer to Fig. 3.26 and consider any closed annulus with inner radius less than one and outer radius greater than one. Observe that the trajectories of the boundary points of the annulus all flow into its interior. Such a region must contain an attracting (limit) set which these trajectories approach as $t\to\infty$. However, the only fixed point of the system (3.81) is the origin. Consequently, the annulus contains a compact limit set with no fixed point. This set must be a closed orbit by Theorem 3.9.1.

The annulus in the above discussion is an example of what is called a **trapping region**. The phase portraits in Fig. 3.30(a), (b) show that, while the existence of a trapping region guarantees the existence of an attracting set, it is not sufficient to ensure that the limit set is a limit cycle. Figure 3.30(c) highlights the fact that the trapping region may contain more than one limit cycle.

We can formalize these ideas as follows.

Definition 3.9.1
A **trapping region** for the system $\dot{\mathbf{x}}=\mathbf{X}(\mathbf{x})$, with flow ϕ_t, is a compact, connected set $D\subset\mathbb{R}^2$ such that $\phi_t(D)\subset D$ for all $t>0$.

In Definition 3.9.1 we have used $\phi_t(D)$ to denote $\{\phi_t(\mathbf{x})\,|\,\mathbf{x}\in D\}$. The following corollary to Theorem 3.9.1 then encapsulates the argument illustrated above.

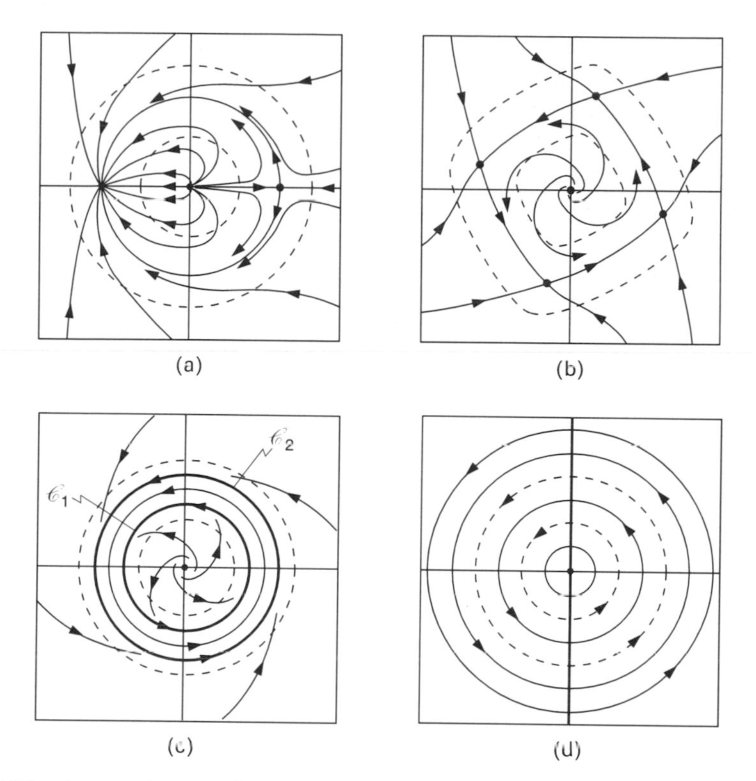

Fig. 3.30. An annular trapping region containing: (a) an attracting fixed point; (b) an attracting set consisting of a union of fixed points and separatrices; (c) two limit cycles $\mathscr{C}_1$ and $\mathscr{C}_2$ and a continuum of closed orbits. (d) The annulus shown is a positively invariant set containing no fixed points but the trajectories of the system do not flow into it as t increases. There are no limit cycles in this annulus.

Corollary

If D is a trapping region for the system $\dot{\mathbf{x}} = \mathbf{X}(\mathbf{x})$ and there are no fixed points in D, then the phase portrait of the system has a limit cycle in D.

In Example 3.8.1 the trajectories of the system are directed into the annulus at every point of its boundary. While this property is sufficient to ensure that a compact, connected set, D, is a trapping region, it is not a necessary one. Of course, it is necessary that no trajectories leave D with increasing time; a set with this property is said to be positively invariant.

Definition 3.9.2

Given the system $\dot{\mathbf{x}} = \mathbf{X}(\mathbf{x})$ with flow ϕ_t, a subset D of $\mathbb{R}^2$ is said to be a **positively invariant set** for the system if, for every point $\mathbf{x}_0$ of D, $\phi_t(\mathbf{x}_0)$ lies in D for all positive t.

The subset D is called an **invariant set** for the system if the conditions of Definition 3.9.2 are satisfied for *all* real t.

For positively invariant D the trajectories of boundary points are not required to enter the interior of D with increasing time. To obtain a trapping region steps must be taken to ensure that $\phi_t(\mathbf{x})$ tends to an attracting set inside D. In particular, we must exclude the possibility that the boundary of D is made up of trajectories of the system (like those in 3.30(d)). Thus, the trajectories of the system must be directed into the set at some, but not necessarily all, of its boundary points. This is sufficient to ensure that $\phi_t(D)$ is a proper subset of D and, consequently, such a positively invariant set can form a trapping region.

Example 3.9.2
Show that the phase portrait of

$$\ddot{x} - \dot{x}(1 - 3x^2 - 2\dot{x}^2) + x = 0$$

has a limit cycle.

Solution
The corresponding first-order system is

$$\dot{x}_1 = x_2, \qquad \dot{x}_2 = -x_1 + x_2(1 - 3x_1^2 - 2x_2^2), \tag{3.90}$$

which becomes

$$\begin{aligned} \dot{r} &= r\sin^2\theta(1 - 3r^2\cos^2\theta - 2r^2\sin^2\theta) \\ \dot{\theta} &= -1 + \tfrac{1}{2}\sin 2\theta(1 - 3r^2\cos^2\theta - 2r^2\sin^2\theta) \end{aligned} \tag{3.91}$$

in polar coordinates. Observe:

1. equation (3.91) with $r = \frac{1}{2}$ gives

$$\dot{r} = \tfrac{1}{4}\sin^2\theta(1 - \tfrac{1}{2}\cos^2\theta) \geqslant 0 \tag{3.92}$$

   with equality only at $\theta = 0$ and π. Thus, $\{\mathbf{x} \mid r \geqslant \frac{1}{2}\}$ is positively invariant;
2. equation (3.91) with $r = 1/\sqrt{2}$ implies

$$\dot{r} = -\frac{1}{2\sqrt{2}}\sin^2\theta\cos^2\theta \leqslant 0, \tag{3.93}$$

   with equality at $\theta = 0$, π, $\pi/2$, $3\pi/2$. Thus $\{\mathbf{x} \mid r \leqslant 1/\sqrt{2}\}$ is positively invariant.

Now 1 and 2 imply that the annular region $\{\mathbf{x} \mid \frac{1}{2} \leqslant r \leqslant 1/\sqrt{2}\}$ is positively invariant. What is more, the circles bounding this annulus are not trajectories of the system because $\dot{r}$ does not vanish identically on them. Consequently the trajectories of points on these circles move into the annulus as time increases. Since the only fixed point of (3.90) is at the origin, we conclude that there is a limit cycle in the annulus. □

The following result gives a condition for there to be no limit cycle in a region D.

Theorem 3.9.2
Let D be a simply connected region of the phase plane in which the vector field $\mathbf{X}(\mathbf{x}) = (X_1(x_1, x_2), X_2(x_1, x_2))$ *has the property that*

$$\frac{\partial X_1}{\partial x_1} + \frac{\partial X_2}{\partial x_2} \tag{3.94}$$

is of constant sign. Then the system $\dot{\mathbf{x}} = \mathbf{X}(\mathbf{x})$ *has no closed trajectories wholly contained in D.*

It will be sufficient for our purpose to recognize that a **simply connected region** of the plane is a region with no 'holes' in it (as illustrated in Fig. 3.31). The theorem follows from Green's theorem in the plane which may be stated as follows:

Let the real-valued functions $P(x_1, x_2)$ *and* $Q(x_1, x_2)$ *have continuous first partial derivatives in a simply connected region* $\mathscr{R}$ *of the* x_1x_2*-plane bounded by a simple closed curve* $\mathscr{C}$. *Then*

$$\oint_{\mathscr{C}} P\mathrm{d}x_1 + Q\mathrm{d}x_2 = \iint_{\mathscr{R}} \left(\frac{\partial Q}{\partial x_1} - \frac{\partial P}{\partial x_2}\right)\mathrm{d}x_1\mathrm{d}x_2. \tag{3.95}$$

where $\oint_{\mathscr{C}}$ *indicates integration along* $\mathscr{C}$ *in an anticlockwise direction.*

To prove Theorem 3.9.2 assume that a limit cycle C of period T exists for the system. Let $P = -X_2$, $Q = X_1$ in (3.95) and obtain

$$\oint_C X_1\mathrm{d}x_2 - X_2\mathrm{d}x_1 = \int_0^T (X_1\dot{x}_2 - X_2\dot{x}_1)\mathrm{d}t \; (=0) \tag{3.96}$$

$$= \iint_{\mathscr{R}} \left(\frac{\partial X_1}{\partial x_1} + \frac{\partial X_2}{\partial x_2}\right)\mathrm{d}x_1\mathrm{d}x_2 \; (\neq 0). \tag{3.97}$$

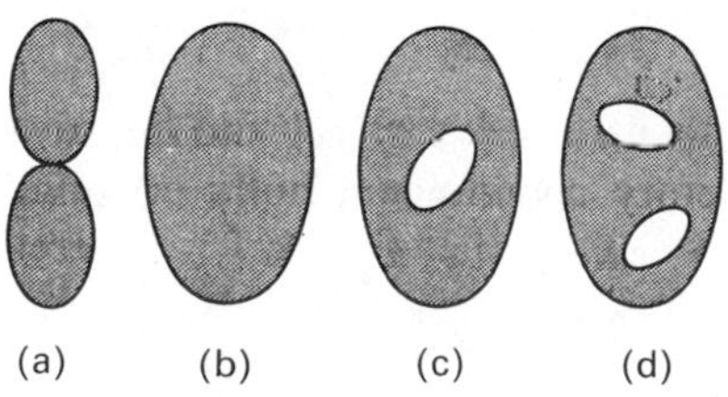

Fig. 3.31. The shaded regions in (a) and (b) have no 'holes' and are simply connected. Those in (c) and (d) have 'holes' and are *not* simply connected.

Equation (3.96) follows because C is a solution curve, while (3.97) follows from (3.94). Hence, the closed trajectory C cannot exist.

Example 3.9.3
Prove that if the system

$$\dot{x}_1 = -x_2 + x_1(1 - x_1^2 - x_2^2), \qquad \dot{x}_2 = x_1 + x_2(1 - x_1^2 - x_2^2) + K, \quad (3.98)$$

where K is a constant, has a closed trajectory, then it will either:

1. encircle the origin; or
2. intersect the circle $x_1^2 + x_2^2 = \frac{1}{2}$.

Solution
The quantity

$$\frac{\partial X_1}{\partial x_1} + \frac{\partial X_2}{\partial x_2} = 2 - 4(x_1^2 + x_2^2) \quad (3.99)$$

is positive inside the circle $x_1^2 + x_2^2 = \frac{1}{2}$ and negative outside it. Thus, any closed trajectory cannot be wholly contained in the simply connected region $\{(x_1, x_2) | x_1^2 + x_2^2 < \frac{1}{2}\}$. Therefore, if a closed trajectory exists it is either contained in $\{(x_1, x_2) | x_1^2 + x_2^2 > \frac{1}{2}\}$ or it will intersect $x_1^2 + x_2^2 = \frac{1}{2}$. If the closed orbit is contained in $\{(x_1, x_2) | x_1^2 + x_2^2 > \frac{1}{2}\}$ then it must encircle the origin, otherwise it will bound a region of constant negative sign of (3.99). □

EXERCISES

Sections 3.1–3.3

3.1 Use the method of isoclines to sketch the global phase portraits of the following systems:
(a) $\dot{x}_1 = x_1 x_2, \quad \dot{x}_2 = \ln x_1, \quad x_1 > 0$;
(b) $\dot{x}_1 = 4x_1(x_2 - 1), \quad \dot{x}_2 = x_2(x_1 + x_1^2)$;
(c) $\dot{x}_1 = x_1 x_2, \quad \dot{x}_2 = x_2^2 - x_1^2$.

3.2 Show that the mapping $(x_1, x_2) \to (f(r)\cos\theta, f(r)\sin\theta)$, where $x_1 = r\cos\theta$, $x_2 = r\sin\theta$ and $f(r) = \tan(\pi r / 2r_0)$, is a continuous bijection of $N = \{(x_1, x_2) | r < r_0\}$ onto $\mathbb{R}^2$. Does the bijection map the set of concentric circles on $\mathbf{0}$ in N onto the set of concentric circles on $\mathbf{0}$ in $\mathbb{R}^2$? What property of local phase portraits of linear systems in the plane does this result illustrate?

3.3 Sketch the local phase portraits of the fixed points in Figs 3.5(a), 3.22(b) and 3.30(a).

3.4 Find the linearizations of the following systems, at the fixed points indicated, by:

(a) introducing local coordinates at the fixed points;
(b) using Taylor's theorem.

(i) $\dot{x}_1 = x_1 + x_1 x_2^3/(1 + x_1^2)^2, \quad \dot{x}_2 = 2x_1 - 3x_2, \quad (0,0);$
(ii) $\dot{x}_1 = x_1^2 + \sin x_2 - 1, \quad \dot{x}_2 = \sinh(x_1 - 1) \quad (1,0);$
(iii) $\dot{x}_1 = x_1^2 - e^{x_2}, \quad \dot{x}_2 = x_2(1 + x_2), \quad (e^{-1/2}, -1).$

State the preferred method (if one exists) for each system.

3.5 Use the linearization theorem to classify, where possible, the fixed points of the systems:
(a) $\dot{x}_1 = x_2^2 - 3x_1 + 2, \quad \dot{x}_2 = x_1^2 - x_2^2;$
(b) $\dot{x}_1 = x_2, \quad \dot{x}_2 = -x_1 + x_1^3;$
(c) $\dot{x}_1 = \sin(x_1 + x_2), \quad \dot{x}_2 = x_2;$
(d) $\dot{x}_1 = x_1 - x_2 - e^{x_1}, \quad \dot{x}_2 = x_1 - x_2 - 1;$
(e) $\dot{x}_1 = -x_2 + x_1 + x_1 x_2, \quad \dot{x}_2 = x_1 - x_2 - x_2^2;$
(f) $\dot{x}_1 = x_2, \quad \dot{x}_2 = -(1 + x_1^2 + x_1^4)x_2 - x_1;$
(g) $\dot{x}_1 = -3x_2 + x_1 x_2 - 4, \quad \dot{x}_2 = x_2^2 - x_1^2.$

3.6 Linearize the system

$$\dot{x}_1 = -x_2, \qquad \dot{x}_2 = x_1 - x_1^5$$

at the origin and classify the fixed point of the linearized system. Show that the trajectories of the non-linear system lie on the family of curves

$$x_1^2 + x_2^2 - x_1^6/3 = C,$$

where C is a constant. Sketch these curves to show that the non-linear system and its linearization have qualitatively equivalent local phase portraits at the origin. Why could this conclusion not be deduced from the linearization theorem?

3.7 Find the principal directions of the fixed points at the origin of the following systems:
(a) $\dot{x}_1 = e^{x_1 + x_2} - 1, \quad \dot{x}_2 = x_2;$
(b) $\dot{x}_1 = -\sin x_1 + x_2, \quad \dot{x}_2 = \sin x_2;$
(c) $\dot{x}_1 = \ln(x_1 + x_2 + 1), \quad \dot{x}_2 = \frac{1}{2}x_1 + x_2, \quad x_1 + x_2 > -1;$
and use them to sketch local phase portraits.

Section 3.4

3.8 Find the family of solution curves which satisfy

$$\frac{dx_2}{dx_1} = \frac{x_2^2 - x_1^3}{2x_1 x_2}, \qquad x_1, x_2 \neq 0,$$

by making the substitution $x_2^2 = u$. Sketch the family of solutions and hence or otherwise sketch the local phase portrait of the non-simple fixed point of

$$\dot{x}_1 = 2x_1x_2, \qquad \dot{x}_2 = x_2^2 - x_1^3.$$

3.9 Are the phase portraits of the systems $\dot{x}_1 = x_1 e^{x_1}$, $\dot{x}_2 = x_2 e^{x_1}$ and $\dot{x}_1 = x_1$, $\dot{x}_2 = x_2$ qualitatively equivalent? If so, state the continuous bijection which exhibits the equivalence.

3.10 Show that the 'straight line' separatrices at the non-simple fixed point of

$$\dot{x}_1 = x_2(3x_1^2 - x_2^2), \qquad \dot{x}_2 = x_1(x_1^2 - 3x_2^2)$$

satisfy $x_2 = kx_1$ where

$$k^2(3 - k^2) = 1 - 3k^2.$$

Hence, or otherwise, find these separatrices and by using isoclines sketch the phase portrait.

3.11 Show that the system

$$\dot{x}_1 = x_1^2 - x_2^3, \qquad \dot{x}_2 = x_1^2(x_1^2 - x_2^3)$$

has a line of fixed points. Furthermore, show that every fixed point on the line is non-simple. Can this conclusion be reached by using the linearization theorem?

Is the above conclusion true for any system with a line of fixed points?

Section 3.5

3.12 Use the method of isoclines to sketch the phase portrait of (3.33), paying particular attention to the manner in which the trajectories cross the x_2-axis. Hence show that, while all trajectories approach $(x_1, x_2) = (0, 0)$ as t tends to infinity, the system is *not* asymptotically stable.

3.13 Show that $V(x_1, x_2) = x_1^2 + x_2^2$ is a strong Liapunov function at the origin for each of the following systems:
(a) $\dot{x}_1 = -x_2 - x_1^3, \quad \dot{x}_2 = x_1 - x_2^3;$
(b) $\dot{x}_1 = -x_1^3 + x_2 \sin x_1, \quad \dot{x}_2 = -x_2 - x_1^2 x_2 - x_1 \sin x_1;$
(c) $\dot{x}_1 = -x_1 - 2x_2^2, \quad \dot{x}_2 = 2x_1x_2 - x_2^3;$
(d) $\dot{x}_1 = -x_1 \sin^2 x_1, \quad \dot{x}_2 = -x_2 - x_2^5;$
(e) $\dot{x}_1 = -(1 - x_2)x_1, \quad \dot{x}_2 = -(1 - x_1)x_2.$

3.14 Find domains of stability at the origin for each of the systems even in Exercise 3.13.

3.15 Show that $V(x_1, x_2) = x_1^2 + x_2^2$ is a weak Liapunov function for the following systems at the origin:
(a) $\dot{x}_1 = x_2, \quad \dot{x}_2 = -x_1 - x_2^3(1 - x_1^2)^2;$
(b) $\dot{x}_1 = -x_1 + x_2^2, \quad \dot{x}_2 = -x_1x_2 - x_1^2;$
(c) $\dot{x}_1 = -x_1^3, \quad \dot{x}_2 = -x_1^2 x_2;$
(d) $\dot{x}_1 = -x_1 + 2x_1x_2^2, \quad \dot{x}_2 = -x_1^2 x_2^3.$
Which of these systems are asymptotically stable?

3.16 Prove that if V is a strong Liapunov function for $\dot{\mathbf{x}} = -\mathbf{X}(\mathbf{x})$, in a neighbourhood of the origin, then $\dot{\mathbf{x}} = \mathbf{X}(\mathbf{x})$ has an unstable fixed point at the origin. Use this result to show that the systems:

(a) $\dot{x}_1 = x_1^3.\quad \dot{x}_2 = x_2^3.$

(b) $\dot{x}_1 = \sin x_1,\quad \dot{x}_2 = \sin x_2;$

(c) $\dot{x}_1 = -x_1^3 + 2x_1^2 \sin x_1,\quad \dot{x}_2 = x_2 \sin^2 x_2;$

are unstable at the origin.

3.17 Prove that the differential equations

(a) $\ddot{x} + \dot{x} - \dot{x}^3/3 + x = 0;$

(b) $\ddot{x} + \dot{x}\sin(\dot{x}^2) + x = 0;$

(c) $\ddot{x} + \dot{x} + x^3 = 0;$

(d) $\ddot{x} + \dot{x}^3 + x^3 = 0,$

have asymptotically stable zero solutions $x(t) \equiv 0$.

3.18 Prove that $V(x_1, x_2) = ax_1^2 + 2bx_1x_2 + cx_2^2$ is positive definite if and only if $a > 0$ and $ac > b^2$. Hence, or otherwise, prove that

$$\dot{x}_1 = x_2,\qquad \dot{x}_2 = -x_1 - x_2 - (x_1 + 2x_2)(x_2^2 - 1)$$

is asymptotically stable at the origin by considering the region $|x_2| < 1$. Find a domain of stability.

3.19 Find domains of stability for the following systems by using an appropriate Liapunov function:

(a) $\dot{x}_1 = x_2 - x_1(1 - x_1^2 - x_2^2)(x_1^2 + x_2^2 + 1)$
$\quad\;\dot{x}_2 = -x_1 - x_2(1 - x_1^2 - x_2^2)(x_1^2 + x_2^2 + 1);$

(b) $\dot{x}_1 = x_2,\quad \dot{x}_2 = -x_2 + x_2^3 - x_1^5.$

3.20 Use $V(x_1, x_2) = (x_1/a)^2 + (x_2/b)^2$ to show that the system

$$\dot{x}_1 = x_1(x_1 - a),\qquad \dot{x}_2 = x_2(x_2 - b),\qquad a, b > 0,$$

has an asymptotically stable origin. Show that all trajectories tend to the origin as $t \to \infty$ in the region

$$\frac{x_1^2}{a^2} + \frac{x_2^2}{b^2} < 1.$$

3.21 Given the system

$$\dot{x}_1 = x_2,\qquad \dot{x}_2 = x_2 - x_1^3$$

show that a positive definite function of the form

$$V(x_1, x_2) = ax_1^4 + bx_1^2 + cx_1x_2 + dx_2^2$$

can be chosen such that $\dot{V}(x_1, x_2)$ is also positive definite. Hence deduce that the origin is unstable.

3.22 Show that the origin of the system

$$\dot{x}_1 = x_2^2 - x_1^2,\qquad \dot{x}_2 = 2x_1x_2$$

is unstable by using

$$V(x_1, x_2) = 3x_1x_2^2 - x_1^3.$$

3.23 Show that the fixed point at the origin of the system

$$\dot{x}_1 = x_1^4, \qquad \dot{x}_2 = 2x_1^2x_2^2 - x_2^2$$

is unstable by using the function

$$V(x_1, x_2) = \alpha x_1 + \beta x_2$$

for a suitable choice of the constants α and β.

Verify the instability at the fixed point by examining the behaviour of the separatrices.

Section 3.6

3.24 Show that the non-linear change of coordinates

$$y_1 = x_1 + x_2^3, \qquad y_2 = x_2 + x_2^2$$

satisfies the requirements of the flow box theorem for the system

$$\dot{x}_1 = -\frac{3x_2^2}{1 + 2x_2}, \qquad \dot{x}_2 = \frac{1}{1 + 2x_2}$$

in the neighbourhood of any point (x_1, x_2) with $x_2 \neq -\frac{1}{2}$.

3.25 Prove that the following systems have no fixed points:

(a) $\dot{x}_1 = e^{x_1 + x_2}, \quad \dot{x}_2 = x_1 + x_2$;

(b) $\dot{x}_1 = x_1 + x_2 + 2, \quad \dot{x}_2 = x_1 + x_2 + 1$;

(c) $\dot{x}_1 = x_2 + 2x_2^3, \quad \dot{x}_2 = 1 + x_2^2$

and sketch their phase portraits.

3.26 Sketch phase portraits consistent with the following information:

(a) two fixed points, a saddle and a stable node;

(b) three fixed points, one saddle and two stable nodes.

3.27 Find the local phase portraits at each of the fixed points of the system

$$\dot{x}_1 = x_1(1 - x_1^2), \qquad \dot{x}_2 = x_2.$$

Use these results to suggest a global phase portrait. Check whether your suggestion is correct by using the method of isoclines.

Section 3.7

3.28 Find first integrals of the following systems together with their domains of definition:

(a) $\dot{x}_1 = x_2, \quad \dot{x}_2 = x_1^2 + 1$;
(b) $\dot{x}_1 = x_1(x_2 + 1), \quad \dot{x}_2 = -x_2(x_1 + 1)$;
(c) $\dot{x}_1 = \sec x_1, \quad \dot{x}_2 = -x_2^2, \quad |x_1| < \pi/2$;
(d) $\dot{x}_1 = x_1(x_1 e^{x_2} - \cos x_2), \quad \dot{x}_2 = \sin x_2 - 2x_1 e^{x_2}$.

3.29 Find a first integral of the system

$$\dot{x}_1 = x_1 x_2 - 3x_1^3, \qquad \dot{x}_2 = x_2^2 - 6x_1^2 x_2 + x_1^4$$

using the substitution $x_2 = ux_1^2$. Sketch the phase portrait.

3.30 How do the phase portraits of the two systems

$$\dot{x}_1 = x_1(x_2^2 - x_1), \qquad \dot{x}_2 = -x_2(x_2^2 - x_1)$$
$$\dot{x}_1 = x_1, \qquad \dot{x}_2 = -x_2$$

differ?

3.31 Find a first integral of the system

$$\dot{x}_1 = x_1 x_2, \qquad \dot{x}_2 = \ln x_1, \; x_1 > 1$$

in the region indicated. Hence, or otherwise, sketch the phase portrait.

3.32 Find first integrals for the linear systems $\dot{\boldsymbol{x}} = \boldsymbol{J}\boldsymbol{x}$, where $\boldsymbol{J}$ is a 2×2 Jordan matrix of node, centre and focus type. State maximal regions on which these first integrals exist.

Is a system which has an asymptotically stable fixed point ever conservative?

3.33 Find a Hamiltonian H for a particle moving along a straight line subject to

$$\ddot{x} = -x + \alpha x^2,$$

$\alpha > 0$, where x is the displacement. Sketch the level curves of the Hamiltonian H in the phase plane. Indicate the regions of the phase plane that contain trajectories which give non-oscillatory motion.

3.34 When expressed in terms of plane polar coordinates (r, θ) given by $x = r \cos\theta$, $p = r\sin\theta$, Hamilton's equations (3.70) take the form

$$\dot{r} = \frac{1}{r}\frac{\partial \tilde{H}}{\partial \theta}, \qquad \dot{\theta} = \frac{-1}{r}\frac{\partial \tilde{H}}{\partial r},$$

where $\tilde{H}(r, \theta) = H(r\cos\theta, r\sin\theta)$. Consider the system with

$$\tilde{H}(r, \theta) = -\mu r^2 + r^4 + r^5 \sin 5\theta,$$

when $0 < \mu \ll \frac{1}{2}$ and $r < \frac{1}{2}$. Show that there are ten fixed points near the circle $r = \sqrt{\mu/2}$ and determine their topological types. Sketch $\tilde{H}(r, \theta)$ as a function of r for $\sin 5\theta = 0, \pm 1$ and verify that the phase portrait takes the form shown below.

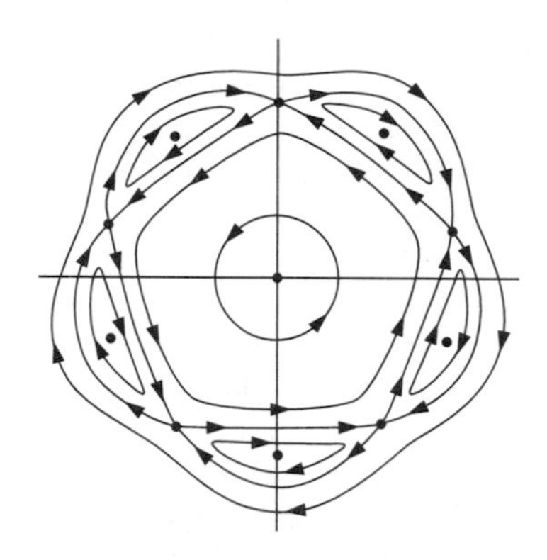

Section 3.8

3.35 Sketch phase portraits for each of the following systems:

(a) $\dot{r} = -r(r-1)^2, \qquad \dot{\theta} = 1;$

(b) $\dot{r} = \begin{cases} r(1-r) & \text{if } r \leqslant 1; \\ 0 & \text{otherwise} \end{cases} \qquad \dot{\theta} = 1;$

(c) $\dot{r} = \begin{cases} 0 & \text{if } r \leqslant 1; \\ r(r-1) & \text{otherwise} \end{cases} \qquad \dot{\theta} = -1.$

In each case, obtain α- and ω-limit sets for all points $\mathbf{x}$ with $|\mathbf{x}| > 0$.

3.36 Show that the Poincaré map, P, defined on the positive x-axis, for the system

$$\dot{r} = ar(1-r), \qquad \dot{\theta} = 1,$$

is given by

$$P(x) = x/[x + (1-x)\exp(-2\pi a)].$$

Verify that $P(x)$ has a stable fixed point at $x = x^* = 1$. The Poincaré map P_1 is defined on some other r-coordinate line; write down an expression for $P_1(r)$.

3.37 Assume that the Poincaré map, P, defined on the x-axis for the system

$$\dot{r} = r(r-1)^2, \qquad \dot{\theta} = 1,$$

satisfies:

(a) $P(x)$ has a fixed point at $x = x^* = 1$;
(b) $[\mathrm{d}P/\mathrm{d}x]_{x=x^*} = 1$;
(c) $[\mathrm{d}^2P/\mathrm{d}x^2]_{x=x^*} = 4\pi$.

Sketch the graph of $P(x)$ for x near x^* and illustrate the iteration $x_{n+1} = P(x_n)$, $n = 0, 1, \ldots$, for $x_0 \neq 1$. Explain how these diagrams change if the signs of $\dot{r}$ and $\dot{\theta}$ are reversed. Confirm that both systems have a semi-stable, circular limit cycle of radius unity and describe the difference between them.

3.38 Consider a planar system with angular equation $\dot{\theta} = 1$ and let P be the Poincaré map defined on the positive x-axis. Suppose:

$$P(x^*) = x^*; \qquad |dP/dx|_{x=x^*} = 1;$$

$$[d^2 P/dx^2]_{x=x^*} = 0; \qquad [d^3 P/dx^3]_{x=x^*} = \varepsilon \neq 0.$$

Draw diagrams to illustrate the form of the iteration $x_{n+1} = P(x_n)$, $n = 0, 1, \ldots$, for x_0 near x^*, when: (i) $\varepsilon > 0$; (ii) $\varepsilon < 0$. Sketch corresponding phase portraits for the planar system on an annular neighbourhood of $r = x^*$.

Section 3.9

3.39 Label and list all the non-empty, closed and bounded limit sets in the phase portrait shown below.

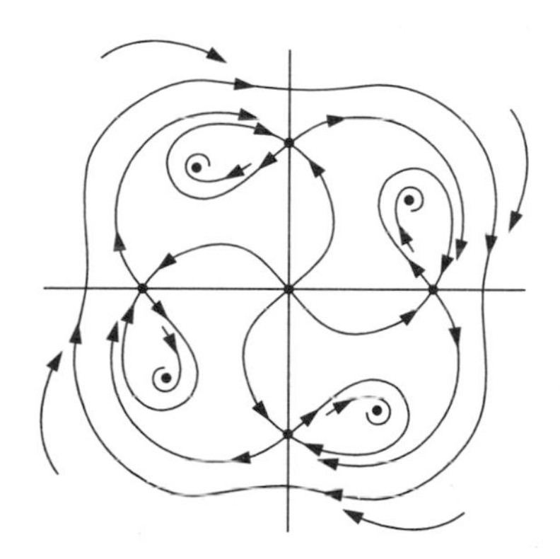

3.40 Sketch phase portraits consistent with the following information:
(a) an unstable limit cycle; three fixed points, one saddle and two stable nodes; and a circular trapping region centred on the origin;
(b) five foci, one unstable and four stable; four saddle points; a non-empty, closed and bounded limit set containing four fixed points; and a circular trapping region centred on the origin.

3.41 Let D be a closed region of the plane bounded by a simple closed curve ∂D. For each $\mathbf{x}$ in ∂D, define $X_\perp(\mathbf{x})$ to be the component of $\mathbf{X}(\mathbf{x})$ along the inward normal to D. Assume that $X_\perp(\mathbf{x})$ satisfies either: (i) $X_\perp(\mathbf{x}) > 0$; or (ii) $X_\perp(\mathbf{x}) \geqslant 0$; for all $\mathbf{x}$ in ∂D. Explain why (i) is a sufficient condition for D to be a trapping region for the flow of $\dot{\mathbf{x}} = \mathbf{X}(\mathbf{x})$ while (ii) is not.

Use the method of isoclines to sketch the phase portrait for the system (3.91). Examine the vector field on the circles $r = \frac{1}{2}$ and $r = 1/\sqrt{2}$

and confirm that the trajectories starting on these circles move into the annulus $\{\mathbf{x} \mid \frac{1}{2} \leqslant r \leqslant 1/\sqrt{2}\}$ as t increases. What feature of systems satisfying condition (ii) does this example illustrate?

3.42 Show that each of the following regions is a positively invariant set for the system given:

(a) the half-plane $x_2 \geqslant 0$ for

$$\dot{x}_1 = 2x_1 x_2, \qquad \dot{x}_2 = x_2^2;$$

(b) the disc $x_1^2 + x_2^2 < 1$ for

$$\dot{x}_1 = -x_1 + x_2 + x_1(x_1^2 + x_2^2), \qquad \dot{x}_2 = -x_1 - x_2 + x_2(x_1^2 + x_2^2);$$

(c) the closed region formed by joining the points $(e^{-2\pi}, 0)$ and $(1, 0)$ by a segment of the x_1-axis and one turn of the spiral with polar form $r = e^{-\theta}$ for

$$\dot{x}_1 = -x_1 - x_2, \qquad \dot{x}_2 = x_1 - x_2;$$

(d) the region inside and on the closed curve that is a subset of $\{(x_1, x_2) \mid 3(x_1^2 + x_2^2) - 2x_1^3 = 1\}$ for

$$\dot{x}_1 = x_2, \qquad \dot{x}_2 = -x_1 + x_1^2;$$

(e) the closed region bounded by the curves $C_{\frac{1}{2}}$ and C_2 for the system

$$\dot{x}_1 = x_1 - x_2 - x_1(x_1^2 + x_2^2)^{\frac{1}{2}}, \qquad \dot{x}_2 = x_1 + x_2 - x_2(x_1^2 + x_2^2)^{\frac{1}{2}},$$

with flow $\phi_t(x_1, x_2)$, where C_x is the closed curve formed by the union of $\{\phi_t(x, 0) \mid 0 \leqslant t \leqslant 2\pi\}$ and the segment of the x_1-axis between $(x, 0)$ and $\phi_{2\pi}(x, 0)$.

Indicate which of these positively invariant sets are trapping regions and specify the attracting set contained within them.

3.43 Prove that there exists a region $R = \{(x_1, x_2) \mid x_1^2 + x_2^2 \leqslant r^2\}$ such that all trajectories of the system

$$\dot{x}_1 = -wx_2 + x_1(1 - x_1^2 - x_2^2), \qquad \dot{x}_2 = wx_1 + x_2(1 - x_1^2 - x_2^2) - F,$$

where w and F are constants, eventually enter R. Show that the system has a limit cycle when $F = 0$.

3.44 Prove that the system

$$\dot{x}_1 = 1 - x_1 x_2, \qquad \dot{x}_2 = x_1$$

has no limit cycles.

3.45 Consider the system

$$\dot{x}_1 = -wx_2 + x_1(1 - x_1^2 - x_2^2) - x_2(x_1^2 + x_2^2),$$
$$\dot{x}_2 = wx_1 + x_2(1 - x_1^2 - x_2^2) + x_1(x_1^2 + x_2^2) - F,$$

where w and F are constants. Prove that if the system has a limit cycle such that all of its points are at a distance greater than $1/\sqrt{2}$ from the origin, then the limit cycle must encircle the origin.

3.46 Suppose that the region $R = \{(x_1, x_2)\}|x_1, x_2 > 0\}$ is positively invariant for the system $\dot{x}_1 = X_1(x_1, x_2), \dot{x}_2 = X_2(x_1, x_2)$ and that

$$\dot{x}_1 \lesseqgtr 0 \qquad \text{for } x_2 \gtreqless -x_1^2 + 3x_1 + 1$$

$$\dot{x}_2 \lesseqgtr 0 \qquad \text{for } x_2 \gtreqless x_1,$$

respectively. Assuming that there are no closed orbits in R prove that the unique fixed point in R is asymptotically stable.

4

Flows on non-planar phase spaces

4.1 FIXED POINTS

In section 2.7 we examined linear differential equations, $\dot{\boldsymbol{x}} = \boldsymbol{A}\boldsymbol{x}$, defined on $\mathbb{R}^n$ with $n \geqslant 3$, and concluded that all the corresponding canonical systems could be solved. However, we did not consider how the increased dimension influences the types of qualitative behaviour that might occur.

4.1.1 Hyperbolic fixed points

If the eigenvalues of the coefficient matrix, $\boldsymbol{A}$, all have non-zero real part, then the system has an isolated hyperbolic fixed point at the origin (cf. section 3.3). Let n_{u} be the number of eigenvalues with real part greater than zero. Then it can be shown that the phase portrait of such a system is qualitatively equivalent (in the sense of Definition 2.4.1) to that of

$$\dot{\boldsymbol{x}}_{\mathrm{u}} = \boldsymbol{x}_{\mathrm{u}}, \qquad \dot{\boldsymbol{x}}_{\mathrm{s}} = -\boldsymbol{x}_{\mathrm{s}}, \tag{4.1}$$

where $\boldsymbol{x}_{\mathrm{u}}(\boldsymbol{x}_{\mathrm{s}})$ lies in $\mathbb{R}^{n_{\mathrm{u}}}$ ($\mathbb{R}^{n_{\mathrm{s}}}$), with $n_{\mathrm{s}} = n - n_{\mathrm{u}}$.

It is easy to check that this result gives the classification shown in Fig. 2.11 when $n = 2$. There are three possibilities for n_{u}, namely $n_{\mathrm{u}} = 0, 1, 2$. If $n_{\mathrm{u}} = 0$ then $n_{\mathrm{s}} = 2$ and the system is qualitatively equivalent to

$$\dot{x}_1 = -x_1, \qquad \dot{x}_2 = -x_2. \tag{4.2}$$

The phase portrait of (4.2) is a stable star node. If $n_{\mathrm{u}} = 1$, then $n_{\mathrm{s}} = 1$ and the equivalent system is

$$\dot{x}_1 = x_1, \qquad \dot{x}_2 = -x_2, \tag{4.3}$$

with a phase portrait of saddle type. Finally, for $n_{\mathrm{u}} = 2, n_{\mathrm{s}} = 0$ the given system is qualitatively equivalent to

$$\dot{x}_1 = x_1, \qquad \dot{x}_2 = x_2, \tag{4.4}$$

which is an unstable star node. These are precisely the topological types of hyperbolic fixed point appearing in Fig. 2.11.

Equation (4.1) shows that, for any dimension n, $n_u = 0$ corresponds to a contraction onto the origin and $n_u = n$ to an expansion away from the origin. For each n, the corresponding phase portraits are n-dimensional star nodes. However, for $n \geqslant 3$ it is important to note that qualitatively distinct fixed points of saddle type occur for $n_u = 1, \ldots, n-1$. The separatrices of the two-dimensional saddle point are replaced by the **unstable** and **stable eigenspaces**, E^u, E^s, of the exponential matrix $\exp(At)$. These are the eigenspaces associated with the eigenvalues of A that have positive and negative real parts, respectively. Thus, $\dim(E^u) = n_u$ and $\dim(E^s) = n_s$. Notice that there are no eigenvalues with zero real part because the fixed point is hyperbolic. Elements of E^u represent points in $\mathbb{R}^n$ whose trajectories approach the origin as $t \to -\infty$, whilst those of E^s correspond to points with trajectories that tend to the origin at $t \to +\infty$.

Example 4.1.1

Sketch phase portrait for $\dot{\boldsymbol{x}} = \boldsymbol{A}\boldsymbol{x}, \boldsymbol{x} \in \mathbb{R}^3$, when $\boldsymbol{A}$ takes the form:

$$\text{(a)} \begin{bmatrix} \lambda_1 & 0 & 0 \\ 0 & -\lambda_2 & 0 \\ 0 & 0 & -\lambda_3 \end{bmatrix}; \quad \text{(b)} \begin{bmatrix} \alpha & -\beta & 0 \\ \beta & \alpha & 0 \\ 0 & 0 & -\lambda \end{bmatrix}; \tag{4.5}$$

where α, β, λ and $\lambda_i, i = 1, 2, 3$, are all positive.

Explain why the phase portraits of (a) and (b) are *not* qualitatively equivalent. Use the matrix in (4.5(a)) to construct a linear system with E^s equal to the plane given by $x_1 = -x_2$ and E^u the straight line through the origin in the direction $(1, 1, 0)$.

Solution

The required phase portraits are shown in Fig 4.1; both are of saddle type. For (4.5(a)) E^s is the x_2x_3-plane and the restriction of the flow to E^s is a node. The matrix A has one positive and two negative eigenvalues so that $n_u = 1$ and $n_s = 2$. It follows that (4.5(a)) is qualitatively equivalent to

$$\dot{x}_1 = x_1, \qquad \dot{x}_2 = -x_2, \qquad \dot{x}_3 = -x_3. \tag{4.6}$$

On the other hand, (4.5(b)) has two complex eigenvalues with positive real part (namely $\alpha \pm i\beta$) and one negative real eigenvalue. Hence $n_u = 2, n_s = 1$ and the system is qualitatively equivalent to

$$\dot{x}_1 = x_1, \qquad \dot{x}_2 = x_2, \qquad \dot{x}_3 = -x_3. \tag{4.7}$$

The systems (4.6) and (4.7) cannot be qualitatively equivalent because any continuous map that takes the x_2x_3-plane in the phase space of (4.6) onto the x_3-axis in that of (4.7) cannot be a bijection.

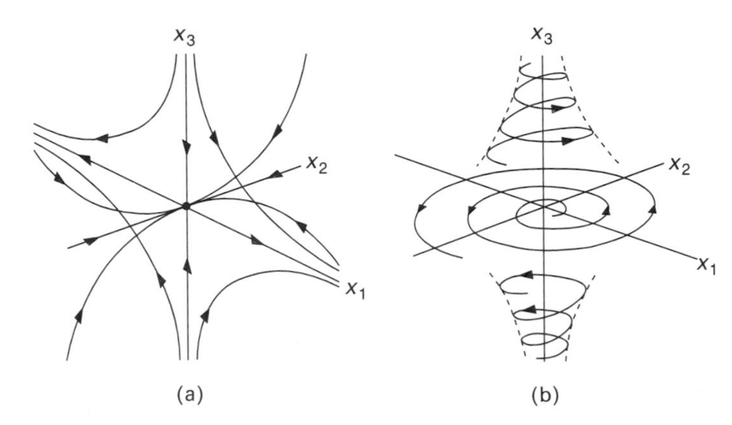

Fig. 4.1. Phase portraits for the linear systems with coefficient matrices given in (4.5). In both (a) and (b) the fixed point at the origin is of saddle type but the systems are not qualitatively equivalent.

A linear system with the desired properties can be obtained from (4.5(a)) as follows. Suppose that coordinates (x_1', x_2', x_3') are related to (x_1, x_2, x_3) by a right-handed rotation through an angle of $\pi/4$ radians about the positive x_3-axis (so that the x_1'-axis lies along the $(1, 1, 0)$-direction in the unprimed coordinates). Then the system $\dot{\boldsymbol{x}}' = \boldsymbol{A}\boldsymbol{x}'$ with $\boldsymbol{A}$ given by (4.5(a)) has the desired properties when expressed in unprimed coordinates. Now, $\boldsymbol{x}' = \boldsymbol{M}\boldsymbol{x}$, where

$$\boldsymbol{M} = \begin{bmatrix} 1/\sqrt{2} & 1/\sqrt{2} & 0 \\ -1/\sqrt{2} & 1/\sqrt{2} & 0 \\ 0 & 0 & 1 \end{bmatrix} \tag{4.8}$$

and since $\dot{\boldsymbol{x}} = \boldsymbol{M}^{-1}\dot{\boldsymbol{x}}' = \boldsymbol{M}^{-1}\boldsymbol{A}\boldsymbol{M}\boldsymbol{x}$, we conclude that the required system is $\dot{\boldsymbol{x}} = \boldsymbol{B}\boldsymbol{x}$ where

$$\boldsymbol{B} = \begin{bmatrix} a & b & 0 \\ b & a & 0 \\ 0 & 0 & -\lambda_3 \end{bmatrix}, \tag{4.9}$$

with $a = \frac{1}{2}(\lambda_1 - \lambda_2)$ and $b = \frac{1}{2}(\lambda_1 + \lambda_2)$. □

The form (4.7) highlights the fact that the fixed point at the origin of the system given by (4.5(b)) is of saddle type in spite of the fact that the flow on E^u is a focus. As we shall see later, such **spiral saddle points** play an important role in the dynamics of the Lorenz equations.

A fixed point, $\mathbf{x}^*$, of a non-linear system, $\dot{\mathbf{x}} = \mathbf{X}(\mathbf{x}), \mathbf{x} \in \mathbb{R}^n$, is said to be **hyperbolic** if the linearized system at $\mathbf{x}^*$ has a hyperbolic fixed point at the origin. In terms of coordinates $x_1, \ldots, x_n$ the linearization, $\dot{\boldsymbol{y}} = \boldsymbol{A}\boldsymbol{y}$, of $\dot{\boldsymbol{x}} = \boldsymbol{X}(\boldsymbol{x})$ at $\boldsymbol{x}^*$ has coefficient matrix

$$A = \begin{bmatrix} \frac{\partial X_1}{\partial x_1} & \cdots & \frac{\partial X_1}{\partial x_n} \\ \frac{\partial X_2}{\partial x_1} & \cdots & \frac{\partial X_2}{\partial x_n} \\ \vdots & \vdots & \vdots \\ \frac{\partial X_n}{\partial x_1} & \cdots & \frac{\partial X_n}{\partial x_n} \end{bmatrix}\Bigg|_{x=x^*}. \tag{4.10}$$

This is the obvious generalization of (3.19). The matrix A (denoted by $DX(x^*)$, cf. (2.1)) represents the derivative $D\mathbf{X}$ at $\mathbf{x}^*$. The linearization theorem holds for hyperbolic fixed points in any dimension (cf. section 3.3). Thus, if $\mathbf{x}^*$ is a hyperbolic fixed point, there is a neighbourhood of $\mathbf{x}^*$ on which the phase portrait of the non-linear system $\dot{\mathbf{x}} = \mathbf{X}(\mathbf{x})$ is qualitatively equivalent to that of its linearization, $\dot{y} = Ay$, on some neighbourhood of the origin.

The reader will recall that in two-dimensions the separatrices of the non-linear system are tangent to those of the linearized system at a hyperbolic fixed point (as discussed in section 3.3). The same is true of their generalization to higher dimensions. The set of phase points whose trajectories approach the fixed point as $t \to \infty (-\infty)$ is called its **stable (unstable) manifold**. These manifolds have the same dimension as E^s and E^u, respectively, and are tangent to them at $\mathbf{x}^*$.

Example 4.1.2

Show that the Lorenz equations

$$\begin{aligned} \dot{x} &= -\sigma(x-y), \\ \dot{y} &= rx - y - xz, \\ \dot{z} &= xy - bz, \end{aligned} \tag{4.11}$$

where σ, r and b are positive constants, have a hyperbolic fixed point of saddle type at the origin for $r > 1$. Obtain E^u and E^s for the linearized system at the origin and deduce that the stable manifold is a surface tangent at the origin to the plane with normal $(-r, 1+\eta, 0)$, where

$$\eta = \tfrac{1}{2}[-(\sigma+1) - \sqrt{(\sigma+1)^2 + 4\sigma(r-1)}]. \tag{4.12}$$

Find trajectories of the linearized system lying in E^s that are also trajectories of the non-linear system. Hence sketch the shape of the stable manifold in the neighbourhood of the origin.

Solution

The Lorenz equations can be written as

$$\begin{bmatrix} \dot{x} \\ \dot{y} \\ \dot{z} \end{bmatrix} = \begin{bmatrix} -\sigma & \sigma & 0 \\ r & -1 & 0 \\ 0 & 0 & -b \end{bmatrix} \begin{bmatrix} x \\ y \\ z \end{bmatrix} + \begin{bmatrix} 0 \\ -xz \\ xy \end{bmatrix}. \tag{4.13}$$

Equation (4.13) is of the form

$$\dot{\boldsymbol{x}} = \boldsymbol{A}\boldsymbol{x} + \boldsymbol{g}(\boldsymbol{x}), \tag{4.14}$$

where the components of $\boldsymbol{g}$ tend to zero faster than $r = (x^2 + y^2 + z^2)^{\frac{1}{2}}$. It follows that the first term in the right-hand side is the linearized system at the origin (cf. (3.2)). The eigenvalues of the coefficient matrix, $\boldsymbol{A}$, of this linearization are given by the roots of the cubic equation

$$(\lambda + b)[\lambda^2 + (\sigma + 1)\lambda + \sigma(1 - r)] = 0. \tag{4.15}$$

For $r > 1, \sigma(1 - r) < 0$ and the quadratic factor in (4.15) has two real roots, $\lambda_\pm$, of opposite sign. More precisely,

$$\lambda_\pm = \tfrac{1}{2}[-(\sigma + 1) \pm \sqrt{(\sigma + 1)^2 + 4\sigma(r - 1)}] \tag{4.16}$$

and, given that σ is positive $\lambda_+ > 0$ and $\lambda_- < 0$. Hence $\boldsymbol{A}$ has eigenvalues $-b, \lambda_-$ that are negative and λ_+ that is positive. Thus the origin is a hyperbolic fixed point of saddle type.

It is easily shown that

$$\boldsymbol{u}_\pm = \begin{bmatrix} 1 + \lambda_\pm \\ r \\ 0 \end{bmatrix}, \qquad \boldsymbol{u}_3 = \begin{bmatrix} 0 \\ 0 \\ 1 \end{bmatrix}, \tag{4.17}$$

are eigenvectors of $\boldsymbol{A}$ with eigenvalues $\lambda_\pm$ and $-b$, respectively. Thus E^{u} is the one-dimensional subspace of $\mathbb{R}^3$ spanned by $\boldsymbol{u}_+$, whilst E^{s} is the two-dimensional subspace spanned by $\boldsymbol{u}_-$ and $\boldsymbol{u}_3$. Geometrically, E^{s} is the plane containing $\boldsymbol{u}_-$ and $\boldsymbol{u}_3$ which has normal $\boldsymbol{u}_3 \wedge \boldsymbol{u}_- = (-r, 1 + \lambda_-, 0) = (-r, 1 + \eta, 0)$ for η given by (4.12). Hence the stable manifold of the non-linear fixed point is tangent to this plane at the origin.

The linearized system in (4.13) occurs in a form in which the motion in the z-direction is decoupled from that in the xy-plane. The linearized equations are obviously satisfied by

$$x(t) \equiv 0, \qquad y(t) \equiv 0, \qquad z(t) = z_0 \exp(-bt). \tag{4.18}$$

It follows that the positive and negative z-axes are trajectories of the linearized system that lie in E^{s}. However, $\boldsymbol{g}(\boldsymbol{x}) \equiv \boldsymbol{0}$ for $\boldsymbol{x}(t)$ given by (4.18) and consequently the positive and negative z-axes are also trajectories of the non-linear system that converge to the origin as time increases. Thus the z-axis lies in the stable manifold of the non-linear fixed point.

We can gain some insight into the shape of the stable manifold of the Lorenz equations near the origin by examining the non-linear terms in (4.13) in the neighbourhood of E^{s}. Observe that the departure of the stable manifold of the non-linear system from E^{s} comes about because of the term $-xz$ in $\dot{y}$. The effect of this term is easier to visualize in the time-reversal of (4.13) in which it becomes $+xz$ and E^{s} becomes the unstable eigenspace at the origin. Let us consider typical trajectories of the time-reversed system that are tangent

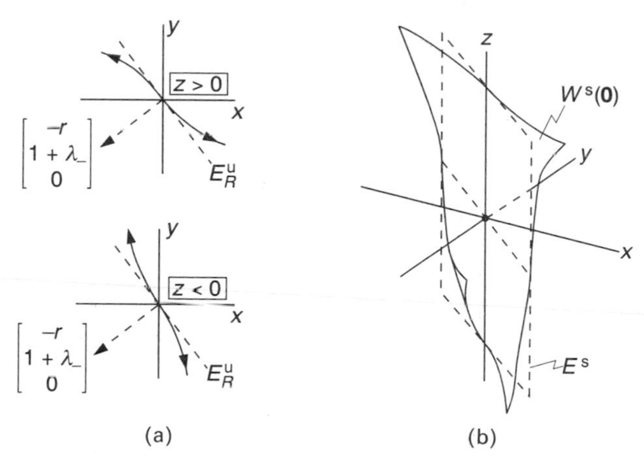

Fig. 4.2. (a) Sketches of the projections onto the (x, y)-plane of typical trajectories of the time-reversal of (4.13) that are tangent at the origin to the unstable eigenspace, E^u_R, of the time-reversed system. The sense in which a trajectory deviates from E^u_R depends on the sign of the non-linear term xz in $\dot{y}$. (b) Suggested form for the stable manifold, $W^s(\mathbf{0})$, of the fixed point at the origin of the Lorenz equations (4.13) in the immediate neighbourhood of the origin. It must be emphasized that $W^s(\mathbf{0})$ can develop into a much more complicated surface as the distance from the origin increases.

to the unstable eigenspace at the origin. The sign of xz suggests that the projections of such trajectories on the xy-plane will take the form shown in Fig. 4.2(a).

These trajectories must lie in the unstable manifold of the time-reversal of (4.13) and, consequently, the same phase curves, with the reverse orientation, lie in the stable manifold of (4.13) itself. We therefore offer Fig. 4.2(b) as a sketch of the stable manifold of the Lorenz equations in the neighbourhood of the origin. □

4.1.2 Non-hyperbolic fixed points

Hyperbolic fixed points are always simple (because no eigenvalue of the linearized system can be zero) but non-hyperbolic fixed points may be either simple or non-simple. Our treatment in Chapter 2 paid particular attention to the simple, non-hyperbolic fixed point known as the centre because it was neutrally stable and corresponded to periodic motion in the state space variables. As in the case of the hyperbolic saddle point, the extension of the discussion to phase spaces of dimension greater than two leads to a proliferation of such neutrally stable possibilities. However, we also encounter a number of new ideas and some subtlely different kinds of qualitative behaviour.

For example, consider the neutrally stable linear system on $\mathbb{R}^3$ with coefficient matrix

$$\begin{bmatrix} 0 & -\beta & 0 \\ \beta & 0 & 0 \\ 0 & 0 & \lambda \end{bmatrix}. \tag{4.19}$$

In cylindrical polar coordinates, (R, θ, z), this linear system takes the form

$$\dot{R} = 0, \qquad \dot{\theta} = \beta, \qquad \dot{z} = \lambda z, \tag{4.20}$$

with solution

$$R(t) \equiv C, \qquad \theta(t) = \beta t + \theta_0, \qquad z = z_0 \exp(\lambda t). \tag{4.21}$$

Thus, for each fixed value of $R = C > 0$, the trajectories of the system are confined to a cylinder of radius C whose axis is the z-axis. The phase portrait for the system is shown in Fig. 4.3 and there are two qualitatively distinct possibilities depending on the sign of λ.

The linear system $\dot{\boldsymbol{x}} = \boldsymbol{A}\boldsymbol{x}$ on $\mathbb{R}^4$ with coefficient matrix

$$\begin{bmatrix} 0 & -\beta_1 & 0 & 0 \\ \beta_1 & 0 & 0 & 0 \\ 0 & 0 & 0 & -\beta_2 \\ 0 & 0 & \beta_2 & 0 \end{bmatrix} \tag{4.22}$$

is an important example. This differential equation is easily solved by introducing polar coordinates in the x_1x_2- and x_3x_4-planes. If the new

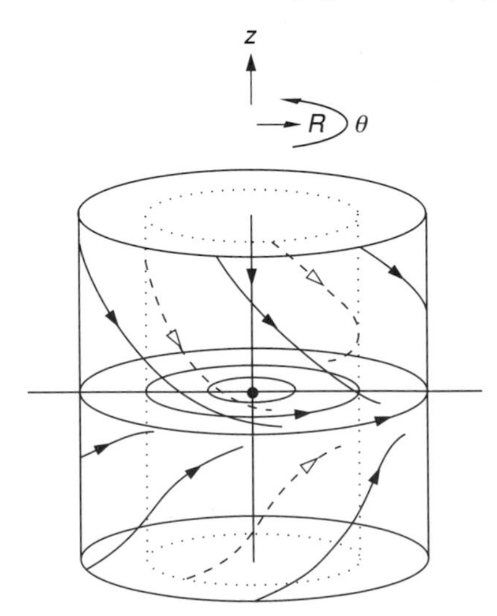

Fig. 4.3. Phase portrait for the neutrally stable linear system with coefficient matrix (4.19) when $\lambda < 0$. For $\lambda > 0$, the trajectories are still confined to cylinders but they expand away from the circular orbits at $z = 0$ rather than contract towards them.

coordinates are (r_1, θ_1) and (r_2, θ_2), respectively, then

$$\begin{aligned} \dot{r}_1 &= 0, \qquad \dot{\theta}_1 = \beta_1, \\ \dot{r}_2 &= 0, \qquad \dot{\theta}_2 = \beta_2. \end{aligned} \tag{4.23}$$

The polar equations (4.23) have solutions

$$\begin{aligned} r_1(t) &\equiv C_1, \qquad \theta_1(t) = \beta_1 t + \theta_1^0, \\ r_2(t) &\equiv C_2, \qquad \theta_2(t) = \beta_2 t + \theta_2^0, \end{aligned} \tag{4.24}$$

where C_1, C_2 are positive constants and θ_1^0, θ_2^0 are the values of θ_1, θ_2 at $t = 0$. Notice that, for $i = 1, 2, \theta_i(t)$ and $\theta_i(t) + 2k\pi, k$ integer, define the same radial direction in the (r_i, θ_i)-plane.

The solutions (4.24) obviously give $x_i, i = 1, 2, 3, 4$, that involve periodic functions but whether or not a particular phase point recurs periodically depends on the relationship between β_1 and β_2.

Definition 4.1.1

Two non-zero numbers β_1 and β_2 are said to be **rationally independent** if

$$k_1 \beta_1 + k_2 \beta_2 = 0, \tag{4.25}$$

with k_1, k_2 integers, is only satisfied by $k_1 = k_2 = 0$. Numbers that are not rationally independent will be called **rationally dependent** numbers.

The point (x_1, x_2, x_3, x_4) with polar coordinates $(C_1, \theta_1^0, C_2, \theta_2^0)$ recurs after time τ if $\theta_1(\tau) = \theta_1^0$ and $\theta_2(\tau) = \theta_2^0$, i.e. if there are integers p, q such that the equations

$$\beta_1 \tau = 2\pi q, \qquad \beta_2 \tau = 2\pi p \tag{4.26}$$

are simultaneously satisfied. Division of the two equations in (4.26) and a minor rearrangement yields

$$p\beta_1 - q\beta_2 = 0. \tag{4.27}$$

Thus if the point $(C_1, \theta_1^0, C_2, \theta_2^0)$ is periodic then β_1 and β_2 are rationally dependent. It is not difficult to show that if β_1 and β_2 are rationally dependent then every phase point is periodic with period

$$\tau = -\frac{2\pi k_2}{\beta_1} = \frac{2\pi k_1}{\beta_2}, \tag{4.28}$$

where k_1, k_2 are the non-zero integers satisfying (4.25). We conclude therefore that a phase point is periodic for the system (4.22) if and only if β_1 and β_2 are rationally dependent.

If β_1 and β_2 are rationally independent then no relation of the form (4.25) exists, therefore there can be no τ, p, q such that (4.26) is satisfied and no periodic points occur. In fact, it can be shown that, in this case, the orbit of any point approaches that point arbitrarily closely infinitely many times as t increases indefinitely and the trajectory is said to correspond to **quasi-periodic motion**.

The importance of this example to our present discussion is that the existence or otherwise of periodic points allows us to prove that the phase portraits for rationally dependent and rationally independent β_1 and β_2 are not qualitatively equivalent. Thus quasi-periodic motion is a new, qualitatively distinct type of behaviour not encountered in lower dimensions. It is perhaps worth emphasizing that the state variables are periodic functions of time whether or not β_1 and β_2 are rationally independent. Rational independence simply means that they do not all return to their initial values simultaneously. In this case, however, the initial state is repeatedly approached arbitrarily closely by its orbit, so that it can be difficult to distinguish between periodic and quasi-periodic motion by straightforward numerical calculations.

Thus far in our discussion of non-planar phase spaces we have focused attention on Euclidean spaces of dimension greater than two. However, the examples discussed above suggest an alternative line of approach. In both cases, the essential features of the dynamics could have been obtained by examining flows on surfaces. In the first example the surface in question is a cylinder and in the second it is a torus (see Fig. 4.4).

In both of these examples the phase space is two-dimensional but it is no longer planar. The fact that the surface closes on itself introduces recurrence, and new types of qualitative behaviour (such as quasi-periodic motion on the torus) can arise. The cylinder, the torus, and the sphere are all examples of what are known as two dimensional **differentiable manifolds**. The plane is itself a simple two-dimensional differentiable manifold which only admits the limited range of dynamics given in the Poincaré–Bendixson theorem.

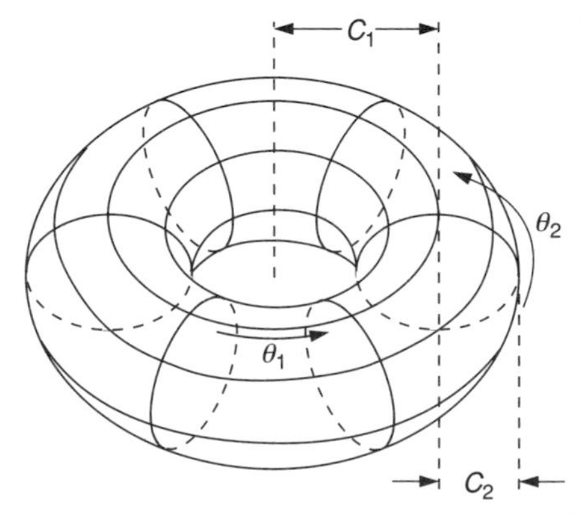

Fig. 4.4. For given values of $C_1 > C_2 > 0$ the solutions (4.24) of the polar equations (4.23) can be interpreted as a flow on the two-dimensional toral surface (or 2-torus), T^2, with radii C_1 and C_2. The 2-torus resembles a tyre inner tube with arbitrarily thin walls.

Our example shows that other types of behaviour can arise on the 2-torus. In fact differentiable manifolds are the natural choice of structure on which to study flows and the reader will find that many important theorems in the literature are stated for phase spaces of this type.

4.2 CLOSED ORBITS

4.2.1 Poincaré maps and hyperbolic closed orbits

We have seen that the qualitative classification of isolated fixed points on non-planar phase spaces involves more types of fixed point than the planar case. In order to investigate the qualitative properties of isolated closed orbits we make use of the corresponding Poincaré map.

Example 4.2.1

Consider the differential equation on $\mathbb{R}^3$ that takes the form

$$\dot{R} = R(1-R), \qquad \dot{\theta} = \beta, \qquad \dot{z} = \lambda z, \tag{4.29}$$

where (R, θ, z) are cylindrical polar coordinates. Assume that (a) $\lambda < 0$; (b) $\lambda > 0$; and sketch the phase portrait of (4.29). Draw diagrams to illustrate the successive intersections of typical trajectories of the flow of (4.29) with the half-plane $\theta = 0$.

Define Cartesian coordinates x, y by $x = R\cos\theta, y = R\sin\theta$ and obtain an explicit expression for the Poincaré map, **P**, of the system defined on the xz-plane with $x > 0$.

Solution

The system (4.29) has a fixed point at the origin. It is decoupled so that the (R, θ)- and z-subsystems can be considered separately. The phase portrait for the former is like that shown in Fig. 3.26, whilst the latter has solution

$$z(t) = z_0 \exp(\lambda t), \tag{4.30}$$

with $z(0) = z_0$. It follows that the restriction of the flow of (4.29) to the plane $z = 0$ has an unstable fixed point at the origin and a circular attracting closed orbit of radius unity centred on the origin. For $z \neq 0$, the flow of (4.29) spirals about the z-axis and contracts towards (expands away from) the plane $z = 0$ for $\lambda < 0 (> 0)$. For $0 < R \neq 1$, the flow contracts towards a cylinder of unit radius with its axis coincident with the z-axis. The trajectories of points on this cylinder remain on it for all t and spiral into (out from) the circular section with the plane $z = 0$ for $\lambda < 0 (> 0)$. The corresponding phase portraits are illustrated in Fig. 4.5.

Intersections of typical trajectories with the half-plane $\theta = 0$ are shown in Fig. 4.6. The trajectory through $(x, z) = (1, 0)$ repeatedly meets the half-plane in that point independently of the sign of λ. However, the intersections of

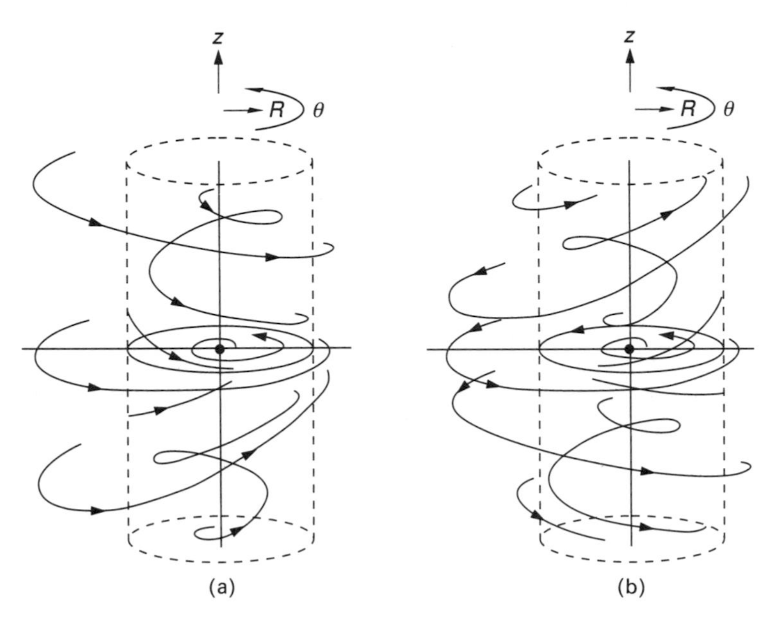

Fig. 4.5. Sketches of the phase portraits for (4.29) with: (a) $\lambda < 0$; (b) $\lambda > 0$.

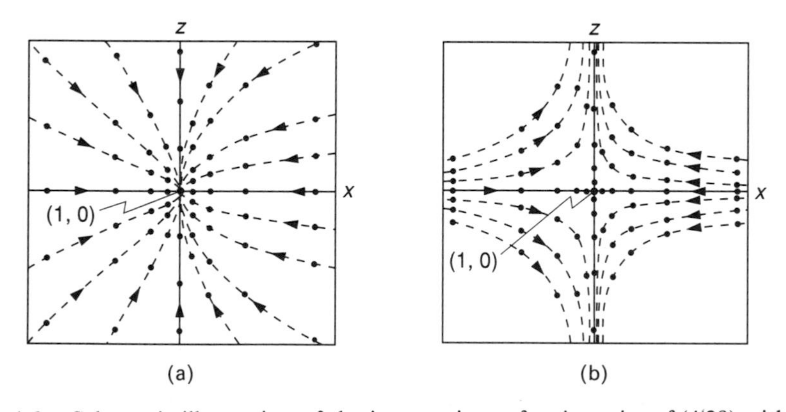

Fig. 4.6. Schematic illustration of the intersections of trajectories of (4.29) with the half-plane $\theta = 0$: (a) $\lambda < 0$; (b) $\lambda > 0$. The trajectory through $(x, z) = (1, 0)$ repeatedly returns to that point. Intersections of trajectories through points near to $(1, 0)$ are sequences of points that lie on recognizable curves. As time increases the points of a sequence appear on its curve in the order indicated by the arrows.

trajectories through other points exhibit what might be described as 'discrete node-like' behaviour for $\lambda < 0$ and 'discrete saddle-like' behaviour for $\lambda > 0$.

We can obtain an explicit expression for the Poincaré map $\boldsymbol{P}(x, z)$ from the flow of the equation (4.29). The system (4.29) is completely decoupled and we can solve each of its three constituent equations independently. We have already shown (in Example 1.5.1) that the differential equation

$$\dot{R} = R(1 - R) \tag{4.31}$$

has solution

$$R(t) = \frac{R_0 \exp(t)}{[R_0 \exp(t) - R_0 + 1]},$$

$$= \frac{R_0}{[R_0 + (1 - R_0)\exp(-t)]}, \tag{4.32}$$

satisfying $R(0) = R_0$ at $t = 0$. Moreover,

$$\theta(t) = \beta t + \theta_0, \tag{4.33}$$

satisfies $\dot{\theta} = \beta$ with $\theta(0) = \theta_0$. Equations (4.30), (4.32) and (4.33) allow us to write the flow of (4.29) in the Cartesian coordinates (x, y, z) defined in the question as

$$\phi_t(x, y, z) = \begin{bmatrix} \dfrac{[x \cos \beta t - y \sin \beta t]}{[(x^2 + y^2)^{\frac{1}{2}} + (1 - (x^2 + y^2)^{\frac{1}{2}})\exp(-t)]} \\ \dfrac{[x \sin \beta t + y \cos \beta t]}{[(x^2 + y^2)^{\frac{1}{2}} + (1 - (x^2 + y^2)^{\frac{1}{2}})\exp(-t)]} \\ z \exp(\lambda t) \end{bmatrix}. \tag{4.34}$$

It is clear from (4.33) that $\theta(t)$ first returns to θ_0 at $t = 2\pi/\beta = \tau$. Moreover, if we consider points that are initially in the xz-plane then $y = 0$ and

$$\phi_\tau(x, 0, z) = \begin{bmatrix} \dfrac{x}{[x + (1 - x)\exp(-\tau)]} \\ 0 \\ z \exp(\lambda \tau) \end{bmatrix}. \tag{4.35}$$

Hence we conclude that

$$\boldsymbol{P}(x, z) = \begin{bmatrix} \dfrac{x}{[x + (1 - x)\exp(-\tau)]} \\ z \exp(\lambda \tau) \end{bmatrix}. \tag{4.36}$$

It is perhaps worth emphasizing that x is to be taken greater than zero in (4.36) to avoid duplication of information from the half-planes $x < 0$ and $x > 0$.

□

Example 4.2.1 shows that the Poincaré map, **P**, has a fixed point at $(x, z) = (1, 0)$. Moreover, if $\mathbf{x}_0 = (x_0, z_0)$ is a point in the xz-plane near to $(1, 0)$ then the successive intersections of $\{\phi_t(x_0, 0, z_0) | t > 0\}$ with the half-plane $\theta = 0$ are given by $\mathbf{P}^k(\mathbf{x}_0), k = 1, 2, \ldots,$. It can be shown that the inverse function $\mathbf{P}^{-1}$ exists and the intersections for $t < 0$ are then given by $\mathbf{P}^k(\mathbf{x}_0), k = -1, -2, \ldots,$. The set of all points $\mathbf{P}^k(\boldsymbol{x}_0)$ for k integer, with $\mathbf{P}^0$ equal to the identity, is known as the **orbit** or **trajectory** of $\mathbf{x}_0$ under **P**. The behaviour of the orbits under **P** of points near to a fixed point determine its qualitative type and the nature of the fixed point (1, 0) in Example 4.2.1 clearly depends on the sign of λ. More precisely, we define the **stable (unstable) manifold,** $W^{s(u)}(\mathbf{x}^*)$, of a fixed point $\mathbf{x}^*$ of $\mathbf{P}$ to be the set of points $\mathbf{x}_0$ for which $\mathbf{P}^k(\mathbf{x}_0) \to \mathbf{x}^*$ as $k \to \infty(-\infty)$. In Example 4.2.1 the dimensions of these manifolds depend on the sign of λ.

In section 3.8, we pointed out that the stability of a closed orbit in the plane is determined by the first derivative of the local Poincaré map, P, provided $|\mathrm{d}P/\mathrm{d}x|_{x=x^*} \neq 1$, where x^* is the fixed point of P corresponding to the closed orbit. What is the analogous result when the domain of the Poincaré map has dimension greater than one? A vital clue to answering this question is the recognition of $\mathrm{d}P/\mathrm{d}x\,|_{x=x^*}$ as the eigenvalue of the linearization of P at $x = x^*$.

Definition 4.2.1
In terms of coordinates $x_1, \ldots, x_m$, the **linearization, $D\mathbf{P}(\mathbf{x})$, of a Poincaré map** $\mathbf{P}: \mathbb{R}^m \to \mathbb{R}^m$ is represented by the $m \times m$ matrix

$$D\boldsymbol{P}(\boldsymbol{x}) = \begin{bmatrix} \dfrac{\partial P_1}{\partial x_1} & \cdots & \dfrac{\partial P_1}{\partial x_m} \\ \vdots & \vdots & \vdots \\ \dfrac{\partial P_m}{\partial x_1} & \cdots & \dfrac{\partial P_m}{\partial x_m} \end{bmatrix}. \tag{4.37}$$

The analogous result referred to above is that: the qualitative nature of the closed orbit, corresponding to the fixed point $\mathbf{x}^*$ of $\mathbf{P}$, is determined by its linearization if every eigenvalue of $D\boldsymbol{P}(\boldsymbol{x}^*)$ has modulus different from unity.

Definition 4.2.2
A fixed point, $\mathbf{x}^*$, of a Poincaré map, $\mathbf{P}$, is said to be **hyperbolic** if no eigenvalue of $D\boldsymbol{P}(\boldsymbol{x}^*)$ has modulus equal to unity.

Thus the nature of the closed orbit is determined if the corresponding fixed point of the Poincaré map is hyperbolic. A closed orbit corresponding to a hyperbolic fixed point of a Poincaré map is said to be a **hyperbolic closed orbit**. However, we must clarify what is meant by 'the nature of the closed orbit is determined'.

4.2.2 Topological classification of hyperbolic closed orbits

Since each closed orbit is associated with a fixed point of a Poincaré map, $\mathbf{P}$, the nature of the closed orbit is given by the qualitative behaviour of the trajectories of $\mathbf{P}$ in the neighbourhood of the fixed point. This problem of the qualitative classification of fixed points of Poincaré maps is usually studied using rather different language. Poincaré maps can be shown to be differentiable maps with differentiable inverses. Such maps are called

diffeomorphisms. In fact, every diffeomorphism is the Poincaré map of a flow known as the **suspension** of the diffeomorphism. Thus we can replace 'Poincaré map' by 'diffeomorphism' in our discussion.

The qualitative classification of the fixed points of diffeomorphisms proceeds along analogous lines to that of flows. The first requirement is an equivalence relation that allows us to decide when two maps have the same qualitative behaviour. The equivalence relation chosen is known as **topological conjugacy**. Two diffeomorphisms are topologically conjugate if there is a continuous bijection between their domains of definition that maps the orbits of one onto those of the other preserving the integer parameter (denoted by k above) that labels the orbit points in terms of iterates of the maps. The essential idea is illustrated schematically in Fig. 4.7.

There is a linearization theorem for diffeomorphisms which relates the local qualitative behaviour at a hyperbolic, non-linear fixed point to that of the linearized system near the origin. More precisely, there are neighbourhoods of the fixed point and the origin, and a continuous bijection between them, that maps the orbits of the diffeomorphism onto those of its linearization in the manner illustrated in Fig. 4.7. For hyperbolic fixed points, therefore, the classification problem reduces to one of classifying the fixed points of hyperbolic linear diffeomorphisms.

Let **A**, **B** be hyperbolic linear diffeomorphisms on $\mathbb{R}^m$ represented by A, B respectively. Define the stable eigenspace of $A(B)$, $E^s_{A(B)}$, to be the subspace spanned by the eigenvectors of $A(B)$ corresponding to eigenvalues with modulus less than unity. Make similar definitions of the unstable eigenspaces of A and B using eigenvalues of modulus greater than unity.

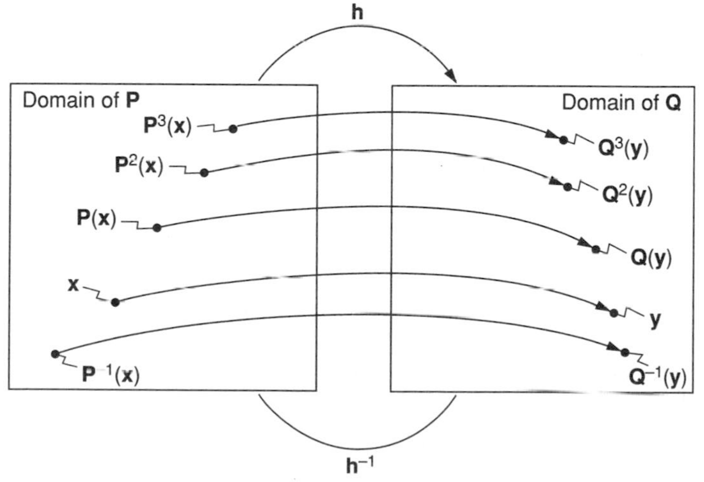

Fig. 4.7. If two diffeomorphisms **P** and **Q** are topologically conjugate then there is a continuous bijection, **h**, taking the domain, $\mathscr{D}_{\mathbf{P}}$, of **P** onto the domain, $\mathscr{D}_{\mathbf{Q}}$, of **Q** such that, for any $\mathbf{x} \in \mathscr{D}$, $\mathbf{h}(\mathbf{P}(\mathbf{x})) = \mathbf{Q}(\mathbf{h}(\mathbf{x}))$. It then follows that **h** maps the orbit of **x** under **P** onto the orbit of $\mathbf{y} = \mathbf{h}(\mathbf{x})$ under **Q** preserving the iterate label k.

Theorem 4.2.1

The hyperbolic linear diffeomorphisms **A** *and* **B** *are topologically conjugate if and only if:* (a) dim $E^s_A = \dim E^s_B$; (b) *for* $i = \mathrm{s}, \mathrm{u}, \boldsymbol{A}_i = \boldsymbol{A} | E^i_A$ *and* $\boldsymbol{B}_i = \boldsymbol{B} | \boldsymbol{E}^i_B$ *are both either orientation-preserving or orientation-reversing.*

A matrix $\boldsymbol{A}$ is said to be **orientation-preserving (reversing)** if det $(\boldsymbol{A})$ is greater than (less than) zero. Figure 4.8 illustrates the orbits of some simple examples of orientation-preserving and orientation-reversing linear diffeomorphisms.

Item (b) in the statement of Theorem 4.2.1 highlights an important difference between the qualitative behaviour of flows and diffeomorphisms. The flow, $\exp(\boldsymbol{A}t)\boldsymbol{x}$, of the hyperbolic linear system $\dot{\boldsymbol{x}} = \boldsymbol{A}\boldsymbol{x}$ is a hyperbolic linear diffeomorphism for all non-zero t but it is always orientation-preserving. For

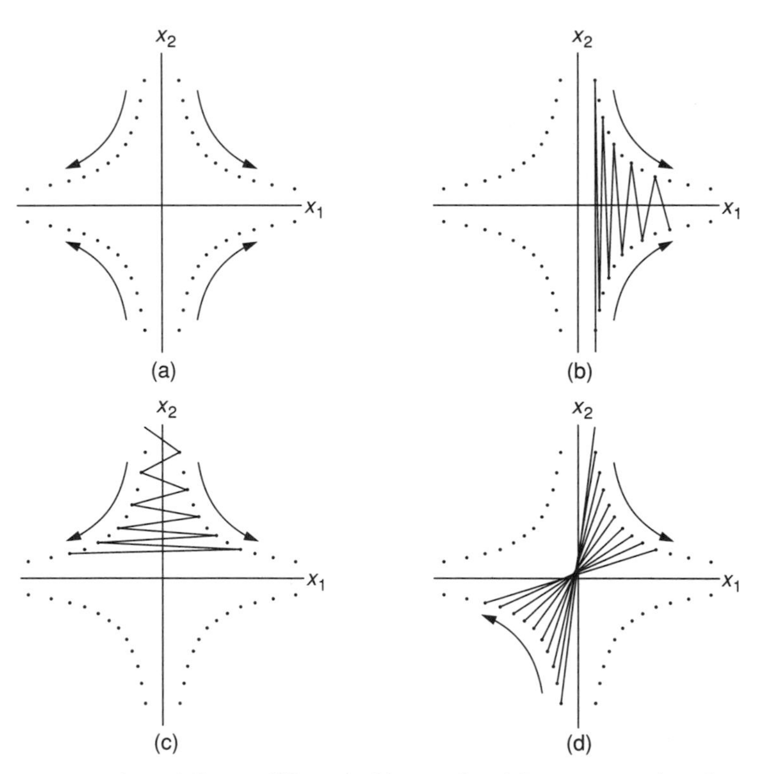

Fig. 4.8. Examples of linear diffeomorphisms of saddle type on the plane. The transformations shown here are $(x_1, x_2) \mapsto (\lambda_1 x_1, \lambda_2 x_2)$ for:

(a) $\lambda_1 > 1 > \lambda_2 > 0$;
(b) $\lambda_1 > 1, \quad -1 < \lambda_2 < 0$;
(c) $\lambda_1 < -1, \quad 0 < \lambda_2 < 1$;
(d) $\lambda_1 < -1 < \lambda_2 < 0$.

Note that (a) and (d) are orientation-preserving, whilst (b) and (c) are orientation-reversing.

if $\boldsymbol{A}$ has eigenvalues, λ_j, $j = 1, \ldots, n$, all of which have non-zero real part, then

$$\det[\exp(\boldsymbol{A}t)] = \prod_{j=1}^{n} \exp(\lambda_j t) > 0 \tag{4.38}$$

and $\exp(\boldsymbol{A}t)$ is orientation-preserving for all t.

Example 4.2.2
Obtain the linearization of the Poincaré map, $\boldsymbol{P}(x, z)$, given in (4.36) at the fixed point $(x, z) = (1, 0)$. Verify that the closed orbit in (4.29) is hyperbolic for all $\lambda \neq 0$. Show that qualitatively distinct types of closed orbit occur for $\lambda > 0$ and $\lambda < 0$.

Solution
The components of $\boldsymbol{P}(x, z)$ are

$$P_1(x, z) = \frac{x}{[x + (1 - x)\exp(-\tau)]}, \qquad P_2(x, z) = z\exp(\lambda\tau). \tag{4.39}$$

Hence (4.37) gives

$$D\boldsymbol{P}(x, z) = \begin{bmatrix} \dfrac{\exp(-\tau)}{[x + (1 - x)\exp(-\tau)]^2} & 0 \\ 0 & \exp(\lambda\tau) \end{bmatrix} \tag{4.40}$$

and

$$D\boldsymbol{P}(1, 0) = \begin{bmatrix} \exp(-\tau) & 0 \\ 0 & \exp(\lambda\tau) \end{bmatrix}. \tag{4.41}$$

The eigenvalues of the diagonal matrix $D\boldsymbol{P}(1, 0)$ are $\exp(-\tau)$ and $\exp(\lambda\tau)$ and, provided $\lambda \neq 0$, both are different from unity. Hence the fixed point of $\boldsymbol{P}$ at $(x, z) = (1, 0)$ is hyperbolic by Definition 4.2.2.

For $\lambda < 0$, the eigenvalues of $D\boldsymbol{P}(1, 0)$ are both positive and less than unity, consequently the stable eigenspace of $D\boldsymbol{P}(1, 0)$ has dimension two. However, when $\lambda > 0, 0 < \exp(-\tau) < 1 < \exp(\lambda\tau)$ and the stable eigenspace of $D\boldsymbol{P}(1, 0)$ has dimension unity. It then follows from Theorem 4.2.1 that the linearizations for $\lambda < 0$ and $\lambda > 0$ are not topologically conjugate. The linearization theorem for diffeomorphisms then assures us that the non-linear fixed point at $(x, z) = (1, 0)$ is qualitatively different for $\lambda < 0$ and $\lambda > 0$. Finally, this means that qualitatively distinct types of hyperbolic closed orbit occur in the two cases. □

The Poincaré map $\boldsymbol{P}(x, z)$ in Example 4.2.2 is atypical in that $P_1(x, z)$ depends only on x and $P_2(x, z)$ depends only on z. Consequently the stable and unstable manifolds at $(1, 0)$ coincide, respectively, with the stable and unstable eigenspaces of $D\boldsymbol{P}(1, 0)$. In general, just as for flows, $W^{s,u}(\boldsymbol{x}^*)$ are only tangent to $E^{s,u}$ at $\boldsymbol{x}^*$.

4.2.3 Periodic orbits and quasi-periodic motion

The closed orbit in Example 4.2.1 closes after a single trip around the z-axis and therefore corresponds to a fixed point of the Poincaré map. However, the freedom afforded by a three-dimensional phase space allows us to envisage a similar system in which the trajectory through x^* fails to close on its first return to the half-plane $\theta = 0$ but does so after a further trip around the z-axis (as illustrated in Fig. 4.9(a)).

In terms of the Poincaré map, $\mathbf{P}$, for such a flow we have $\mathbf{P}(\mathbf{x}^*) \neq \mathbf{x}^* = \mathbf{P}(\mathbf{P}(\mathbf{x}^*)) = \mathbf{P}^2(\mathbf{x}^*)$. Thus $\mathbf{x}^*$ is a fixed point of $\mathbf{P}^2$ but it is not a fixed point of $\mathbf{P}$. The following definition provides the terminology used to describe such cases.

Definition 4.2.3
A point $\mathbf{x}^*$ is said to be a **periodic point** of $\mathbf{P}$ if $\mathbf{P}^q(\mathbf{x}^*) = \mathbf{x}^*$ for some integer $q \geqslant 1$.

The smallest value of q satisfying Definition 4.2.3 is referred to as the **period** of $\mathbf{x}^*$ and $\{\mathbf{x}^*, \mathbf{P}(\mathbf{x}^*), \ldots, \mathbf{P}^{q-1}(\mathbf{x}^*)\}$ is said to be a **periodic orbit of period** q or a q-**cycle** of $\mathbf{P}$. Notice that fixed points are included as periodic points of period one. In the illustration discussed above, $\mathbf{x}^*$ is a periodic point of $\mathbf{P}$ of period two and $\{\mathbf{x}^*, \mathbf{P}(\mathbf{x}^*)\}$ is the corresponding 2-cycle.

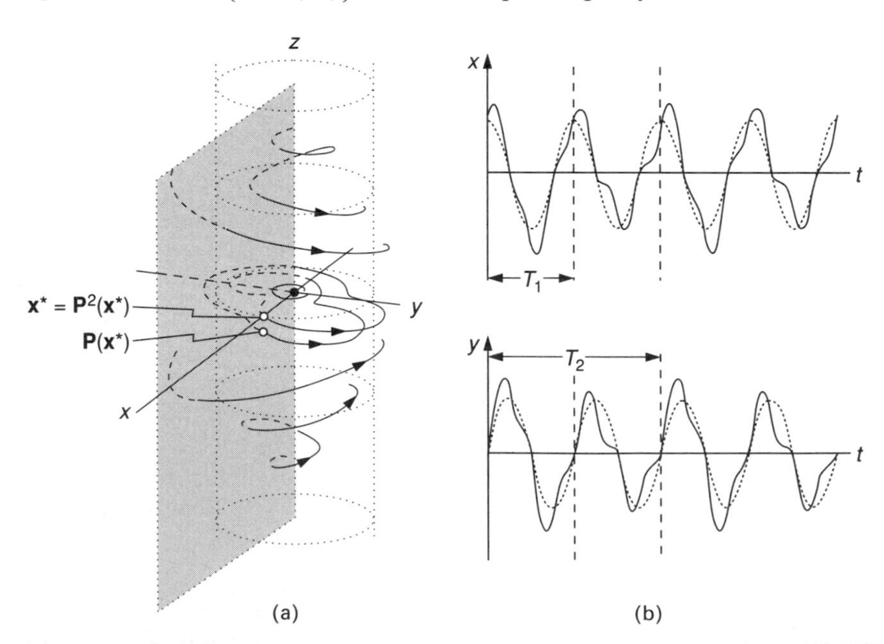

Fig. 4.9. (a) Modification of the phase portrait of (4.29) in which the closed orbit of period $T_1 = 2\pi/\beta$ is replaced by one of period $T_2 = 4\pi/\beta$, giving rise to a 2-cycle in the Poincaré map $\mathbf{P}$. (b) Schematic comparison of the time dependence of state variables for closed orbits associated with period-1 (dashed) and period-2 (solid) points in the Poincaré maps of (4.29) and its modification as suggested in (a).

The oscillations of the state variables with time when such a closed orbit occurs in the phase portrait of a system differ significantly from those arising from a fixed point of **P**. For example, if we assume that the equation $\dot{\theta} = \beta$ in (4.29) is preserved, then the closed orbit in Fig. 4.9(a) (associated with a 2-cycle of **P**) is such that $x(t)$ and $y(t)$ have period $T_2 = 4\pi/\beta$ in contrast to the period $T_1 = \tau = 2\pi/\beta$ of the circular closed orbit in (4.29) (associated with a 1-cycle of **P**). What is more, if the orbit is not confined to the cylinder $R = 1$, then the dependence of these state variables on t may no longer be given by a cosine or sine function. The result, as Fig. 4.9(b) suggests, has the appearance of an irregular fluctuation in x and y that is regularly repeated. If the closed orbit winds around the z-axis $q > 2$ times before closing, then the irregular fluctuation extends over a period T_q (equal to $2\pi q/\beta$ in our example) before repetition occurs. If q is large it can be difficult to determine T_q by examining plots of $x(t)$ and $y(t)$.

All the fixed points of $\mathbf{P}^q$ lying on a q-cycle of **P** are of the same topological type and this determines the type of the cycle itself. In particular, the **stable (unstable) manifold of a periodic orbit** is the union of the stable (unstable) manifolds of the fixed points of $\mathbf{P}^q$ that constitute it.

Example 4.2.3

Consider the differential equation

$$\dot{\theta}_1 = \beta_1, \qquad \dot{\theta}_2 = \beta_2, \tag{4.42}$$

defined on the two-dimensional torus shown in Fig. 4.4. Obtain the Poincaré map, P, for the flow of (4.42) on the section consisting of the circle given by $\theta_1 = \theta_1^0$. Show that P has periodic points if and only if β_1 and β_2 are rationally dependent. Interpret this result in terms of the behaviour of the flow of (4.42).

Solution

The system of equations (4.42) has solution

$$\theta_1(t) = \beta_1 t + \theta_1^0, \qquad \theta_2(t) = \beta_2 t + \theta_2^0, \tag{4.43}$$

where, for $i = 1, 2$, $\theta_i(t)$ and $\theta_i(t) + 2k\pi$, k integer, are to be identified to obtain the angular coordinate on the torus. The orbit of a point which, at $t = 0$, is on the circle, $\mathscr{C}$, given by $\theta_1 = \theta_1^0$, first returns to $\mathscr{C}$, having made one complete trip around the torus in the θ_1-direction, when $\beta_1 t = 2\pi$. Hence, if the θ_2-coordinate of the point is initially θ_2^0, then at first return it is given by

$$P(\theta_2^0) = \theta_2\left(\frac{2\pi}{\beta_1}\right) = 2\pi\left(\frac{\beta_2}{\beta_1}\right) + \theta_2^0, \tag{4.44}$$

reduced mod 2π. Clearly, the Poincaré map, P, is simply a rotation through $2\pi(\beta_2/\beta_1)$ radians. Equation (4.44) implies

$$P^q(\theta_2^0) = \theta_2^0 + 2\pi q\left(\frac{\beta_2}{\beta_1}\right), \tag{4.45}$$

reduced mod 2π, and therefore $P^q(\theta_2^0) = \theta_2^0$ if and only if

$$q\left(\frac{\beta_2}{\beta_1}\right) = p, \tag{4.46}$$

where p is an integer. However, (4.46) means there are non-zero integers p, q such that

$$p\beta_1 - q\beta_2 = 0, \tag{4.47}$$

and β_1 and β_2 must be rationally dependent.

When β_1 and β_2 are rationally dependent there exist non-zero integers p, q satisfying (4.47). Any point θ_2^0 on $\mathscr{C}$ is a periodic point of period q of the Poincaré map, P, and the trajectory of (4.42) through any point (θ_1^0, θ_2^0) on the torus closes after making q trips around the torus in the θ_1-direction. During this process, the points of intersection of this trajectory with $\mathscr{C}$ make p trips around that circle. This means that while the trajectory of (4.42) is winding q times around the torus in the θ_1-sense, it is, at the same time, wrapping itself p times around the torus in the θ_2-direction. Each point of the circle $\mathscr{C}$ gives rise to a distinct trajectory of (4.42) and the set of all such trajectories completely fills out the torus.

When β_1 and β_2 are rationally independent there are no non-zero integers p, q satisfying (4.47) and P has no periodic points. Thus, the trajectory of (4.42) through (θ_1^0, θ_2^0) winds around the torus indefinitely in both the θ_1- and θ_2-senses without closing. However, it can be shown (Arnold, 1973) that the orbit of any point, $\theta_2 = \theta_2^0$ say, of $\mathscr{C}$ under P is dense in $\mathscr{C}$ and therefore visits any neighbourhood of $\theta_2 = \theta_2^0$ (however small) infinitely often. The corresponding trajectory of (4.42) forms a dense subset of the torus and comes arbitrarily close to closing infinitely many times. Thus the motion can be reasonably described as almost- or quasi-periodic. □

Consider the following problem. Let the closed orbit in Fig. 4.9(a) be replaced by a torus with the dynamics considered in Example 4.2.3 but with no knowledge of whether or not β_1 and β_2 are rationally independent. Given that we compute numerical values for the state variable y as a function of t for $0 < t < t_0$ and find no sign of periodicity in y, can we conclude that β_1 and β_2 are rationally independent? Of course, the answer is 'no' because they could satisfy the relation (4.47) with q so large that the period, $2\pi q/\beta_1$, of the closed orbit is greater than t_0. Thus it can be difficult to distinguish between quasi-periodic motion and some kinds of periodic motion. In terms of Poincaré maps this idea corresponds to it being difficult to distinguish between an aperiodic point and a q-periodic point of long period.

4.3 ATTRACTING SETS AND ATTRACTORS

Attracting sets and limit sets of planar phase portraits were discussed in sections 3.8 and 3.9. Recall that all attracting sets are limit sets and, according to the Poincaré–Bendixson theorem, the compact limit sets of planar flows consist of fixed points, closed orbits or unions of trajectories containing fixed points. Attracting sets of this kind are found in flows on non-planar phase spaces. However, from a topological point of view, the plane is rather special. For example, it has the property that any closed curve disconnects the points inside it from those outside it. This property, together with the fact that the trajectories of a flow do not intersect, restricts the kind of limit sets that can occur. The additional topological freedom enjoyed by flows on non-planar phase spaces means that more complicated attracting sets can occur.

Example 4.3.1
Characterize the phase space of the system

$$\dot{r} = r(1-r), \qquad \dot{\theta} = \beta, \qquad \dot{\varphi} = \gamma, \tag{4.48}$$

where (r, θ) are plane polar coordinates and φ is a second angular coordinate of period 2π. Describe the phase portrait for (4.48) and find the ω-limit set, $L_\omega(\mathbf{x})$, for any point, $\mathbf{x}$, with non-zero radial coordinate. Discuss how sensitive the nature of the dynamics on $L_\omega(\mathbf{x})$ is to the relative values of β and γ.

Solution
The phase space of the (r, θ)-subsystem of (4.48) is obviously the plane. In adding the new state variable φ, we are adding a third dimension but, by insisting that φ is an angular variable, we are demanding that this dimension behave like a circle. In order to clarify the nature of this phase space, let us first consider a similar system in which the radial coordinate, r, must be less than some upper limit, R. Now there is no problem: the phase space of the r, θ-subsystem is an open disc, D_R, of radius R that is swept out into a solid torus, $D_R \times S^1$, as the angular coordinate, φ, moves around the circle S^1 (as

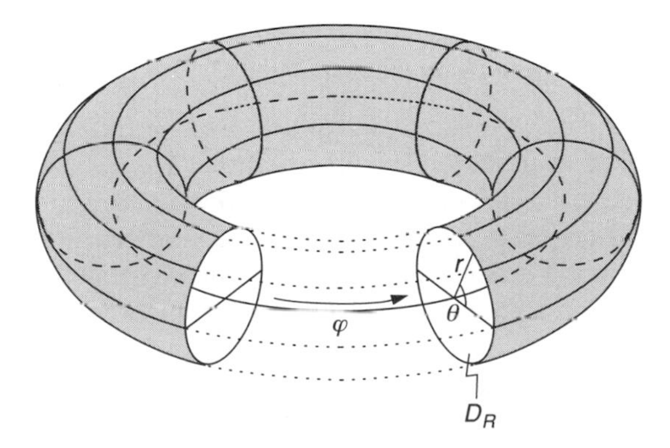

Fig. 4.10. Illustration of the solid torus $D_R \times S^1$. This set of points is topologically the same for all non-zero values of the radius R of the disc D_R. Notice that the radius of the ϕ-coordinate circle with $r = 0$ is always taken to be greater than R.

shown in Fig. 4.10). Here '$\times$' denotes the Cartesian product of two sets. However, from a topological point of view, the open disc, D_R, is completely equivalent to the plane, $\mathbb{R}^2$. Observe that the mapping $r \to R[1 - \exp(-r)]$ takes $[0, \infty)$ onto $[0, R)$ and allows us to construct a continuous bijection between $\mathbb{R}^2$ and D_R. Thus, the phase space, $\mathbb{R}^2 \times S^1$, is, topologically, just a solid torus. The (r, θ)- and φ-subsystems in (4.48) are decoupled and may therefore be solved separately. The phase portrait for the former is like that shown in Fig. 3.26 and, as time increases, the trajectories of this subsystem are swept around the circle defined by φ to produce the phase portrait of (4.48). For example, the fixed point at the origin in Fig. 3.26 becomes a closed orbit in the phase portrait of (4.48) and the limit cycle in Fig. 3.26 is drawn out into an invariant 2-torus. In the phase portrait of (4.48) trajectories of points not in either of these two sets spiral around the solid torus, $\mathbb{R}^2 \times S^1$, in both the θ- and φ-senses drawing closer and closer to the invariant 2-torus as time increases. Since the (r, θ)- and φ-motions are independent of one another, a similar construction to that used in Example 3.8.1 provides, for each value of φ, a sequence of times satisfying Definition 3.8.1 and confirming that the 2-torus is the ω-limit set, $L_\omega(\mathbf{x})$, of any point, $\mathbf{x}$, with non-zero radial coordinate.

The restriction of the flow of (4.48) to the invariant 2-torus is the same as that considered in Example 4.2.3. If β/γ is a rational number, p/q say, then (4.47) is satisfied, every point on the 2-torus lies on a closed orbit and the motion will be periodic; if not, there are no closed orbits on the 2-torus and the motion is quasi-periodic. If we think of β/γ as a parameter that can be varied through some interval of real values, then the sets of parameter values for which periodic and quasi-periodic motion occur are intermingled in the same way as the rational and irrational numbers. Viewed in this way, the nature of the flow on the 2-torus depends delicately on the relative values of β and γ. □

4.3.1 Trapping regions for Poincaré maps

It is important to emphasize that the closed orbit in Example 3.8.1 and the two-dimensional torus in Example 4.3.1 are rather special ω-limit sets in that they are both examples of attracting sets. We pointed out in section 3.9 that the existence of a trapping region, D, in the phase portrait of a flow, ϕ_t, is sufficient to ensure the existence of an attracting set, A, within D but how is the attracting set itself characterized? With Definition 3.9.1 of a trapping region in mind, the key idea is that the image of D under the flow must collapse onto A as time increases, i.e. $\phi_{t_2}(D) \subset \phi_{t_1}(D)$ if $t_2 > t_1$ and $\phi_t(D)$ is a trapping region for all $t > 0$.

Definition 4.3.1
A closed, invariant set, A, is said to be **attracting** for a flow, ϕ_t, if within any

neighbourhood of A there is a trapping region, T, such that

$$A = \bigcap_{t>0} \phi_t(T). \tag{4.49}$$

It is not difficult to verify that the closed orbit in Example 3.8.1 and the 2-torus in Example 4.3.1 are attracting sets in the sense of Definition 4.3.1.

Example 4.3.1 also illustrates the need to distinguish between an attracting set and an attractor. The restriction of the flow of (4.48) to the attracting 2-torus is the same as that considered in Example 4.2.3. When β and γ are rationally independent the orbit of any point spreads throughout the 2-torus, approaching every point arbitrarily closely, i.e. it is dense in the attracting set. An attracting set that contains a dense orbit is called an **attractor**. The existence of such an orbit ensures that the attracting set behaves as a single unit that cannot be decomposed into a collection of disjoint subsets that are themselves attractors. If β and γ are rationally dependent then each point of the 2-torus lies on a closed orbit winding finitely many times around the torus before closing on itself. It follows that there is no point on the 2-torus whose orbit is dense in the attracting set and the 2-torus is not an attractor.

The existence of an attracting set A for a flow circulating around $\mathbb{R}^2 \times S^1$ is reflected in the behaviour of the Poincaré map, $\mathbf{P}$, defined, for example, on the $\varphi = 0$ plane. Typically, the intersection of the trapping region D for ϕ_t with the $\varphi = 0$ plane defines a region, $T \subset \mathbb{R}^2$, such that

$$\mathbf{P}(T) \subseteq \text{int}(T), \tag{4.50}$$

where $\text{int}(T)$ is the interior of T. A closed and bounded set, T, in the domain of definition of a Poincaré map, $\mathbf{P}$, that satisfies (4.50) is said to be a **trapping region** for $\mathbf{P}$. It can be shown that (4.50), together with the fact that $\mathbf{P}$ is a diffeomorphism, means that

$$\mathbf{P}^n(T) \subseteq \text{int}(\mathbf{P}^{n-1}(T)), \tag{4.51}$$

for all positive integers n. Observe that application of $\mathbf{P}$ to both sides of (4.50) gives

$$\mathbf{P}^2(T) \subseteq \mathbf{P}(\text{int}(T)) = \text{int}(\mathbf{P}(T)). \tag{4.52}$$

The equality in (4.52) holds because $\mathbf{P}$ is a diffeomorphism. Thus $\mathbf{P}(T) \subseteq \text{int}(T)$ is also a trapping region for $\mathbf{P}$. It follows that the images of T under $\mathbf{P}^n$ contract onto

$$A_{\mathbf{P}} = \bigcap_{n>0} \mathbf{P}^n(T). \tag{4.53}$$

as n increases to infinity. This attracting set for $\mathbf{P}$ is the intersection of A with the $\varphi = 0$ plane.

Example 4.3.2

Find the flow of (4.48) in terms of the coordinates (r, θ, φ) and obtain the

Poincaré map, **P**, defined on the plane $\varphi = \varphi_0$. Prove that the circle $\mathscr{C} = \{(r, \theta) | r = 1, 0 \leqslant \theta < 2\pi\}$ is an attracting set for **P** and show that the restriction of **P** to $\mathscr{C}$ is a rotation through an angle $2\pi\beta/\gamma$. Under what circumstances is $\mathscr{C}$ an attractor? Explain your answer.

Solution

The differential equation (4.48) is completely decoupled when expressed in terms of the curvilinear coordinates (r, θ, φ). The flow ϕ_t is obtained by writing down solutions to the three independent equations. The θ-, φ-equations can be integrated directly and the solution to the r-equation has already been discussed in Example 1.5.1 and Example 4.2.1. We conclude that

$$\phi_t(r, \theta, \varphi) = \begin{bmatrix} \dfrac{r}{[r + (1 - r)\exp(-t)]} \\ \beta t + \theta \\ \gamma t + \varphi \end{bmatrix}. \tag{4.54}$$

Equation (4.54) shows that a trajectory passing through the plane $\varphi = \varphi_0$ at $t = 0$, first returns to that plane when $\gamma t = 2\pi$. Moreover, if the point of intersection of trajectory and plane at $t = 0$ has coordinates (r, θ) then the point at which the trajectory first returns to the plane has coordinates given by the first two components of (4.54) with $t = 2\pi/\gamma$. Thus

$$\mathbf{P}(r, \theta) = \begin{bmatrix} \dfrac{r}{[r + (1 - r)\exp(-2\pi/\gamma)]} \\ \theta + 2\pi\beta/\gamma \end{bmatrix}. \tag{4.55}$$

Notice that (4.55) is independent of φ_0. This is a result of the fact that the (r, θ)- and φ-subsystems of (4.48) are decoupled.

The effect of the map **P** on the point (r, θ) consists of anticlockwise rotation through an angle $2\pi\beta/\gamma$ followed by a change of radial coordinate which is independent of θ. Thus, in terms of the set of concentric circles centred on the origin, **P** maps the circle of radius r onto the circle of radius, r_1, where

$$\frac{r_1}{r} = \frac{1}{[r(1 - \alpha) + \alpha]}, \tag{4.56}$$

with $0 < \alpha = \exp(-2\pi/\gamma) < 1$. Hence $r_1 > r$ for $r < 1$ and $r_1 < r$ when $r > 1$. If $r = 1$ then $r_1 = 1$ and, consequently, **P** maps the circle $\mathscr{C}$ onto itself. We say that **P** has an **invariant circle** of unit radius centred on the origin.

Consider two circles centred on the origin, one of radius $a < 1$ and the other of radius $b > 1$. Let T be the closed annular region bounded by these two circles. Equation (4.56) shows that T is a trapping region for **P** and that the images of T under $\mathbf{P}^n$ contract onto the circle $\mathscr{C}$ as n tends to infinity. Hence $\mathscr{C}$ is an attracting set for **P**.

The restriction of **P** to the invariant circle $\mathscr{C}$ is obtained by substituting

$r = 1$ in (4.55). The result is

$$\boldsymbol{P}(1, \theta) = \begin{bmatrix} 1 \\ \theta + 2\pi\beta/\gamma \end{bmatrix}, \tag{4.57}$$

which is an anticlockwise rotation through an angle $2\pi\beta/\gamma$. The invariant circle is an attractor provided β and γ are rationally independent. If β/γ is an irrational number, then the orbit of any point on the unit circle is everywhere dense on that circle (Arnold, 1973). If β and γ are rationally dependent, with $\beta/\gamma = p/q$ in lowest terms, then every point of the invariant circle is a periodic point of period q. There is no point whose orbit is dense in the circle and the attracting set is not an attractor. □

4.3.2 Saddle points in attracting sets

Example 4.3.2 is particularly simple in that the system (4.48) is completely decoupled, the Poincaré map can be obtained explicitly and both trapping region and attracting set are easily recognized. A more substantial example, which illustrates the importance of stable and unstable manifolds in the study of attracting sets, is provided by the following system. The phase space is once again the solid torus $\mathbb{R}^2 \times S^1$, the plane $\mathbb{R}^2$ is described by Cartesian coordinates (x, y) with dynamics given by

$$\dot{x} = y, \qquad \dot{y} = x - x^3 - \delta y. \tag{4.58}$$

The circular coordinate θ satisfies

$$\dot{\theta} = \omega, \tag{4.59}$$

so the x, y- and θ-subsystems are decoupled but explicit expressions for the flow of (4.58) are not available.

We can sketch the phase portrait of (4.58) using the ideas developed in Chapter 3. For $\delta = 0$ (4.58) is a Hamiltonian system with Hamiltonian

$$H(x, y) = \frac{1}{2}y^2 - \frac{1}{2}x^2 + \frac{1}{4}x^4, \tag{4.60}$$

and its phase curves are given by the level sets of $H(x, y)$. The surface $z = H(x, y)$ and the resulting phase portrait are illustrated in Fig. 4.11.

For $\delta > 0$ the positions of the fixed points remain unchanged and the linearization theorem shows that the centres at $(x, y) = (\pm 1, 0)$ are stable foci if $\delta < 2\sqrt{2}$. The global effect of the term $-\delta y$ is to destroy the closed orbits shown in Fig. 4.11(b). To see this, let $\mathbf{X}_\delta(x, y)$ be the vector field defined by the right-hand side of (4.58). Then $\mathbf{X}_0(x, y)$ is tangent, at (x, y), to the closed trajectory in Fig. 4.11(b) passing through (x, y). It follows that $\mathbf{n}(x, y) = (x - x^3, -y)$ lies along the inward normal to the trajectory at (x, y) and that

$$\begin{aligned} \mathbf{n}(x, y) \cdot \mathbf{X}_\delta(x, y) &= (x - x^3)y - y(x - x^3 - \delta y) \\ &= \delta y^2. \end{aligned} \tag{4.61}$$

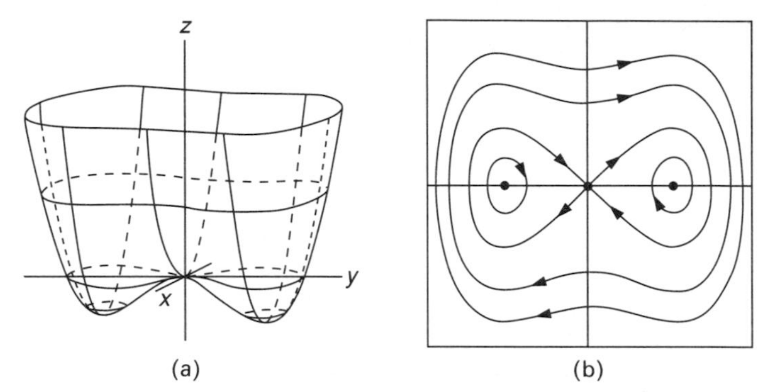

Fig. 4.11. (a) The surface $z = H(x, y)$ for the Hamiltonian given in (4.60). (b) The phase curves of the planar system (4.58) with $\delta = 0$ are the intersections of the surface shown in (a) with the planes $z = C$, for $-\frac{1}{4} \leqslant C < \infty$.

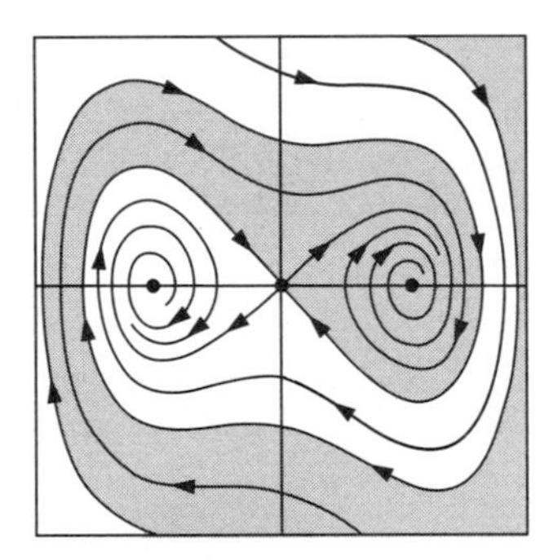

Fig. 4.12. Phase portrait for (4.58) with $\delta = 0.25$. The shading distinguishes the basins of attraction of the foci at $(\pm 1, 0)$.

Equation (4.61) shows that $\mathbf{X}_\delta, \delta > 0$, has a positive component along the inward normal to the trajectories of $\mathbf{X}_0(x, y)$ at all points (x, y) for which $y \neq 0$. Hence the trajectories of (4.58) with $\delta > 0$ are tilted inward relative to those of the phase portrait illustrated in Fig. 4.11(b) at all points not on the x-axis. We conclude that the phase portrait for (4.58) with $0 < \delta < 2\sqrt{2}$ must take the form shown in Fig. 4.12.

Example 4.3.3

Consider the system defined by (4.58) and (4.59) with $0 < \delta < 2\sqrt{2}$. Obtain the Poincaré map, $\mathbf{P}$, defined on the $\theta = \theta_0$ plane in terms of the flow ϕ_t of (4.58) and verify that it is independent of θ_0. Let T_C be the closed set of points in the $\theta = 0$ plane bounded by the level curve, L_C, defined by $H(x, y) = C$ with $C > 0$. Show that T_C is a trapping region for $\mathbf{P}$ and describe the attracting set lying within it. Explain why this attracting set is not an attractor.

Solution

Equation (4.59) has solution $\theta = \omega t + \theta_0$ satisfying $\theta = \theta_0$ at $t = 0$. Since $\theta = \theta_0$ and $\theta = \theta_0 + 2\pi$ are identified, a trajectory of (4.58) and (4.59) starting in the $\theta = \theta_0$ plane at $t = 0$ first returns to that plane when $t = \tau = 2\pi/\omega$, for then $\theta = \theta_0 + \omega\tau = \theta_0 + 2\pi$.

The (x, y)- and θ-motions are independent of one another and the evolution of phase points in the (x, y)-subsystem is given by the flow, ϕ_t, of (4.58). In other words, ϕ_t is the projection of the flow of (4.58) and (4.59) onto any plane of constant θ. Therefore, if the above trajectory of (4.58) and (4.59) starts at the point (x, y) of the $\theta = \theta_0$ plane, its projection onto that plane will have evolved to the point $\phi_\tau(x, y)$ at $t = \tau$. Thus, the trajectory of (4.58) and (4.59) through the point (x, y) of the $\theta = \theta_0$ plane first returns to that plane at the point $\phi_\tau(x, y)$, where ϕ_t is the flow of (4.58). Hence

$$\mathbf{P}(x, y) = \phi_\tau(x, y), \tag{4.62}$$

which is independent of θ_0.

We have already shown that the vector field, $\mathbf{X}_\delta$, of (4.58) with $\delta > 0$ points into T_C at every point of its boundary, L_C, for which $y \neq 0$ and is tangent to it when $y = 0$. Since L_C meets the x-axis in only two points, we conclude that all the trajectories of (4.58) that encounter L_C pass into the interior of T_C as time increases. Hence

$$\phi_\tau(T_C) = \mathbf{P}(T_C) \subseteq \text{int}(T_C) \tag{4.63}$$

and T_C is a trapping region for $\mathbf{P}$.

We can recognize the attracting set lying within T_C by examining how the images $\mathbf{P}(T_C)$ and $\mathbf{P}^2(T_C)$ of T_C are formed. The effect of $\mathbf{P}$ on T_C can be obtained from the phase portrait of (4.58). Observe that (4.62) means that the fixed points of $\mathbf{P}$ coincide with those of ϕ_t. What is more, the stable and unstable separatrices of the saddle point at the origin in Fig. 4.12 consist of points that approach that fixed point under forward and reverse iterations of $\mathbf{P}$, respectively. Consequently, they form the stable and unstable manifolds of the fixed point of $\mathbf{P}$. Equation (4.62) also shows that the image of any point of T_C lies on the forward trajectory of (4.58) passing through that point. Therefore the images of interior points of T_C must be interior points of the image of T_C, and it is sufficient to consider the effect of $\mathbf{P}$ on the boundary curve L_C.

The two points of intersection of the stable manifold of the saddle point at the origin with L_C are of special interest. The images of these points under $\mathbf{P}^n$ approach the origin as n tends to infinity. However, as Fig. 4.12 shows, they separate points of L_C that are attracted to different stable fixed points. It follows that a small segment of L_C such as AB in Fig. 4.13(a) will ultimately approach the saddle point from below and be stretched along the underside of the unstable manifold of that fixed point. A similar segment, CD in

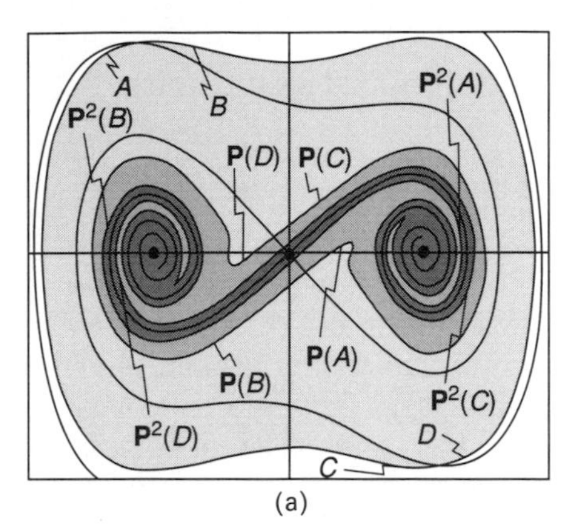

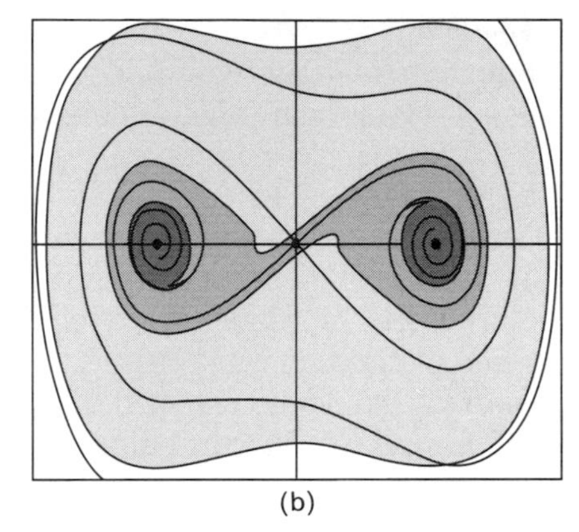

Fig. 4.13. (a) Schematic diagram illustrating the formation of $\mathbf{P}(T_C)$ and $\mathbf{P}^2(T_C)$, $C > 0$. (b) Plots of $\mathbf{P}(T_1)$ and $\mathbf{P}^2(T_1)$ for the Poincaré map, $\mathbf{P}$, of (4.58) and with $\omega = 1$ and $\delta = 0.25$. The diagram in (a) is a topologically faithfully representation of the numerical data shown in (b).

Fig. 4.13(a), spanning the other stable separatrix, is ultimately stretched along the upperside of the unstable manifold of the saddle point at the origin.

Since $\mathbf{P}$ is a diffeomorphism, the end-points of each pair of mapped intervals must be joined up to form a closed curve and, in view of (4.51), successive images of T_C must lie inside each other. However, it is important to note that some points must lie in *all* images of T_C. If $S \subset T_C$ is such that

$$\mathbf{P}(S) = S \tag{4.64}$$

then

$$\mathbf{P}^n(S) = S \subset \mathbf{P}^n(T_C), \tag{4.65}$$

for all positive integers n. Subsets S of T_C satisfying (4.64) are the fixed points $(0, 0)$, $(\pm 1, 0)$ and the unstable manifold of the saddle point at the origin. Fig. 4.13(a) shows schematically how the first two images of T_C under $\mathbf{P}$ meet the above requirements. Moreover, it is a faithful topological representation of the numerical data shown in Fig. 4.13(b), where, for the parameter values used, $\mathbf{P}^2(AB)$ and $\mathbf{P}^2(CD)$ are pressed so close to the unstable manifold that all three curves appear to merge into one.

Subsequent images of T_C are formed in a similar way. At the nth application of $\mathbf{P}$ two small intervals of $\mathbf{P}^{n-1}(L_C)$ that span the stable manifold just above and below the saddle point are stretched along the unstable manifold and moved closer to it. The segments of $\mathbf{P}^{n-1}(L_C)$ joining the ends of the small intervals lie in the domains of attraction of the foci. Since their images must lie inside the $(n-1)$th image itself, they extend deeper in around the foci. As n increases to infinity, the images of the trapping boundary collapse onto the three fixed points and the unstable manifold connecting them. This is the attracting set of $\mathbf{P}$ that lies within T_C. It is concisely described as the closure of the unstable manifold of the saddle fixed point of $\mathbf{P}$.

Excluding the fixed points themselves, the orbit of any point in the above attracting set must lie wholly in one of two disjoint subsets, namely the two branches of the unstable manifold between the origin and the points $(\pm 1, 0)$.

Consequently, there is no point whose orbit is dense in the attracting set and it is therefore not an attractor. □

The system (4.58) and (4.59) is a special case of the Duffing equation. In Chapter 6, we show how the ideas developed in Example 4.3.3 are relevant to the formation of chaotic attracting sets in the Duffing problem.

4.4 FURTHER INTEGRALS

We discussed the value of first integrals in the construction of planar phase portraits in section 3.7. In that case a single scalar function, f, is constant on the trajectories of the system and the level curves of f coincide with the phase curves of the differential equation.

First integrals exist for differential equations defined on Euclidean phase spaces of higher dimension. A function $f_1 : D(\subseteq \mathbb{R}^n) \to \mathbb{R}$ is a first integral for the system $\dot{\mathbf{x}} = \mathbf{X}(\mathbf{x})$ on D if

$$\frac{\mathrm{d}f_1}{\mathrm{d}t} = \frac{\partial f_1}{\partial x_1} X_1 + \frac{\partial f_1}{\partial x_2} X_2 + \cdots + \frac{\partial f_1}{\partial x_n} X_n = 0, \tag{4.66}$$

for all $\mathbf{x} \in D$ (cf. (3.57)–(3.59)). Typically, however, for $n \geqslant 3$, the level sets defined by $f_1(\mathbf{x}) = C, C$ constant, only provide a constraint on the phase curves of the system. For example,

$$f_1(\mathbf{x}) = \tfrac{1}{2}(x_1^2 + x_2^2) \tag{4.67}$$

is a first integral for the linear system with coefficient matrix (4.19) but $\{(x_1, x_2, x_3) | \tfrac{1}{2}(x_1^2 + x_2^2) = C > 0\}$ is a cylinder in $\mathbb{R}^3$. The fact that $f_1(\phi_t(\mathbf{x}))$, $t \in \mathbb{R}$, is constant for each $\mathbf{x}$, means that each trajectory of the system is confined to one of a family of circular cylinders with axes along the x_3-axis of phase space.

If further integrals, $f_2, f_3, \ldots$, satisfying (4.66) can be found then it may be possible to use the intersection of the corresponding level sets to construct phase curves for the system. Assuming that the level sets of the first integral are $(n-1)$ dimensional and that each new integral reduces the dimension of the cumulative intersection of the level sets by unity, then $n-1$ integrals would be required to achieve this. However, more commonly, integrals are used to reduce the dimension of the differential equation that need be considered. For example, the linear system on $\mathbb{R}^4$ with coefficient matrix (4.22) has two integrals,

$$\begin{aligned} f_1(\mathbf{x}) &= \tfrac{1}{2}(x_1^2 + x_2^2), \\ f_2(\mathbf{x}) &= \tfrac{1}{2}(x_3^2 + x_4^2), \end{aligned} \tag{4.68}$$

which allow us to focus attention on the two angular equations in (4.23). This idea receives more systematic attention in the study of Hamiltonian systems.

4.4.1 Hamilton's equations

A system of differential equations on $\mathbb{R}^{2n}$ is said to be a **Hamiltonian system with n degrees of freedom** if it can be written in the form

$$\dot{q}_i = \frac{\partial H}{\partial p_i}, \qquad \dot{p}_i = -\frac{\partial H}{\partial q_i}, \tag{4.69}$$

for $i = 1, \ldots, n$, where $H(\mathbf{x}) = H(q_1, \ldots, q_n, p_1, \ldots, p_n)$ is a twice continuously differentiable function. The dimension of the phase space is $2n$ and the state vector $\mathbf{x}$ consists of n **generalized coordinates**, $\{q_i\}_{i=1}^n$, and n **generalized momenta**, $\{p_i\}_{i=1}^n$. The function, H, is called the **Hamiltonian** for the system and the equations (4.69) are referred to as **Hamilton's equations**.

Substitution of the vector field in (4.69) into (4.66) with f_1 replaced by H gives

$$\sum_{i=1}^{n} \left(\frac{\partial H}{\partial q_i} \frac{\partial H}{\partial p_i} - \frac{\partial H}{\partial p_i} \frac{\partial H}{\partial q_i} \right) \equiv 0. \tag{4.70}$$

Thus H is a (first) integral for (4.69), but how can further integrals be found?

Let us suppose we can find a change of coordinates,

$$(q_1, \ldots, q_n, p_1, \ldots, p_n) \mapsto (Q_1, \ldots, Q_n, P_1, \ldots, P_n)$$

with the following properties:

1. the transformed system of differential equations is a Hamiltonian system with Hamiltonian function $\tilde{H}$ given by
$$\tilde{H}(Q_1, \ldots, P_n) = H(q_1(Q_1, \ldots, P_n), \ldots, p_n(Q_1, \ldots, P_n)); \tag{4.71}$$
2. $\tilde{H}$ is independent of one of the new generalized coordinates, Q_n say.

Assuming that 1 and 2 are satisfied, the transformed differential equation for P_n is

$$\dot{P}_n = -\frac{\partial \tilde{H}}{\partial Q_n} = 0. \tag{4.72}$$

Hence P_n is a constant of the motion and takes a fixed value, I_n say, independent of t. Moreover, the equation for Q_n can be written

$$\dot{Q}_n = \left. \frac{\partial \tilde{H}}{\partial P_n} \right|_{P_n = I_n}. \tag{4.73}$$

Now the t-dependence of the right-hand side of (4.73) arises only from $Q_1, \ldots, Q_{n-1}$ and $P_1, \ldots, P_{n-1}$. Given that the remaining $2(n-1)$ equations (in which the constant value I_n appears as a parameter) can be solved for these variables, (4.73) can, in principle, be integrated to obtain $Q_n(t)$. It follows that a change of variables, with the properties 1 and 2, effectively allows us to reduce the dimension of the Hamiltonian system by two, focusing attention on the first $2(n-1)$ equations of the transformed Hamiltonian system.

Further reductions in dimension occur if $\tilde{H}$ is independent of more than one generalized coordinate, corresponding to the existence of more than one new integral. The maximum reduction of dimension that can be achieved in the above scheme occurs when $\tilde{H}$ is independent of all of the generalized coordinates $Q_1, \ldots, Q_n$. In this case

$$\dot{P}_i = -\frac{\partial \tilde{H}}{\partial Q_i} = 0, \tag{4.74}$$

$i = 1, \ldots, n$ and each P_i is a constant $(= I_i)$ of the motion or an integral of the system. Furthermore, for each i,

$$\begin{aligned} \dot{Q}_i &= \left.\frac{\partial \tilde{H}}{\partial P_i}\right|_{P_1 = I_1, \ldots, P_n = I_n}, \\ &= \omega_i(I_1, \ldots, I_n). \end{aligned} \tag{4.75}$$

The right-hand sides of the equations (4.75) are independent of t and they can be integrated to obtain

$$Q_i(t) = \omega_i t + C_i. \tag{4.76}$$

A system for which such a maximal reduction is possible is said to be **integrable** and the transformed Hamiltonian system

$$\dot{P}_i = 0, \qquad \dot{Q}_i = \omega_i(P_1, \ldots, P_n), \tag{4.77}$$

$i = 1, \ldots, n$, is referred to as its **normal form**. For recurrent systems (Arnold and Avez, 1968, pp. 210–14), the variables Q_i are cyclic and (4.77) consists of a collection of n subsystems of the type (2.51) with $\alpha = 0$, where P_i corresponds to the radial coordinate r and Q_i to the angular coordinate θ. Indeed the variables exhibiting the normal form are called **action–angle variables**. For each $i = 1, \ldots, n$, the action P_i takes the constant value I_i independent of time and the angle Q_i increases linearly with t at a constant rate $\omega_i(I_1, \ldots, I_n)$. In these terms, each subsystem exhibits the oscillatory behaviour appearing in (2.53) and an integrable Hamiltonian system can be likened to a collection of n independent oscillators.

It should be noted that $\tilde{H}(I_1, \ldots, I_n)$ is also a constant of the motion but it is obviously determined by the choice of the n actions $P_1, \ldots, P_n$. This highlights the fact that the action variables are not unique and alternative choices involving functions of $P_1, \ldots, P_n$ are possible. The important point is that there are only n independent integrals and if, for example, we wish to include $\tilde{H}$ as one of the actions then one of $P_1, \ldots, P_n$ must be discarded in such a way that the new set of n action variables are independent of one another.

Example 4.4.1
Consider the Hamiltonian system with Hamiltonian function

$$H(q_1, q_2, p_1, p_2) = \frac{1}{2}\sum_{i=1}^{2} (p_i^2 + \omega_i^2 q_i^2). \tag{4.78}$$

Let

$$\omega_i q_i = \rho_i \cos\theta_i, \qquad p_i = \rho_i \sin\theta_i, \tag{4.79}$$

$i = 1, 2$, and show that the transformed Hamiltonian, $\tilde{H}$, is independent of both θ_1 and θ_2.

Write down Hamilton's equations for the Hamiltonian H given in (4.78) and use (4.79) to obtain differential equations for $\rho_i, \theta_i, i = 1, 2$. Compare your result with Hamilton's equations for $\tilde{H}$. What conclusion can be drawn about the transformation (4.79)?

Solution
The transformed Hamiltonian, $\tilde{H}$, is given by using (4.78) and (4.79) in (4.71). Observe that

$$\begin{aligned}\tilde{H}(\theta_1, \theta_2, \rho_1, \rho_2) &= \frac{1}{2}\sum_{i=1}^{2} \rho_i^2(\sin^2\theta_i + \cos^2\theta_i),\\ &= \frac{1}{2}(\rho_1^2 + \rho_2^2),\end{aligned} \tag{4.80}$$

which is independent of both θ_1 and θ_2.

Hamilton's equations for $H(q_1, q_2, p_1, p_1)$ are given by substituting (4.78) into (4.69). We obtain

$$\begin{aligned}\dot{q}_i &= \frac{\partial H}{\partial p_i} = p_i,\\ \dot{p}_i &= -\frac{\partial H}{\partial q_i} = -\omega_i^2 q_i,\end{aligned} \tag{4.81}$$

Differentiation of (4.79) gives

$$\begin{bmatrix}\omega_i \dot{q}_i\\ \dot{p}_i\end{bmatrix} = \begin{bmatrix}-\rho_i \sin\theta_i & \cos\theta_i\\ \rho_i \cos\theta_i & \sin\theta_i\end{bmatrix}\begin{bmatrix}\dot{\theta}_i\\ \dot{\rho}_i\end{bmatrix}, \tag{4.82}$$

so that

$$\begin{bmatrix}\dot{\theta}_i\\ \dot{\rho}_i\end{bmatrix} = -\frac{1}{\rho_i}\begin{bmatrix}\sin\theta_i & -\cos\theta_i\\ -\rho_i\cos\theta_i & -\rho_i\sin\theta_i\end{bmatrix}\begin{bmatrix}\omega_i \dot{q}_i\\ \dot{p}_i\end{bmatrix}. \tag{4.83}$$

Hamilton's equations (4.81) together with the change of variable (4.79) give

$$\omega_i \dot{q}_i = \omega_i \rho_i \sin\theta_i, \qquad \dot{p}_i = -\omega_i \rho_i \cos\theta_i. \tag{4.84}$$

Thus

However, Hamilton's equations for $\tilde{H}$ take the form

$$\dot{\theta}_i = \frac{\partial \tilde{H}}{\partial \rho_i} = \rho_i, \qquad \dot{\rho}_i = -\frac{\partial \tilde{H}}{\partial \theta_i} = 0. \tag{4.86}$$

Comparison of equations (4.85) and (4.86) shows that the transformed Hamilton's equations for H are not Hamilton's equations for $\tilde{H}$. Therefore, while (4.79) is a transformation that eliminates θ_1 and θ_2 from the transformed Hamiltonian (cf. property 2 above), it fails to satisfy the requirement that the transformed equations be Hamilton's equations for the transformed Hamiltonian function (cf. property 1 above). □

A change of variable for which property 1 holds is called a **symplectic** or **canonical transformation**. The following definition, which provides an algebraic characterization of such a transformation, is based on the observation that Hamilton's equations can be expressed in the matrix form

$$\dot{\boldsymbol{x}}^{\mathrm{T}} = D H(\boldsymbol{x}) \boldsymbol{S}, \tag{4.87}$$

where

$$\dot{\boldsymbol{x}}^{\mathrm{T}} = [\dot{q}_1 \cdots \dot{p}_n], \tag{4.88}$$

$$DH(\boldsymbol{x}) = \left[\frac{\partial H}{\partial q_1} \cdots \frac{\partial H}{\partial p_n}\right] \tag{4.89}$$

and

$$\boldsymbol{S} = \begin{bmatrix} \boldsymbol{O}_n & -\boldsymbol{I}_n \\ \boldsymbol{I}_n & \boldsymbol{O}_n \end{bmatrix}. \tag{4.90}$$

In (4.90), $\boldsymbol{O}_n$ is the $n \times n$ null matrix and $\boldsymbol{I}_n$ is the $n \times n$ unit matrix. Let us take

$$\mathbf{y} = \mathbf{h}(\mathbf{x}), \tag{4.91}$$

where $\mathbf{y} \in \mathbb{R}^{2n}$ has components $Q_1, \ldots, P_n$, and focus attention on differentiable changes of coordinate with differentiable inverses so that $\mathbf{h}$ is a diffeomorphism.

Definition 4.4.1
A diffeomorphism $\mathbf{h}: U(\subseteq \mathbb{R}^{2n}) \to \mathbb{R}^{2n}$ is said to be **symplectic** if

$$[D\boldsymbol{h}(\boldsymbol{x})]^{\mathrm{T}} \boldsymbol{S} D\boldsymbol{h}(\boldsymbol{x}) = \boldsymbol{S}, \tag{4.92}$$

for all $\mathbf{x}$ in U.

The matrix $\boldsymbol{S}$ has some intriguing properties (explored in Exercise 4.27). Moreover the block structure of $\boldsymbol{S}$ can be used to simplify the process of checking (4.92) by expressing the matrix products in terms of $n \times n$ matrices. For instance, the task of verifying that the transformation (4.79) is not symplectic is facilitated by noting that the corresponding blocks in the derivative of (4.79) are 2×2 diagonal matrices (as shown in Exercise 4.28).

In a similar way, the alternative transformation

$$q_i = \left(\frac{2\tau_i}{\omega_i}\right)^{\frac{1}{2}} \cos\theta_i, \qquad p_i = -(2\tau_i\omega_i)^{\frac{1}{2}} \sin\theta_i, \tag{4.93}$$

can be shown to satisfy (4.92).

If (4.93) is used instead of (4.79) in Example 4.4.1. then (4.85) is replaced by

$$\begin{bmatrix} \dot{\theta}_i \\ \dot{\tau}_i \end{bmatrix} = \begin{bmatrix} -\dfrac{\omega_i}{\eta_i}\sin\theta_i & -\dfrac{1}{\eta_i}\cos\theta_i \\ \eta_i\cos\theta_i & -\dfrac{\eta_i}{\omega_i}\sin\theta_i \end{bmatrix} \begin{bmatrix} -\eta_i\sin\theta_i \\ -\omega_i\eta_i\cos\theta_i \end{bmatrix} = \begin{bmatrix} \omega_i \\ 0 \end{bmatrix}, \tag{4.94}$$

where $\eta_i = (2\omega_i\tau_i)^{\frac{1}{2}}$. In this case

$$\tilde{H}(\theta_1, \theta_2, \tau_1, \tau_2) = \omega_1\tau_1 + \omega_2\tau_2 \tag{4.95}$$

and Hamilton's equations are

$$\dot{\theta}_i = \frac{\partial \tilde{H}}{\partial \tau_i} = \omega_i, \qquad \dot{\tau}_i = -\frac{\partial \tilde{H}}{\partial \theta_i} = 0, \tag{4.96}$$

as required.

For each set of fixed values, $\{P_i = I_i\}_{i=1}^n$, for the actions, the flow of the normal form of an integrable Hamiltonian system is confined to an n-torus in the phase space defined by the action–angle variables. In other words, each set $\{I_1, \ldots, I_n\}$ defines an **invariant torus** for the flow of the normal form. For example, in (4.96) fixed values of the actions τ_1, τ_2 define the radii of a 2-torus the points of which are generated as θ_1, θ_2 are varied (cf. equation (4.23) and Fig. 4.4). What is more, the nature of the flow on the 2-torus depends on whether or not ω_1 and ω_2 are rationally independent (cf. Example 4.2.3).

4.4.2 Poincaré maps of Hamiltonian flows

It can be shown that all Hamiltonian systems with one degree of freedom and analytic Hamiltonian functions are integrable. Moreover, most of the systems with two or more degrees of freedom that are encountered in traditional courses on classical mechanics are integrable, for it is essentially only for such systems that closed-form solutions are available. However, integrability is not a typical property of Hamiltonian systems with more than one degree of freedom. In order to observe phenomena associated with non-integrability, we must consider systems with at least two degrees of freedom, which means we must cope with phase spaces of dimension four at least. The phase portraits of such systems are difficult to visualize and a lower-dimensional representation of the qualitative behaviour of the system must be sought.

Let us consider a system with two degrees of freedom that is a non-integrable perturbation of an integrable system which is in normal form with action–

angle variables (Q_1, Q_2, P_1, P_2). Given that the Hamiltonian function, $H(Q_1, Q_2, P_1, P_2)$, for the perturbed system is known, its trajectories are confined to three-dimensional solid tori on which H is constant. By focusing attention on one such torus, $H = C$ say, we can reduce the dimension of the phase space under consideration by unity. This is equivalent to eliminating one of the actions, say P_2, from the discussion. Since Q_2 is an angle-variable, the trajectories of the system on the solid torus repeatedly pass through the $Q_2 = 0$ plane as time increases. The Poincaré map, **P**, defined on this plane gives a two-dimensional geometrical representation of the dynamics on the solid torus. The following example illustrates these ideas by using the integrable problem considered in Example 4.4.1 for which explicit solutions can be written down. Notice that it is not essential for the system to be in normal form.

Example 4.4.2

Consider the Hamiltonian system discussed in Example 4.4.1. Given that (4.81) has solutions

$$\begin{aligned} q_i &= A_i \cos(\omega_i t + \eta_i), \\ p_i &= -\omega_i A_i \sin(\omega_i t + \eta_i), \end{aligned} \tag{4.97}$$

$i = 1, 2$, obtain the Poincaré map, **P**, defined on the $q_2 = 0$ plane. Show that **P** leaves the quantity $p_1^2 + \omega_1^2 q_1^2$ invariant and, hence or otherwise, describe the qualitative behaviour of **P**.

Solution

Assume that p_2 is to be determined by the condition $H = C$. Equation (4.97) shows that q_2 is periodic with period $2\pi/\omega_2$. Thus if a trajectory passes through the $q_2 = 0$ plane at $t = t_0$ then it first returns to that plane at $t = t_0 + 2\pi/\omega_2$. The corresponding values of q_1, p_1 are given by (4.97) and we conclude that

$$\begin{bmatrix} q_1(t_0 + 2\pi/\omega_2) \\ p_1(t_0 + 2\pi/\omega_2) \end{bmatrix} = \begin{bmatrix} A_1 \cos(\omega_1 t_0 + 2\pi\mu + \eta_1) \\ -\omega_1 A_1 \sin(\omega_1 t_0 + 2\pi\mu + \eta_1) \end{bmatrix}, \tag{4.98}$$

where $\mu = \omega_1/\omega_2$. The right-hand side of (4.98) can be rewritten in terms of $q_1(t_0)$ and $p_1(t_0)$ to yield

$$\begin{bmatrix} \cos(2\pi\mu) & \omega_1^{-1}\sin(2\pi\mu) \\ -\omega_1 \sin(2\pi\mu) & \cos(2\pi\mu) \end{bmatrix} \begin{bmatrix} q_1(t_0) \\ p_1(t_0) \end{bmatrix} = \boldsymbol{P} \begin{bmatrix} q_1(t_0) \\ p_1(t_0) \end{bmatrix}, \tag{4.99}$$

which gives an explicit expression for **P**.

Let the right-hand side of (4.98) be $[\tilde{q}_1, \tilde{p}_1]^{\mathrm{T}}$, then

$$\tilde{p}_1^2 + \omega_1^2 \tilde{q}_1^2 = (\omega_1 A_1)^2 = p_1(t_0)^2 + \omega_1^2 q_1(t_0)^2. \tag{4.100}$$

Therefore $p_1^2 + \omega_1^2 q_1^2$ is invariant under **P**. The equation

$$p_1^2 + \omega_1^2 q_1^2 = C_1, \tag{4.101}$$

where $0 < C_1 < 2C$, represents an ellipse centred on the origin in the

(q_1, p_1)-plane. The matrix form for $\mathbf{P}$ in (4.99) rotates the points of such an ellipse about the origin by moving them along the curve itself. □

Hamiltonian flows have the property that they preserve volumes in phase space (**Liouville's theorem**). This result follows from the special form taken by Hamilton's equations. Let ϕ_t be the flow of the differential equation $\dot{\mathbf{x}} = \mathbf{X}(\mathbf{x})$ defined on $\mathbb{R}^m$. A region R of phase space defined at $t = 0$ has image $\phi_t(R)$ at time t. Its volume is then given by

$$\Omega(t) = \int_{\phi_t(R)} \mathrm{d}x_1 \cdots \mathrm{d}x_m, \tag{4.102}$$

where $x_1, \ldots, x_m$ are Cartesian coordinates. A short time h later we have

$$\begin{aligned} \Omega(t+h) &= \int_{\phi_{t+h}(R)} \mathrm{d}x'_1 \cdots \mathrm{d}x'_m, \\ &= \int_{\phi_h(\phi_t(R))} \mathrm{d}x'_1 \cdots \mathrm{d}x'_m, \\ &= \int_{\phi_t(R)} |J| \mathrm{d}x_1 \cdots \mathrm{d}x_m, \end{aligned} \tag{4.103}$$

where J is the Jacobian of the coordinate transformation $\mathbf{x}' = \phi_h(\mathbf{x})$. For small h,

$$\phi_h(\mathbf{x}) = \mathbf{x} + h\mathbf{X}(\mathbf{x}) + O(h^2) \tag{4.104}$$

and (cf. (4.10))

$$D\phi_h(\boldsymbol{x}) = \boldsymbol{I} + hD\boldsymbol{X}(\boldsymbol{x}) + O(h^2). \tag{4.105}$$

It follows (from Exercise 4.30) that

$$\det(D\phi_h(\boldsymbol{x})) = 1 + h\,\mathrm{tr}(D\boldsymbol{X}(\boldsymbol{x})) + O(h^2), \tag{4.106}$$

and, for sufficiently small h, $|J|$ is given by (4.106). Therefore,

$$\Omega(t+h) = \Omega(t) + h \int_{\phi_t(R)} \mathrm{tr}(D\boldsymbol{X}(\boldsymbol{x}))\mathrm{d}x_1 \cdots \mathrm{d}x_m + O(h^2)$$

and

$$\dot{\Omega}(t) = \int_{\phi_t(R)} \mathrm{tr}(D\boldsymbol{X}(\boldsymbol{x}))\mathrm{d}x_1 \cdots \mathrm{d}x_m. \tag{4.107}$$

When $\mathbf{X}$ is given by (4.69), $m = 2n$, $(x_1 \cdots x_m) = (q_1, \ldots q_n, p_1, \ldots p_n)$ and

$$\begin{aligned} \mathrm{tr}\,(D\boldsymbol{X}(\boldsymbol{x})) &= \sum_{i=1}^{m} \frac{\partial X_i}{\partial x_i}, \\ &= \sum_{j=1}^{m} \left\{ \frac{\partial}{\partial q_j}\left(\frac{\partial H}{\partial p_j}\right) + \frac{\partial}{\partial p_j}\left(-\frac{\partial H}{\partial q_j}\right) \right\} \equiv 0. \end{aligned} \tag{4.108}$$

Thus $\dot{\Omega}(t) \equiv 0$ and Ω is independent of t.

It can also be shown that the Poincaré map of a recurrent Hamiltonian system with two degrees of freedom, constructed in the manner of Example 4.4.2, preserves areas in the surface of the section on which it is defined. This result is valid for both integrable and non-integrable systems and is an important constraint on the qualitative phenomena that can arise from non-integrability (see section 6.7). A diffeomorphism $\mathbf{Q}$ is area-preserving if $|\det(D\boldsymbol{Q}(\boldsymbol{x}))| \equiv 1$. If $\mathscr{C}$ is a closed curve in the domain of $\mathbf{Q}$, then the area lying within its image, $\mathbf{Q}(\mathscr{C})$, is

$$\int_{\mathbf{Q}(\mathscr{C})} \mathrm{d}x'\mathrm{d}y' = \int_{\mathscr{C}} |\det(D\boldsymbol{Q}(\boldsymbol{x}))|\mathrm{d}x\mathrm{d}y = \int_{\mathscr{C}} \mathrm{d}x\mathrm{d}y, \tag{4.109}$$

which is the area within $\mathscr{C}$. If we let $\mathbf{Q}$ be the linear Poincaré map in Example 4.4.2, then $\boldsymbol{Q}(\boldsymbol{x}) = \boldsymbol{P}\boldsymbol{x}$, where $\boldsymbol{P}$ is the matrix in (4.99). We find $D\boldsymbol{Q}(\boldsymbol{x}) = \boldsymbol{P}$ and $\det \boldsymbol{P} = 1$ as required.

EXERCISES

Section 4.1

4.1 Find the eigenvalues of the system $\dot{\boldsymbol{x}} = \boldsymbol{A}\boldsymbol{x}$ for the following matrices $\boldsymbol{A}$:

$$\text{(a)} \begin{bmatrix} -1 & 2 & 1 \\ 1 & -2 & 3 \\ 3 & -2 & 1 \end{bmatrix}; \qquad \text{(b)} \begin{bmatrix} 5 & 1 & 2 \\ 3 & 3 & 2 \\ 3 & -1 & 6 \end{bmatrix}.$$

Write down the equivalent system given by (4.1) for each case.

4.2 Consider the linear system

$$\dot{\boldsymbol{x}} = \begin{bmatrix} \alpha & -\beta & 0 \\ \beta & \alpha & 0 \\ 0 & 0 & \lambda \end{bmatrix} \boldsymbol{x},$$

where λ, α are non-zero and β is positive. Explain why this system can exhibit four distinct types of hyperbolic behaviour. Sketch a typical phase portrait for each type and indicate the region of the $\alpha\lambda$-plane in which it occurs.

4.3 Find the topological type of the fixed point $\boldsymbol{x} = \boldsymbol{0}$ for the system $\dot{\boldsymbol{x}} = \boldsymbol{A}\boldsymbol{x}$, where

$$\boldsymbol{A} = \begin{bmatrix} 0 & 0 & 1 \\ 1 & -1 & 1 \\ 1 & 0 & 0 \end{bmatrix}.$$

Find the eivenvectors of $\boldsymbol{A}$ and construct a conjugating matrix $\boldsymbol{M}$ such that $\boldsymbol{M}^{-1}\boldsymbol{A}\boldsymbol{M} = \boldsymbol{D}$ is a diagonal matrix. Hence sketch the phase portrait of $\dot{\boldsymbol{x}} = \boldsymbol{A}\boldsymbol{x}$ in $\boldsymbol{x}$-space.

4.4 Find the fixed points of the system

$$\dot{x} = x - x^2 - xz, \qquad \dot{y} = y - z, \qquad \dot{z} = z - z^2.$$

Use the linearization theorem, when it is applicable, to determine the topological type of each fixed point.

4.5 Verify that the Lorenz equations (4.11) with $r > 1$ have non-trivial fixed points at

$$(x, y, z) = (\pm\sqrt{b(r-1)}, \pm\sqrt{b(r-1)}, r-1).$$

Prove that the linearized systems at both of these fixed points have eigenvalues satisfying the cubic equation

$$f(\lambda) = \lambda^3 + (1 + b + \sigma)\lambda^2 + b(\sigma + r)\lambda + 2\sigma b(r-1) = 0$$

Assume that this equation has only one real root $\lambda = \lambda_1 < 0$. Show that the non-trivial fixed points are spiral saddles if $|\lambda_1| > (1 + b + \sigma)$.

4.6 The **centre eigenspace**, E^c, of a non-hyperbolic, linear system $\dot{\boldsymbol{x}} = \boldsymbol{A}\boldsymbol{x}$ is the eigenspace of $\boldsymbol{A}$ associated with those eigenvalues that have zero real part. Obtain E^c for the systems:

(a) $\dot{x} = 0, \quad \dot{y} = -y, \quad \dot{z} = -z;$
(b) $\dot{x} = -y, \quad \dot{y} = x, \quad \dot{z} = -z;$
(c) $\dot{x} = -x, \quad \dot{y} = 0, \quad \dot{z} = z;$
(d) $\dot{x} = y, \quad \dot{y} = 0, \quad \dot{z} = -z.$

Sketch phase portraits of (a)–(d) paying particular attention to the manner in which trajectories of points not on E^c approach E^c as $t \to \pm\infty$. Comment on the role played by E^c in these phase portraits.

4.7 Show that the fixed point at the origin of each of the following systems is non-hyperbolic:

(a) $\dot{x} = x^2, \quad \dot{y} = -y, \quad \dot{z} = -z;$
(b) $\dot{x} = yz, \quad \dot{y} = xz, \quad \dot{z} = xy.$

Sketch phase portraits for both systems and compare each one with the phase portrait of the linearized system at the origin.

4.8 Consider the system (4.23) with $\beta_1 = 2$ and $\beta_2 = 1$. Let the circular coordinates θ_1, θ_2 be represented by a square of side 2π with opposite sides identified. What surface is obtained in this way? Exhibit the solution curves (4.24) on the θ_1, θ_2-square and describe their behaviour on the surface. Explain why the values of $C_1 > C_2 > 0$ in (4.24) are not important from a qualitative point of view.

Section 4.2

4.9 Locate the positions of the eigenvalues in the complex plane and list the dimensions of the stable and unstable eigenspaces of each of the following hyperbolic linear diffeomorphisms on $\mathbb{R}^2$:

(a) $\begin{bmatrix} 2 & 3 \\ -1 & 1 \end{bmatrix}$; (b) $\begin{bmatrix} 7 & 5 \\ 1 & 1 \end{bmatrix}$;

(c) $\begin{bmatrix} 2 & 0 \\ 0 & -2 \end{bmatrix}$; (d) $\begin{bmatrix} 1 & 2 \\ -1 & -3 \end{bmatrix}$;

(e) $\begin{bmatrix} 1/3 & -1/2 \\ 1/2 & 1/3 \end{bmatrix}$; (f) $\begin{bmatrix} -4 & 1 \\ -2 & -1 \end{bmatrix}$;

(g) $\begin{bmatrix} 1 & 1 \\ -3/4 & -1 \end{bmatrix}$; (h) $\begin{bmatrix} -3 & 0 \\ 0 & 1/2 \end{bmatrix}$.

Determine which of the matrices (a)–(h) are order-preserving and arrange them in groups of the same topological type.

4.10 Find bases for the stable and unstable eigenspaces of the hyperbolic linear diffeomorphism **A** represented by

$$A = \begin{bmatrix} 0 & 2 & -1 \\ 1 & 1 & -1 \\ 2 & -1 & 2 \end{bmatrix}$$

Describe the qualitative behaviour of $\mathbf{A}|E^s$, $\mathbf{A}|E^u$ and **A**.

4.11 Consider the set, $\mathscr{S}$, of all diffeomorphisms defined on $\mathbb{R}^n$. Let **P** and **Q** be two such diffeomorphisms and write $\mathbf{P} \sim \mathbf{Q}$ if **P** is topologically conjugate to **Q**. Show that:

(a) $\mathbf{P} \sim \mathbf{Q}$ implies $\mathbf{Q} \sim \mathbf{P}$; (b) $\mathbf{P} \sim \mathbf{Q}$ and $\mathbf{Q} \sim \mathbf{R}$ implies $\mathbf{P} \sim \mathbf{R}$.

What is the significance of these results for the set $\mathscr{S}$?

4.12 Use Theorem 4.2.1 to find the number of distinct topological types of hyperbolic linear diffeomorphism on: (a) $\mathbb{R}$; (b) $\mathbb{R}^2$; (c) $\mathbb{R}^m$.

4.13 Let $\boldsymbol{B} = \exp(\boldsymbol{A}t)$ be the evolution operator of the hyperbolic linear system $\dot{\boldsymbol{x}} = \boldsymbol{A}\boldsymbol{x}$. If λ_j is an eigenvalue of $\boldsymbol{A}$, show that $\exp(\lambda_j t)$ is an eigenvalue of $\boldsymbol{B}$. Hence prove that:

(a) $\boldsymbol{B}$ is orientation-preserving for all real t;
(b) $\boldsymbol{B}$ is a hyperbolic linear diffeomorphism for $t \neq 0$.

Why is $t = 0$ excluded from (b)?

4.14 Draw a diagram (cf. Fig. 4.8) illustrating the nature of the orbits of the following hyperbolic linear diffeomorphisms near the origin:

(a) $\begin{bmatrix} 4 & 1 \\ -2 & 1 \end{bmatrix}$; (b) $\begin{bmatrix} 2 & 2 \\ 3 & -3 \end{bmatrix}$;

(c) $\begin{bmatrix} 1/4 & -1 \\ 1/2 & 3/4 \end{bmatrix}$; (d) $\begin{bmatrix} -3/2 & -1 \\ 2 & 3/2 \end{bmatrix}$.

Explain why all 2×2 real matrices with complex eigenvalues are orientation-preserving.

Confirm that the examples (a)–(d), together with those shown in Fig. 4.8, exhaust the possible types of hyperbolic linear diffeomorphisms on $\mathbb{R}^2$.

4.15 Consider the linear diffeomorphism defined by

$$x_1 = \lambda_1 x, \qquad y_1 = \lambda_2 y.$$

Find the value of α for which this mapping has invariant curves of the form $y = ax^\alpha$. Hence sketch invariant curves for the linear diffeomorphism $DP(1, 0)$ in (4.41) for: (a) $\lambda < -1$; (b) $\lambda = -1$; (c) $-1 < \lambda < 0$; (d) $\lambda > 0$.

4.16 Let $\mathbf{x}^*$ be a periodic point of period q of the diffeomorphism $\mathbf{P}$. Show that $\mathbf{P}(\mathbf{x}^*), \mathbf{P}^2(\mathbf{x}^*), \ldots, \mathbf{P}^{q-1}(\mathbf{x}^*)$ are fixed points of $\mathbf{P}^q$ of the same topological type as $\mathbf{x}^*$.

Let ϕ_t be the flow with phase portrait shown below.

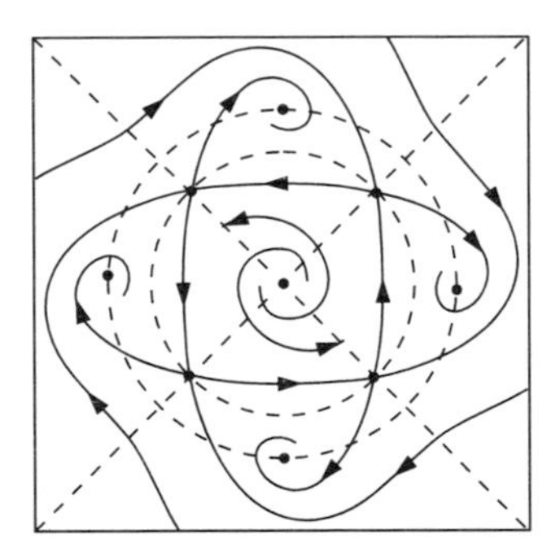

Define $\mathbf{Q}(x, y) = \phi_{2\pi}(\mathbf{R}_{\pi/2}(x, y))$, where $\mathbf{R}_{\pi/2}$ is an anticlockwise rotation through an angle of $\pi/2$ about the origin. Show that $\boldsymbol{Q}$ has: (a) a stable; (b) a saddle-type; four-cycle. Copy the phase portrait of ϕ_t and label the stable and unstable manifolds of both cycles. Comment on the resulting diagram.

Section 4.3

4.17 Consider the flow $\dot{r} = 0$, $\dot{\theta} = 1$ on the cylinder $\{(r, \theta) | -1 \leqslant r \leqslant 1, 0 \leqslant \theta < 2\pi\}$. Obtain the Poincaré map $P_1(r)$ defined on the section $\theta = \theta_0$.

Suppose the cylinder is cut along the generator with $\theta = 0$, one of the cut edges is rotated through π radians and then the edges are glued together (thereby identifying $(r, 0)$ with $(-r, 2\pi)$) to form a Möbius strip. Interpreting (r, θ) as coordinates on this strip, obtain the Poincaré map $P_2(r)$ of $\dot{r} = 0, \dot{\theta} = 1$ on any section on which θ is constant. Could $P_2(r)$ be the Poincaré map of a flow on the cylinder? Which of the following transformations could be the Poincaré map of a flow on the solid torus:

(a) $(r, \theta) \mapsto (r, (\theta + \alpha) \bmod 2\pi)$; (b) $(r, \theta) \mapsto (r, (-\theta) \bmod 2\pi)$.

4.18 Use the Poincaré map $P(x)$ obtained in Exercise 3.36, to show that the circle $r = 1$ is an attracting set for the flow of

$$\dot{r} = r(1 - r), \qquad \dot{\theta} = 1.$$

4.19 Verify that $\mathbf{P}(x, y) = (\lambda x, \lambda y)$ is a diffeomorphism of the x, y-plane for any non-zero λ. For each of the following closed sets $\mathscr{S}$, identify $\text{int}(\mathscr{S})$ and verify that $\mathbf{P}(\text{int}(\mathscr{S})) = \text{int}(\mathbf{P}(\mathscr{S}))$:

(a) $\{(x, y) | x^2 + y^2 \leqslant 1\}$;
(b) $\{(x, y) | 1 \leqslant x^2 + y^2 \leqslant 2\}$;
(c) $\{(x, y) | (x - 2)^2 + y^2 \leqslant 1\}$.

Under what circumstances does one of the above sets $\mathscr{S}$ form a trapping region T for $\mathbf{P}$? Find $\mathbf{P}^n(T)$ for any positive integer n, confirm (4.51) and obtain the attracting set given by (4.53).

4.20 Let $\mathbf{P}(x, y)$ be a planar diffeomorphism and let T be a closed, connected region in the plane whose boundary is a simple closed curve. Given that T is a trapping region for $\mathbf{P}$, which of the following statements is true:

(a) if (x, y) is boundary point of T then $\mathbf{P}(x, y)$ must be a boundary point of $\mathbf{P}(T)$;
(b) if (x, y) is a boundary point of T then $\mathbf{P}(x, y)$ is a boundary point of T;
(c) if (x, y) is an interior point of T then $\mathbf{P}(x, y)$ is an interior point of $\mathbf{P}(T)$;
(d) the image of the boundary of T is not necessarily a simple closed curve;
(e) $\mathbf{P}(T)$ must have non-empty interior;
(f) $\bigcap_{n=1}^{\infty} \mathbf{P}^n(T)$ must have non-empty interior.

Explain your answers.

4.21 Consider the function $H(x, y)$ given in (4.60). Sketch representatives of the level sets given by $H(x, y) = C$ when $C < 0$. Why is such a set not the boundary of a trapping region for the flow, ϕ_t, of (4.58) with $\delta > 0$? Obtain trapping regions for ϕ_t from the level sets sketched above and state whether or not the attracting sets lying within them are attractors?

4.22 Verify that (4.58) has fixed points at $(x, y) = (0, 0), (\pm 1, 0)$ for all $\delta > 0$. Use the linearization theorem to show that the non-trivial fixed points are: (a) foci for $0 < \delta < 2\sqrt{2}$; (b) nodes for $\delta > 2\sqrt{2}$.

4.23 Let $\mathbf{P}$ be a diffeomorphism defined on the plane and suppose $S \subseteq T \subseteq \mathbb{R}^2$. Show that $\mathbf{P}(S) \subseteq \mathbf{P}(T)$.

Equation (4.64) holds if and only if $\mathbf{P}(S) \subseteq S$ and $\mathbf{P}^{-1}(S) \subseteq S$, i.e. if S is invariant under both $\mathbf{P}$ and $\mathbf{P}^{-1}$. For the problem considered in Example 4.3.3, show that (4.64) holds if S is taken to be any phase curve of the planar system (4.58). Explain why the fixed points and the unstable manifold of the saddle point at the origin are the only such sets to lie in all iterates of the trapping region.

Section 4.4

4.24 A non-linear system defined on $\mathbb{R}^4$, with Cartesian coordinates x_1, x_2, x_3, x_4 takes the form

$$\dot{r}_1 = 0, \qquad \dot{\theta}_1 = 1,$$

$$\dot{r}_2 = 5r_2^4 \cos 5\theta_2, \qquad \dot{\theta}_2 = 2\mu - 4r_2^2 - 5r_2^3 \sin 5\theta_2,$$

where $x_1 = r_1 \cos\theta_1, x_2 = r_1 \sin\theta_1, x_3 = r_2 \cos\theta_2, x_4 = r_2 \sin\theta_2$ and $0 < \mu \ll \frac{1}{2}$. Confirm that

$$f_1(r_1, \theta_1, r_2, \theta_2) = \frac{1}{2} r_1^2$$

is a constant of the motion for the system and show that trajectories satisfying $f_1 = C_1 > 0$ are confined to a solid torus embedded in $\mathbb{R}^4$. Verify that

$$f_2(r_1, \theta_1, r_2, \theta_2) = -\mu r_2^2 + r_2^4 + r_2^5 \sin 5\theta_2$$

is another integral for the system and use the results obtained in Exercise 3.34 to describe how f_2 further constrains the trajectories lying inside the solid torus when $0 \leqslant r_2 < \frac{1}{2}$.

4.25 Write down Hamilton's equations (4.69) for a Hamiltonian system with one degree of freedom. Define plane polar coordinates (r, θ) on the q, p-plane by $q = r\cos\theta, p = r\sin\theta$ and obtain the transformed system equations.

Use the Hamiltonian $H(q, p) = \frac{1}{2}(q^2 + p^2)$ to verify that the transformed system equations are not the same as Hamilton's equations for the equations for the transformed Hamiltonian. Show that plane polar coordinates fail to satisfy the condition (4.92).

4.26 Repeat Exercise 4.25 using canonical polar coordinates defined by $q = \sqrt{2\tau}\cos\theta, p = \sqrt{2\tau}\sin\theta$. Show that these coordinates satisfy the symplectic condition (4.92).

4.27 (a) Verify that the matrix $\mathbf{S}$ defined in (4.90) satisfies $\boldsymbol{S}^{\mathrm{T}} = -\boldsymbol{S} = \boldsymbol{S}^{-1}$ and prove that $\det(\boldsymbol{S}) = 1$.

(b) Let $\mathbf{h}: \mathbb{R}^{2n} \to \mathbb{R}^{2n}$ be a symplectic diffeomorphism. Prove that:

(i) $\mathbf{h}$ preserves volumes in $\mathbb{R}^{2n}$;

(ii) $\mathbf{h}^{-1}$ is symplectic;

(iii) $[D\boldsymbol{h}(\boldsymbol{x})]^{-1} = \boldsymbol{S}^T [D\boldsymbol{h}(\boldsymbol{x})]^T \boldsymbol{S}$.

4.28 Consider Definition 4.4.1 when $\mathbf{h}$: $U(\subseteq \mathbb{R}^4) \to \mathbb{R}^4$. Given that

$$D\boldsymbol{h}(\boldsymbol{x}) = \begin{bmatrix} \boldsymbol{A} & \boldsymbol{B} \\ \boldsymbol{C} & \boldsymbol{D} \end{bmatrix},$$

where $A, \ldots, D$ are $n \times n$ matrices, show that (4.92) can be written as

$$S = \begin{bmatrix} C^{\mathrm{T}}A - A^{\mathrm{T}}C & C^{\mathrm{T}}B - A^{\mathrm{T}}D \\ D^{\mathrm{T}}A - B^{\mathrm{T}}C & D^{\mathrm{T}}B - B^{\mathrm{T}}D \end{bmatrix}.$$

Hence show that (4.93) is symplectic, whilst (4.79) is not.

4.29 Solve Hamilton's equations (4.96) and repeat the calculation of Example 4.4.2 using these solutions. In particular, assume τ_2 is fixed and obtain the Poincaré map $\mathbf{P}$ defined on the $\theta_2 = 0$ plane. Characterize the invariant curves of $\mathbf{P}$ and describe its qualitative behaviour.

4.30 If A is a real, $n \times n$ matrix with distinct eigenvalues $\lambda_i, i = 1, \ldots, n$, then:

(a) $\mathrm{tr}(A) = \sum_i \lambda_i$; (b) $\det(A) = \prod_i \lambda_i$.

Use these results to show that

$$\det(D\phi_h(x)) = 1 + h\,\mathrm{tr}(DX(x)) + O(h^2),$$

where $D\phi_h(x)$ is given by (4.105).

4.31 Consider the area-preserving linear transformation $\mathbf{P}$ of the plane given by Px, where

$$x = \begin{bmatrix} x \\ y \end{bmatrix} \text{ and } P = \begin{bmatrix} \lambda & 0 \\ 0 & \lambda^{-1} \end{bmatrix}$$

with $\lambda > 1$. Draw a diagram showing the images under $\mathbf{P}$ and $\mathbf{P}^{-1}$ of the square $S = \{(x, y) | 1 \leqslant x, y \leqslant 2\}$. Verify explicitly that the areas within $\mathbf{P}(S)$ and $\mathbf{P}^{-1}(S)$ are both unity. Prove that $\mathbf{P}^n$, n integer, also preserves area and describe how $\mathbf{P}^n(S)$ varies as: (a) $n \to \infty$; (b) $n \to -\infty$.

4.32 Show that a transformation $\mathbf{h}: \mathbb{R}^2 \to \mathbb{R}^2$ is symplectic if and only if it is both area- and orientation-preserving.

5

Applications I: planar phase spaces

The results of Chapters 2 and 3 play an important role in the construction and analysis of models of real time-dependent systems. In this chapter, we illustrate these applications by presenting a number of models from several different areas of study.

In each model the vector $\mathbf{x}(t)$ gives the state of the system at time t (cf. section 1.2.2). The time development of the states is governed by the **dynamical equations** (or equations of motion) $\dot{\mathbf{x}} = \mathbf{X}(\mathbf{x})$ and the qualitative behaviour of the evolution of the states is given by the corresponding phase portrait.

5.1 LINEAR MODELS

A model is said to be linear if it has linear dynamical equations. As we have seen in Chapter 2, such equations only show certain kinds of qualitative behaviour. For example, in the plane the qualitative behaviour is confined to the classification given in section 2.4.

5.1.1 A mechanical oscillator

Consider a mass m supported on a vertically mounted spring as shown in Fig. 5.1. The mass is constrained to move only along the axis of the spring. The mass is also attached to a piston that moves in a cylinder of fluid contained in the spring. The piston resists any motion of the mass. This arrangement is an idealization of the shock absorbers that can be seen on most motorcycles.

Let x be the displacement of the mass m below its equilibrium position at rest on the spring. Assume that:

1. the spring obeys Hooke's law so that it exerts a restoring force of $Kx (K > 0)$ on the mass;
2. the force exerted by the piston, which opposes the motion of the mass, is $2k (k \geqslant 0)$ times the momentum p of the mass.

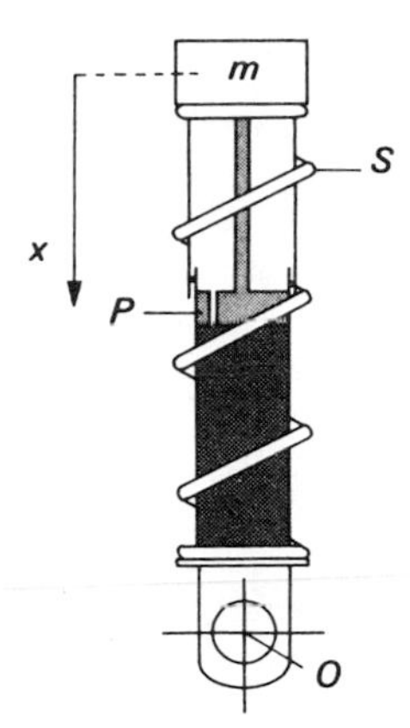

Fig. 5.1. Mass m supported on a vertically mounted coiled spring S rigidly fixed at its lower end O. The mass is also attached to the piston P. P moves in a fluid-filled cylinder and this impedes the motion of m.

With these assumptions, the dynamical equations which model the motion of the mass are linear. They can be constructed by observing that:

1. the momentum p of the mass is given by $m\dot{x} = p$;
2. the rate of change of linear momentum of the mass is equal to the force applied to it (i.e. Newton's second law of motion).

The second observation gives

$$p = -K(l + x) - 2kp + mg, \tag{5.1}$$

where l is the compression of the spring at equilibrium. However, at equilibrium $\dot{p} = p = x = 0$, so $Kl = mg$ and (5.1) can be written as

$$\dot{p} = m\ddot{x} = -Kx - 2km\dot{x}.$$

Thus, the position of the mass satisfies the linear second-order equation

$$\ddot{x} + 2k\dot{x} + \omega_0^2 x = 0, \tag{5.2}$$

where $\omega_0^2 = K/m > 0$ and $k \geqslant 0$. An equivalent first-order system is given by putting $x_1 = x$ and $x_2 = \dot{x}$ (cf. Exercise 1.28) to obtain

$$\dot{x}_1 = x_2, \qquad \dot{x}_2 = -\omega_0^2 x_1 - 2kx_2. \tag{5.3}$$

The linear system (5.3) has coefficient matrix

$$A = \begin{bmatrix} 0 & 1 \\ -\omega_0^2 & -2k \end{bmatrix}, \tag{5.4}$$

with $\mathrm{tr}(A) = -2k \leqslant 0$ and $\det(A) = \omega_0^2 > 0$. It follows that the fixed point at the origin of the phase portrait of (5.3) is always stable (asymptotically for $k > 0$). As Fig. 5.2 shows, for each fixed value of ω_0^2 (i.e. the spring constant

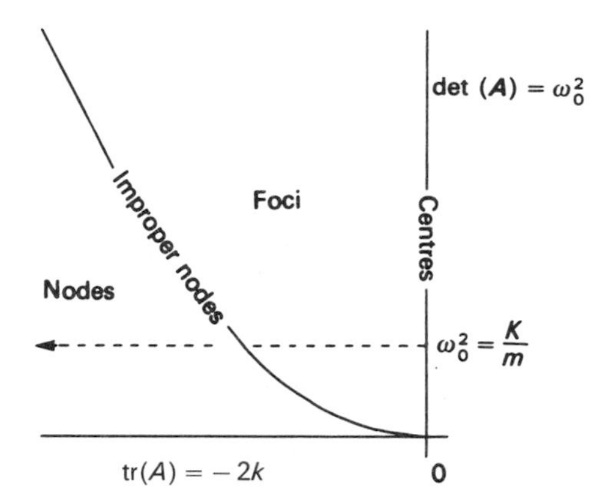

Fig. 5.2. Quadrant of the tr(A)–det(A) plane reached by A in (5.4); tr$(A) \leqslant 0$ implies that all phase portraits are stable (cf. Fig. 2.7).

K), the phase portrait passes through the sequence: centre, foci, improper nodes, nodes, as k increases through the interval $0 \leqslant k < \infty$.

Each of the above types of phase portrait corresponds to a qualitatively different motion of the mass m. There are four cases:

(a) $k = 0$

The eigenvalues of A are $\lambda_1 = -\lambda_2 = i\omega_0$. The corresponding canonical system $\dot{y} = Jy$ has solution (2.53) with $\beta = \omega_0$ and $\dot{x} = Ax$ has trajectories

$$(x_1(t), x_2(t)) = (R\cos(\omega_0 t + \theta), -\omega_0 R \sin(\omega_0 t + \theta)), \tag{5.5}$$

(as shown in Exercise 5.1). Notice that for this problem $x_2(t) = \dot{x}_1(t)$. The phase portrait is therefore as shown in Fig. 5.3. The relationship between the oscillations in x_1 and the trajectories is also illustrated in this figure.

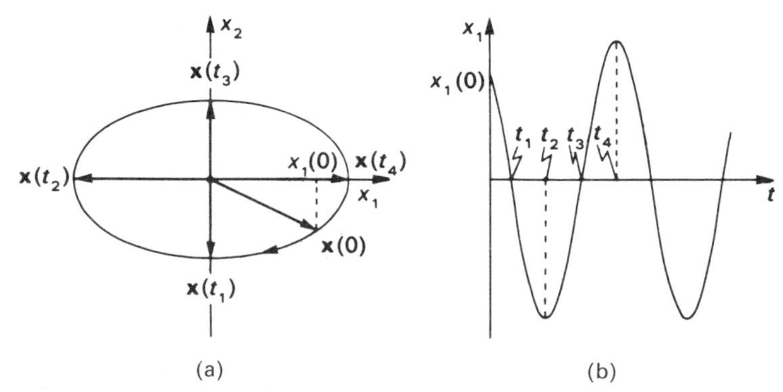

Fig. 5.3. (a) Phase portrait for $\dot{x}_1 = x_2$, $\dot{x}_2 = -\omega_0^2 x_1$ consists of a continuum of concentric ellipses. The trajectory (5.5) through $\mathbf{x}(0) = (R\cos\theta, -\omega_0 R \sin\theta)$ corresponds to the oscillations in x_1 shown in (b).

The motion of the mass consequently consists of persistent oscillations of both position and velocity with the same period $T_0 = 2\pi/\omega_0$. The mass is said to execute **free** (no external applied forces), **undamped** ($k = 0$) oscillations at the **natural frequency** $\omega_0 = (K/m)^{1/2}$ of the system.

(*b*) $0 < k < \omega_0$
The eigenvalues of $\boldsymbol{A}$ are now $-k \pm \mathrm{i}(\omega_0^2 - k^2)^{1/2}$ and (2.52) implies that

$$x_1(t) = Re^{-kt}\cos(\beta t + \theta), \tag{5.6}$$

where $\beta = (\omega_0^2 - k^2)^{1/2}$ (as shown in Exercise 5.1).
The motion is, therefore, modified in two ways:

1. the amplitude of the oscillation decays with increasing t; and
2. the period of the oscillations is $T = 2\pi/\beta > T_0$.

The system is said to be **underdamped** and to undergo **damped**, free oscillations. These oscillations, and the corresponding phase portrait, are shown in Fig. 5.4.

(*c*) $k = \omega_0$
As $k \to \omega_{0-}$ the period $T = 2\pi/(\omega_0^2 - k^2)^{1/2}$ increases indefinitely and finally at $k = \omega_0$ the oscillations disappear. The eigenvalues of $\boldsymbol{A}$ are real and both equal to $-k$. We have reached the line of stable improper nodes in Fig. 5.2. The system is said to be **critically damped**, in so much as the oscillations of the underdamped ($0 < k < \omega_0$) system have just disappeared.
In this case we have from (2.48)

$$\mathbf{x}(t) = \mathrm{c}^{-kt}(a + bt, b - k(a + bt)), \tag{5.7}$$

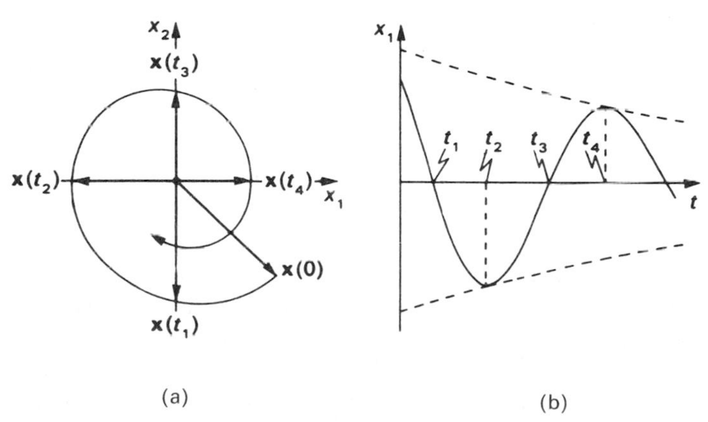

Fig. 5.4. Phase portrait, (a), and $x_1(t)$ versus t, (b), for $\dot{x}_1 = x_2$, $\dot{x}_2 = -\omega_0^2 x_1 - 2kx_2$, $0 < k < \omega_0$. The phase portrait is a stable focus and the envelope of the oscillations—shown dashed in (b)—is Re^{-kt}.

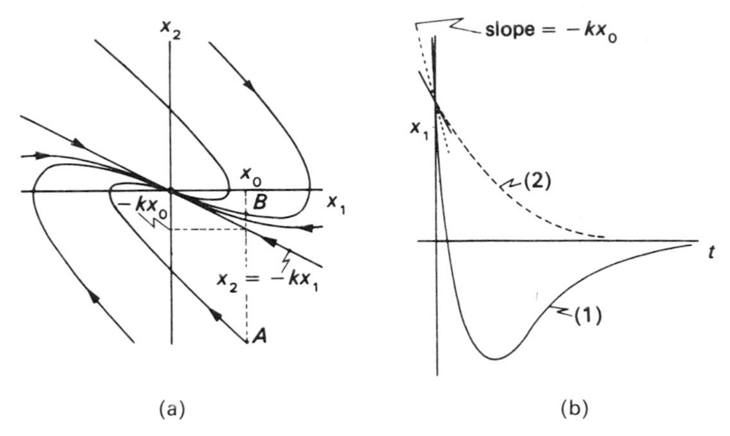

Fig. 5.5. (a) Phase portrait for $\dot{x}_1 = x_2, \dot{x}_2 = -k^2 x_1 - 2kx_2$: the origin is an improper node. The trajectory through A gives $x_1(t)$ shown by curve (1) in (b), while the trajectory through B gives that shown in curve (2).

see with $a, b \in \mathbb{R}$ (see Exercise 5.1). The motion of the mass depends on the initial conditions as the phase portrait in Fig. 5.5(a) shows.

Suppose the mass is projected with velocity $x_2 = -s_0$ (speed $s_0 > 0$) from $x_1 = x_0 > 0$. The phase portrait shows that if $s_0 > kx_0$ (e.g. point A in Fig. 5.5(a)) then the mass overshoots the equilibrium position once and then approaches $x_1 = 0$. If $s_0 < kx_0$ (e.g. point B) then no overshoot occurs. The two kinds of behaviour of x_1 are illustrated in Fig. 5.5(b).

(d) $k > \omega_0$

The eigenvalues of $\boldsymbol{A}$ are no longer equal, but both are negative (i.e. $\lambda_1 = -k + (k^2 - \omega_0^2)^{1/2}, \lambda_2 = -k - (k^2 - \omega_0^2)^{1/2}$) and so the origin is a stable node. It can be shown that

$$x_1(t) = a e^{\lambda_1 t} + b e^{\lambda_2 t}, \qquad x_2(t) = \dot{x}_1(t) \tag{5.8}$$

and the phase portraits look like those in Fig. 5.6.

As $k \to \infty$, $\lambda_1 \to 0$ and $\lambda_2 \to -\infty$ so the principal directions at the fixed point move to coincide with the coordinate axes as indicated in Fig. 5.6. The trajectories steepen and the speed, $|x_2|$, decreases rapidly from large values while the mass moves through a relatively short distance. The area of the plane from which overshoot can occur diminishes and the system is said to be **overdamped**.

The different motions of the mass described in (a)–(d) are easily recognizable in the context of vehicle suspensions. The overdamped case would be a 'very hard' suspension transmitting shocks almost directly to the vehicle. The highly underdamped case would allow the vehicle to 'wallow'. These situations would both be unsatisfactory. The critically damped case, with at most one

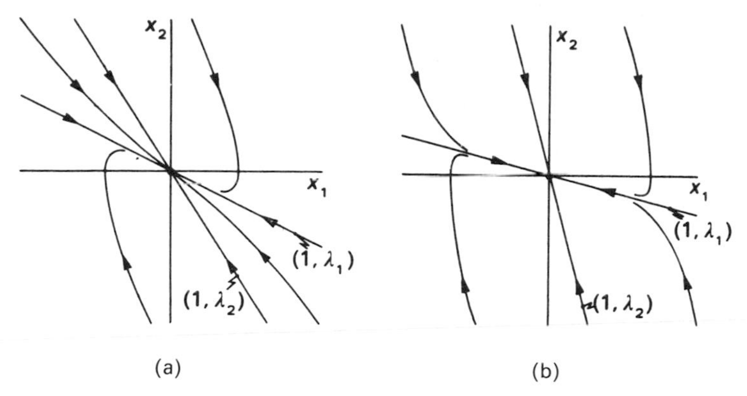

Fig. 5.6. Phase portraits for $\dot{x}_1 = x_2, \dot{x}_2 = -\omega_0^2 x_1 - 2kx_2$ with (a) $k = k_1 > \omega_0$; (b) $k = k_2 > k_1 > \omega_0$. The principal directions are given by $(1, \lambda_1)$, $(1, \lambda_2)$ (see Exercise 5.2).

overshoot, would obviously be most reasonable. Indeed, this is usually the case as the reader may verify on the suspension of any motor cycle or car.

The equations (5.2) or (5.3) are arguably the most frequently used linear dynamical equations. They appear in models of all kinds of time-dependent systems (see sections 5.1.2 and 5.1.3). Their solutions are characterized by trigonometric (or harmonic) oscillations, which are damped when $k > 0$. As a result (5.2) and (5.3) are known as the **damped harmonic oscillator** equations.

5.1.2 Electrical circuits

Electrical circuit theory is a rich source of both linear and non-linear dynamical equations. We will briefly review the background for readers who are not familiar with the subject.

An electrical circuit is a collection of 'circuit elements' connected in a network of closed loops. The notation for some typical circuit elements is given in Fig. 5.7, along with the units in which they are measured.

Differences in **electrical potential** (measured in volts) cause **charge** to move through a circuit. This flow of charge is called a **current**, j (measured in amperes). We can think of an electrical circuit as a set of 'nodes' or 'terminals'

Resistor	Inductor	Capacitor	Battery	Generator
(ohms)	(henrys)	(farads)	(volts)	(volts)
R	L	C	E	$E(t)$

Fig. 5.7. The values R, L and C of the resistor, inductor and capacitor are always non-negative (unless otherwise stated) and independent of t. Batteries and generators are sources of electrical potential difference.

with circuit elements connected between them. If a circuit element is connected between node n and node m of a circuit we associate with it:

1. a potential difference or **voltage** v_{nm};
2. a current j_{nm}.

The voltage v_{nm} is the potential difference between node n and node m, so that $v_{nm} = -v_{mn}$. Equally j_{nm} measures the flow of charge from n to m and therefore $j_{nm} = -j_{mn}$.

The currents and voltages associated with the elements in a circuit are related in a number of ways. The potential differences satisfy **Kirchhoff's voltage law**:

The sum of the potential differences around any closed loop in a circuit is zero; (5.9)

and the currents satisfy **Kirchhoff's current law**:

The sum of the currents flowing into a node is equal to the sum of the currents flowing out of it. (5.10)

Apart from these fundamental laws, the current flowing through a resistor, inductor or capacitor is related to the voltage across it. If a resistor is connected between nodes n and m then

$$v_{nm} = j_{nm} R. \tag{5.11}$$

This is **Ohm's law** and such a resistor is said to be **ohmic**. More generally, the relationship is non-linear with

$$v_{nm} = f(j_{nm}), \tag{5.12}$$

Unless otherwise stated we will assume (5.11) to be valid.

The inductor and capacitor provide relationships involving time derivatives which, in turn, lead to dynamical equations. With the notation defined in Fig. 5.8, these relations are:

$$v = L\frac{\mathrm{d}j}{\mathrm{d}t}; \tag{5.13}$$

Fig. 5.8. In (5.13) and (5.14), $j \equiv j_{nm}$ and $v \equiv v_{nm}$; the potential difference is taken in the direction of j.

and

$$C\frac{dv}{dt}=j. \tag{5.14}$$

Example 4.1.1

Find the dynamical equations of the *LCR* circuit shown in Fig. 5.9. Show that, with $x_1=v_{23}$ and $x_2=\dot{v}_{23}$, these equations can be written in the form (5.3).

Solution

Kirchhoff's current law is already satisfied by assigning a current, j, to the single closed loop as shown, while the voltage law implies

$$v_{12}+v_{23}+v_{31}=0. \tag{5.15}$$

The circuit element relations give

$$v_{12}=jR, \tag{5.16}$$

$$v_{31}=L\frac{dj}{dt}. \tag{5.17}$$

and

$$C\frac{dv_{23}}{dt}=j. \tag{5.18}$$

Substituting (5.15) and (5.16) into (5.17) and writing $v_{23}\equiv v$ gives

$$\frac{dv}{dt}=\frac{j}{C},\qquad \frac{dj}{dt}=-\frac{R}{L}j-\frac{v}{L}. \tag{5.19}$$

Let $x_1=v$ and $x_2=\dot{v}=j/C$ and (5.19) becomes

$$\dot{x}_1=x_2,\qquad \dot{x}_2=-\omega_0^2x_1-2kx_2. \tag{5.20}$$

with $\omega_0^2=1/LC>0$ and $2k=R/L\geqslant 0$. Equation (5.20) is precisely the same system as (5.3) for the mechanical oscillator. □

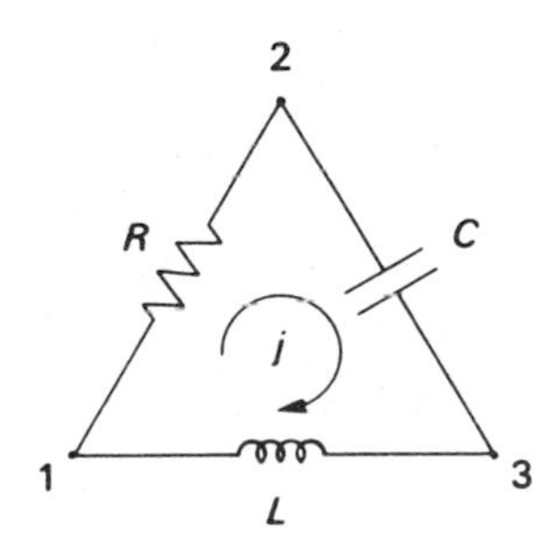

Fig. 5.9. The 'series' *LCR* circuit.

The fact that (5.20) and (5.3) are the same means that the series *LCR* circuit is an electrical analogue of the mechanical oscillator. The potential difference $v(=x_1)$ across the capacitor behaves in exactly the same way with time as the displacement of the mass on the spring. Clearly, (5.20) implies

$$\ddot{v} + 2k\dot{v} + \omega_0^2 v = 0, \tag{5.21}$$

which is exactly the same as (5.2).

The phase portraits in Figs 5.3–5.6 can be re-interpreted in terms of $x_1 = v$ and $x_2 = j/C$. For given L and C, the capacitor voltage may: oscillate without damping ($R = 0$); execute damped oscillations $[0 < R < 2(L/C)^{1/2}]$; be critically damped $[R = 2(L/C)^{1/2}]$; or overdamped $[R > 2(L/C)^{1/2}]$.

Observe that the sign of j determines the sense in which charge flows around the circuit. If $j > 0$, the flow is clockwise; if $j < 0$ the flow is counter clockwise (cf. Fig. 5.9). Consider the case when $R = 0$, for example, and suppose $v(0) = v_0 > 0$, $j(0) = 0$. Equation (5.5) implies that

$$\left(v(t), \frac{j(t)}{C}\right) = \left(v_0 \cos\left(\frac{t}{\sqrt{(LC)}}\right), -\frac{v_0}{\sqrt{(LC)}} \sin\left(\frac{t}{\sqrt{(LC)}}\right)\right). \tag{5.22}$$

Charge flows around the circuit in a counter-clockwise sense until $t = \pi\sqrt{(LC)}$, when

$$v(\pi\sqrt{(LC)}) = -v_0 \quad \text{and} \quad j(\pi\sqrt{(LC)}) = 0. \tag{5.23}$$

As t increases further, $v(t)$ increases and $j(t)$ becomes positive. Charge now flows around the circuit in a clockwise sense until $t = 2\pi\sqrt{(LC)}$ when

$$v(2\pi\sqrt{(LC)}) = v_0 \quad \text{and} \quad j(2\pi\sqrt{(LC)}) = 0. \tag{5.24}$$

The state of the circuit is now the same as it was at $t = 0$ and the oscillations continue. This behaviour is easily recognized in the elliptical trajectories of Fig. 5.3(a).

5.1.3 Economics

An economy is said to be **closed** if all its output is either consumed or invested within the economy itself. There are no exports, imports or external injections of capital. Thus, if Y, C and I are, respectively, the output, consumption and investment for a closed economy at time t, then $Y = C + I$. When external capital injections, such as government expenditure G, are included the economy is no longer closed and output is increased so that

$$Y = C + I + G. \tag{5.25}$$

Furthermore, consumption increases with output; we take

$$C = dY = (1 - s)\, Y, \tag{5.26}$$

where $d, s > 0$ are, respectively, the marginal propensities to consume and to save (Hayes, 1975).

Let us consider an economy which operates with a constant level of government spending G_0. At any particular time t, the economy is subject to a demand $D(t)$ which measures the desired level of consumption and investment. The aim is to balance the economy so that output matches demand, i.e. $D(t) \equiv Y(t)$. However, in practice output is unable to respond instantaneously to demand. There is a response time, τ, associated with development of new plant, etc.

To achieve balance in the presence of lags we must plan ahead and adjust output to meet predicted demand, by taking

$$D(t) = (1 - s)\, Y(t - \tau) + I(t) + G_0. \tag{5.27}$$

In (5.27) we have assumed that investment does not change significantly during the period τ so that $I(t - \tau) = I(t)$. If we now take

$$Y(t - \tau) = Y(t) - \tau \dot{Y}(t) + O(\tau^2), \tag{5.28}$$

we see that (5.27) means that balance is achieved to first order in τ provided

$$(1 - s)\tau \dot{Y}(t) = - sY(t) + I(t) + G_0, \tag{5.29}$$

for every real t.

Although $I(t)$ does not change significantly in periods of order τ, it is not a constant. Investment will respond to trends in output. One plausible investment policy—the 'accelerator principle'—claims that $I(t) \equiv a\dot{Y}(t)$, $a > 0$, is suitable. Time lags in implementing investment decisions prevent this relation from being satisfied; however, we always move towards this ideal if we take

$$\dot{I}(t) = b(a\dot{Y}(t) - I(t)), \qquad b > 0. \tag{5.30}$$

Equations (5.29) and (5.30) are the basis of the dynamics of the economy. They can be cast into more familiar form by differentiating (5.29) to obtain

$$(1 - s)\tau \ddot{Y} + s\dot{Y} = \dot{I} = b(a\dot{Y} - I). \tag{5.31}$$

Substituting for I from (5.29) finally gives

$$(1 - s)\tau \ddot{y} + (s - ba + (1 - s)\tau b)\dot{y} + sby = 0, \tag{5.32}$$

where $y = Y - (G_0/s)$. Thus the output y, relative to G_0/s, satisfies (5.2) with

$$k = (s - ba + (1 - s)\tau b)/2\tau(1 - s) \quad \text{and} \quad \omega_0^2 = sb/\tau(1 - s) > 0. \tag{5.33}$$

When k is non-negative, the phase plane analysis of section 5.1.1 applies, with suitable re-interpretation, to the output of the economy. In particular, the oscillations in Y (about G_0/s) correspond to the 'booms' and 'depressions' experienced by many economies.

A new feature in this model is that the 'damping constant', k, given in

(5.33) can be negative. This means that the equivalent first-order system

$$\dot{x}_1 = x_2, \qquad \dot{x}_2 = -\omega_0^2 x_1 + 2|k|x_2 \tag{5.34}$$

has an unstable fixed point at the origin of its phase plane. It has the qualitative behaviour characteristic of the positive quadrant of the tr(A)–det(A) plane given in Fig. 2.7. Under such circumstances, the peaks of the booms increase with t; as do the depths of the depressions. External intervention is then necessary to prevent output from eventually reaching zero.

5.1.4 Coupled oscillators

The decoupling of linear systems with dimension > 2 into subsystems of lower dimension (see section 2.7) is the basis of the treatment of several models involving interacting oscillators. Small oscillations are usually assumed in order to ensure that the dynamical equations are linear.

Consider two identical simple pendula connected by a light spring as illustrated in Fig. 5.10. We will examine the motion of this system in the vertical plane through OO'.

The dynamical equations can be obtained from the components of force perpendicular to the pendula rods. For small oscillations we obtain

$$\begin{aligned} \ddot{\theta} &= -\frac{g}{a}\theta - \frac{\lambda l}{a}(\theta - \phi), \\ \ddot{\phi} &= -\frac{g}{a}\phi + \frac{\lambda l}{a}(\theta - \phi). \end{aligned} \tag{5.35}$$

If the unit of time is chosen so that $t\sqrt{(g/a)}$ is replaced by t and we take

$$x_1 = \theta, \qquad x_2 = \dot{\theta}, \qquad x_3 = \phi \text{ and } x_4 = \dot{\phi}, \tag{5.36}$$

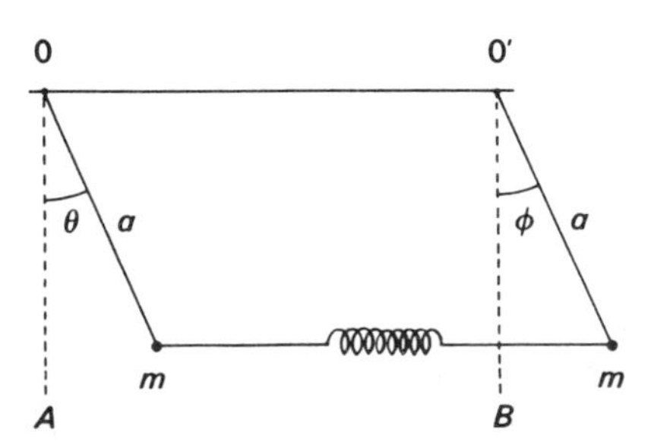

Fig. 5.10. Two identical pendula, A and B, of length a are suspended at points O and O' and connected by a spring of natural length $l = OO'$ and spring constant $m\lambda$. The angles of deflection from the vertical are θ and ϕ as shown.

(5.35) can be written as the first-order linear system,

$$\begin{aligned} \dot{x}_1 &= x_2, & \dot{x}_2 &= -x_1 - \alpha(x_1 - x_3), \\ \dot{x}_3 &= x_4, & \dot{x}_4 &= -x_3 + \alpha(x_1 - x_3), \end{aligned} \tag{5.37}$$

where $\alpha = \lambda l/g > 0$.

The system (5.37) can be decoupled into two-dimensional subsystems by the linear change of variable

$$\begin{aligned} x_1' &= x_1 + x_3, & x_2' &= x_2 + x_4, \\ x_3' &= x_1 - x_3, & x_4' &= x_2 - x_4. \end{aligned} \tag{5.38}$$

The transformed system has coefficient matrix

$$\left[\begin{array}{cc|cc} 0 & 1 & 0 & 0 \\ -1 & 0 & 0 & 0 \\ \hline 0 & 0 & 0 & 1 \\ 0 & 0 & -(1+2\alpha) & 0 \end{array}\right]. \tag{5.39}$$

The diagonal blocks correspond to a pair of undamped harmonic oscillators, one with $\omega_0 = 1$ and the other with $\omega_0 = \sqrt{(1+2\alpha)}$. Of course, (5.39) is not a Jordan form; a further change

$$\begin{aligned} x_1' &= y_1, & x_2' &= -y_2, \\ x_3' &= y_3, & x_4' &= -\sqrt{(1+2\alpha)}\,y_4, \end{aligned} \tag{5.40}$$

is necessary to obtain the canonical system $\dot{\boldsymbol{y}} = \boldsymbol{J}\boldsymbol{y}$ with

$$\boldsymbol{J} = \left[\begin{array}{cc|cc} 0 & -1 & 0 & 0 \\ 1 & 0 & 0 & 0 \\ \hline 0 & 0 & 0 & -\sqrt{(1+2\alpha)} \\ 0 & 0 & \sqrt{(1+2\alpha)} & 0 \end{array}\right]. \tag{5.41}$$

The solutions of (5.37) are linear combinations of those of $\dot{\boldsymbol{y}} = \boldsymbol{J}\boldsymbol{y}$ and a number of interesting features appear for different initial conditions. For example:

(*a*) $\theta = \phi = 0, \dot{\theta} = \dot{\phi} = v$ at $t = 0$
Equations (5.36), (5.38) and (5.40) imply that $y_1 = y_3 = y_4 = 0$ and $y_2 = -2v$ at $t = 0$; therefore (cf. (2.53)),

$$y_1 = 2v\sin(t), \qquad y_2 = -2v\cos(t), \qquad y_3 \equiv y_4 \equiv 0. \tag{5.42}$$

Transformation back to the natural variables gives

$$\theta = v\sin(t), \qquad \phi = v\sin(t). \tag{5.43}$$

Thus $\theta \equiv \phi$ and the pendula oscillate in phase with period 2π; the spring remains unstretched at all times;

(*b*) $\theta = \phi = 0, \dot{\theta} = -\dot{\phi} = v$ at $t = 0$
In this case, we find

$$y_1 \equiv y_2 \equiv 0, \qquad y_3 = \frac{2v}{\sqrt{(1+2\alpha)}} \sin(\sqrt{(1+2\alpha)}t),$$

$$y_4 = \frac{-2v}{\sqrt{(1+2\alpha)}} \cos(\sqrt{(1+2\alpha)}t) \tag{5.44}$$

and

$$\theta = -\phi = \frac{v}{\sqrt{(1+2\alpha)}} \sin(\sqrt{(1+2\alpha)}t), \tag{5.45}$$

so that the pendula oscillate with period $2\pi/\sqrt{(1+2\alpha)}$ and a phase difference of π. They are always symmetrically placed relative to the vertical bisecting OO'.

These two special forms of oscillation are called the **normal modes** of the coupled pendula. They correspond to the solutions of the canonical system (5.41) in which oscillations take place in one of the subsystems while the other has solutions which are identically zero. In (a) one **normal coordinate** $y_1 = \theta + \phi$ oscillates while the other $y_3 = \theta - \phi \equiv 0$. In (b) these roles are reversed and $y_1 \equiv 0$.

Another kind of motion is obtained if the initial conditions are $\theta = \phi = \dot{\theta} = 0$ and $\dot{\phi} = v$ when $t = 0$, so that $y_1(0) = y_3(0) = 0, y_2(0) = -v$ and $y_4(0) = v/\sqrt{(1+2\alpha)}$. In this case,

$$y_1 = v\sin(t) \text{ and } y_3 = \frac{-v}{\sqrt{(1+2\alpha)}} \sin(\sqrt{(1+2\alpha)}t), \tag{5.46}$$

so that

$$\theta = \frac{v}{2}\left\{\sin(t) - \frac{\sin(\sqrt{(1+2\alpha)}t)}{\sqrt{(1+2\alpha)}}\right\},$$

$$\phi = \frac{v}{2}\left\{\sin(t) + \frac{\sin(\sqrt{(1+2\alpha)}t)}{\sqrt{(1+2\alpha)}}\right\}. \tag{5.47}$$

Suppose, now, that the spring constant λ is sufficiently small for $0 < \alpha \ll 1$, so that the coupling is light. Since

$$\sin(t) \pm \frac{\sin(\beta t)}{\beta} = \sin(t) \pm \sin(\beta t) \pm \left(\frac{1}{\beta} - 1\right)\sin(\beta t) \tag{5.48}$$

for any β, we can uniformly approximate θ and ϕ in (5.47) by

$$\theta = \frac{v}{2}\{\sin(t) - \sin(\sqrt{(1+2\alpha)}t)\},$$

$$\phi = \frac{v}{2}\{\sin(t) + \sin(\sqrt{(1+2\alpha)}t)\}, \tag{5.49}$$

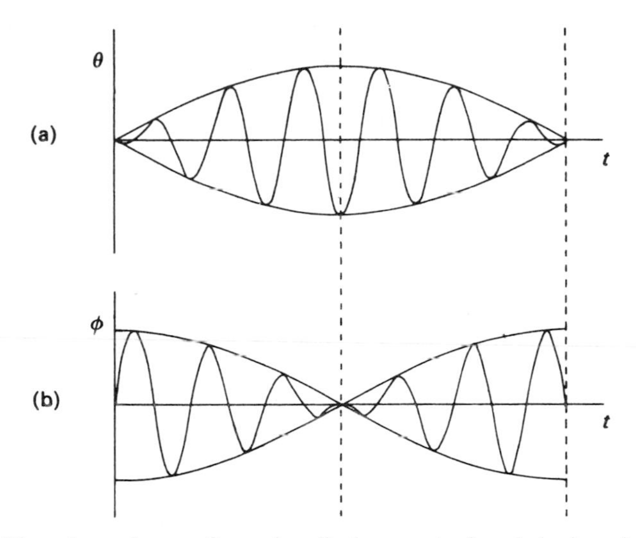

Fig. 5.11. Time dependence of angular displacements θ and ϕ given in (5.51) and (5.52) for coupled pendula. The envelopes (a) $\sin(\alpha t/2)$ for θ and (b) $\cos(\alpha t/2)$ for ϕ are out of phase by $\pi/2$ so that the amplitude of the oscillations in θ is a maximum when ϕ is almost zero.

incurring, at worst, an error of magnitude

$$1 - \frac{1}{\sqrt{(1+2\alpha)}} = \alpha + O(\alpha^2). \tag{5.50}$$

Finally, (5.49) can be approximated by

$$\theta \doteq -v\cos(t)\sin(\alpha t/2) \tag{5.51}$$

and

$$\phi \doteq v\sin(t)\cos(\alpha t/2). \tag{5.52}$$

The time dependence of θ and ϕ is shown in Fig. 5.11. For t near zero, pendulum B oscillates strongly while pendulum A shows only small amplitude oscillations. As t increases, the amplitude of the oscillations of B decays, while A oscillates with increasing vigour. At $t = \pi/\alpha$, pendulum B is stationary and A is oscillating with its maximum amplitude. The roles of A and B are now the reverse of that at $t = 0$ and the oscillations of A decline while those of B grow as t increases to $t = 2\pi/\alpha$. This phenomenon, which corresponds to a repeated exchange of energy between the pendula via the spring, is known as **beats**. It is a common feature of the motion of lightly coupled systems.

5.2 AFFINE MODELS

In this section we consider models whose dynamical equations take the affine form,

$$\dot{\boldsymbol{x}} = \boldsymbol{A}\boldsymbol{x} + \boldsymbol{h}(t). \tag{5.53}$$

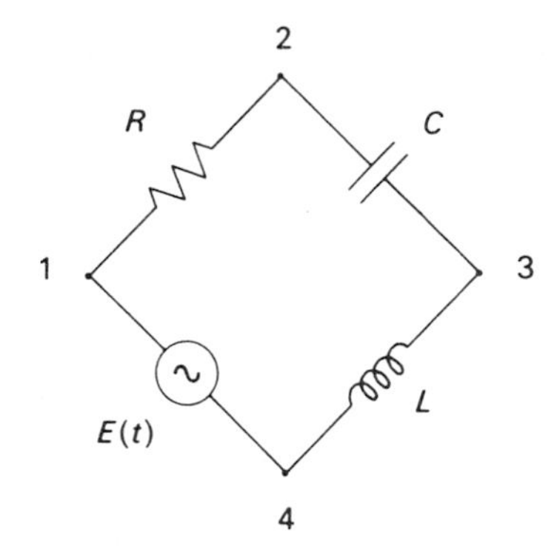

Fig. 5.12. Introduction of the generator as shown above replaces the linear dynamical equations (5.20) by an affine form.

Such models can represent physical systems like those in section 5.1 in the presence of time-dependent external disturbances.

The source of the disturbance or **forcing term**, $\boldsymbol{h}(t)$, depends on the physical system being modelled. For example, if the mass on the spring in section 5.1.1 is subjected to an externally applied downward force $F(t)$ per unit mass, (5.3) becomes

$$\dot{x}_1 = x_2, \qquad \dot{x}_2 = -\omega_0^2 x_1 - 2kx_2 + F(t) \tag{5.54}$$

and

$$\boldsymbol{h}(t) = \begin{bmatrix} 0 \\ F(t) \end{bmatrix}. \tag{5.55}$$

Similarly, the inclusion of a generator in the LCR circuit of section 5.1.2 (illustrated in Fig. 5.12) leads to

$$\boldsymbol{h}(t) = \begin{bmatrix} 0 \\ E(t)/LC \end{bmatrix}. \tag{5.56}$$

Let us consider the effect of a periodic forcing term

$$\boldsymbol{h}(t) = \begin{bmatrix} 0 \\ h_0 \cos \omega t \end{bmatrix} \tag{5.57}$$

on the damped harmonic oscillator discussed in sections 5.1.1 and 5.1.2.

5.2.1 The forced harmonic oscillator

Equation (2.92) gives

$$\boldsymbol{x}(t) = \mathrm{e}^{At}\boldsymbol{x}_0 + \mathrm{e}^{At}\int_0^t \mathrm{e}^{-As}\boldsymbol{h}(s)\mathrm{d}s \tag{5.58}$$

for the solution of (5.53) which satisfies the initial condition $\boldsymbol{x}(0)=\boldsymbol{x}_0$. For $\boldsymbol{h}(t)$ given by (5.57), the integral in (5.58) can be evaluated by using the matrix equivalent of integration by parts (derived in Exercise 5.12). The result is

$$(\omega^2\boldsymbol{I}+\boldsymbol{A}^2)\int_0^t \mathrm{e}^{-\boldsymbol{A}s}\boldsymbol{h}(s)\mathrm{d}s - \boldsymbol{A}\boldsymbol{h}(0) - \mathrm{e}^{-\boldsymbol{A}t}\{\dot{\boldsymbol{h}}(t)+\boldsymbol{A}\boldsymbol{h}(t)\}. \tag{5.59}$$

Thus

$$\boldsymbol{x}(t)=\boldsymbol{x}_{\mathrm{T}}(t)+\boldsymbol{x}_{\mathrm{S}}(t), \tag{5.60}$$

where

$$\boldsymbol{x}_{\mathrm{T}}(t)=\mathrm{e}^{\boldsymbol{A}t}\{\boldsymbol{x}_0+(\omega^2\boldsymbol{I}+\boldsymbol{A}^2)^{-1}\boldsymbol{A}\boldsymbol{h}(0)\},$$
$$\boldsymbol{x}_{\mathrm{S}}(t)=-(\omega^2\boldsymbol{I}+\boldsymbol{A}^2)^{-1}\{\dot{\boldsymbol{h}}(t)+\boldsymbol{A}\boldsymbol{h}(t)\}).$$

The first term $\boldsymbol{x}_{\mathrm{T}}(t)$ in (5.60) is a particular solution of the free (undisturbed) harmonic oscillator

$$\dot{\boldsymbol{x}}=\boldsymbol{A}\boldsymbol{x}, \qquad \text{with } \boldsymbol{A}=\begin{bmatrix} 0 & 1 \\ -\omega_0^2 & -2k \end{bmatrix}. \tag{5.61}$$

In section 5.1.1, we observed that the fixed point at the origin of (5.61) is asymptotically stable for all $k>0$, so $\boldsymbol{x}_{\mathrm{T}}(t)\to\boldsymbol{0}$ as $t\to\infty$. We say that $\boldsymbol{x}_{\mathrm{T}}(t)$ is the **transient** part of the solution. The second term, $\boldsymbol{x}_{\mathrm{S}}(t)$, in (5.60) is persistent and is said to represent the **steady state solution**; this remains after the 'transients' have died away.

5.2.2 Resonance

This phenomenon occurs when the frequency, ω, of the forcing term is near a special frequency determined by ω_0 and k. To see how this comes about we must examine $\boldsymbol{x}_{\mathrm{S}}(t)$ more closely.

It is straightforward to show that

$$\omega^2\boldsymbol{I}+\boldsymbol{A}^2=\begin{bmatrix} \omega^2-\omega_0^2 & -2k \\ 2k\omega_0^2 & \omega^2-\omega_0^2+4k^2 \end{bmatrix}. \tag{5.62}$$

Observe that

$$\begin{aligned}\det(\omega^2\boldsymbol{I}+\boldsymbol{A}^2)&=(\omega^2-\omega_0^2)^2+4k^2\omega^2\\ &\neq 0 \qquad \text{for } k>0,\end{aligned} \tag{5.63}$$

so an inverse exists. This is given by

$$(\omega^2\boldsymbol{I}+\boldsymbol{A}^2)^{-1}=\frac{1}{(\omega^2-\omega_0^2)^2+4k^2\omega^2}\begin{bmatrix} \omega^2-\omega_0^2+4k^2 & +2k \\ -2k\omega_0^2 & \omega^2-\omega_0^2 \end{bmatrix}. \tag{5.64}$$

Finally we can conclude that

$$\boldsymbol{x}_S(t) = \begin{bmatrix} x_{S1}(t) \\ x_{S2}(t) \end{bmatrix}$$

$$= \frac{h_0}{(\omega^2 - \omega_0^2)^2 + 4k^2\omega^2} \begin{bmatrix} (\omega_0^2 - \omega^2)\cos\omega t + 2k\omega\sin\omega t \\ 2k\omega^2\cos\omega t - (\omega_0^2 - \omega^2)\omega\sin\omega t \end{bmatrix}. \tag{5.65}$$

Observe $x_{S2}(t) = \dot{x}_{S1}(t)$, as indeed it must since $\boldsymbol{x}(t) \to \boldsymbol{x}_S(t)$ as $t \to \infty$; so let us examine $x_{S1}(t)$ alone. This is simply a linear combination of cos ωt and sin ωt and it can therefore be written as

$$x_{S1}(t) = G(\omega)\cos(\omega t - \eta(\omega)), \tag{5.66}$$

where

$$G(\omega) = \frac{h_0}{[(\omega^2 - \omega_0^2)^2 + 4k^2\omega^2]^{1/2}}, \tag{5.67}$$

and

$$\eta(\omega) = \tan^{-1}\left[\frac{2k\omega}{\omega_0^2 - \omega^2}\right]. \tag{5.68}$$

For a given physical system (i.e. fixed ω_0 and k), $G(\omega)$ has a maximum at $\omega = \omega_R$ given by

$$\omega_R = (\omega_0^2 - 2k^2)^{1/2}, \tag{5.69}$$

provided $\omega_0^2 > 2k^2$. What is more,

$$G(\omega_R) = h_0/2k(\omega_0^2 - k^2)^{1/2} \tag{5.70}$$

and $G(\omega_R)/h_0 \to \infty$ as $k \to 0$. This means that if the physical system is sufficiently lightly damped, then the amplitude, $G(\omega_R)$, of the response is large, compared with that of the forcing term, h_0 (as illustrated in Fig. 5.13).

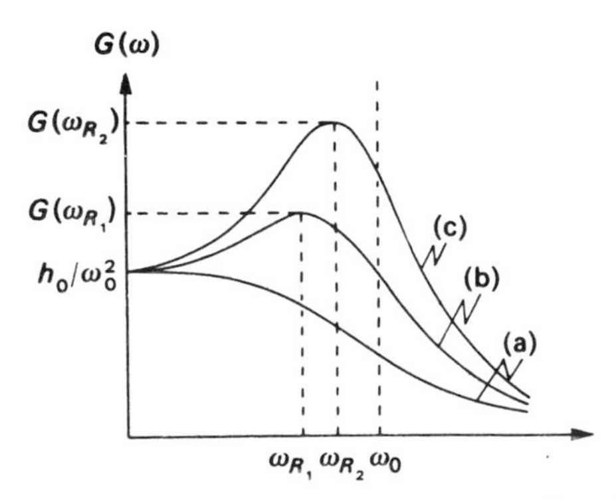

Fig. 5.13. Plots of $G(\omega)$ versus ω for: (a) $k = k_0$ where $\omega_0^2 < 2k_0^2$; (b) $k = k_1$ where $\omega_0^2 > 2k_1^2$ with a maximum at $\omega_{R_1} < \omega_0$; (c) $k = k_2 < k_1$ with a maximum at ω_{R_2} where $\omega_{R_1} < \omega_{R_2} < \omega_0$. Observe the sharpening and shift of the maxima as k decreases.

The system is said to **resonate** with the disturbance and ω_R is called the **resonant frequency**.

Resonances do have very important applications particularly in electronics where tuning devices are in common use. We can illustrate this idea by considering the series LCR circuit shown in Fig. 5.12 with $E(t) = E_0 \cos \omega t$. We have already observed in (5.20) and (5.56) that for this case

$$\omega_0^2 = 1/LC, \qquad 2k = R/L \quad \text{and} \quad \boldsymbol{h}(t) = \begin{bmatrix} 0 \\ (E_0/LC)\cos \omega t \end{bmatrix}. \tag{5.71}$$

The current flowing through the circuit in the steady state is $C x_{S2}(t)$, i.e.

$$j(t) = -C\omega G(\omega)\sin(\omega t - \eta(\omega)) = j_0 \cos(\omega t + [\pi/2 - \eta(\omega)]). \tag{5.72}$$

This has amplitude

$$j_0 = C\omega G(\omega) = \frac{E_0 \omega}{L\{[\omega^2 - (1/LC)]^2 + (R^2/L^2)\omega^2\}^{1/2}}$$
$$= \frac{E_0}{\{R^2 + [\omega L - (1/\omega C)]^2\}^{1/2}}. \tag{5.73}$$

The ratio $Z = E_0/j_0$ is called the **impedance** of the circuit; so for the series LCR circuit

$$Z = \{R^2 + [\omega L - (1/\omega C)]^2\}^{1/2}. \tag{5.74}$$

In this case the resonant frequency is given by

$$\omega_R L - \frac{1}{\omega_R C} = 0 \qquad \text{or} \qquad \omega_R^2 = 1/LC. \tag{5.75}$$

At this frequency the impedance of the circuit is a minimum ($= R$) and for frequencies in this neighbourhood the impedance is small if R is small. For other frequencies the impedance increases with their separation from ω_R and the amplitudes of the associated currents would be correspondingly smaller. The circuit is said to be **selective**, favouring response in the neighbourhood of ω_R.

If, for example, the capacitance C can be varied then the resonant frequency can be changed. It is then possible to tune the circuit to provide strong current response at a desired frequency.

5.3 NON-LINEAR MODELS

In spite of the simplicity and undoubted success of linear models, they do have limitations. For example, the qualitative behaviour of a linear model in the plane must be one of a finite number of types. Linear models can have only one isolated fixed point; they can only produce persistent oscillations of a harmonic (or trigonometric) kind; such oscillations always come from centres in their phase portraits, etc. If the qualitative behaviour of the physical system to be modelled is not consistent with these limitations then a linear model will be inadequate.

In this section, we consider some problems from population dynamics for which non-linear models are required.

5.3.1 Competing species

Two similar species of animal compete with each other in an environment where their common food supply is limited. There are several possible outcomes to this competition:

1. species 1 survives and species 2 becomes extinct;
2. species 2 survives and species 1 becomes extinct;
3. the two species coexist;
4. both species become extinct.

Each of these outcomes can be represented by an equilibrium state of the populations x_1 and x_2 of the two species. The differential equations used to model the dynamics of x_1 and x_2 are therefore required to have four isolated fixed points.

Consider the following non-linear dynamical equations:

$$\dot{x}_1 = (a - bx_1 - \sigma x_2)x_1, \qquad \dot{x}_2 = (c - \nu x_1 - dx_2)x_2, \tag{5.76}$$

where $a, b, c, d, \sigma, \nu > 0$. Observe that the per capita growth rate $\dot{x}_1/x_1 = (a - bx_1 - \sigma x_2)$ consists of three terms: the growth rate, a, of the isolated population; the 'intra'-species competition, $-bx_1$ (cf. section 1.2.2); and the 'inter'-species competition, $-\sigma x_2$. A similar interpretation can be given for the terms c, $-dx_2$ and $-\nu x_1$ in $\dot{x}_2/x_2$.

A necessary condition for coexistence of the two species is that (5.76) has a fixed point with both populations greater than zero. Such a fixed point can only arise in (5.76) if the linear equations

$$bx_1 + \sigma x_2 = a, \qquad \nu x_1 + dx_2 = c \tag{5.77}$$

have a solution with x_1, $x_2 > 0$. We will assume that (5.77) has a unique solution in the positive quadrant of the x_1x_2-plane. The fixed point is then given by

$$\left(\frac{ad - \sigma c}{bd - \nu\sigma}, \frac{bc - a\nu}{bd - \nu\sigma} \right), \tag{5.78}$$

where either:

$$bd < \nu\sigma \qquad \text{with } ad < \sigma c \quad \text{and} \quad bc < a\nu; \tag{5.79}$$

or

$$bd > \nu\sigma \qquad \text{with } ad > \sigma c \quad \text{and} \quad bc > a\nu. \tag{5.80}$$

The geometrical significance of these inequalities is illustrated in Fig. 5.14.

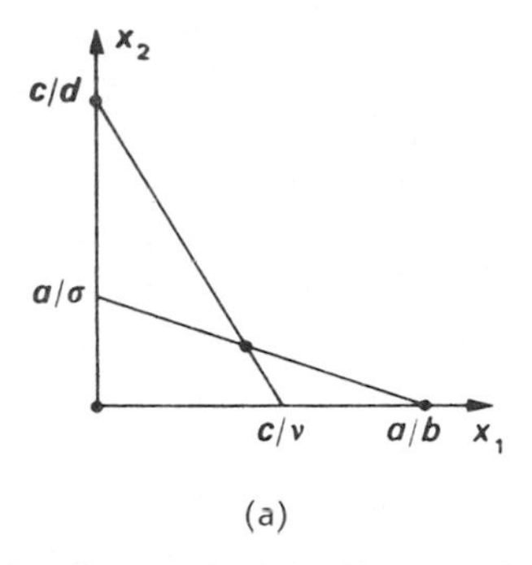

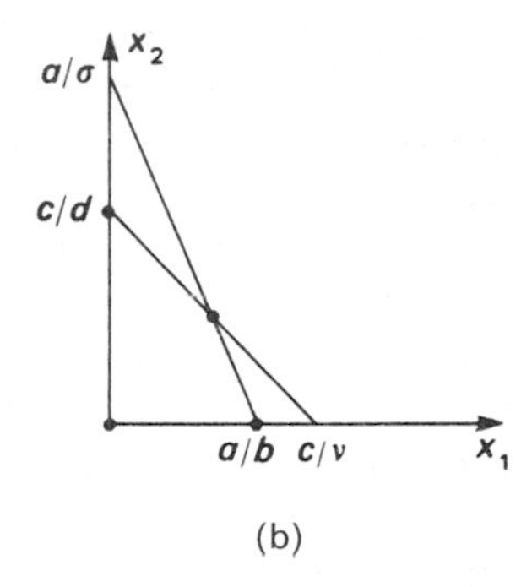

Fig. 5.14. Geometrical significance of (a) (5.79) and (b) (5.80). The fixed points of (5.76) are denoted by •.

We will now consider the dynamics of x_1 and x_2 when (5.79) is satisfied; some other possibilities are dealt with in Exercise 5.16. We begin by examining the linearizations at the four fixed points. Let each linearized system be denoted by $\dot{\boldsymbol{y}} = \boldsymbol{W}\boldsymbol{y}$ where $\boldsymbol{y}$ are local coordinates at the fixed point. Then each fixed point and its corresponding $\boldsymbol{W}$ are given by:

$$(0, 0), \qquad \begin{bmatrix} a & 0 \\ 0 & c \end{bmatrix}; \tag{5.81}$$

$$(0, c/d), \qquad \begin{bmatrix} a - \sigma c/d & 0 \\ -vc/d & -c \end{bmatrix}; \tag{5.82}$$

$$(a/b, 0), \qquad \begin{bmatrix} -a & -\sigma a/b \\ 0 & c - va/b \end{bmatrix}; \tag{5.83}$$

$$\left(\frac{\sigma c - ad}{v\sigma - bd}, \frac{av - bc}{v\sigma - bd}\right), \qquad \frac{1}{bd - v\sigma}\begin{bmatrix} b(\sigma c - ad) & \sigma(\sigma c - ad) \\ v(av - bc) & d(av - bc) \end{bmatrix}. \tag{5.84}$$

The fixed points are all simple and their nature is determined by the eigenvalues of $\boldsymbol{W}$. For (5.81) these are $a > 0$ and $c > 0$ and, by the linearization theorem, the origin is an unstable node. The eigenvalues of $\boldsymbol{W}$ in (5.82) are also given by inspection, since $\boldsymbol{W}$ is triangular, but both are negative (cf. (5.79)). The same is true of (5.83) and in both cases the linearization theorem allows us to conclude that these fixed points are stable nodes or improper nodes (note that $a - \sigma c/d = -c$ or $c - va/b = -a$ is possible). Finally, the fixed point (5.84) is a saddle point, because

$$\det(\boldsymbol{W}) = \frac{(\sigma c - ad)(av - bc)}{bd - v\sigma} < 0, \tag{5.85}$$

by (5.79), and the eigenvalues of $\boldsymbol{W}$ must have opposite sign (cf. Fig. 2.7).

It is now apparent that coexistence is extremely unlikely in this model since only the two stable separatrices approach the saddle point as $t \to \infty$.

The fixed points at $(0, c/d)$ and $(a/b, 0)$ are stable and correspond, respectively, to either species 1 or species 2 becoming extinct. The origin is an unstable node so there is no possibility of both populations vanishing since all trajectories move away from this point. Assuming that all initial states in the positive quadrant are equally likely, then by far the most likely outcome of the competition is that one or other of the species will die out. This result agrees with the 'Law of Competitive Exclusion'; namely, that when situations of this competitive kind occur in nature one or other of the species involved becomes extinct.

Further details of the evolution of the two species can be obtained by sketching the phase portrait for the model. Observe that the straight lines given in (5.77) are respectively the $\dot{x}_1 = 0$ and $\dot{x}_2 = 0$ isoclines. The sense of the vector field $((a - bx_1 - \sigma x_2)x_1, (c - vx_1 - dx_2)x_2)$ can be obtained by recognizing that $\dot{x}_1$ and $\dot{x}_2$ take positive and negative values as shown in

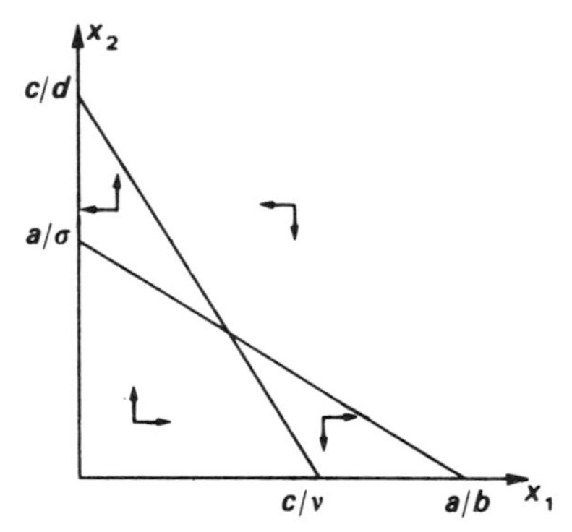

Fig. 5.15. The signs of $\dot{x}_1$ and $\dot{x}_2$ given by (5.76) in the positive quadrant of the x_1x_2-plane. The notation is as follows: ↑→ means $\dot{x}_1, \dot{x}_2 > 0$; ↓→ is $\dot{x}_1 > 0$, $\dot{x}_2 < 0$; ←↑ is $\dot{x}_1 < 0$, $\dot{x}_2 > 0$ and ←↓ is $\dot{x}_1, \dot{x}_2 > 0$.

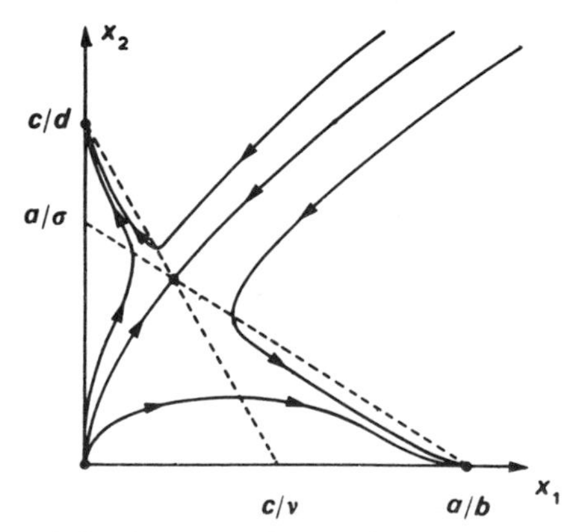

Fig. 5.16. Phase portrait for the competing species model (5.76) with $bd < v\sigma$, $ad < \sigma c$, $bc < av$.

Fig. 5.15. This information, along with the nature of the fixed points, is sufficient to construct a sketch of the phase portrait shown in Fig. 5.16.

Details such as whether trajectories leave the origin tangential to the x_1- or x_2-axis depend upon actual values of the eigenvalues. For instance, in Fig. 5.16 we have assumed $c < a$ so that trajectories are drawn tangential to the x_2-axis. Further such detail is dealt with in Exercise 5.15.

5.3.2 Volterra–Lotka equations

The most general oscillatory form that can arise from linear dynamical equations in the plane is

$$x_1(t) = R_1 \cos(\beta t + \theta_1), \qquad x_2(t) = R_2 \cos(\beta t + \theta_2), \tag{5.86}$$

corresponding to elliptical closed orbits in the phase portrait. The harmonic oscillations in (5.86) are symmetric and, while R_i and θ_i $(i = 1, 2)$ allow variation of amplitude and shift of origin, their shape remains unchanged.

In population dynamics, there are many examples of populations exhibiting oscillatory behaviour. However, these oscillations are markedly different from harmonic forms and non-linear dynamical equations are required to model them. The Volterra–Lotka equations are a particularly well-known example.

Consider a two-species model in which one species is a predator that preys upon the other. Let x_1 and x_2 be the populations of prey and predator, respectively, and assume that there is no competition between individuals in either species. The per capita growth rate $\dot{x}_1/x_1$ of prey is taken as $a - bx_2, a, b > 0$; a is the growth rate in the absence of predators and $-bx_2$ allows for losses due to their presence. The predator population would decline in the absence of their food supply (i.e. the prey), so when $x_1 = 0$, $\dot{x}_2/x_2 = -c$, $c > 0$. However, successful hunting offsets this decline and we take $\dot{x}_2/x_2 = -c + dx_1$, $d > 0$, when $x_1 > 0$. Thus

$$\dot{x}_1 = (a - bx_2)x_1, \qquad \dot{x}_2 = (-c + dx_1)x_2, \tag{5.87}$$

where a, b, c, $d > 0$. These are the Volterra–Lotka equations.

The system (5.87) has two fixed points (0, 0) and $(c/d, a/b)$. The former is a saddle point with separatrices which coincide with the x_1- and x_2-axes, the latter being stable. The linearization at the non-trivial fixed point, $(c/d, a/b)$, is a centre and therefore the linearization theorem is unable to determine its nature.

The system (5.87) has a first integral

$$f(x_1, x_2) = x_1^c \mathrm{e}^{-dx_1} x_2^a \mathrm{e}^{-bx_2} = g(x_1)h(x_2) \tag{5.88}$$

(cf. Example 3.7.3). The functions $g(x_1)$ and $h(x_2)$ have the same form; each is positive for all arguments in $(0, \infty)$ and has a single maximum in this interval. These functions are plotted for typical values of their parameters in Fig. 5.17.

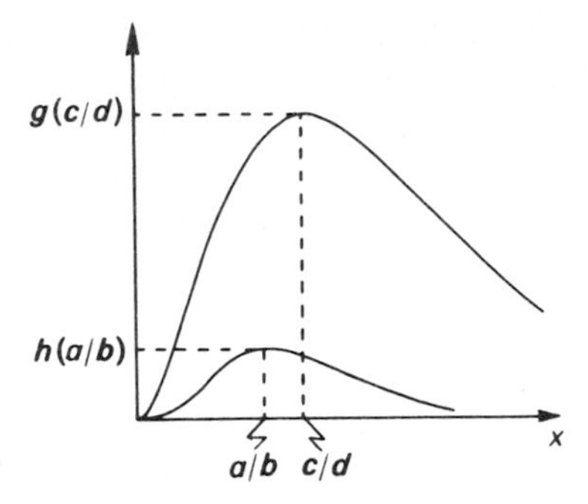

Fig. 5.17. Graphs of $g(x)$ and $h(x)$ for $a = 4.0$, $b = 2.5$, $c = 2.0$ and $d = 1.0$. Maxima occur at $x = c/d$ and $x = a/b$, respectively.

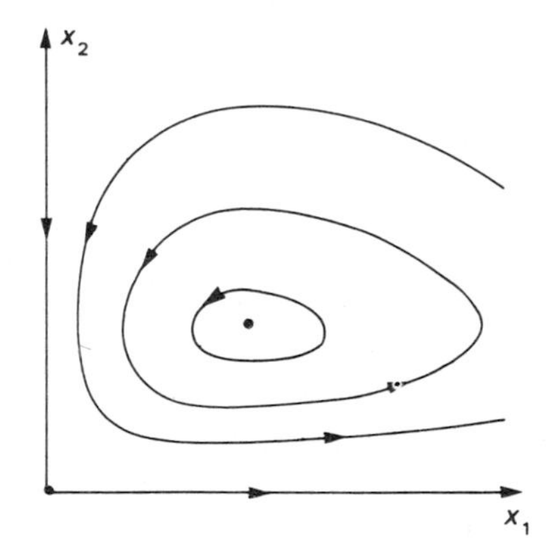

Fig. 5.18. Phase portrait for the Volterra–Lotka equations $\dot{x}_1 = (a - bx_2)x_1$, $\dot{x}_2 = (-c + dx_1)x_2$ with $a = 4.0$, $b = 2.5$, $c = 2.0$, $d = 1.0$.

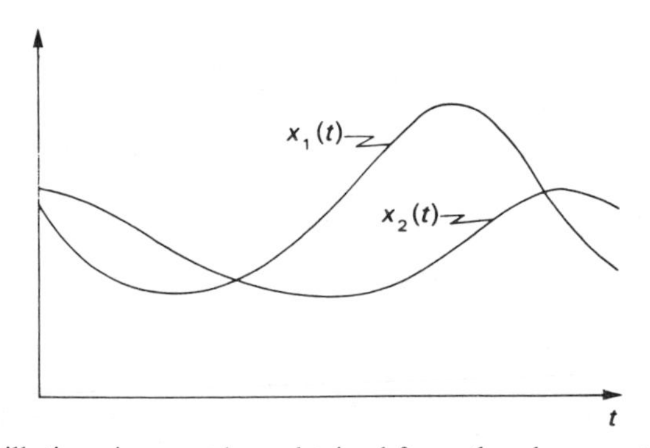

Fig. 5.19. Oscillations in x_1 and x_2 obtained from the phase portrait in Fig. 5.18.

The maxima occur at $x_1 = c/d$ and $x_2 = a/b$, respectively, and thus $f(x_1, x_2)$ has a maximum of $g(c/d)h(a/b)$ at $(x_1, x_2) = (c/d,\ a/b)$. The level curves of $f(x_1, x_2)$ are therefore closed paths surrounding $(c/d,\ a/b)$. The trajectories of (5.87) coincide with these curves and clearly $(c/d,\ a/b)$ is a centre.

The functions $g(x_1)$ and $h(x_2)$ are not symmetric about their extreme points and consequently the closed trajectories are certainly not ellipses. The sharp rise followed by slower fall shown by both g and h (see Fig. 5.17) means that the trajectories take the form shown in Fig. 5.18.

The non-elliptic form of the trajectories of the non-linear centre is reflected in the non-harmonic nature of the oscillations in the populations (as illustrated in Fig. 5.19).

5.3.3 The Holling–Tanner model

Although persistent non-harmonic oscillations in dynamical variables can be modelled by systems with centres in their phase portraits, the stability of such oscillations to perturbations of the model is suspect. We saw in section 5.1 that the centre for the harmonic oscillator was destroyed by even the weakest of damping forces becoming a focus. The same is true of the simple pendulum (discussed in Exercise 5.25). Similarly, suppose the Volterra–Lotka equations (5.87) are modified to include the additional effects of competition within the species. The resulting system no longer has a centre in its phase portrait (see Exercise 5.17) and the population oscillations decay. This tendency to be very easily destroyed is inherent in the make-up of centres. They are said to lack **structural stability**.

An alternative way for persistent oscillations to occur in non-linear systems is by the presence of a stable limit cycle in their phase portraits (cf. section 3.8). The limit cycle is structurally stable and is consequently a 'more permanent' feature of a phase portrait; it is not likely to disappear as a result of relatively small perturbations of the model. Models that are insensitive to perturbations are said to be **robust**. Since most models are idealizations which focus attention on certain central variables and interactions, this kind of stability is very important.

The two-species prey–predator problem is obviously an idealization of the kind mentioned above and the Volterra–Lotka model is not robust. It is therefore questionable whether (5.87) contains the true mechanism for population oscillations.

A model which provides structurally stable population oscillations for the prey–predator problem is the Holling–Tanner model. The dynamical equations are

$$\begin{aligned} \dot{x}_1 &= r(1 - x_1 K^{-1})x_1 - w x_1 x_2 (D + x_1)^{-1}, \\ \dot{x}_2 &= s(1 - J x_2 x_1^{-1}) x_2, \end{aligned} \tag{5.89}$$

with $r,\ s,\ K,\ D,\ J > 0$.

The rate of growth of prey $\dot{x}_1$ is the difference of two terms.

1. $r(1 - x_1K^{-1})x_1$ gives the growth of the prey population in the absence of predators. This includes competition between prey for some limiting resource (cf. section 1.2.2).
2. $wx_1x_2(D + x_1)^{-1}$ describes the effect of the predators.

To explain the form of 2 it is convenient to think of the effect of predators on $\dot{x}_1$ in terms of a **predation rate**. This is the number of prey killed per predator per unit time. In (5.87) the predation rate is bx_1. This means that the number of prey killed per predator per unit time increases indefinitely with prey population.

A more reasonable assumption is that there is an upper limit to the predation rate, e.g. when the predator's appetite is satisfied. This is taken into account in 2 where the predation rate is $wx_1(D + x_1)^{-1}$ (illustrated in Fig. 5.20).

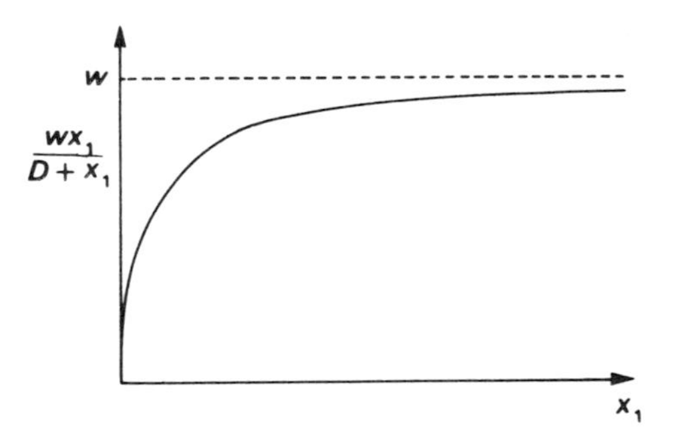

Fig. 5.20. The predation rate for the Holling–Tanner model. The slope at $x_1 = 0$ is w/D and the saturation value is w.

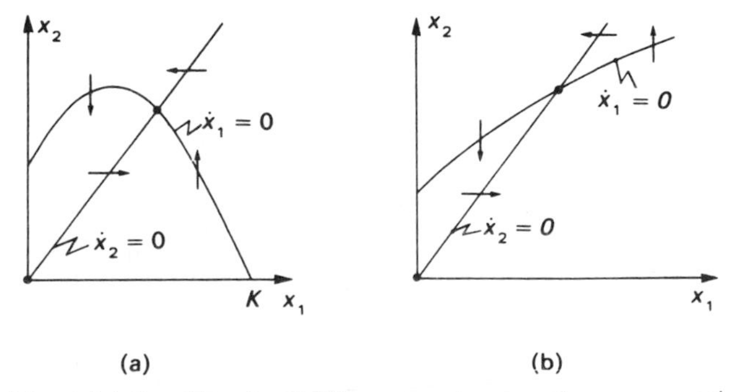

Fig. 5.21. Non-trivial isoclines for (5.89) are given by: $\dot{x}_1 = 0$ on $x_2 = rw^{-1}(1 - x_1K^{-1})(D + x_1)$, which has a maximum at $x_1 = (K - D)/2$; $\dot{x}_2 = 0$ on $x_2 = J^{-1}x_1$. Intersection of these isoclines can take place with (a) $x_1 > (K - D)/2$ or (b) $x_1 < (K - D)/2$.

The rate of growth of the predator population, $\dot{x}_2$, is obtained by viewing the prey as a scarce resource. We will assume that the number of prey required to support one predator is J. Thus if the prey population is x_1 it can support no more than x_1/J predators. We can ensure that the predator population does not exceed its environmental limits by taking (cf. (1.18))

$$\dot{x}_2 = x_2\left(s - \frac{sJ}{x_1}x_2\right).$$

This is the equation given in (5.89).

The $\dot{x}_1 = 0$ and $\dot{x}_2 = 0$ isoclines of (5.89) are shown in Fig. 5.21. As can be seen, there is a fixed point with x_1 and x_2 both greater than zero, at $(x_1, x_2) = (x_1^*, x_2^*)$, say. To determine the nature of this fixed point it is convenient to scale the variables in (5.89) by x_1^*. We take $y_1 = x_1/x_1^*$ and $y_2 = x_2/x_1^*$ and obtain

$$\begin{aligned} \dot{y}_1 &= r(1 - y_1 k^{-1})y_1 - w y_1 y_2 (d + y_1)^{-1}, \\ \dot{y}_2 &= s(1 - J y_2 y_1^{-1}) y_2. \end{aligned} \tag{5.90}$$

where $k = K/x_1^*$ and $d = D/x_1^*$. It follows that the fixed point in the $y_1 y_2$-plane has coordinates

$$\begin{aligned} (y_1^*, y_2^*) &= (1, rw^{-1}(1 - k^{-1})(1 + d)) \\ &= (1, J^{-1}). \end{aligned} \tag{5.91}$$

The coefficient matrix of the linearized system at (y_1^*, y_2^*) is then

$$\boldsymbol{W} = \begin{bmatrix} r(-k^{-1} + w(rJ)^{-1}(1 + d)^{-2}) & -w(1 + d)^{-1} \\ sJ^{-1} & -s \end{bmatrix}. \tag{5.92}$$

Observe that

$$\det(\boldsymbol{W}) = rs(k^{-1} + wd(rJ)^{-1}(1 + d)^{-2}) > 0 \tag{5.93}$$

so that (y_1^*, y_2^*) is never a saddle point. However, substituting for $w(rJ)^{-1}$ from (5.91),

$$\begin{aligned} \operatorname{tr}(\boldsymbol{W}) &= r(w(rJ)^{-1}(1 + d)^{-2} - k^{-1}) - s \\ &= r(k - d - 2)(k(1 + d))^{-1} - s, \end{aligned} \tag{5.94}$$

which can be either positive or negative.

Let ϕ_t be the evolution operator for (5.90) and consider $\phi_t(k, 0_+)$ for $t > 0$. As Fig. 5.22 shows, this trajectory must move around the fixed point and intersect the $\dot{x}_1 = 0$ isocline at A. The set S enclosed by this trajectory and the part of the $\dot{x}_1 = 0$ isocline between A and $(k, 0)$ is positively invariant.

If $\operatorname{tr}(\boldsymbol{W}) > 0$, so that (y_1^*, y_2^*) is an unstable fixed point then the linearization theorem ensures that there is a neighbourhood N of (y_1^*, y_2^*) in S such that $S\backslash N$ is a trapping region. However, there are no fixed points in $S\backslash N$ and by the corollary to Theorem 3.9.1 this set must contain a closed orbit. Thus,

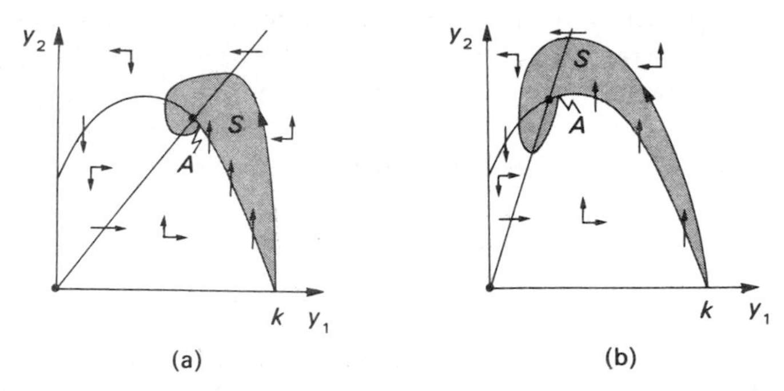

Fig. 5.22. Positively invariant sets S containing the fixed point (y_1^*, y_2^*), bounded by the trajectory $\{\phi_t(k, 0_+) | t > 0\}$ and part of the $\dot{x}_1 = 0$ isocline: (a) $x_1^* > (K - D)/2$; (b) $x_1^* < (K - D)/2$.

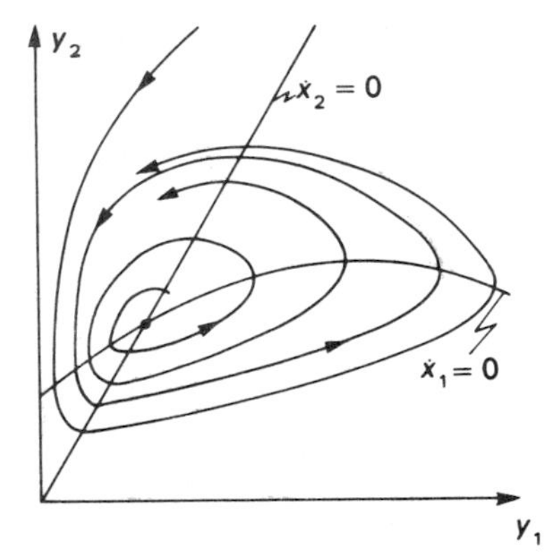

Fig. 5.23. Phase portrait for (5.90) with $r = 1.0$, $s = 0.2$, $k = 7.0$, $d = 1.0$, showing the stable limit cycle of the Holling–Tanner model.

for those sets of parameters (r, s, k, d) for which

$$s < \frac{r(k - d - 2)}{k(1 + d)}, \tag{5.95}$$

the phase portrait of (5.90) contains a stable limit cycle (shown in Fig. 5.23). In Exercise 5.19, it is shown that this can only occur if (y_1^*, y_2^*) lies to the left of the peak in the $\dot{x}_1 = 0$ isocline. Thus the phase portrait corresponding to Fig. 5.22(a) has no limit cycle; (y_1^*, y_2^*) is simply a stable focus.

5.4 RELAXATION OSCILLATIONS

5.4.1 Van der Pol oscillator

Consider the modification of the series LCR circuit shown in Fig. 5.24. Suppose that the 'black box' B is a circuit element (or collection of circuit

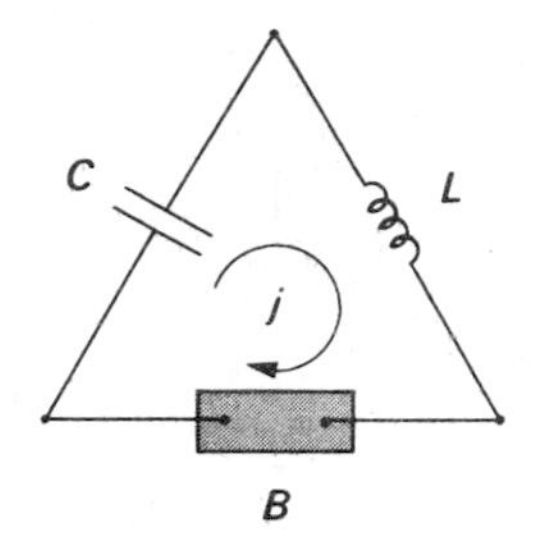

Fig. 5.24. A modification of the series LCR circuit.

elements) with a voltage–current relationship like that shown in Fig. 5.25(a). The black box is said to be a non-linear resistor with the cubic **characteristic**,

$$v_{\mathrm{B}} = f(j_{\mathrm{B}}) = j_{\mathrm{B}}(\tfrac{1}{3}j_{\mathrm{B}}^2 - 1). \tag{5.96}$$

Let v_L, v_{B} and $-v_C$ be the potential differences across the elements in Fig. 5.24 in the direction of the current j. The dynamical equations for the circuit are

$$L\frac{\mathrm{d}j}{\mathrm{d}t} = v_C - f(j), \qquad C\frac{\mathrm{d}v_C}{\mathrm{d}t} = -j, \tag{5.97}$$

where we have used Kirchhoff's voltage law to eliminate v_L. Now let $L^{-1}t \to t$, $x_1 = j$, $x_2 = v_C$, $L/C = \eta$, to obtain

$$\dot{x}_1 = x_2 - f(x_1), \qquad \dot{x}_2 = -\eta x_1. \tag{5.98}$$

The system (5.98) has a fixed point at $x_1 = x_2 = 0$ and the isoclines shown in Fig. 5.26.

Suppose now that η is so small (L small compared with C) that $\dot{x}_2$ can be neglected, relative to $\dot{x}_1$, at all points of the phase plane that are not in the immediate neighbourhood of the $\dot{x}_1 = 0$ isocline. This means that the vector

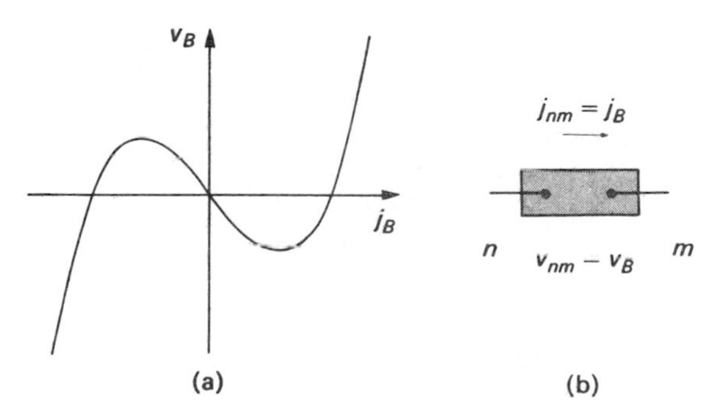

Fig. 5.25. The voltage–current 'characteristic' (a) of the black box in Fig. 5.24. With the notation in (b) $v_{\mathrm{B}} = f(j_{\mathrm{B}}) = j_{\mathrm{B}}(\tfrac{1}{3}j_{\mathrm{B}}^2 - 1)$.

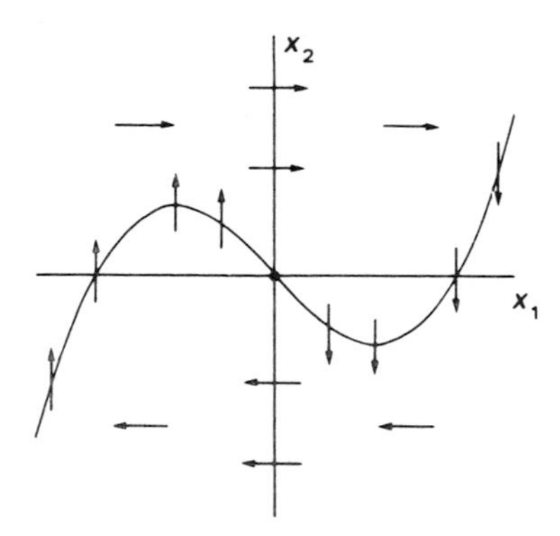

Fig. 5.26. Isoclines for (5.98): $\dot{x}_1 = 0$ on $x_2 = f(x_1)$; $\dot{x}_2 = 0$ on x_2-axis. The sense of the vector field is shown for small η.

field $(x_2 - f(x_1), -\eta x_1)$ is directed essentially horizontally in Fig. 5.26 except near the curve

$$x_2 = f(x_1) = \frac{x_1^3}{3} - x_1. \tag{5.99}$$

It follows that the phase portrait must be as shown in Fig. 5.27. As can be seen, all initial values of (x_1, x_2) lead to trajectories which home in on the limit cycle *ABCD* at the first opportunity. At points away from the cubic (5.99) $\dot{x}_1$ is comparatively large so that the phase point moves rapidly to the neighbourhood of the characteristic and slowly follows it.

The fast movement on *AB*, *CD* and the slow movement on *BC*, *DA* leads to an 'almost discontinuous' wave form for the persistent oscillations of the current $x_1(t)$ (as shown in Fig. 5.28). Oscillations of this kind are called **relaxation oscillations**.

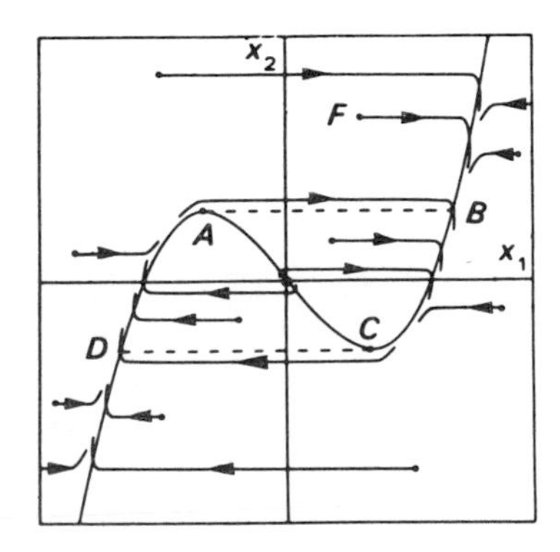

Fig. 5.27. Phase portrait for (5.98) when $\eta \to 0$. All trajectories rapidly home in on the limit cycle *ABCD*.

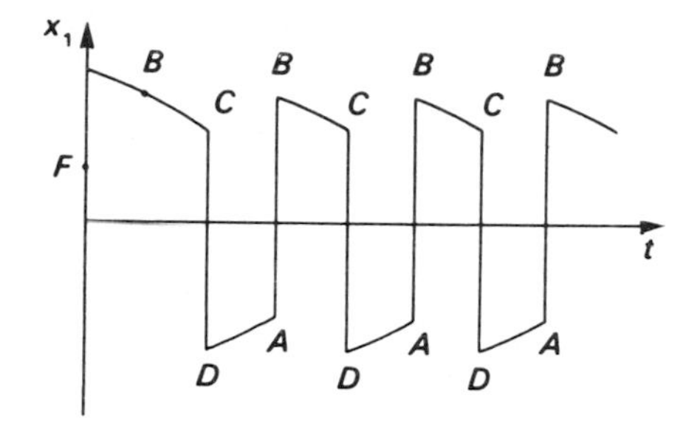

Fig. 5.28. Relaxation oscillations from the trajectory starting at F in Fig. 5.27. Sections marked CD and AB are almost vertical corresponding to rapid motion between these points in the phase plane.

The system (5.98) does not only have a limit cycle when $\eta \to 0$. For example, when $\eta = 1$ it is a special case of the **Van der Pol equation**

$$\ddot{x} + \varepsilon(x^2 - 1)\dot{x} + x = 0, \qquad \varepsilon > 0, \tag{5.100}$$

when $\varepsilon = 1$. This equation has the form

$$\ddot{x} + g(x)\dot{x} + h(x) = 0, \tag{5.101}$$

which is equivalent to the system

$$\dot{x}_1 = x_2 - \int^{x_1} g(u)\,du, \qquad \dot{x}_2 = -h(x_1) \tag{5.102}$$

(cf. Exercise 1.27). Equations of the form (5.101) are sometimes called **Liénard equations** and the coordinates in (5.102) are said to define the **Liénard plane**.

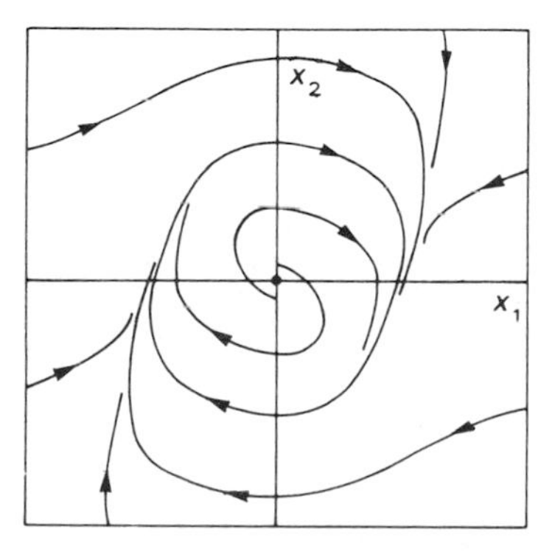

Fig. 5.29. Limit cycle of Van der Pol oscillator in the Liénard plane: $\varepsilon = 1.0$.

The Liénard representation of the Van der Pol equation is

$$\dot{x}_1 = x_2 - \varepsilon\left(\frac{x_1^3}{3} - x_1\right), \qquad \dot{x}_2 = -x_1, \tag{5.103}$$

which is (5.98) when $\varepsilon = \eta = 1$.

The Van der Pol equation can be thought of as an extension of the harmonic oscillator equation (5.2). The difference is that it has a non-linear damping coefficient, $\varepsilon(x^2 - 1)$, instead of the constant $2k$ in (5.2).

The Van der Pol oscillator has a unique stable limit cycle for all values of ε. When $\varepsilon \to \infty$, this limit cycle takes the shape shown in Fig. 5.27 and relaxation oscillations result (see Exercise 5.28). When $\varepsilon \to 0$, the limit cycle is approximately circular with radius 2 (see Exercise 5.27). As ε increases from zero the limit cycle distorts smoothly from one of these extremes to the other (cf. Fig. 5.29).

5.4.2 Jumps and regularization

If we let $L = 0$ in the circuit equations (5.97) the differential equation

$$L\frac{\mathrm{d}j}{\mathrm{d}t} = v_C - f(j) \tag{5.104}$$

is replaced by the algebraic equation

$$v_C = f(j). \tag{5.105}$$

Thus the system of two dynamical equations (5.97) with $(j, v_C) \in \mathbb{R}^2$ is replaced by a single dynamical equation

$$C\frac{\mathrm{d}v_C}{\mathrm{d}t} = C\dot{v}_C = -j \tag{5.106}$$

defined on $\{(j, v_C) | v_C = f(j)\}$.

Geometrically, (5.106) means that the dynamics of the circuit are confined to the curve (5.105) in the j, v_C-plane. Equation (5.106) implies that

$$\dot{v}_C \lesseqgtr 0 \qquad \text{for } j \gtreqless 0, \tag{5.107}$$

and so the states of the circuit must evolve along the cubic characteristic as shown in Fig. 5.30. For any starting point on $v_C = f(j)$ the state of the circuit evolves to one of the two points A or C. These points are not fixed points; clearly $\dot{v}_C = -j/C \neq 0$ at both. However they are special in that

$$\frac{\mathrm{d}j}{\mathrm{d}t} = -j\left(C\frac{\mathrm{d}f}{\mathrm{d}j}\right)^{-1} \tag{5.108}$$

is undefined (since $\mathrm{d}f/\mathrm{d}j$ vanishes).

Without the analysis of (5.97) given in section 5.4.1, we would have difficulty

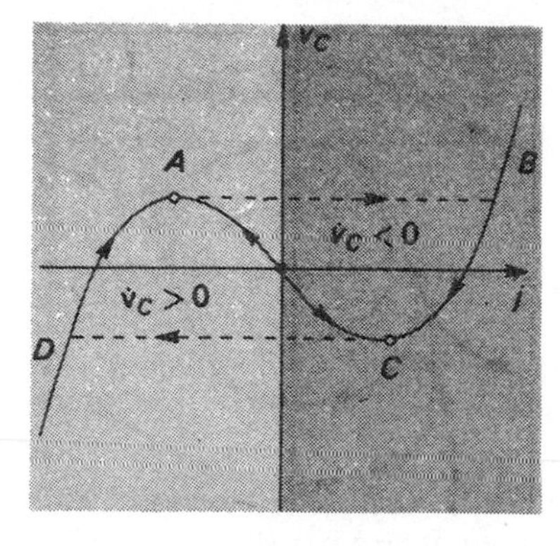

Fig. 5.30. Equation (5.107) means that v_C is increasing for $j < 0$ and decreasing for $j > 0$. Thus, for any initial point on $v_C = f(j)$, the state of the circuit evolves either to point A or to point C.

in deciding what happens when A or C is reached. As it is, we can reasonably assume that j changes discontinuously, instantaneously 'jumping' to B or D, respectively, while v_C remains constant. Thus, relaxation oscillations occur in $j(t)$, but the changes are now genuinely discontinuous.

When a constrained system of dynamical equations is asymptotically related to an unconstrained system of higher dimension, the unconstrained system is called the **regularization** of the constrained one. Thus, (5.97) is the regularization of (5.106).

Sometimes constrained dynamical equations and assumptions of 'jumps' in the variables occur in the natural formulation of a model. In such cases, the regularization can represent a refinement of the model. We can illustrate this by the following example involving an electrical circuit containing a 'neon tube'.

A neon tube, or 'neon', is essentially a glass envelope containing an inert gas (neon) which is fitted with two electrodes. As the potential difference between these electrodes is increased from zero, no current flows until a threshold voltage v_1 is reached. At this voltage the tube suddenly conducts (the inert gas ionizes) and the current jumps to a non-zero value. Thereafter the current increases essentially linearly with increasing voltage. This is the portion $0 \to A \to B \to E$ of the voltage current characteristic for the neon shown in Fig. 5.31.

When the potential difference is reduced from its value at E, however, the tube does not stop conducting when voltage v_1 is reached. Instead, the current continues to decrease with voltage until $v_2 (< v_1)$ is reached; here the current falls sharply to zero. This is the path $E \to B \to C \to D$ in Fig. 5.31. This behaviour is completely reproducible if the voltage is again increased.

There is also physical evidence for a section of characteristic connecting C and A, shown dotted in Fig. 5.31. This does not appear in the simple steady

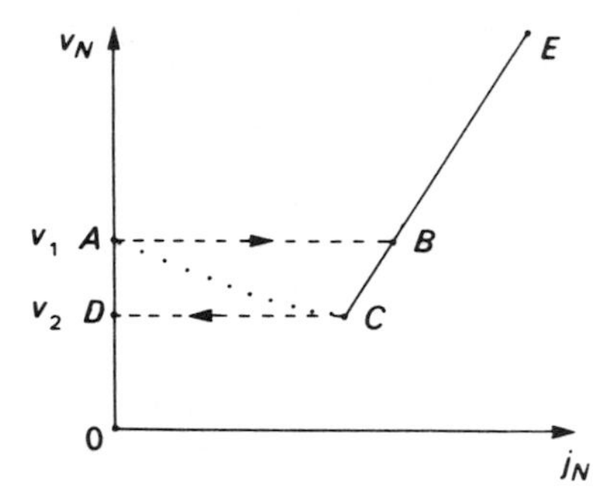

Fig. 5.31. An idealized voltage–current characteristic of a neon tube. Observe the similarity between this and the cubic characteristic; both are folded curves.

voltage experiment described above and need not concern us greatly here. However, it does help to emphasize the 'folded' nature of the neon characteristic.

Example 5.4.1

Show that the dynamics of the circuit shown in Fig. 5.32 are governed by a constrained dynamical equation describing the time dependence of the voltage, v_C, across the capacitor.

Suppose that a small inductor, L, is introduced between nodes 2 and 3 in the figure, what effect does this have on the dynamical equations?

Solution

Let $v_{12} = v_N, v_{13} = v_C$ and $v_{34} = v_R$. Then, with the notation given in Fig. 5.32, we have

$$v_R + v_C = \mathscr{E} = jR + v_C \tag{5.109}$$

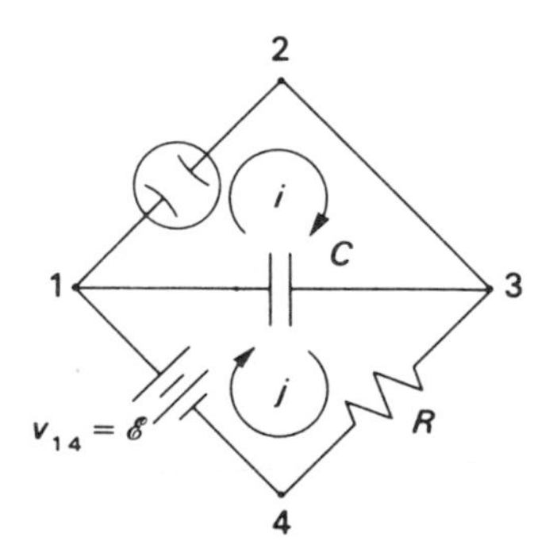

Fig. 5.32. A rudimentary time-base circuit (see section 5.5.1).

and

$$v_C = v_N = f(i), \tag{5.110}$$

where $f(i)$ is the voltage–current characteristic of the neon tube. Furthermore,

$$C\dot{v}_C = j - i. \tag{5.111}$$

Thus the behaviour of the circuit is determined by

$$C\dot{v}_C = \frac{\mathscr{E} - v_C}{R} - i \qquad \text{with } v_C = f(i), \tag{5.112}$$

which is a constrained dynamical equation.

If the inductor, L, is introduced between nodes 2 and 3, Equation (5.110) becomes

$$v_L = v_C - v_N, \tag{5.113}$$

where $v_L = v_{23}$. It then follows that the constraint $v_C = f(i)$ is replaced by the new dynamical equation

$$L\frac{\mathrm{d}i}{\mathrm{d}t} = v_C - f(i). \tag{5.114}$$

This equation, along with the differential equation in (5.112), forms an unconstrained system which is the regularization of (5.112). □

Jumps are built into the dynamics of the capacitor voltage in Example 5.4.1 by the characteristic of the neon. Suppose, v_C is initially zero; the battery, $\mathscr{E}$, charges the capacitor, C, through the resistor, R. The voltage across the neon tube, v_N, is equal to v_C. The current, i, through the neon is zero until $v_N = v_C = v_1$ ($0 \to D \to A$ in Fig. 5.31) when the neon conducts (i.e. there is a jump in i). Provided R is sufficiently large, the neon represents a low resistance path and the capacitor discharges through it. This means that v_C falls. However, the neon continues to conduct until $v_C = v_N = v_2$ when i jumps back to zero ($B \to C \to D$ in Fig. 5.31). The battery then charges the capacitor again. The jumps arise because v_C must be equal to $f(i)$.

The small inductance, L, changes all this. The dynamics are defined on the whole i, v_C-plane and the jumps are smoothed out as $ABCD$ in Fig. 5.31 becomes a limit cycle.

5.5 PIECEWISE MODELLING

In section 5.4 we have considered models in which the dynamical equations are constrained by functional relationships between the dynamical variables. Another, apparently different, type of constraint arises in models which involve different dynamical equations in certain intervals of time or regions of the dynamical variables. Such models can often be thought of as being constructed in pieces and frequently give rise to piecewise continuous or piecewise differentiable time dependences in the variables.

5.5.1 The jump assumption and piecewise models

At first sight these models might appear to be completely separate from the models with constraints considered in section 5.4. However, this is not always the case, as the following example shows.

Example 5.5.1
Show that the constrained dynamical equation (5.112), i.e.

$$C\dot{v}_C = \frac{\mathscr{E} - v_C}{R} - i, \qquad \text{with } v_C = f(i),$$

obtained in Example 5.4.1, is equivalent to a model involving two different dynamical equations in different intervals of time.

Refer to the neon tube characteristic shown in Fig. 5.31 and assume that on CE $v_N = rj_N$. Use the dynamical equations you have developed to find oscillatory solutions for $v_C(t)$ if $v_C = v_2$ and $i = 0$ when $t = 0$.

Solution
Consider the significance of the jump assumption associated with the dynamics of (5.112). If the portion CE of the characteristic shown in Fig. 5.31 is given by

$$j_N = \phi(v_N), \tag{5.115}$$

then the jump assumption implies that

$$i = \begin{cases} 0, & \text{on } OA; \\ \phi(v_C), & \text{on } CE. \end{cases} \tag{5.116}$$

Thus, when the state of the circuit corresponds to a point on OA the dynamical equation is

$$C\dot{v}_C = \frac{\mathscr{E} - v_C}{R}. \tag{5.117}$$

However, when the state corresponds to a point on CE, the dynamical equation is

$$C\dot{v}_C = \frac{\mathscr{E} - v_C}{R} - \phi(v_C). \tag{5.118}$$

If $v_C(0) = v_2$ and $i(0) = 0$, the initial state of the system is at D on OA and the dynamics are given by (5.117). This is a linear (strictly affine) equation with solution

$$v_C(t) = \mathscr{E}[1 - \exp(-t/CR)] + v_2 \exp(-t/CR). \tag{5.119}$$

The voltage $v_C = v_N$ therefore increases towards $\mathscr{E}$.

For oscillations to take place we require $\mathscr{E} > v_1$, so that after time T_1, say, the neon begins to conduct. The state of the circuit jumps to $B \in CE$ and the dynamical equation switches to (5.118). If we take $\phi(v_C) = v_C/r$, then (5.118) becomes

$$\dot{v}_C + \frac{v_C}{C\bar{R}} = \frac{\mathscr{E}}{CR}, \tag{5.120}$$

where $\bar{R}^{-1} = R^{-1} + r^{-1}$. This linear differential equation has solution

$$v_C = \frac{\mathscr{E}\bar{R}}{R}[1 - \exp(-\tau/C\bar{R})] + v_1 \exp(-\tau/C\bar{R}), \tag{5.121}$$

where $\tau = t - T_1$.

For oscillations to occur $\mathscr{E}\bar{R}/R$ must be smaller than v_2 so that v_C now falls as t increases, finally reaching v_2 at $\tau = T_2$, say (as shown in Fig. 5.33). At this voltage the neon stops conducting and the state jumps again to D where (5.117) takes over again and v_C rises once more. □

The voltage across the capacitor in Example 5.5.1 increases to v_1 at a rate determined by R (see (5.119)) and 'relaxes' back to v_2 in a way determined by $\bar{R}$ (see (5.121)). If $R \gg r$, $\bar{R} \simeq r$ and the relaxation takes place much more rapidly than the growth of v_C. If T_2 is sufficiently small compared with T_1 the return to v_2 is essentially instantaneous. Furthermore if $v_1 - v_2$

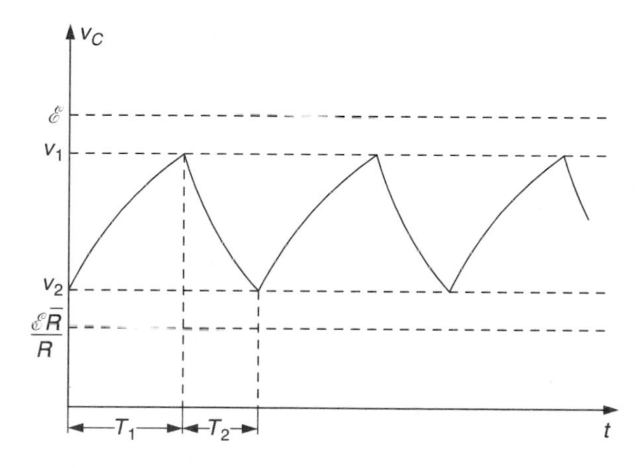

Fig. 5.33. Non-linear 'saw-tooth' oscillations obtained from the linear differential equations (5.117) and (5.118).

is small enough compared with $\mathscr{E}$, v_C is essentially linear for the rise from v_2 to v_1.

These are just the properties required for the time base of an oscilloscope. The x-plates are connected across the capacitor and the signal to be displayed is applied across the y-plates. The electron beam is therefore drawn steadily across the screen and returned rapidly to its starting point, once in every cycle. Synchronization of the period of v_C and the signal allows the latter to be displayed as a stationary wave form.

This example is particularly interesting because the non-harmonic, saw-tooth oscillations are clearly a non-linear phenomenon and yet they are modelled successfully by using the essentially linear differential equations (5.117) and (5.118) in the appropriate time intervals.

The two types of solution are joined continuously at $t = T_1, t = T_1 + T_2$, etc., although the slope $\dot{v}_C$ is discontinuous at these points. This discontinuity arises because of the jumps which take place in the phase plane shown in Fig. 5.31. The equations (5.117) and (5.118) are defined on the separated sets DA and BC of this figure and it is this separation which gives rise to the jumps and hence the discontinuity of $\dot{v}_C$.

5.5.2 A limit cycle from linear equations

There are many models in which non-linear phenomena are constructed by piecing together linear dynamical equations. Perhaps the most intriguing example of this kind is the construction of a limit cycle from two damped, harmonic oscillator equations (Andronov and Chaikin, 1949).

The system is defined by:

$$\ddot{x} + 2k\dot{x} + \omega_0^2 x = \omega_0^2 g, \qquad \dot{x} > 0; \tag{5.122}$$

$$\ddot{x} + 2k\dot{x} + \omega_0^2 x = 0, \qquad \dot{x} \leqslant 0; \tag{5.123}$$

where g is a real constant. We have already examined the possible solutions to (5.123) in section 5.1. For our purpose here it will be necessary to assume that $0 < k < \omega_0$ so that the system undergoes damped, free oscillations of the form

$$x(t) = R\mathrm{e}^{-kt}\cos(\beta t + \theta), \tag{5.124}$$

where $\beta = (\omega_0^2 - k^2)^{1/2}$. In the phase plane defined by $x_1 = x, x_2 = \dot{x}$ the phase portrait has a fixed point at $(0, 0)$ which is a stable focus.

The solutions to (5.122) are easily obtained from the same information by recognizing that this equation can be written as

$$\ddot{y} + 2k\dot{y} + \omega_0^2 y = 0, \tag{5.125}$$

with $y = x - g$. It follows that (5.122) has solutions of the form

$$x(t) = g + \bar{R}\mathrm{e}^{-kt}\cos(\beta t + \bar{\theta}) \tag{5.126}$$

and its phase portrait in the x_1x_2-plane has a stable focus at $(g, 0)$.

In order to show that the system defined by (5.122) and (5.123) has a limit cycle in its phase portrait, we will first prove that a closed trajectory can be found. Let us suppose that at some time $t = t_1$ the phase point is at $(x_1(t_1), 0)$, with $x_1(t_1) > 0$; it moves according to (5.123) and spirals about the origin. At time $t_1 + \pi/\beta$ it again encounters the x_1-axis on the other side of the origin (i.e. it follows the segment of trajectory from P to Q in the lower half-plane of Fig. 5.34). Equation (5.124) gives

$$x_1(t_1) = R\exp(-kt_1)\cos(\beta t_1 + \theta) \tag{5.127}$$

and

$$\begin{aligned} x_1\left(t_1 + \frac{\pi}{\beta}\right) &= R\exp\left[-k\left(t_1 + \frac{\pi}{\beta}\right)\right]\cos(\beta t_1 + \pi + \theta) \\ &= -R\exp(-kt_1)\exp\left(-\frac{k\pi}{\beta}\right)\cos(\beta t_1 + \theta). \end{aligned} \tag{5.128}$$

Thus

$$x_1\left(t_1 + \frac{\pi}{\beta}\right) = -\rho x_1(t_1), \tag{5.129}$$

with $\rho = \exp(-k\pi/\beta) < 1$.

For $t > t_1 + \pi/\beta$ the phase point moves into the upper half-plane in Fig. 5.34 and its motion is governed by (5.122). It consequently executes part of a spiral centred on $(g, 0)$, until at $t = t_1 + 2\pi/\beta$ the positive x_1-axis is encountered at N. Equation (5.126) gives

$$x_1\left(t_1 + \frac{\pi}{\beta}\right) - g = \bar{R}\exp\left[-k\left(t_1 + \frac{\pi}{\beta}\right)\right]\cos(\beta t_1 + \pi + \bar{\theta})$$

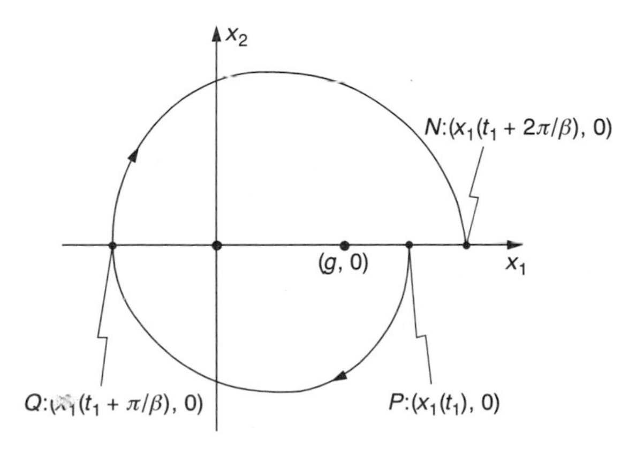

Fig. 5.34. The phase plane of the system defined by (5.122) and (5.123) with $x_1 = x$ and $x_2 = \dot{x}$. The trajectory evolving from $(x_1(t_1), 0)$ is shown. For $x_2 \leqslant 0$, this trajectory spirals about $(0, 0)$; for $x_2 > 0$ it spirals about $(g, 0)$.

and

$$x_1\left(t_1+\frac{2\pi}{\beta}\right)-g=-\bar{R}\exp\left[-k\left(t_1+\frac{\pi}{\beta}\right)\right]\rho\cos(\beta t_1+\pi+\bar{\theta}). \quad (5.130)$$

Thus, using (5.129) we have

$$x_1\left(t_1+\frac{2\pi}{\beta}\right)-g=-[-\rho x_1(t_1)-g]\rho. \quad (5.131)$$

If a closed trajectory exists, there must be a value of $x_1(t_1)$ such that N coincides with P. This will be so provided there is a positive solution to (5.131) with $x_1(t_1+2\pi/\beta)=x_1(t_1)$. It is easily seen that there is only one such solution, namely

$$x_1(t_1)=\frac{g}{1-\rho}=x_1^{(c)}. \quad (5.132)$$

Consequently there is a single closed trajectory.

We may confirm that the closed orbit is an attracting limit cycle by arguing as follows. Consider the trajectory starting from an arbitrary point in the x_1x_2-plane and let $\bar{x}_1 \neq x_1^{(c)}$ be the x_1-coordinate of its first intersection with the positive x_1-axis. Equation (5.131) then implies that the x_1-coordinates of intersections with the positive x_1-axis are given by the sequence

$$\{\bar{x}_1, g(1+\rho)+\rho^2\bar{x}_1, g(1+\rho+\rho^2+\rho^3)+\rho^4\bar{x}_1,\ldots,$$
$$g(1+\rho+\rho^2+\cdots+\rho^{2m-1})+\rho^{2m}\bar{x}_1,\ldots\}, \quad (5.133)$$

where m is the number of complete revolutions of the trajectory about the origin. Since $\rho<1$ the limit of the sequence (5.133) exists and is equal to the sum of the geometric series

$$g\sum_{i=0}^{\infty}\rho^i=\frac{g}{1-\rho}=x_1^{(c)}. \quad (5.134)$$

Thus all trajectories 'home in' on the closed one and we conclude that the system has a unique, attracting limit cycle.

Observe that the above analysis amounts to a calculation of the Poincaré map, $P(x_1)$, of the system (5.122) and (5.123) defined on the positive x_1-axis. The image of a point $x_1=x_1(t_1)$ under this map is $P(x_1)=x_1(t_1+2\pi/\beta)$ given by (5.131). It is easily shown that $x_1^{(c)}=P(x_1^{(c)})$ and $\mathrm{d}P/\mathrm{d}x=\rho^2$. Hence $x_1^{(c)}$ is an asymptotically stable fixed point of P for $\rho<1$, corresponding to an attracting limit cycle in the flow of (5.122) and (5.123).

Example 5.5.2

Consider the electrical circuit shown in Fig. 5.35 (Andronov and Chaikin, 1949). There are two unfamiliar circuit elements involved here. The **triode valve**, T, consists of an evacuated glass envelope with three electrodes which

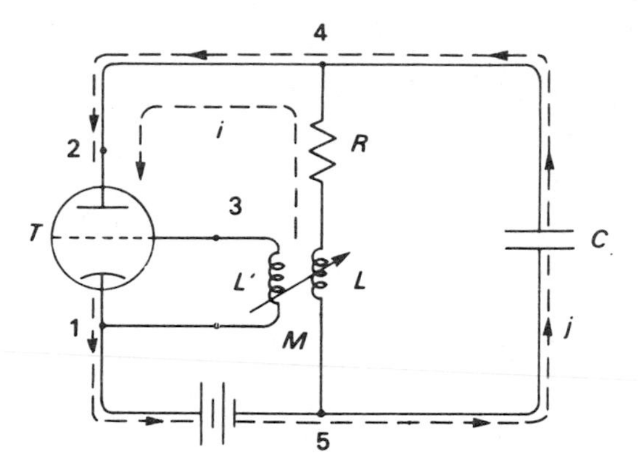

Fig. 5.35. An electrical circuit which provides a realization of the dynamical equations (5.122) and (5.123).

are connected to nodes 1, 2 and 3 of the network. When a potential difference is applied between node 2 and node 1 a current flows through the triode. However the magnitude of this current is a function, f say, of the potential difference between node 3 and node 1. This function is called the **mutual characteristic** of T. Taking currents j and i as shown to satisfy Kirchhoff's current law, we have

$$i + j = f(v_{31}). \tag{5.135}$$

The origin of v_{31} involves the second new circuit element. This is the **mutual inductance**, M. This consists of two coils L and L' wound together; a change of current, i, in L produces a potential difference of $M\,\mathrm{d}i/\mathrm{d}t$ across L', i.e.

$$v_{31} = M\,\mathrm{d}i/\mathrm{d}t, \tag{5.136}$$

where M is a positive constant. Find the dynamical equation governing the current through the resistor R.

Solution

If v_L, v_R, v_C are the potential differences across the corresponding elements, in the direction of the currents i and j, then Kirchhoff's voltage law and the properties of L, C and R give:

$$v_L + v_R = v_C; \qquad L\frac{\mathrm{d}i}{\mathrm{d}t} = v_L; \qquad C\frac{\mathrm{d}v_C}{\mathrm{d}t} = j; \qquad v_R = iR. \tag{5.137}$$

Equations (5.135) and (5.136) imply

$$i + j = f\left(M\frac{\mathrm{d}i}{\mathrm{d}t}\right),$$

while (5.137) gives

$$\frac{\mathrm{d}v_C}{\mathrm{d}t} = \frac{j}{C} = \frac{\mathrm{d}v_L}{\mathrm{d}t} + \frac{\mathrm{d}v_R}{\mathrm{d}t} = L\frac{\mathrm{d}^2 i}{\mathrm{d}t^2} + R\frac{\mathrm{d}i}{\mathrm{d}t}.$$

Thus

$$\frac{\mathrm{d}^2 i}{\mathrm{d}t^2} + \frac{R}{L}\frac{\mathrm{d}i}{\mathrm{d}t} + \frac{i}{LC} = \frac{1}{L} f\left(M\frac{\mathrm{d}i}{\mathrm{d}t}\right). \tag{5.138}$$

□

Now suppose $f(v)$ is simply a step function, i.e.

$$f(v) = \begin{cases} 0, & v \leqslant 0; \\ I_0/C, & v > 0; \end{cases}$$

then, defining $2k = R/L$ and $\omega_0^2 = 1/LC$, we have

$$\frac{\mathrm{d}^2 i}{\mathrm{d}t^2} + 2k\frac{\mathrm{d}i}{\mathrm{d}t} + \omega_0^2 i = \begin{cases} \omega_0^2 I_0, & \mathrm{d}i/\mathrm{d}t > 0; \\ 0, & \mathrm{d}i/\mathrm{d}t \leqslant 0. \end{cases} \tag{5.139}$$

We conclude therefore that the triode valve with a discontinuous mutual characteristic can provide a realization of the dynamical equations (5.122) and (5.123).

EXERCISES

Section 5.1

5.1 For the harmonic oscillator equations

$$\dot{x}_1 = x_2, \qquad \dot{x}_2 = -\omega_0^2 x_1 - 2kx_2,$$

write down the canonical system $\dot{\boldsymbol{y}} = \boldsymbol{J}\boldsymbol{y}$ when:
(a) $k = 0$; (b) $0 < k < \omega_0$; (c) $k = \omega_0$; (d) $k > \omega_0$;
and its solutions. Without finding the transformation matrix $\boldsymbol{M}$ explicitly, show that the solutions in the x_1, x_2 variables take the form given in Equations (5.5)–(5.8).

5.2 Sketch the phase portrait for the canonical system $\dot{\boldsymbol{y}} = \boldsymbol{J}\boldsymbol{y}$ corresponding to the critically damped harmonic oscillator

$$\dot{x}_1 = x_2, \qquad \dot{x}_2 = -k^2 x_1 - 2kx_2.$$

Show that $\boldsymbol{x} = \boldsymbol{M}\boldsymbol{y}$ where

$$\boldsymbol{M} = \begin{bmatrix} 1 & 0 \\ -k & 1 \end{bmatrix}$$

and write down the principal directions at the origin. Use these directions and the method of isoclines to sketch the phase portrait in the $x_1 x_2$-plane. Is this the same as Fig. 5.5(a)?

5.3 Consider an overdamped harmonic oscillator

$$\dot{x}_1 = x_2, \qquad \dot{x}_2 = -\omega_0^2 x_1 - 2kx_2, \qquad k > \omega_0.$$

Write down the canonical system $\dot{\boldsymbol{y}} = \boldsymbol{J}\boldsymbol{y}$. Find the principal directions at the origin and investigate their behaviour as $k \to \infty$. Show that for large k the trajectories in the x_1x_2-plane are essentially vertical except near to the x_1-axis.

5.4 Consider a circuit with three elements, an inductor L, capacitor C and a linear resistor R connected in parallel between two nodes. Show that the current through the inductor satisfies a damped harmonic oscillator equation. Identify the damping k and natural frequency ω_0. Assuming L and R are fixed, find a condition on C for the inductor current to execute damped oscillations.

5.5 Consider the circuit shown below.

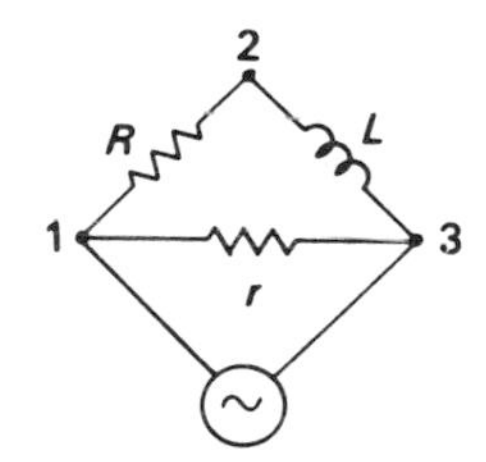

Suppose that

$$v_{13} = E(t) = \begin{cases} 0, & t \leqslant 0; \\ E_0, & t > 0. \end{cases}$$

Show that the current flowing through the resistor R at time $t > 0$ is

$$j_R = \frac{E_0}{R}(1 - e^{-Rt/L}).$$

5.6 An electrical circuit consists of a capacitor C and a linear resistor R joined in parallel between two nodes. Prove that the potential difference across the resistor decays exponentially to zero regardless of its initial value.

5.7 A model for price adjustment in relation to stock level is given as follows. The rate of change of stock (q) is assumed to be proportional to the difference between supply (s) and demand (u), i.e.

$$\dot{q} = k(s - u), \qquad k > 0.$$

The rate of change of price (p) is taken to be proportional to the amount by which stock falls short of a given level q_0 and so

$$\dot{p} = -m(q - q_0), \qquad m > 0.$$

If both s and u are assumed to be affine functions of p, find a second-order differential equation for p. Show that if $u > s$ when $p = 1$ and $\mathrm{d}s/\mathrm{d}p > \mathrm{d}u/\mathrm{d}p$, then the price oscillates with time.

5.8 A cell population consists of 2-chromosome and 4-chromosome cells. The dynamics of the population are modelled by

$$\dot{D} = (\lambda - \mu)D, \qquad \dot{U} = \mu D + \nu U,$$

where D and U are the numbers of 2- and 4-chromosome cells, respectively. Assume that $D = D_0$ and $U = U_0$ at $t = 0$ and find the proportion of 2-chromosome cells present in the population as a function of t. Show that this tends to a saturation level independent of D_0 and U_0 providing $\lambda > \mu + \nu$.

5.9 The motion of a particle P moving in the plane with coordinates x and y is governed by the differential equations

$$\ddot{x} = -\omega^2 x, \qquad \ddot{y} = -y.$$

Plot the parametrized curve $(x(t), y(t))$ in the xy-plane when $x(0) = 0$, $\dot{x}(0) = 1$, $y(0) = 1$, $\dot{y}(0) = 0$ for $\omega = 1, 2, 3$. (These dynamical equations are realized in the **biharmonic oscillator** (Arnold, 1978).)

5.10 Three springs (each of natural length l and spring constant k) and two masses, m, are arranged as shown below on a smooth horizontal table The ends A and B are held fixed and the masses are displaced along the line of the springs and released.

Let x_1 and x_2, respectively, be the displacements (measured in the same sense) of the masses from their equilibrium positions. Show that the equations of motion of the masses are

$$m\ddot{x}_1 = k(-2x_1 + x_2), \qquad m\ddot{x}_2 = k(x_1 - 2x_2).$$

Write these equations in the matrix form $\ddot{\boldsymbol{x}} = \boldsymbol{A}\boldsymbol{x}$, where $\boldsymbol{x} = \begin{bmatrix} x_1 \\ x_2 \end{bmatrix}$,

and find a linear change of variables, $\boldsymbol{x} = \boldsymbol{M}\boldsymbol{y}$, such that $\ddot{\boldsymbol{y}} = \boldsymbol{D}\boldsymbol{y}$ with $\boldsymbol{D}$ diagonal. Hence, find normal coordinates and describe the normal modes of oscillation of the two masses.

Section 5.2

5.11 In a simple model of a national economy, $\dot{I} = I - \alpha C, \dot{C} = \beta(I - C - G)$, where I is the national income, C is the rate of consumer spending and G is the rate of government expenditure. The model is restricted to its natural domain $I \geqslant 0, C \geqslant 0, G \geqslant 0$ and the constants α and β satisfy $1 < \alpha < \infty, 1 \leqslant \beta < \infty$.

(a) Show that if the rate of government expenditure $G = G_0$, a constant, then there is an equilibrium state. Classify the equilibrium state when $\beta = 1$ and show that then the economy oscillates.

(b) Assume government expenditure is related to the national income by the rule $G = G_0 + kI$, where $k > 0$. Find the upper bound A on k for which an equilibrium state exists in the natural domain of this model. Describe both the position and the behaviour of this state for $\beta > 1$, as k tends to the critical value A.

5.12 Suppose $\boldsymbol{U}(s)$ and $\boldsymbol{V}(s)$ are matrices with elements $u_{ij}(s)$ and $v_{ij}(s)$ $(i, j = 1, \ldots, n)$. Verify the relation

$$\frac{\mathrm{d}}{\mathrm{d}s}(\boldsymbol{U}\boldsymbol{V}) = \frac{\mathrm{d}\boldsymbol{U}}{\mathrm{d}s}\boldsymbol{V} + \boldsymbol{U}\frac{\mathrm{d}\boldsymbol{V}}{\mathrm{d}s} = \dot{\boldsymbol{U}}(s)\,\boldsymbol{V}(s) + \boldsymbol{U}(s)\,\dot{\boldsymbol{V}}(s)$$

and show that

$$\int_0^t \boldsymbol{U}(s)\,\dot{\boldsymbol{V}}(s)\,\mathrm{d}s = [\boldsymbol{U}(s)\,\boldsymbol{V}(s)]_0^t - \int_0^t \dot{\boldsymbol{U}}(s)\,\boldsymbol{V}(s)\,\mathrm{d}s.$$

Let

$$\boldsymbol{P} = \int_0^t \mathrm{e}^{-\boldsymbol{A}s}\begin{bmatrix} 0 \\ \cos(\omega s) \end{bmatrix}\mathrm{d}s$$

and

$$\boldsymbol{Q} = \int_0^t \mathrm{e}^{-\boldsymbol{A}s}\begin{bmatrix} 0 \\ \sin(\omega s) \end{bmatrix}\mathrm{d}s.$$

Use (1) to obtain:

(a) $\omega\boldsymbol{P} = \mathrm{e}^{-\boldsymbol{A}t}\begin{bmatrix} 0 \\ \sin(\omega t) \end{bmatrix} + \boldsymbol{A}\boldsymbol{Q}$;

(b) $\omega\boldsymbol{Q} = \left\{-\mathrm{e}^{-\boldsymbol{A}s}\begin{bmatrix} 0 \\ \cos(\omega s) \end{bmatrix}\right\}_0^t - \boldsymbol{A}\boldsymbol{P}$.

Hence show that

$$(\omega^2\boldsymbol{I} + \boldsymbol{A}^2)\boldsymbol{P} = \boldsymbol{A}\begin{bmatrix} 0 \\ 1 \end{bmatrix} + \mathrm{e}^{-\boldsymbol{A}t}\left\{\begin{bmatrix} 0 \\ \omega\sin(\omega t) \end{bmatrix} - \boldsymbol{A}\begin{bmatrix} 0 \\ \cos(\omega t) \end{bmatrix}\right\}$$

and obtain Equation (5.59).

5.13 Consider the electrical circuit shown below, where $u(t)$ is an 'input' voltage applied at A and $y(t)$ is the corresponding 'output' voltage measured across the smaller capacitor at B.

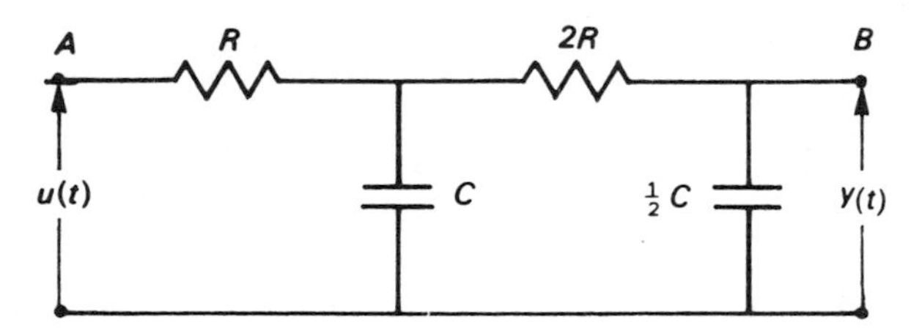

Obtain the dynamical equation

$$R^2C^2\,\ddot{y}(t) + \tfrac{5}{2}RC\,\dot{y}(t) + y(t) = u(t)$$

and derive the equivalent first-order system

$$\frac{dx_1}{d\tau} = x_2, \qquad \frac{dx_2}{d\tau} = -x_1 - \tfrac{5}{2}x_2 + u(RC\tau)$$

where $RC\tau = t$ and $x_1 = y$. Hence show that the steady-state output of the circuit is given by

$$y_S(t) = \frac{2}{3RC}\int_{t_0}^{t} [e^{(s-t)/2RC} - e^{2(s-t)/RC}]u(s)ds$$

for any initial values of $y(t_0)$ and $\dot{y}(t_0)$.

5.14 Consider the electrical circuit shown below.

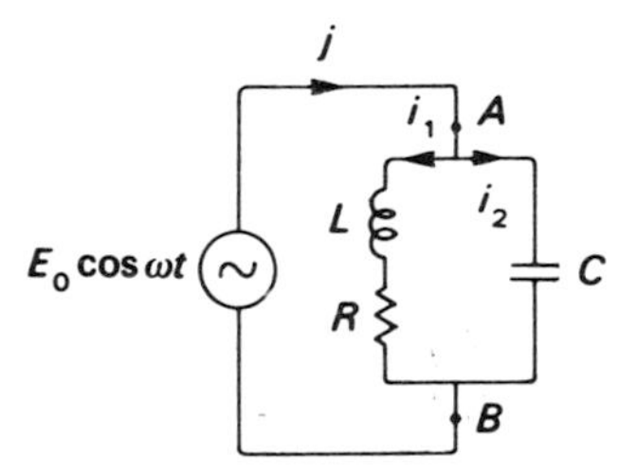

Obtain differential equations for the currents i_1 and i_2 and hence find their steady-state values. Calculate the amplitude, j_0, of the current j. Show that the impedance $Z(= E_0/j_0)$ of the LCR circuit between A and B is given by

$$Z = \frac{L/C}{[R^2 + (\omega L - 1/\omega C)^2]^{1/2}},$$

where R is small compared with ωL. Sketch Z as a function of ω and obtain the resonant frequency of the circuit.

Section 5.3

5.15 Investigate the nature of the fixed points of the competing species model

$$\dot{x}_1 = (2 - x_1 - 2x_2)x_1, \qquad \dot{x}_2 = (2 - 2x_1 - x_2)x_2$$

and indicate their position, together with the $\dot{x}_1 = 0, \dot{x}_2 = 0$ isoclines, in the $x_1 x_2$-plane. Find the principal directions at the fixed points, sketch the phase portrait and interpret it in terms of species behaviour.

5.16 Examine the behaviour of the fixed points of the competing species model

$$\dot{x}_1 = (1 - x_1 - x_2)x_1, \qquad \dot{x}_2 = (v - x_2 - 4v^2 x_1)x_2, \qquad x_1, x_2 > 0,$$

as v varies through positive values. Show that changes in the number and the nature of the fixed points occur at $v = \frac{1}{4}$ and $v = 1$. Sketch typical phase portraits for v in the intervals $(0, \frac{1}{4})$, $(\frac{1}{4}, 1)$ and $(1, \infty)$.

5.17 Consider the prey–predator equations with 'logistic' corrections

$$\dot{x}_1 = x_1(1 - x_2 - \alpha x_1), \qquad \dot{x}_2 = -x_2(1 - x_1 + \alpha x_2),$$

where $0 \leqslant \alpha < 1$. Show that, at the non-trivial fixed point, the centre that exists for $\alpha = 0$ changes into a stable focus for $0 < \alpha < 1$. Sketch the phase portrait.

5.18 Let $(x_1(t), x_2(t))$ be a periodic solution of the prey–predator equations

$$\dot{x}_1 = x_1(a - bx_2), \qquad \dot{x}_2 = -x_2(c - dx_1).$$

Define the average value, $\bar{x}_i$, of x_i by

$$\bar{x}_i = \frac{1}{T}\int_0^T x_i(t)\mathrm{d}t$$

where T is the period of the solution. Show that $\bar{x}_1 = c/d$ and $\bar{x}_2 = a/b$.

Suppose the dynamical equations are modified by the addition of 'harvesting' terms, $-\varepsilon x_i (\varepsilon > 0)$, to $\dot{x}_i$ for $i = 1, 2$. Such terms correspond, for example, to the effects of fishing on fish populations or chemical sprays on insect populations. What effect do these harvesting terms have on the average populations, $\bar{x}_1$ and $\bar{x}_2$?

5.19 Show that the fixed point $(1, J^{-1})$ of the Holling–Tanner model (5.90) is stable if the $\dot{y}_2 = 0$ isocline intersects the parabola $\dot{y}_1 = 0$ to the right of its peak. Hence show that if a phase portrait for this model contains only one limit cycle, which is stable, then the $\dot{y}_2 = 0$ isocline must intersect the parabola to the left of its peak.

5.20 An age-dependent population model is given by

$$\dot{P} = -\mu(P)P + B, \qquad \dot{B} = [\gamma - \mu(P)]B, \qquad \gamma > 0,$$

where P is the total population and B is the birth rate. Prove that $B = \gamma P$ is a union of trajectories for all choices of the function $\mu(P)$. Investigate the phase portrait when $\mu(P) = b + cP$ where $b < 0$ and $c > 0$. Show that for all positive initial values of the variables, both population and birth rate stabilize at non-zero values.

5.21 Find a first integral for the general epidemic model

$$\dot{x} = -2xy, \qquad \dot{y} = 2xy - y,$$

where x is the number of susceptibles and y is the number of infectives, suitably scaled. Hence, or otherwise, sketch the phase portrait in the region $x, y \geqslant 0$. Show that the number of infectives reaches a peak of

$$c_0 - \tfrac{1}{2}(1 + \ln 2)$$

when $x = \frac{1}{2}$, where c_0 is the total number of susceptibles and infectives when $x = 1$. How does the epidemic evolve?

5.22 In Exercise 5.21, suppose the stock of susceptibles is being added to at a constant rate so that

$$\dot{x} = -2xy + 1, \qquad \dot{y} = 2xy - y.$$

Show that the new system has a stable fixed point in the region $x, y > 0$. What are the implications for the development of this epidemic?

5.23 The system

$$\dot{S} = -rIS, \qquad \dot{I} = rIS - \gamma I, \qquad R = 1 - S - I$$

$(r, \gamma > 0)$ models how a disease, which confers permanent immunity, spreads through a population. Let S, I and R be the fractions of the population which are, respectively, susceptible, infected and immune. Define $\sigma = r/\gamma$ and assume initial values $S = S_0$, $I = I_0$ and $R = 0$. Prove that:

(a) if $\sigma S_0 \leqslant 1$, then $I(t)$ decreases to zero;
(b) if $\sigma S_0 > 1$, then $I(t)$ increases to a maximum value of $1 - (1 + \ln(\sigma S_0))/\sigma$ and then decreases to zero.

Show that in both (a) and (b) the population $S(t)$ approaches S_L as $t \to \infty$, where S_L is the unique root of the equation $S_L = (1/\sigma)\ln(S_L/S_0) + 1$ in the interval $(0, 1/\sigma)$.

5.24 A simple model of the molecular control mechanism in cells involves the quantity X of messenger ribonucleic acid and the quantity Y of a related enzyme. The dynamical equations are given by

$$\dot{X} = \frac{a}{A + kY} - b, \qquad \dot{Y} = \alpha X - \beta, \qquad a > bA,$$

where $A, k, a, b, \alpha, \beta$ are positive constants. Prove that both X and Y exhibit persistent oscillations in time.

5.25 Investigate the fixed points of the equations of motion

$$\dot{x}_1 = x_2, \qquad \dot{x}_2 = -\omega_0^2 \sin x_1$$

of the simple pendulum where $\omega_0^2 = g/l$. Here, g is the acceleration due to gravity and l is the length of the pendulum. Find a first integral and sketch the phase portrait. Suppose a damping term $-2kx_2, k > 0$, is added to $\dot{x}_2$. Find the nature of the fixed points of the damped system when k is small and sketch its phase portrait.

Interpret both of the above phase portraits in terms of the motions of the pendulum.

5.26 The behaviour of a simple disc dynamo is governed by the system

$$\dot{x} = -\mu x + xy, \qquad \dot{y} = 1 - \nu y - x^2; \qquad \mu, \nu > 0,$$

where x is the output current of the dynamo and y is the angular velocity of the rotating disc. Prove that for $\mu\nu > 1$ there is one stable fixed point A at $(0, \nu^{-1})$ but for $\mu\nu < 1$, A becomes a saddle point and stable fixed points occur at $(\pm\sqrt{(1-\mu\nu)}, \mu)$.

Section 5.4

5.27 Verify that the system

$$\dot{x}_1 = x_2, \qquad \dot{x}_2 = -x_1 - \varepsilon(x_1^2 - 1)x_2, \qquad \varepsilon \geqslant 0, \tag{1}$$

is equivalent to:

(a) the Van der Pol equation $\varepsilon > 0$;
(b) the undamped harmonic oscillator when $\varepsilon = 0$.

Obtain a system equivalent to (1) in polar coordinates, where $x_1 = r\cos\theta$ and $x_2 = r\sin\theta$, and derive an expression for $\mathrm{d}r/\mathrm{d}\theta$.

Assume ε is small compared with unity and suppose the trajectory passing through $(r, \theta) = (r_0, 0)$ evolves to $(r_1, -2\pi)$. Show that

$$\Delta r = r_1 - r_0 = \int_0^{2\pi} \varepsilon r_0 \sin^2\theta(1 - r_0^2\cos^2\theta)\mathrm{d}\theta$$

to first order in ε. Evaluate this integral and explain why the result implies the existence of a stable limit cycle with approximate radius 2.

5.28 (a) Show that the Van der Pol equation is obtained by differentiating the Rayleigh equation,

$$\ddot{x} + \varepsilon(\tfrac{1}{3}\dot{x}^3 - \dot{x}) + x = 0,$$

with respect to time and setting $y = \dot{x}$. Show also that the two equations can be represented by the same first-order system.

(b) Consider the Liénard representation

$$\dot{x}_1 = x_2 - \varepsilon(\tfrac{1}{3}x_1^3 - x_1), \qquad \dot{x}_2 = -x_1$$

of the Van der Pol equation. Show that the dependence of the characteristic on the parameter ε can be removed by rescaling the variable x_2 and letting $x_2 = \varepsilon\omega$. Sketch the phase portrait of the Van der Pol oscillator in the $x_1\omega$-plane as $\varepsilon \to \infty$.

5.29 Let a resistor R, with a current–voltage characteristic of $j = v^3 - v$, be connected to an inductor L to form a single loop. With the notation in the

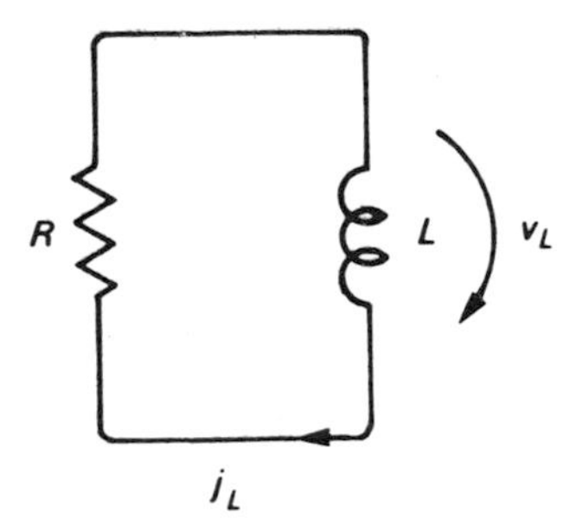

figure, show that the dynamics of the circuit are given by

$$L\frac{\mathrm{d}j_L}{\mathrm{d}t} = v_L$$

for $j_L = v_L - v_L^3$.

Regularize this circuit by introducing a small capacitor in an appropriate way and deduce that the regularized dynamical equations can oscillate.

Section 5.5

5.30 A model for a population which becomes susceptible to epidemics is constructed as follows. The population is originally governed by

$$\dot{p} = ap - bp^2 \tag{1}$$

and grows to a certain value $Q < a/b$. At this population the epidemic strikes and the population is governed by

$$\dot{p} = Ap - Bp^2 \tag{2}$$

where $Q > A/B$. The population falls to the value q, where $A/B < q < Q$, at which point the epidemic ceases and the population is again controlled by (1) and so on. Sketch curves in the p–t plane to illustrate the fluctuations in population with time. Show the time T_1 for the population to increase from q to Q is given by

$$T_1 = \frac{1}{a}\ln\left[\frac{Q(a-bq)}{q(a-bQ)}\right].$$

Find the time T_2 taken for the population to fall from Q to q under the influence of (2) and deduce the period of a typical population cycle.

5.31 A model of a trade cycle is given by

$$\ddot{Y} - \phi(\dot{Y}) + k\dot{Y} + Y = 0, \qquad 0 < k < 2,$$

where Y is the output. The function ϕ satisfies $\phi(0) = 0$,

$$\phi(x) \to L \text{ as } x \to \infty \qquad \text{and} \qquad \phi(x) \to -M \text{ as } x \to -\infty,$$

where M is the scrapping rate of existing stock and L is the net amount of capital-goods trade over and above M. If the function ϕ is idealized to be:

$$\phi(0) = 0; \quad \phi(y) = L, \ y > 0; \quad \phi(y) = -M, \ y < 0;$$

sketch the phase portrait of the differential equation. Show that any non-trivial output $Y(t)$, eventually oscillates with a fixed amplitude and period.

5.32 A model of economic growth is given by $\ddot{Y} + 2\dot{Y} + Y = G(t)$, where $Y(t)$ is the output and $G(t)$ is government spending. The function $G(t)$ has the form $G(t) = 0, 0 \leqslant t < 1$ and $G(t) = G_0, t \geqslant 1$.

Show that there exists an output curve $Y_1(t)$ with the following features:

(a) $Y_1(0) = 0$;
(b) $Y_1(t)$ increases for $t \in [0, 1]$ to a maximum value G_0 where $\dot{Y}_1(1) = 0$;
(c) $Y_1(t) = G_0, t \geqslant 1$.

Investigate the long-term effect on the output $Y(t)$ if, with the same initial conditions ((a) and (b)) as for $Y_1(t)$, the onset of government spending occurs at a time later than $t = 1$.

5.33 Consider a block of mass m whose motion on a horizontal conveyor belt is constrained by a light spring as shown below.

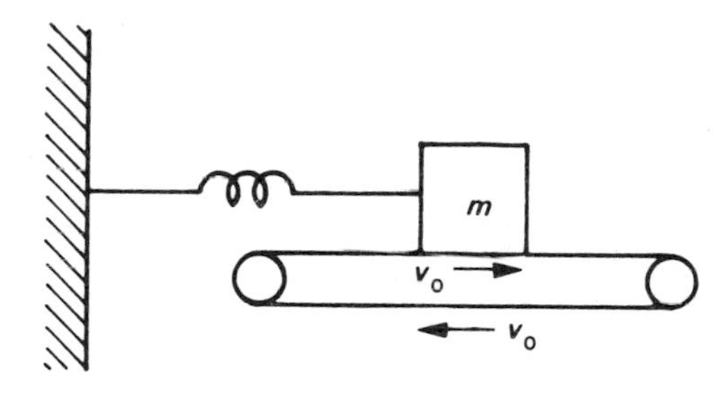

Suppose the belt is driven at a constant speed v_0 and let $x(t)$ be the extension of the spring at time t. The frictional force F exerted on the block by the belt is taken to be

$$F = \begin{cases} F_0, & \dot{x} < v_0 \\ -kx, & \dot{x} = v_0 \\ -F_0, & \dot{x} > v_0, \end{cases}$$

i.e. 'dry' or 'Coulomb' friction.

Show that the equation of motion of the block is

$$m\ddot{x} + kx = -F$$

and sketch the phase portrait in the $x\dot{x}$-plane. Describe the possible motions of the block.

6

Applications II: non-planar phase spaces, families of systems and bifurcations

In this chapter we present some applications of the ideas set out in Chapter 4. However, we also take a wider view of the role that qualitative considerations can play in the modelling process. In particular, most models of dynamical systems involve a family of differential equations rather than an individual equation as the examples in Chapter 5 might suggest. Consequently, it is important to extend our discussion of qualitative analysis to include families of differential equations and the changes of topological behaviour, or **bifurcations**, that can occur within them.

6.1 THE ZEEMAN MODELS OF HEARTBEAT AND NERVE IMPULSE

These models (Zeeman, 1973) are examples of the geometrical approach to modelling with differential equations. They are constructed from a purely qualitative description of the dynamics of the biological mechanisms. The particular differential equations chosen are merely the 'simplest' ones having the required dynamics. There are no specific conditions imposed on their form by any mechanisms whereby the dynamics arise. Only a variety of qualitative features of phase portraits are invoked in order to produce a mathematical description of the heartbeat and nerve impulse. This is in direct contrast to the modelling procedure in section 6.4 where assumptions about the mechanisms involved completely determine the form of the dynamical equations.

The heart is predominantly in one of two states; relaxed (diastole) or contracted (systole). Briefly, in response to an electrochemical trigger each muscle fibre contracts rapidly, remains contracted for a short period and then relaxes rapidly to its stable relaxed state and so on. In contrast, nerve impulses have different dynamic behaviour. The part of the nerve cell which transmits messages is the axon. The operative quantity here is the electrochemically stimulated potential between the inside and outside of the axon. In the absence of stimuli the axon potential remains at a constant rest potential. When a message is being transmitted the axon potential changes sharply and then returns slowly to its rest potential.

These actions have three qualitative features in common and they form the basis of the models. They are:

1. the existence of a stable equilibrium, to which the system returns periodically,
2. a mechanism for triggering the action; and
3. a return to equilibrium after the action is completed.

In this context the main difference between the heartbeat and nerve impulse results from the way in which item 3 takes place.

To construct a model of heartbeat with the above qualities we need to be clear about their interpretation in terms of phase portrait behaviour. Property 1 is interpreted as a stable fixed point in the phase portrait. Property 2 assumes there is a device for periodically moving the state of the system from the fixed point to some other state point. The trajectory through this nearby 'threshold' state then executes the fast action followed by the fast return to equilibrium as required in 3. To show how this can be done we discuss a sequence of models of increasing complexity.

We will use Zeeman's (1973) notation and consider

$$\dot{x} = -\lambda x, \qquad \dot{b} = -b, \tag{6.1}$$

where λ is much larger than 1 (see Fig. 6.1). All trajectories, other than those on the b-axis, are almost parallel to the principal direction of $-\lambda$, the fast eigenvalue, and eventually move in a direction closely parallel to the principal direction of the slow eigenvalue -1. The equations $\dot{x} = -\lambda x$, $\dot{b} = -b$ are called the fast and slow equations respectively. Another example of this type of behaviour is

$$\varepsilon \dot{x} = x - b, \qquad \dot{b} = x, \tag{6.2}$$

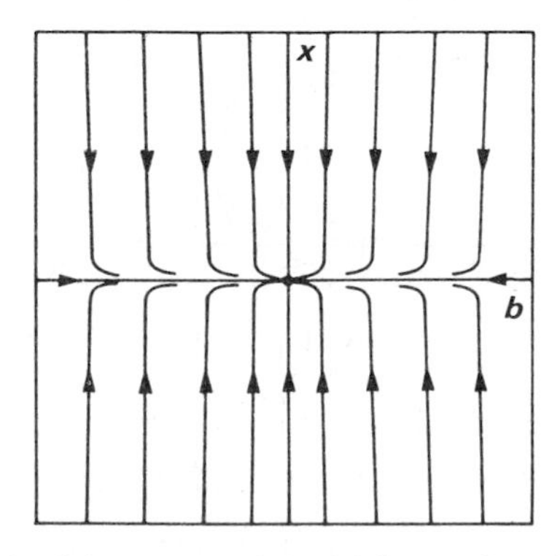

Fig. 6.1. The stable node of the system (6.1) with fast $(-\lambda)$ and slow (-1) eigenvalues.

where ε is positive and much smaller than 1. The eigenvalues are approximately $1/\varepsilon$ and 1 and the principal directions are approximately along $b=0$ and $b=(1-\varepsilon)x$. The fast equation is $\varepsilon\dot{x}=x-b$ as this contains the small term ε which gives the fast eigenvalue $1/\varepsilon$. The phase portrait of system (6.2) is sketched in Fig. 6.2. As in the previous example the fast and slow movements are almost parallel to the principal directions of the node.

Now consider the system

$$\varepsilon\dot{x}=x-x^3-b, \qquad \dot{b}=x, \tag{6.3}$$

for ε positive and small compared to 1. Equation (6.3) can be regarded as a modification of (6.2) with the linear term x replaced by the cubic function $x-x^3$. In fact the linearization of (6.3) at the fixed point $(0,0)$ is just (6.2). The $\dot{x}=0$ isocline for the non-linear system (6.3) is now a cubic curve $b=x-x^3$ as illustrated in Fig. 6.3. The fast movement is almost parallel to

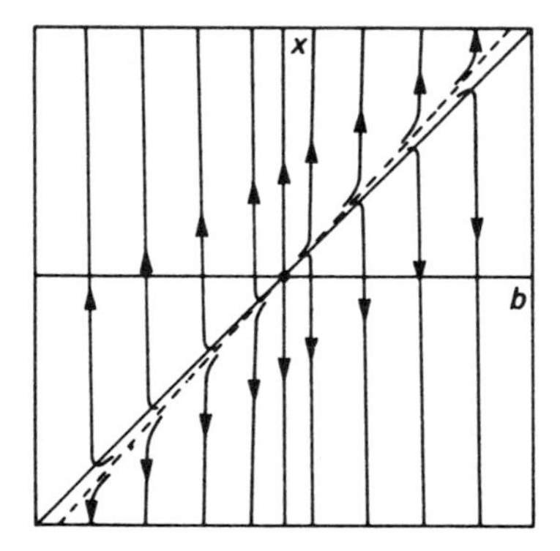

Fig. 6.2. The stable node $\varepsilon\dot{x}=x-b, \dot{b}=x$ with fast movement parallel to the x-axis and slow movement close to $b=(1-\varepsilon)x$.

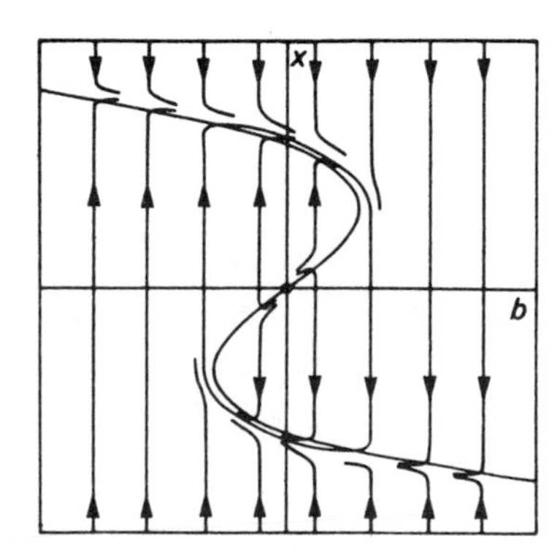

Fig. 6.3. The phase portrait of the Liénard system $\varepsilon\dot{x}=x-x^3-b$, $\dot{b}=x$, where $0<\varepsilon\ll 1$.

the x-direction and the slow movement occurs close to the characteristic curve $b = x - x^3$. The system (6.3) is of Liénard type as discussed in section 5.4.1. The shape of the limit cycle and some neighbouring trajectories are sketched in Fig. 6.3.

Zeeman (1973) makes the adjustment

$$\varepsilon \dot{x} = x - x^3 - b, \qquad \dot{b} = x - x_0 \tag{6.4}$$

to the Liénard system (6.3) to obtain a model of heartbeat. The value of x_0 is taken to be greater than $1/\sqrt{3}$, the value of x at A in Fig. 6.4. The system (6.4) then has a unique fixed point $E = (x_0, b_0)$, where $b_0 = x_0 - x_0^3$, on the upper fold of the characteristic $b = x - x^3$. The linearization of (6.4) at E is

$$\begin{bmatrix} \dot{x} \\ \dot{b} \end{bmatrix} = \begin{bmatrix} 1 - 3x_0^2 & -1 \\ 1 & 0 \end{bmatrix} \begin{bmatrix} x \\ b \end{bmatrix}$$

and therefore E is stable.

The variable x is interpreted as the change in length of a muscle fibre and b is an electrochemical control. The trigger mechanism moves the heart muscle from its equilibrium state E to the nearby state A (as illustrated in Fig. 6.4). The muscle fibre then contracts rapidly as x decreases along the trajectory from A to B. A rapid relaxation of the muscle fibre occurs on the segment CD of the trajectory before returning to its relaxed state E. This behaviour is then repeated periodically by the action of the trigger to mimic the behaviour of the muscle fibres in a beating heart.

To obtain the slow return involved in a nerve impulse a model in $\mathbb{R}^3$ is required. The need for a higher-dimensional system of differential equations can be seen as follows. The fast action of a model like (6.4) arises from a folded characteristic curve. This essentially means that if there is a periodic orbit there must be a second fold and, consequently, a fast return. The introduction of a further dimension avoids this difficulty because the return

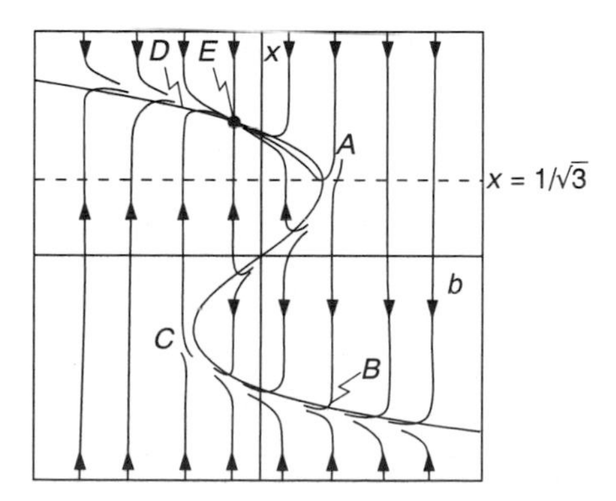

Fig. 6.4. The phase portrait of $\varepsilon \dot{x} = x - x^3 - b$, $\dot{b} = x - x_0$, where $0 < \varepsilon \ll 1$.

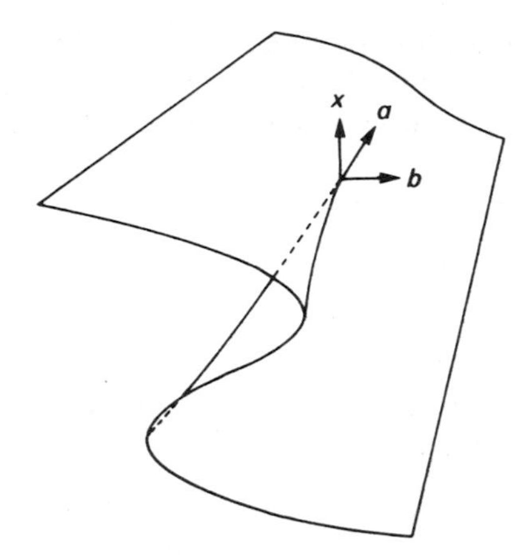

Fig. 6.5. The folded surface $x^3 + ax + b = 0$.

does not have to take place in the vicinity of a fold. We are led, therefore, to models involving a characteristic surface in $\mathbb{R}^3$.

The folded surface M around which the dynamics are constructed is given in (x, a, b)-space by

$$x^3 + ax + b = 0, \tag{6.5}$$

and is illustrated in Fig. 6.5.

To see why Equation (6.5) gives the surface depicted here, it is instructive to consider various $a =$ constant planes in $\mathbb{R}^3$ and see how the surface M intersects these planes as a increases from negative values through zero to positive values. For $a =$ constant, the equation $x^3 + ax + b = 0$ is a cubic curve in the x, b-coordinates. Figure 6.6 illustrates how the fold in the surface for $a < 0$ disappears for $a > 0$.

Differential equations which model fast action with slow return are given by

$$\begin{aligned} \varepsilon\dot{x} &= -(x^3 + ax + b), \\ \dot{a} &= -2x - 2a, \\ \dot{b} &= -a - 1, \end{aligned} \tag{6.6}$$

where ε is a small positive constant. As usual the fast equation gives large value of $\dot{x}$ for points not close to the surface M. This gives rise to fast

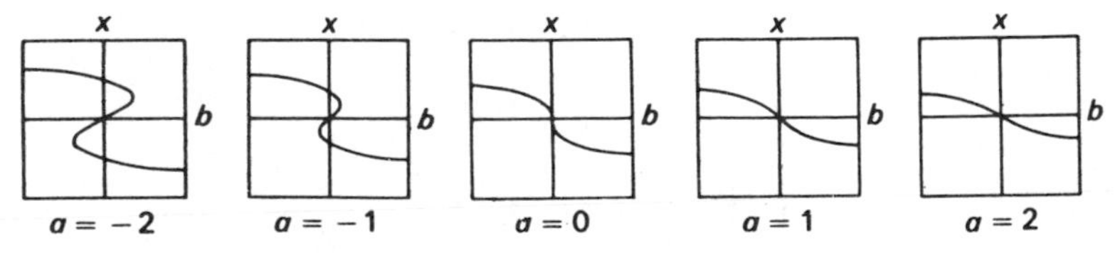

Fig. 6.6. Various $a =$ constant sections of the surface M.

movement in the x-direction in Fig. 6.8. For the slow return we shall see that the equations enable movement from the bottom to the top sheet of M with no abrupt changes in any of the variables.

There is a unique fixed point E for system (6.6) at $x = 1$, $a = -1$, $b = 0$ on the surface M with linearization

$$\begin{bmatrix} \dot{x} \\ \dot{a} \\ \dot{b} \end{bmatrix} = \begin{bmatrix} -2/\varepsilon & -1/\varepsilon & -1/\varepsilon \\ -2 & -2 & 0 \\ 0 & -1 & 0 \end{bmatrix} \begin{bmatrix} x \\ a \\ b \end{bmatrix}. \tag{6.7}$$

This linear system has eigenvalues $\frac{1}{2}(-1 \pm i\sqrt{3})$ and $-2/\varepsilon$, to the leading order in ε, and so E is a stable fixed point. The fast eigenvalue gives rapid contraction towards E in the x-direction at points away from the surface M whilst the slow eigenvalues give spiralling towards E in the vicinity of M. A sketch of the phase portrait of (6.6) viewed from above M is given in Fig. 6.7, the line OF being the top fold of the surface M.

With the help of Fig. 6.8, we can now describe the 'fast action–slow return' dynamic of system (6.6). We assume that the triggering process entails b increasing from zero at the equilibrium point E to some $b_c \geqslant 2/(3\sqrt{3})$ while x and a remain constant. The constraint on b_c ensures that the point $A = (1, -1, b_c)$ is to the right of the fold line OF in Fig. 6.8. The dynamic given by (6.6) quickly changes x from 1 at A to the x-value at B which is almost vertically below A on the lower sheet of M. The trajectory through B then lies close to M and this ensures its relatively slow return to the top sheet before spiralling into E. The whole cycle can then be repeated.

To model the nerve impulse the variables x, a, b have to be suitably interpreted. During the transmission of a nerve impulse the ability of sodium ions to conduct through the cell undergoes a substantial and rapid increase and so sodium conductance is associated with $-x$ (recall x decreases rapidly). The trigger for this abrupt change was taken to be a small change in b. If b

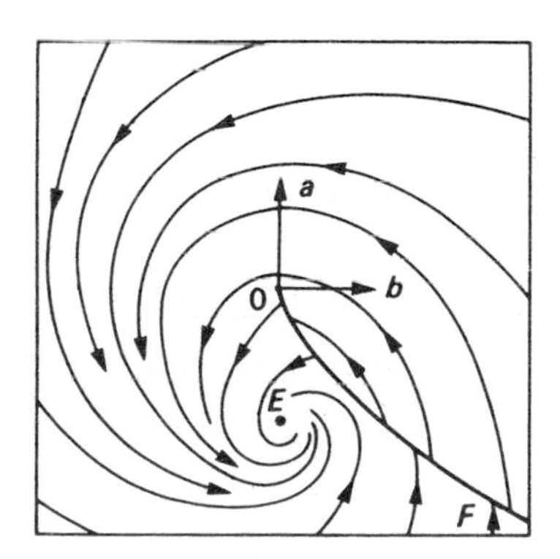

Fig. 6.7. A sketch of the phase portrait of (6.6) near to the surface M as viewed from a point on the positive x-axis.

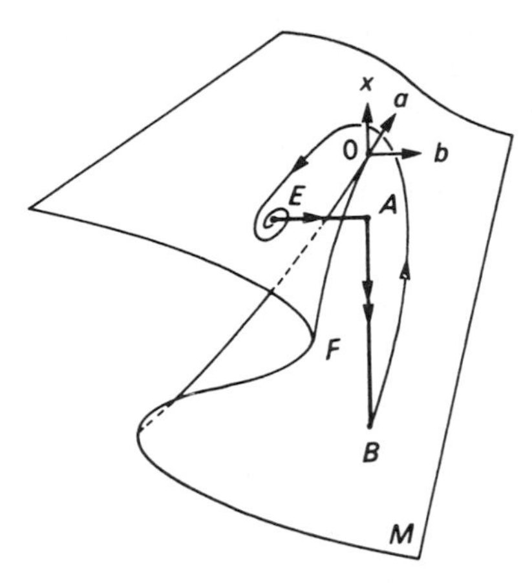

Fig. 6.8. The essential features of the dynamic on M.

is taken to represent the potential across the cell membranes then a small increase in this potential is the nerve impulse triggering mechanism. Finally, the variable a represents potassium conductance. After the action, the conductance of potassium ions does appear to follow changes consistent with the trajectory through B. A slow rise in a is followed by a slow fall to equilibrium at $a = -1$, as the trajectory from A to E swings around O.

6.2 A MODEL OF ANIMAL CONFLICT

Suppose we wish to model the ritualized conflicts which occur within a species when, for example, there is competition for mates, territory or dominance. Conflict occurs when two individuals confront one another and we will suppose that it consists of three possible elements:

1. display;
2. escalation of a fight; or
3. running away.

The population to be modelled is taken to consist of individuals who respond to confrontation in one of a finite number of ways. Suppose that each individual adopts one of the strategies given in the following table.

Index i	Strategy	Initial tactic	Tactic if opponent escalates
1	Hawk (H)	Escalate	Escalate
2	Dove (D)	Display	Run away
3	Bully (B)	Escalate	Run away

An individual playing strategy i against an opponent playing j receives a 'pay-off' a_{ij}. This pay-off is taken to be related to the individual's capability to reproduce (i.e the greater the pay-off, the greater the number of off-spring). Assuming that only pure strategies are played (i.e. an individual is always true to type and always plays the same strategy) and that individuals breed true (i.e. off-spring play the same strategies as their parent), the model is able to determine the evolution of the three sections of the population.

Let x_i be the proportion of the population playing strategy i. It follows that

$$\sum_{i=1}^{3} x_i = 1 \tag{6.8}$$

and

$$x_i \geqslant 0, \qquad i = 1, 2, 3. \tag{6.9}$$

The pay-off to an individual playing i against the rest of the population is

$$\sum_j a_{ij} x_j = (\boldsymbol{A}\boldsymbol{x})_i, \tag{6.10}$$

where $\boldsymbol{A}$ is the 'pay-off matrix'. The average pay-off to an individual is

$$\sum_i x_i (\boldsymbol{A}\boldsymbol{x})_i = \boldsymbol{x}^{\mathrm{T}} \boldsymbol{A}\boldsymbol{x}. \tag{6.11}$$

The 'advantage' of playing i is therefore

$$(\boldsymbol{A}\boldsymbol{x})_i - \boldsymbol{x}^{\mathrm{T}} \boldsymbol{A}\boldsymbol{x}. \tag{6.12}$$

The per capita growth rate of the section of the population playing i is taken to be proportional to this advantage. A suitable choice of units of time then gives

$$\dot{x}_i = x_i((\boldsymbol{A}\boldsymbol{x})_i - \boldsymbol{x}^{\mathrm{T}} \boldsymbol{A}\boldsymbol{x}), \tag{6.13}$$

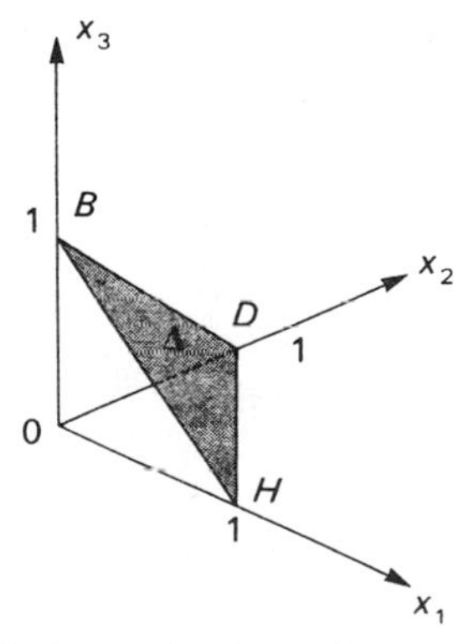

Fig. 6.9. The dynamics of the animal conflict model (6.13) are confined to $\Delta = \{(x_1, x_2, x_3) \mid \Sigma_{i=1}^{3} x_i = 1, x_i \geqslant 0, i = 1, 2, 3\}$. This is the plane triangular 'constraint surface' BDH.

$i = 1, 2, 3$. These equations only represent the state of the population for points in $\mathbb{R}^3$ which satisfy (6.8) and (6.9), i.e. on the region, Δ, in Fig. 6.9.

We can obtain a pay-off matrix by awarding scores, at each confrontation, according to: win $= 6$, lose $= 0$, injury $= -10$, time wasted $= -1$. The actual values chosen here are not important; it is their signs and the order of their magnitudes that are significant. A hawk meeting a dove or a bully simply wins so $a_{12} = a_{13} = 6$. If two hawks meet they fight until one is injured. Each hawk has an equal probability of winning and so the expected gain is $\frac{1}{2}(6 - 10) = -2 = a_{11}$. A dove meeting either a hawk or a bully loses so that $a_{21} = a_{23} = 0$, but two doves continue to display for some time before one gives up. Thus $a_{22} = \frac{1}{2}(6 + 0) - 1 = 2$. Finally, bullies lose against hawks ($a_{31} = 0$), win against doves ($a_{32} = 6$) and have a 50% chance of winning against bullies [$a_{33} = \frac{1}{2}(6 + 0) = 3$]. Thus

$$\boldsymbol{A} = \begin{bmatrix} -2 & 6 & 6 \\ 0 & 2 & 0 \\ 0 & 6 & 3 \end{bmatrix}. \tag{6.14}$$

It is also useful to note that the advantage of a strategy is unchanged by the addition of a constant to a column of $\boldsymbol{A}$. Thus $\boldsymbol{A}$ can be simplified by reducing its diagonal elements to zero by such column changes without changing the dynamical equation (6.13). Therefore we take

$$\boldsymbol{A} = \begin{bmatrix} 0 & 4 & 3 \\ 2 & 0 & -3 \\ 2 & 4 & 0 \end{bmatrix}. \tag{6.15}$$

Example 6.2.1

Show that the dynamical equations (6.13), with $\boldsymbol{A}$ given by (6.15), have a fixed point at $(x_1, x_2, x_3) = (\frac{3}{5}, 0, \frac{2}{5})$. Use the function

$$V(\mathbf{x}) = x_1^{3/5} x_3^{2/5}, \tag{6.16}$$

to show that this fixed point is asymptotically stable, with domain of stability

$$\mathring{\Delta} = \{(x_1, x_2, x_3) | x_1 + x_2 + x_3 = 1;\ x_1, x_2, x_3 > 0\}.$$

Find the remaining fixed points of (6.13), determine their nature and sketch the phase portrait on Δ. What happens to a population consisting entirely of hawks and doves if a mutant bully appears?

Solution

To check that $\mathbf{x} = (\frac{3}{5}, 0, \frac{2}{5})$ is a fixed point of (6.13) we note that $\boldsymbol{x}^{\mathrm{T}}\boldsymbol{A}\boldsymbol{x} = \frac{6}{5}$. For $i = 1$ and 3, $(\boldsymbol{A}\boldsymbol{x})_i = \frac{6}{5}$ and hence $\dot{x}_1 = \dot{x}_3 = 0$. When $i = 2$, $\dot{x}_2 = 0$ since $x_2 = 0$.

We show that the point $(\frac{3}{5}, 0, \frac{2}{5})$ is asymptotically stable on $\mathring{\Delta}$ by using an argument of the Liapunov type. The level surfaces $V(x_1, x_2, x_3)$, cut the $x_2 = 0$

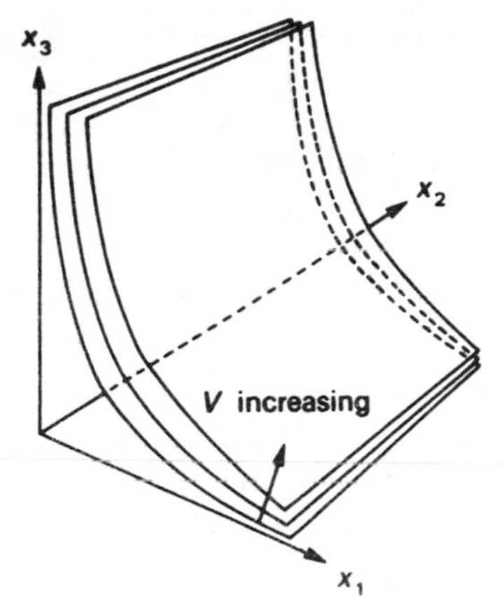

Fig. 6.10. The level surfaces of $V(x_1, x_2, x_3) = x_1^{3/5} x_3^{2/5}$ are generated by translating the hyperbolae $x_3 = Cx_1^{-3/2}$, C constant, in the x_2-direction.

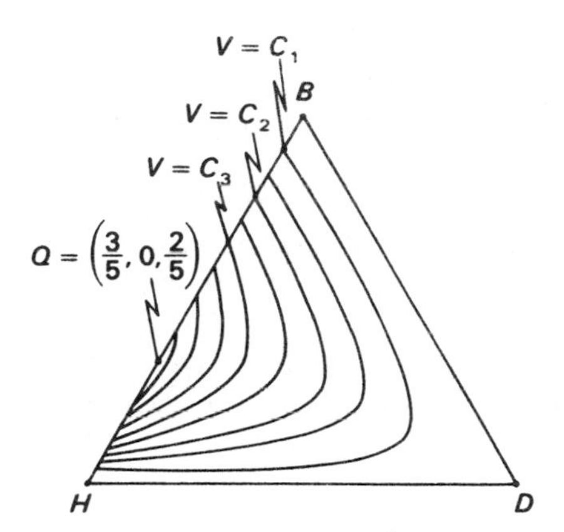

Fig. 6.11. The level curves of $V = x_1^{3/5} x_3^{2/5}$ on $\mathring{\Delta}$ obtained by intersecting the level surfaces of V with $\mathring{\Delta}$. V takes its unique maximum value on Δ at Q so that $C_3 > C_2 > C_1$.

plane in hyperbolae and are invariant under translation parallel to the x_2-axis. A sketch is given in Fig. 6.10. The triangular surface Δ intersects these level surfaces in a system of curves illustrated in Fig. 6.11.

On $\mathring{\Delta}$ the derivative of V along the trajectories of (6.13) is

$$\begin{aligned} \dot{V}(\mathbf{x}) &= V(\mathbf{x}) \left(\frac{3\dot{x}_1}{5x_1} + \frac{2\dot{x}_3}{5x_3} \right) \\ &= V(\mathbf{x}) [(\tfrac{3}{5}, 0, \tfrac{2}{5}) \boldsymbol{Ax} - \boldsymbol{x}^{\mathrm{T}} \boldsymbol{Ax}] \\ &= V(\mathbf{x}) [(1 - x_1 - x_3)(\tfrac{11}{5} - x_1 - x_3) + 5(x_1 - \tfrac{3}{5})^2]. \end{aligned} \tag{6.17}$$

Hence, $\dot{V}(\mathbf{x})$ is positive for $\mathbf{x} \in \mathring{\Delta}$ and V increases along trajectories of (6.13) as t increases.

Using a similar argument to that given in Theorem 3.5.1, we can conclude that along any trajectory in $\mathring{\Delta}$ the function V increases to its maximum at Q (whereas V decreases to a minimum in the proof of Theorem 3.5.1). Thus all trajectories in $\mathring{\Delta}$ approach Q as t increases. It also follows that $\mathring{\Delta}$ is a subset of the domain of stability of Q and therefore cannot contain any fixed points. All the fixed points of (6.13) therefore appear on the boundary of Δ. On $HB(x_2 = 0)$, the equations $\dot{x}_1 = 0$, $\dot{x}_3 = 0$ are given by $x_1(3x_3 - 5x_1x_3) = 0$ and $x_3(2x_1 - 5x_1x_3) = 0$, respectively. Thus, besides Q, there are fixed points at $H = (1, 0, 0)$ and $B = (0, 0, 1)$. Similarly, on $BD(x_1 = 0)$ and $HD(x_3 = 0)$ we find that $D = (0, 1, 0)$ and $P = (\frac{2}{3}, \frac{1}{3}, 0)$ are the only other fixed points on Δ.

We can obtain the behaviour of the trajectories on the boundary of Δ by observing that:

1. on HB, $\dot{x}_1 > 0$ for $x_1 < \frac{3}{5}$, and $\dot{x}_1 < 0$ for $x_1 > \frac{3}{5}$;
2. on BD, $\dot{x}_3 > 0$;
3. on HD, $\dot{x}_2 > 0$ for $x_2 < \frac{1}{3}$, and $\dot{x}_2 < 0$ for $x_2 > \frac{1}{3}$.

Finally, we obtain the phase portrait shown in Fig. 6.12.

Suppose a mutant bully appears in a population consisting originally of only hawks and doves. The population mix is represented by a phase point in $\mathring{\Delta}$ near to HD. Since all trajectories in $\mathring{\Delta}$ approach Q as t increases, we conclude that the state of the population evolves to $x = (\frac{3}{5}, 0, \frac{2}{5})$ where the doves become extinct. □

The dynamical equations (6.13) are only part of a more substantial model described by Zeeman (1979). The complete model involves individuals who

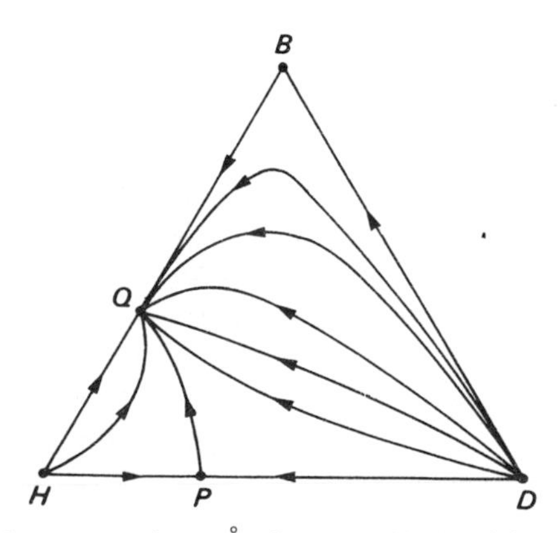

Fig. 6.12 The phase portrait on $\mathring{\Delta}$ of system (6.13) with A given by (6.15).

adopt a fourth strategy namely 'retaliation'. The initial tactic of a retaliator is to display (like a dove) however if its opponent escalates a fight, then the retaliator escalates also. The phase space of this extended problem is three-dimensional, there are four subgames like the one considered above and the dynamics can be represented on a solid tetrahedron. Zeeman's clear account of this model and the perturbations necessary to ensure the existence of an 'evolutionary stable strategy' (i.e. one that is not sensitive to mutations) is a compelling introduction to models whose dynamics make use of game theory.

6.3 FAMILIES OF DIFFERENTIAL EQUATIONS AND BIFURCATIONS

6.3.1 Introductory remarks

The dynamical equations of a model frequently involve time-independent quantities as well as dynamical variables. The values of these parameters are fixed for a particular application of the equations. For example:

1. the per capita growth rate, a, in the population equation

$$\dot{x} = ax; \tag{6.18}$$

2. the quantity ε in the Van der Pol equation

$$\ddot{x} + \varepsilon(x^2 - 1)\dot{x} + x = 0; \tag{6.19}$$

3. the natural frequency, ω_0, and the damping constant, k, in the harmonic oscillator

$$\ddot{x} + 2k\dot{x} + \omega_0^2 x = 0; \tag{6.20}$$

are all parameters. For a particular population described by (6.18) there is a fixed value for a, for a particular mass-on-spring system the values of k and ω_0 are determined; and so on.

However, there are circumstances in which it is advantageous to think of a parameter as a continuous variable which is independent of time. The result is a family of dynamical equations indexed by the parameter. For example, much of the value of modelling is that it allows us to group similar applications together. If we consider all the populations that can be described by (6.18), then each one has a per capita growth rate that is a real number, i.e. their dynamical equations all lie in the family of differential equations defined by (6.18) with real a. Alternatively, the results of an experiment to evaluate oils of different viscosities in the shock absorber shown in Fig. 5.1, would be modelled by a one-parameter family of dynamical equations given by (6.20) with $k > 0$ and ω_0 fixed by the stiffness of the spring. If a number of springs and oils were evaluated then the two-parameter family with both k and ω_0^2 taking positive values would be required.

The qualitative analysis of a family of differential equations involves recognizing the topologically distinct types of phase portrait exhibited by its members. The parameter values at which changes of type take place are called **bifurcation points** of the family. For example, the family given by (6.18) with a real has an attractor at $x = 0$ for $a < 0$ and a repellor at $x = 0$ for $a > 0$. The qualitative behaviour of the members of the family changes as a passes through zero and $a = 0$ is a bifurcation point for this family. The characteristic feature of a bifurcation point is that *every* neighbourhood of it in parameter space contains points giving rise to topologically distinct phase portraits.

Just as examining the fixed points of a differential equation is a useful way to start the task of sketching its phase portrait, so an investigation of the bifurcation points of a family of differential equations gives valuable information about its qualitative behaviour. Indeed, we can carry this analogy further, defining types of bifurcation and setting up criteria for recognizing when they occur. However, we must first find our bifurcation point. The clue as to how this can be done lies in the nature of the bifurcation point itself. The phase portrait at a bifurcation point must be such that an arbitrarily small change in the parameters can result in qualitatively distinct behaviour; in other words, it must be **structurally unstable** (cf. section 5.3.3). It follows that any structurally unstable feature of a phase portrait such as a non-hyperbolic fixed point, a non-hyperbolic closed orbit, or a saddle-connection (cf. Example 3.6.1) can lie at the heart of a corresponding bifurcation.

Consider the one-parameter family of planar systems given by

$$\dot{x}_1 = x_1, \qquad \dot{x}_2 = \mu - x_2^2, \tag{6.21}$$

where μ is real. The differential equation obtained by setting $\mu = 0$ has a fixed point at $(x_1, x_2) = (0, 0)$ with linearization $\dot{x}_1 = x_1, \dot{x}_2 = 0$. For this equation, the origin is a non-hyperbolic fixed point because one of the eigenvalues of the linearized system is zero. Phase portraits for this family are shown in Fig. 6.13 and $\mu = 0$ is clearly a bifurcation point. For $\mu < 0$ there are no fixed points, at $\mu = 0$ a non-hyperbolic fixed point appears at the origin and, as μ increases above zero, this separates into two fixed points: a saddle and a node. The family is said to undergo a **saddle–node bifurcation** at $\mu = 0$ because of the nature of the non-hyperbolic fixed point occurring at the bifurcation point.

Another example is provided by the family

$$\dot{x}_1 = \mu x_1 - x_2 - x_1(x_1^2 + x_2^2), \qquad \dot{x}_2 = x_1 + \mu x_2 - x_2(x_1^2 + x_2^2), \tag{6.22}$$

where μ is a real parameter. For $\mu = 0$ the linearization of (6.22) at $(x_1, x_2) = (0, 0)$ has pure imaginary eigenvalues so that the origin is a non-hyperbolic fixed point. In polar coordinates (6.22) takes the form

$$\dot{r} = r(\mu - r^2), \qquad \dot{\theta} = 1, \tag{6.23}$$

from which it is easy to construct the sequence of phase portraits shown in Fig. 6.14. As μ increases from negative values, the fixed point at the origin changes stability at $\mu = 0$ and a stable limit cycle appears for $\mu > 0$. This family undergoes what is called a **Hopf bifurcation** at $\mu = 0$.

Observe that in the saddle–node and Hopf bifurcations the phase portrait changes that occur are initiated in the neighbourhood of the non-hyperbolic fixed point around which the bifurcation is centred. As a consequence they are referred to as **local bifurcations** and they can be detected by local analysis of the families involved.

The reader may have noticed that Figs 6.13 and 6.14 resemble the phase portraits of differential equations defined on $\mathbb{R}^3$. This is indeed the case. The families (6.21) and (6.22) can both be represented by a single differential equation involving three variables. This equation is obtained by replacing μ

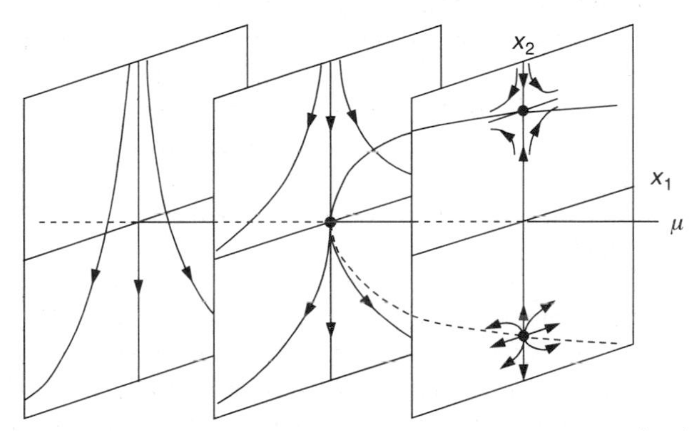

Fig. 6.13. Illustration of the saddle–node bifurcation that occurs in the family of differential equations (6.21) (after Arrowsmith and Place, 1990).

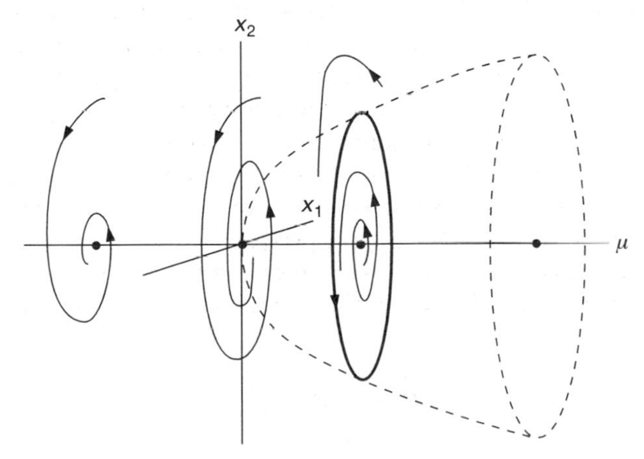

Fig. 6.14. Illustration of the Hopf bifurcation that occurs in the family of differential equations (6.22) (after Arrowsmith and Place, 1990).

by z in (6.21) or (6.22) and adding the third equation $\dot{z}=0$. Clearly, the new equation has solutions $z=\mu$, where μ is a real constant, and the trajectories of the three-dimensional system are confined to the planes of constant z.

6.3.2 The saddle–node bifurcation

The problem of detecting a saddle–node bifurcation in the dynamical equations of a model can be more difficult than our discussion of the family (6.21) suggests. For example, the bifurcation point may not be at the origin of the parameter space, the saddle–node itself may not be at the origin of the phase plane and the whole picture may be obscured by parameters that are not relevant to this bifurcation. Consequently, we must be clear about what 'symptoms' to look for.

In a saddle–node bifurcation either a single fixed point appears and separates into two fixed points, which move apart, or two fixed points move together, coalesce into one and disappear (consider $\mu \mapsto -\mu$ in (6.21)). Technically, however, the distinguishing feature is the nature of the non-hyperbolic fixed point that occurs at the bifurcation point. The linearized system at this fixed point must have one zero and one non-zero eigenvalue. It follows from (2.20) that if $\boldsymbol{A}$ is the coefficient matrix of the linearization then $\det(\boldsymbol{A})=0$ and $\operatorname{tr}(\boldsymbol{A}) \neq 0$.

These conditions, together with the necessary fixed point behaviour, are not sufficient to ensure that a saddle–node bifurcation occurs. This is because the saddle–node bifurcation is characterized by the appearance of quadratic terms in the expression for $\dot{x}_2$ in (6.21). For example, similar behaviour would occur if x_2^2 in (6.21) were replaced by x_2^4 but it would represent only part of the more complicated bifurcation that can arise from the system $\dot{x}_1 = x_1$, $\dot{x}_2 = -x_2^4$. Further discussion of this point is beyond the scope of this book (Arrowsmith and Place, 1990, Chapter 4) but, in practice, unless particular symmetries are in play, it is unlikely that the dynamical equations of a model will fail to contain the necessary quadratic terms. We say that the occurrence of a saddle–node bifurcation is a **generic property** of families of differential equations exhibiting the 'symptoms' described above.

A model of the interaction between a plant population, p, and its animal pollinator population, a, proposed by Soberon *et al.* (1981) exhibits a saddle–node bifurcation. It would be inappropriate to consider the complete model here and we therefore confine attention to a much simplified (and possibly unrealistic) special case for which the dynamical equations take the form

$$\dot{a} = a\left[(K-a)+\frac{p}{1+p}\right],$$

$$\dot{p} = -\frac{p}{2}+\frac{ap}{1+p}, \tag{6.24}$$

where K is a positive parameter. These equations capture the essence of how the saddle–node bifurcation occurs in the full model.

The non-trivial fixed points of (6.24) (i.e. those for which both populations are non-zero) lie at the intersection of the curves

$$p = \frac{a - K}{K + 1 - a}, \tag{6.25}$$

$$p = 2a - 1, \tag{6.26}$$

on which, respectively, $\dot{a}$ and $\dot{p}$ are zero. Substitution of (6.26) into (6.25) yields the quadratic equation

$$2a^2 - 2(K + 1)a + 1 = 0, \tag{6.27}$$

for the a-coordinates of these fixed points. Equation (6.27) has solutions

$$a = \tfrac{1}{2}[(K + 1) \pm \sqrt{(K + 1)^2 - 2}]. \tag{6.28}$$

Since these solutions are complex for $K < K^* = \sqrt{2} - 1$, it follows that:

1. there are no non-trivial fixed points for $K < K^*$;
2. there is one non-trivial fixed point for $K = K^*$;
3. there are two non-trivial fixed points for $K > K^*$.

Equation (6.28) with $K = K^*$ gives the a-coordinate of the single non-trivial fixed point as $a = a^* = 1/\sqrt{2}$ and substitution into (6.26) yields $p = p^* = \sqrt{2} - 1$ for its p-coordinate. The linearization of (6.24) at this fixed point has coefficient matrix, $\boldsymbol{A}$, given by

$$\begin{bmatrix} K^* - 2a + p(1+p)^{-1} & a(1+p)^{-2} \\ p(1+p)^{-1} & -\frac{1}{2} + a(1+p)^{-2} \end{bmatrix}_{(a^*, p^*)}$$

$$= \frac{1}{2\sqrt{2}} \begin{bmatrix} -2 & 1 \\ 2K^* & -K^* \end{bmatrix}. \tag{6.29}$$

Equation (6.29) gives $\det(\boldsymbol{A}) = 0$ and $\operatorname{tr}(\boldsymbol{A}) = -(1 + \sqrt{2})/2\sqrt{2} \neq 0$. Thus we can expect (6.24) to undergo a saddle–node bifurcation at $(a, p) = (a^*, p^*)$ when K increases through K^*.

The family (6.24) also allows us to illustrate a useful geometrical indicator for the occurrence of a saddle–node bifurcation. Figure 6.15 shows the curves (6.25) and (6.26) for K less than, equal to and greater than K^*. As $K < K^*$ increases, the two curves move together, touch at $K = K^*$ and move on to intersect in two points for $K > K^*$. The single fixed point for $K = K^*$ occurs at a tangential intersection of the $\dot{a} = 0$ and $\dot{p} = 0$ isoclines. This fact alone is sufficient to prove that the determinant of the linearized system at such a fixed point is zero.

Consider the planar system

$$\dot{x} = f(x, y), \qquad \dot{y} = g(x, y). \tag{6.30}$$

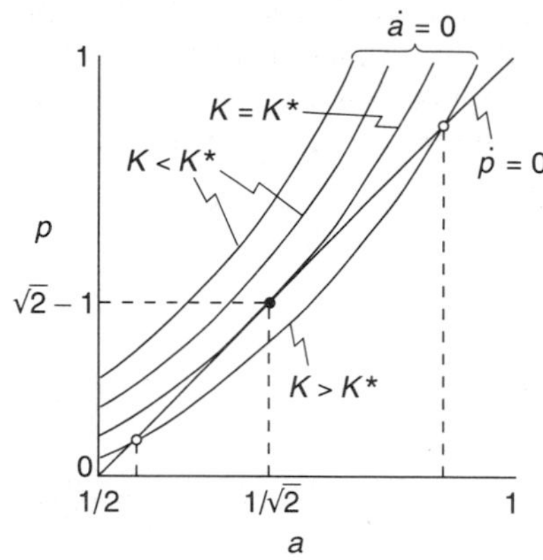

Fig. 6.15. Intersections of the $\dot{a}=0$ and $\dot{p}=0$ isoclines of the plant-pollinator equations (6.24) for: $K<K^*$; $K=K^*$; and $K>K^*$.

Suppose that the equations

$$f(x,y)=0, \qquad g(x,y)=0, \tag{6.31}$$

define curves that meet tangentially at (x^*, y^*). Since the curves have a common tangent at (x^*, y^*) differentiation of (6.31) with respect to x gives

$$\left.\frac{f_x}{f_y}\right|_{(x^*,y^*)} = \left.\frac{g_x}{g_y}\right|_{(x^*,y^*)}, \tag{6.32}$$

where

$$f_x = \frac{\partial f}{\partial x}, \qquad f_y = \frac{\partial f}{\partial y}, \qquad g_x = \frac{\partial g}{\partial x}, \qquad g_y = \frac{\partial g}{\partial y}, \tag{6.33}$$

which implies

$$(f_x g_y - f_y g_x)|_{(x^*,y^*)} = 0. \tag{6.34}$$

However, the left-hand side of (6.34) is the determinant of the linearization of (6.30) evaluated at (x^*, y^*).

It must be emphasized that, while the tangency of these isoclines means that a fixed point with zero determinant occurs, it need not be a saddle–node. For example, as we shall see in section 6.4, if the trace of the linearized system is also zero then a more complicated bifurcation occurs.

6.3.3 The Hopf bifurcation

This bifurcation is characterized by a change in stability of a fixed point accompanied by the creation of a limit cycle. The following theorem gives explicit conditions for such a bifurcation to occur. A proof of the theorem is beyond the scope of our discussion (see Marsden and McCracken, 1976) but we have already covered the techniques needed to use it.

Theorem 6.3.1 (Hopf bifurcation theorem)
Suppose the parametrized system

$$\dot{x}_1 = X_1(x_1, x_2, \mu), \qquad \dot{x}_2 = X_2(x_1, x_2, \mu), \tag{6.35}$$

has a fixed point at the origin for all values of the real parameter μ. *Furthermore suppose the eigenvalues of the linearized system,* $\lambda_1(\mu)$ *and* $\lambda_2(\mu)$, *are purely imaginary when* $\mu = \mu^*$. *If the real part of the eigenvalues* $\mathrm{Re}[\lambda_1(\mu)]$ $(= \mathrm{Re}[\lambda_2(\mu)])$, *satisfies* $(d/d\mu)\{\mathrm{Re}[\lambda_1(\mu)]\}|_{\mu=\mu^*} > 0$ *and the origin is an asymptotically stable fixed point when* $\mu = \mu^*$, *then:*

1. $\mu = \mu^*$ *is a bifurcation point of the system;*
2. *for* $\mu \in (\mu_1, \mu^*)$, *some* $\mu_1 < \mu^*$, *the origin is a stable focus;*
3. *for* $\mu \in (\mu^*, \mu_2)$, *some* $\mu_2 > \mu^*$, *the origin is an unstable focus surrounded by a stable limit cycle, whose size increases with* μ.

To apply this result it is necessary to check that the origin is asymptotically stable when μ is at the bifurcation point μ^*. The linearization theorem can never determine the nature of this non-linear fixed point, because the linearized system is always a centre. Sometimes a strong Liapunov function can be found but, failing this, there is an 'index' that may be capable of detecting the required stability. This index is calculated by the following procedure.

1. Find the linearization $\dot{\boldsymbol{x}} = \boldsymbol{A}\boldsymbol{x}$ of the system (6.35) at the origin of coordinates when $\mu = \mu^*$.
2. Find a non-singular matrix $\boldsymbol{M}$ such that

$$\boldsymbol{M}^{-1}\boldsymbol{A}\boldsymbol{M} = \begin{bmatrix} 0 & \omega^* \\ -\omega^* & 0 \end{bmatrix}, \tag{6.36}$$

   where the eigenvalues of $\boldsymbol{A}$ are $\pm i\omega^*$, with $\omega^* > 0$.
3. Transform the system $\dot{x}_1 = X_1(x_1, x_2, \mu^*)$, $\dot{x}_2 = X_2(x_1, x_2, \mu^*)$ by the change of variables $\boldsymbol{x} = \boldsymbol{M}\boldsymbol{y}$ into $\dot{y}_1 = Y_1(y_1, y_2)$, $\dot{y}_2 = Y_2(y_1, y_2)$.
4. Calculate

$$\begin{aligned} I = \omega^*(Y^1_{111} + Y^1_{122} + Y^2_{112} + Y^2_{222}) \\ + (Y^1_{11}Y^2_{11} - Y^1_{11}Y^1_{12} + Y^2_{11}Y^2_{12} \\ + Y^2_{22}Y^2_{12} - Y^1_{22}Y^1_{12} - Y^1_{22}Y^2_{22}), \end{aligned} \tag{6.37}$$

where

$$Y^i_{jk} = \frac{\partial^2 Y_i}{\partial y_j \partial y_k}(0,0) \quad \text{and} \quad Y^i_{jkl} = \frac{\partial^3 Y_i}{\partial y_j \partial y_k \partial y_l}(0,0),$$

If the index I is negative, then the origin is asymptotically stable.

Example 6.3.1
Show that the equation

$$\ddot{x} + (x^2 - \mu)\dot{x} + 2x + x^3 = 0 \tag{6.38}$$

has a bifurcation point at $\mu = 0$ and that x is oscillatory for some $\mu > 0$.

Solution
The corresponding first-order system is

$$\dot{x}_1 = x_2, \qquad \dot{x}_2 = -(x_1^2 - \mu)x_2 - 2x_1 - x_1^3, \tag{6.39}$$

which has a fixed point at the origin. The eigenvalues of the linearization are given by $\lambda = [\mu \pm \sqrt{(\mu^2 - 8)}]/2$; at $\mu = 0$ they are purely imaginary and $(\mathrm{d}/\mathrm{d}\mu)[\mathrm{Re}(\lambda)]|_{\mu=0} = \frac{1}{2}$. The linearization of (6.39) at the origin is

$$\begin{bmatrix} \dot{x}_1 \\ \dot{x}_2 \end{bmatrix} = \begin{bmatrix} 0 & 1 \\ -2 & \mu \end{bmatrix} \begin{bmatrix} x_1 \\ x_2 \end{bmatrix} \tag{6.40}$$

and so the coefficient matrix is not in the form required to calculate the index I.

A matrix $\boldsymbol{M}$, with the property required in 2 above, is given by $\boldsymbol{M} = [\boldsymbol{u} \,\vdots\, \boldsymbol{v}]$, where $\boldsymbol{u} + \mathrm{i}\boldsymbol{v}$ is an eigenvector of $\boldsymbol{A}$ with eigenvalue $\mathrm{i}\omega^*$. Thus, for example, the matrix

$$\boldsymbol{M} = \begin{bmatrix} 1 & 0 \\ 0 & \sqrt{2} \end{bmatrix}$$

is such that $\boldsymbol{M}^{-1}\boldsymbol{A}\boldsymbol{M}$, where $\boldsymbol{A}$ is obtained from (6.40) with $\mu = 0$, takes the form (6.36) with $\omega^* = \sqrt{2}$. The change of variable $\boldsymbol{x} = \boldsymbol{M}\boldsymbol{y}$ converts the system (6.39) with $\mu = 0$ into

$$\dot{y}_1 = \sqrt{2}y_2, \qquad \dot{y}_2 = -\sqrt{2}y_1 - y_1^2 y_2 - y_1^3/\sqrt{2}, \tag{6.41}$$

and I can now be calculated to be $-2\sqrt{2}$.

Thus as μ increases through 0 the system (6.39) bifurcates to a stable limit cycle surrounding an unstable fixed point at the origin. The system (6.39) is the phase plane representation of the equation (6.38) and the existence of a stable closed orbit implies that $x(t)$ is oscillatory for some $\mu > 0$. □

Lefever and Nicolis (1971) discuss a simple model of oscillatory phenomena in chemical systems that involves a Hopf bifurcation.

They consider the set of chemical reactions

$$\begin{aligned} A &\to X \\ B + X &\to Y + D \\ 2X + Y &\to 3X \\ X &\to E \end{aligned} \tag{6.42}$$

where the inverse reactions are neglected and the initial and final product concentrations A, B, D, E are assumed constant.

The resulting chemical kinetic equations are

$$\dot{X} = a - (b+1)X + X^2 Y, \qquad \dot{Y} = bX - X^2 Y, \tag{6.43}$$

for some positive parameters a and b. There is a unique fixed point P at $(a, b/a)$. The linearization of (6.43) at P has coefficient matrix

$$\left.\begin{bmatrix} 2XY - b - 1 & X^2 \\ b - 2XY & -X^2 \end{bmatrix}\right|_{(a,b/a)} = \begin{bmatrix} b-1 & a^2 \\ -b & -a^2 \end{bmatrix}. \tag{6.44}$$

The determinant of this matrix is a^2 and so the stability of P is determined by the trace. The fixed point is stable for $a^2 + 1 > b$ and unstable for $a^2 + 1 < b$.

To use the Hopf bifurcation theorem we introduce local coordinates $x_1 = X - a$, $x_2 = Y - b/a$ to obtain

$$\begin{aligned} \dot{x}_1 &= (b-1)x_1 + a^2 x_2 + 2a x_1 x_2 + \frac{b}{a} x_1^2 + x_1^2 x_2, \\ \dot{x}_2 &= -bx_1 - a^2 x_2 - 2a x_1 x_2 - \frac{b}{a} x_1^2 - x_1^2 x_2. \end{aligned} \tag{6.45}$$

We now interpret (6.45) as a system parameterized by b with a remaining fixed. The eigenvalues of (6.44) are complex with real part $\frac{1}{2}(b - a^2 - 1)$ for $(a-1)^2 < b < (a+1)^2$ and so

$$\frac{\mathrm{d}}{\mathrm{d}b}\left[\tfrac{1}{2}(b - a^2 - 1)\right] = \tfrac{1}{2}$$

at the bifurcation value $b = a^2 + 1$.

All that needs to be checked now is the stability of (6.45) when $b = a^2 + 1$. For this value of b, the matrix

$$\boldsymbol{M} = \begin{bmatrix} a^2 & 0 \\ -a^2 & a \end{bmatrix}$$

satisfies

$$\boldsymbol{M}^{-1}\begin{bmatrix} b-1 & a^2 \\ -b & -a^2 \end{bmatrix}\boldsymbol{M} = \begin{bmatrix} 0 & a \\ -a & 0 \end{bmatrix} \tag{6.46}$$

and the transformation $\boldsymbol{x} = \boldsymbol{M}\boldsymbol{y}$ converts (6.45) into

$$\begin{aligned} \dot{y}_1 &= a y_2 + (1 - a^2) a y_1^2 + 2a^2 y_1 y_2 - a^4 y_1^3 + a^3 y_1^2 y_2, \\ \dot{y}_2 &= -a y_1. \end{aligned} \tag{6.47}$$

The stability index (6.37) can now be calculated and, since only Y^1_{111} and $Y^1_{11} Y^1_{12}$ are non-zero, $I = -2a^5 - 4a^3$. It follows that system (6.43) bifurcates to an attracting limit cycle surrounding P as b increases through the critical value $1 + a^2$. An example of a typical phase portrait is given in Fig. 6.16.

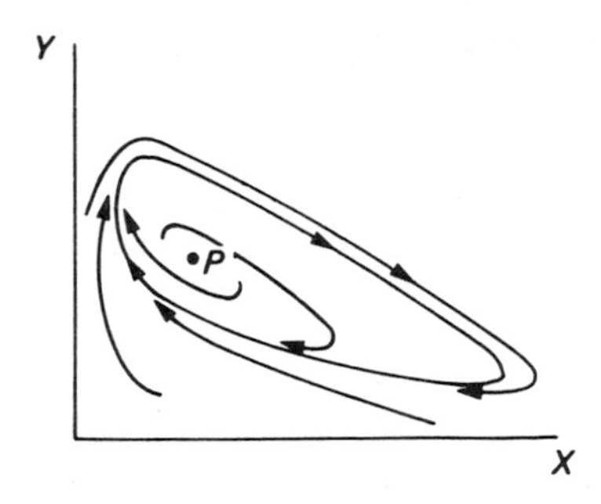

Fig. 6.16. The limit cycle of system (6.43) when $a = 1$, $b = 3$.

6.4 A MATHEMATICAL MODEL OF TUMOUR GROWTH

In this section, we consider a model dealing with the response of an animal's immune system to foreign tissue, in this case a tumour. The model (Rescigno and De Lisi, 1977) is described at length in a survey by Swan (1977); we only sketch the background here.

6.4.1 Construction of the model

Tumour cells contain substances (antigens) which cause an immune response in the host animal. This consists of the production of cells (lymphocytes) which attack and destroy the tumour cells.

The following variables are involved in the model (in each case they refer to the size of the cell population described):

1. L–free lymphocytes on the tumour surface;
2. C–tumour cells in and on the tumour;
3. C_s–tumour cells on the surface of the tumour;
4. $\bar{C}$–tumour cells on the surface of the tumour not bound by lymphocytes;
5. C_f–tumour cells in and on the tumour not bound by lymphocytes.

It follows immediately from these definitions that

$$C = C_f - \bar{C} + C_s. \tag{6.48}$$

The tumour is assumed to be spherical at all times, so that

$$C_s = K_1 C^{2/3}, \tag{6.49}$$

where K_1 is constant, and interactions take place only on the surface of the tumour. Not all tumour cells are susceptible to attack and destruction by lymphocytes and only a proportion of free tumour cell–free lymphocyte interactions result in binding. An equilibrium relation

$$C_s - \bar{C} = K_2 \bar{C} L, \tag{6.50}$$

where K_2 is constant, is assumed between the numbers of free and bound lymphocytes, so that (6.48) and (6.49) imply that

$$C_f = C - K_1 K_2 L C^{2/3}/(1 + K_2 L) \tag{6.51}$$

and

$$\bar{C} = K_1 C^{2/3}/(1 + K_2 L). \tag{6.52}$$

This means that L and C can be taken as the basic variables of the model.

The specific growth rate, $\dot{L}/L$, of the free lymphocyte population is assumed to consist of two terms:

1. a constant death rate λ_1;
2. a stimulation rate $\alpha'_1 \bar{C}(1 - L/L_M)$.

Item 2 shows that, while for small L the stimulation of free lymphocytes increases linearly with $\bar{C}$, there is a maximum population L_M at which the stimulation rate becomes zero. Thus, L satisfies

$$\dot{L} = -\lambda_1 L + \alpha'_1 \bar{C} L(1 - L/L_M). \tag{6.53}$$

The growth rate of the tumour cell population C is given by

$$\dot{C} = \lambda_2 C_f - \alpha'_2 \bar{C} L. \tag{6.54}$$

The first term in (6.54) describes the growth of tumour cells unaffected by lymphocytes, while the second takes account of free tumour cell–free lymphocyte interactions on the tumour surface.

On substituting for $\bar{C}$ and C_f from (6.51) and (6.52), equations (6.53) and (6.54) can be written as

$$\begin{aligned} \dot{x} &= -\lambda_1 x + \alpha_1 x y^{2/3}\left(1 - \frac{x}{c}\right)/(1 + x) \\ \dot{y} &= \lambda_2 y - \alpha_2 x y^{2/3}/(1 + x), \end{aligned} \tag{6.55}$$

where

$$x = K_2 L, \qquad c = K_2 L_M, \qquad y = K_2 C$$

and λ_1, λ_2, α_1, α_2 are positive parameters. Since x and y are populations they must be non-negative, but x cannot exceed c because L is bounded above by L_M.

We now turn to the qualitative implications of the dynamical equations (6.55) and characterize the various phase portraits that can occur, together with their associated parameter ranges.

6.4.2 An analysis of the dynamics

It can be seen at once that the system (6.55) has a saddle point at the origin, for all values of the parameters, by the linearization theorem. The positive

x- and y-axes are trajectories of the system and form the separatrices of the saddle.

To investigate the non-trivial fixed points, however, we write (6.55) in the form

$$\dot{x} = xf(x, y), \qquad \dot{y} = y^{2/3}g(x, y), \tag{6.56}$$

where

$$f(x, y) = -\lambda_1 + \alpha_1 y^{2/3}(1 - \frac{x}{c})/(1 + x) \tag{6.57}$$

and

$$g(x, y) = \lambda_2 y^{1/3} - \alpha_2 x/(1 + x). \tag{6.58}$$

The fixed point equations

$$f(x, y) = 0, \qquad g(x, y) = 0$$

give

$$y^{2/3} = \frac{\lambda_1}{\alpha_1}\left(\frac{1 + x}{1 - (x/c)}\right) = \left(\frac{\alpha_2}{\lambda_2}\frac{x}{1 + x}\right)^2,$$

so that the x-coordinates of the non-trivial fixed points satisfy

$$\psi(x) = \frac{x^2(1 - (x/c))}{(1 + x)^3} = \frac{\lambda_1 \lambda_2^2}{\alpha_1 \alpha_2^2}. \tag{6.59}$$

The function $\psi(x)$ has a unique global maximum at $x^* = 2c/(c + 3)$, with $\psi(x^*) = 4c^2/27(c + 1)^2$. Now define

$$\mu_1 = \frac{\lambda_1 \lambda_2^2}{\alpha_1 \alpha_2^2} - \frac{4c^2}{27(c + 1)^2}, \tag{6.60}$$

when it follows that (6.59) has:

(a) no real roots for $\mu_1 > 0$;
(b) exactly one root at $x = x^*$ for $\mu_1 = 0$; and
(c) exactly two roots, x_1^* and x_2^*, where $0 < x_1^* < 2c/(c + 3) < x_2^* < c$, when $\mu_1 < 0$.

Geometrically, (a)–(c) correspond to the curves $f(x, y) = 0$ and $g(x, y) = 0$ having the intersections shown in Fig. 6.17.

We now consider the phase portraits for each of these situations.

(a) $\mu_1 > 0$

There are no non-trivial fixed points. Recalling that the origin is a saddle point, the signs of $\dot{x}$ and $\dot{y}$ are sufficient to construct a sketch of the phase portrait as in Fig. 6.18.

Observe that $y \to \infty$ as $t \to \infty$, for all initial states of the populations. This corresponds to uncontrolled growth of the tumour.

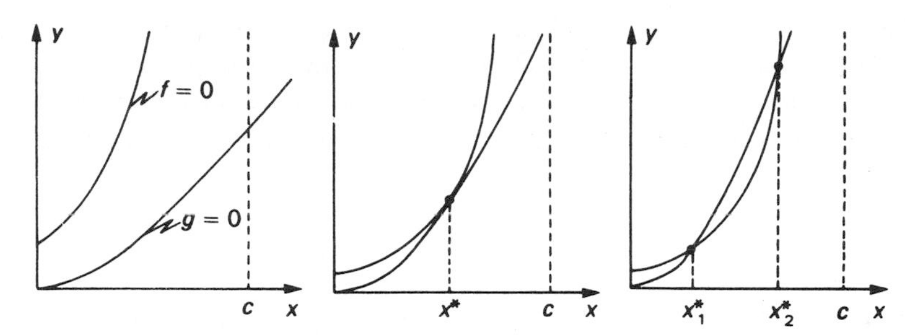

Fig. 6.17. The equation $f(x, y) = 0$ defines a curve which is concave and has asymptote $x = c$, while the curve $g(x, y) = 0$ has asymptote $y = (\alpha_2/\lambda_2)^3$. The possible configurations of these curves are shown: (a) $\mu_1 > 0$; (b) $\mu_1 = 0$; (c) $\mu_1 < 0$.

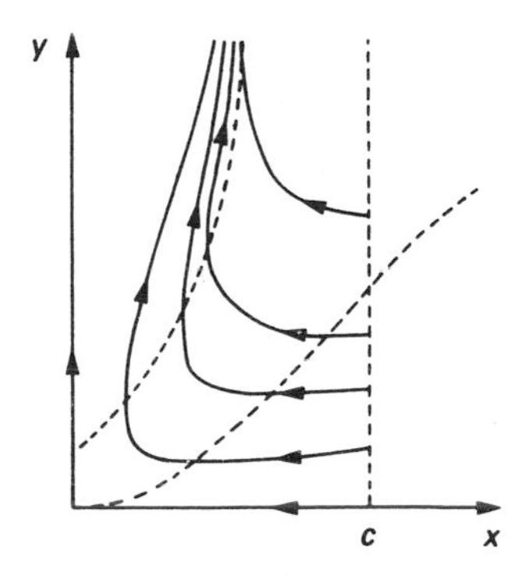

Fig. 6.18. Phase portrait for (6.55) with $\mu_1 > 0$. All trajectories approach $x = c$ asymptotically as $t \to \infty$.

(*b*) $\mu_1 = 0$

There is a single non-trivial fixed point (x^*, y^*), where $x^* = 2c/(c + 3)$. The coefficient matrix of the linearized system is

$$W = \begin{bmatrix} xf_x & xf_y \\ y^{2/3}g_x & y^{2/3}g_y \end{bmatrix}\Bigg|_{(x^*, y^*)}, \tag{6.61}$$

since $f(x^*, y^*) = g(x^*, y^*) = 0$. We have used the notation defined in (6.33). It follows that

$$\det(W) = x^* y^{*2/3}(f_x g_y - f_y g_x)|_{(x^*, y^*)}. \tag{6.62}$$

However, as Fig. 6.17(b) shows, when $\mu_1 = 0$ the slopes of the curves $f(x, y) = 0$ and $g(x, y) = 0$ are the same at (x^*, y^*). This implies that

$$\left.\frac{f_x}{f_y}\right|_{(x^*, y^*)} = \left.\frac{g_x}{g_y}\right|_{(x^*, y^*)}, \tag{6.63}$$

so that det $(W) = 0$. Thus (x^*, y^*) is a non-hyperbolic fixed point and linear analysis cannot determine its nature.

We will return to the problem of the precise nature of the fixed point (x^*, y^*) later on. However, the global behaviour of the phase portrait for $\mu_1 = 0$ is clearly such that uncontrolled tumour growth occurs for the majority of initial states (the area of the domain of stability of (x^*, y^*) is finite).

(c) $\mu_1 < 0$
In this case, there are two non-trivial fixed points, $P_1 = (x_1^*, y_1^*)$ and $P_2 = (x_2^*, y_2^*)$ with $0 < x_1^* < x^* < x_2^* < c$ and $x^* = 2c/(c+3)$.

If $W_i (i = 1, 2)$ is the coefficient matrix of the linearization at (x_i^*, y_i^*), then (6.57), (6.58) and (6.62) give

$$\det(W_i) = \left\{ \alpha_1 \lambda_2 x y^{2/3} \left(\frac{2c}{x} - c - 3 \right) / 3c(1+x)^2 \right\} \Big|_{(x_i^*, y_i^*)}. \tag{6.64}$$

Thus, $\det(W_1)$ is positive and $\det(W_2)$ is negative. We can immediately conclude that P_2 is a saddle point, but the stability of P_1 is determined by $\operatorname{tr}(W_1)$. The matrix (6.6.1), evaluated at (x_1^*, y_1^*), gives

$$\begin{aligned} \operatorname{tr}(W_1) &= (x f_x + y^{2/3} g_y)|_{(x_1^*, y_1^*)} \\ &= \frac{\lambda_2}{3} \left\{ 1 - 3\left(\frac{\lambda_1}{\lambda_2}\right)\left(\frac{1+c}{c}\right) \frac{x}{(1+x)(1-x/c)} \right\} \Big|_{x_1^*}, \\ &= \eta(x_1^*) \end{aligned} \tag{6.65}$$

where (6.57) has been used to eliminate y. The function $x\{(1+x)(1-x/c)\}^{-1}$ is strictly increasing on $(0, c)$ and therefore, $\eta(x)$ is a strictly decreasing function of x on $(0, x^*)$ (recall $0 < x^* < c$). There are several possibilities for $\operatorname{tr}(W_1) = \eta(x_1^*)$ as shown in Fig. 6.19.

Observe that $\eta(x^*) \geqslant 0$ implies that $\eta(x_1^*) > 0$, for any x_1^*, (curves (a) and (b)) whereas if $\eta(x^*) < 0$ then $\eta(x_1^*)$ may be positive (curve (c)), zero (curve (d)) or negative (curve (e)). We can express this result in terms of the

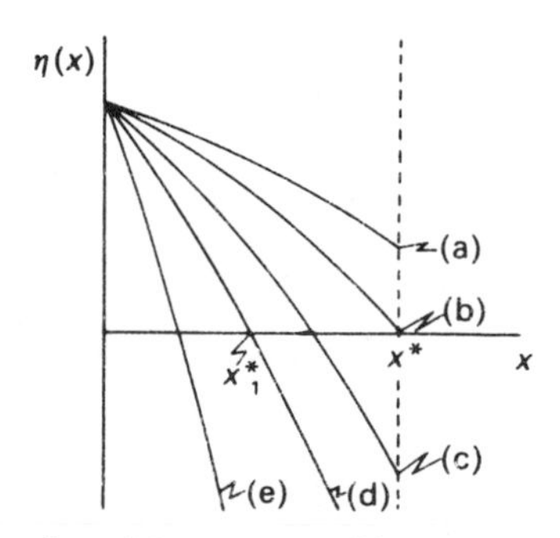

Fig. 6.19. Possible forms for $\eta(x)$ versus x. Observe $\operatorname{tr}(W_1) = \eta(x_1^*)$ is positive for (a)–(c), zero for (d) and negative for (e).

parameters of the model in the following way. When $x = x^* = 2c/(c+3)$,

$$\eta(x^*) = \frac{\lambda_2}{3}\left\{1 - \left(\frac{\lambda_1}{\lambda_2}\right)\left(\frac{2(c+3)}{(c+1)}\right)\right\}, \tag{6.66}$$

which is zero (i.e. curve (b) in Fig. 6.19) if $\lambda_1/\lambda_2 = (c+1)/2(c+3)$. Now define

$$\mu_2 = \frac{\lambda_1}{\lambda_2} - \frac{(c+1)}{2(c+3)} \tag{6.67}$$

when it follows that if:

(a) $\mu_2 \leqslant 0$ then $\eta(x^*) \geqslant 0$ and $\mathrm{tr}(\boldsymbol{W}_1) - \eta(x_1^*) > 0$ and hence P_1 is unstable;
(b) $\mu_2 > 0$ then sign of $\mathrm{tr}(\boldsymbol{W}_1)$ is not determined.

When $\mu_2 > 0$, we can say that $\eta(x)$ moves through the sequence of curves (c), (d) and (e) in Fig. 6.19 as μ_2 increases from zero. Correspondingly, the sign of $\mathrm{tr}(\boldsymbol{W}_1) = \eta(x_1^*)$ is positive, zero and then negative. Thus, as μ_2 increases through $(0, \infty)$ (with μ_1 constant) the fixed point P_1 changes from being unstable at $\mu_2 = 0$ to being stable at sufficiently large μ_2.

The results obtained above are summarized in Fig. 6.20. The μ_1, μ_2-plane is shown, divided into three major regions: A, where there are no non-trivial fixed points; B, where P_1 is unstable; and C, where P_1 is stable. We have not determined the shape of the boundary between B and C in the figure and so it is shown schematically as a broken straight line.

The local behaviour shown in Fig. 6.20 is qualitatively equivalent to that of the simpler two-parameter local family

$$\dot{x} = -(\bar{\mu}_1 + y^2), \qquad \dot{y} = -(x + \bar{\mu}_2 y + y^2), \tag{6.68}$$

investigated by Takens (1974). The bifurcations occurring in (6.68) are illustrated in Figs 6.21 and 6.22. It is not difficult to confirm that (6.68) has no fixed points for $\bar{\mu}_1 > 0$, one for $\bar{\mu}_1 = 0$ and two for $\bar{\mu}_1 < 0$. When $\bar{\mu}_1$ is negative, one of the two fixed points is a saddle and the other is a node/focus. What is more, the stability of the latter changes on the curve $\bar{\mu}_2 = 2\sqrt{-\bar{\mu}_1}, \bar{\mu}_1 < 0$. Observe also that the local phase portraits at the fixed points are the same in Fig. 6.22(b) and (c). Thus, from the point of view of local behaviour B_1' and B_2' are equivalent and we can identify A with A', B with $B' = B_1' \cup B_2'$ and C with C'.

It is the nature of the non-hyperbolic fixed point at the origin of (6.68) with $\bar{\mu}_1 = \bar{\mu}_2 = 0$ that is responsible for the form of Fig. 6.21. It can be seen that both the trace and the determinant of the linearized system at this fixed point are zero. The local phase portrait, shown in Fig. 6.23, is characterized by a pair of separatrices which form a cuspoidal curve at the fixed point and the bifurcation exemplified by (6.68) is consequently called a **cusp bifurcation**.

The relationship between (6.55) and (6.68) extends beyond local behaviour; in fact, every global phase portrait presented by Swan (1977) for (6.55) has

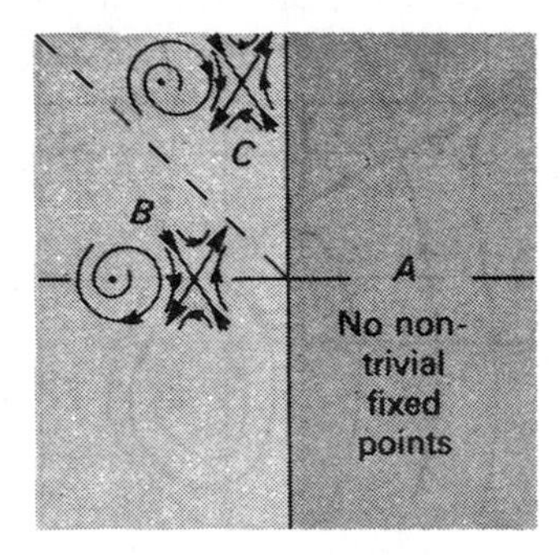

Fig. 6.20. Summary of results of linear stability analysis of (6.55). The nature of the fixed points on $\mu_1 = 0$ and the boundary between B and C are not revealed by this analysis.

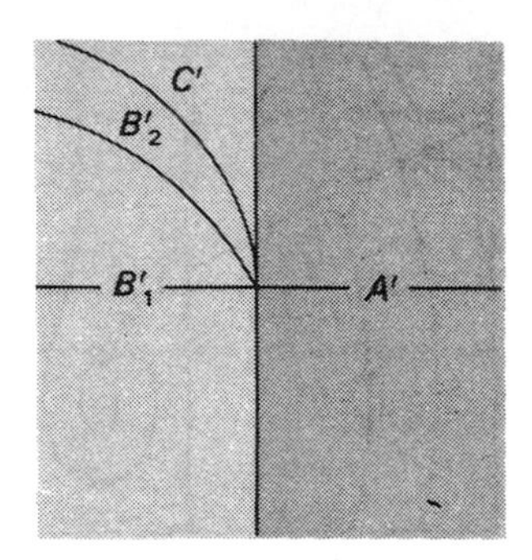

Fig. 6.21. Summary of global analysis of (6.68). The $\bar{\mu}_1\bar{\mu}_2$-plane is divided into four regions A', B'_1, B'_2, C' in which the phase portraits are as shown in Fig. 6.22.

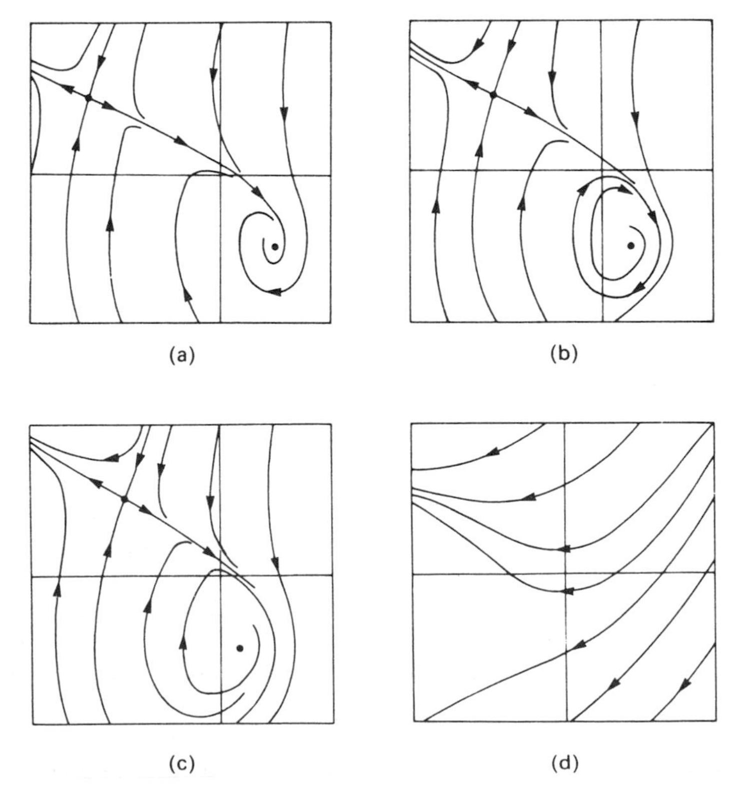

Fig. 6.22. Phase portraits for (6.68) when $(\bar{\mu}_1, \bar{\mu}_2)$ belongs to: (a) C'; (b) B'_2; (c) B'_1; (d) A'.

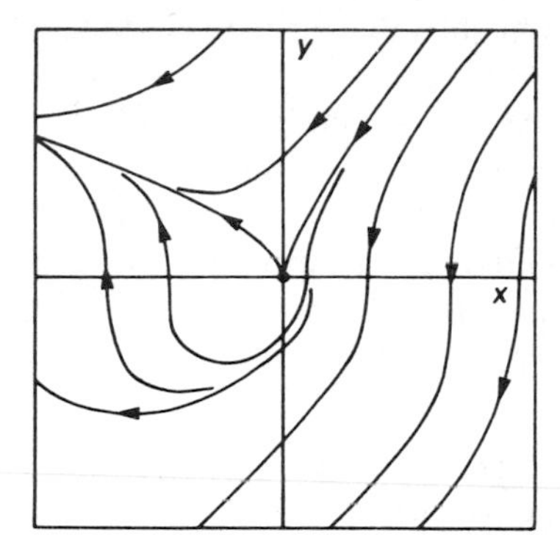

Fig. 6.23. The non-simple fixed point of (6.68) with $\bar{\mu}_1 = \bar{\mu}_2 = 0$.

a qualitatively equivalent counterpart in Takens' treatment of (6.68). More precisely, it can be shown that (6.55) and (6.68) are **qualitatively equivalent families** in the sense that all the bifurcational behaviour exhibited by one must also occur in the other. Takens has given a complete global analysis of (6.68) and this can therefore be used to guide the study of (6.55).

For example, for $\bar{\mu}_1 < 0$, the boundary between B' and C' in Fig. 6.21 is a Hopf bifurcation in (6.68). The reader will recall that (6.55) has purely imaginary eigenvalues on the CB-boundary ($\mathrm{tr}(\boldsymbol{W}_1) = 0$) and there is a change in stability of P_1. These are the symptoms of a Hopf bifurcation and a limit cycle consistent with this appears in Swan's treatment.

As $\bar{\mu}_1$ decreases through zero in (6.68) with $\bar{\mu}_2 \neq 0$, the pair of fixed points appear at a saddle–node bifurcation. The equivalence of the families (6.55) and (6.68) means that the same type of bifurcation occurs as μ_1 decreases through zero with $\mu_2 \neq 0$. Recall that our earlier analysis showed that a single non-trivial fixed point of (6.55) is present when $\mu_1 = 0$ and the determinant of the linearized system at this point is zero. This is consistent with the fixed point being a saddle–node.

There is no mention in Swan's work of the bifurcation corresponding to the $B'_2 B'_1$-boundary in Fig. 6.21 but family equivalence of (6.55) and (6.68) implies its existence. As μ_2 decreases, at fixed $\mu_1 < 0$, the limit cycle created at the CB-boundary expands until it reaches the saddle point P_2 and a saddle connection forms when (μ_1, μ_2) is on a curve corresponding to the $B'_2 B'_1$-boundary. Of course, such a configuration is structurally unstable and a **saddle-connection bifurcation** (in which the phase portrait changes from one like Fig. 6.22(b) to one like Fig. 6.22(c)) takes place as μ_2 moves through this boundary. The formation and subsequent destruction of the saddle-connection results in the sudden disappearance of the attracting limit cycle at the bifurcation point. This means that the bounded oscillations arising from initial states within the limit cycle in (b) turn into the uncontrolled growth exhibited by the flow in (c).

It is apparent that uncontrolled growth dominates the behaviour of the family (6.55). In parameter space, only points (μ_1, μ_2) in C and a relatively small part of B, where the limit cycle persists, offer a possibility of remission or controlled cycling of the tumour. What is more, such an outcome is only possible for initial states lying in the basin of attraction of P_1 or the limit cycle. The problem is to find realistic modifications of the model which enlarge these favourable regions of state and parameter space. We must refer the reader to the excellent review by Swan (1977) for further details.

6.5 SOME BIFURCATIONS IN FAMILIES OF ONE-DIMENSIONAL MAPS

6.5.1 The fold bifurcation

This bifurcation occurs in families of one-dimensional maps and it is associated with a non-hyperbolic fixed point with eigenvalue equal to $+1$.

Consider the one-parameter family of differential equations with polar form

$$\dot{r} = r[(r-1)^2 - \mu r], \qquad \dot{\theta} = 1, \tag{6.69}$$

where $\mu \in (-\mu_0, \mu_0)$, with $0 < \mu_0 \ll 1$. The trajectories of this system repeatedly intersect the positive x-axis and we can define the Poincaré map $P_\mu(x)$ by

$$P_\mu(x) = \phi_{2\pi}(x), \tag{6.70}$$

where $\phi_t(r)$ is the μ-dependent flow of the radial equation in (6.69).

The fixed points of P_μ correspond to the closed orbits of (6.69) and the latter are given by the non-trivial zeros of $\dot{r}$. Figure 6.24 shows how $\dot{r}$ changes as μ increases through zero. For $\mu < 0$, $\dot{r} = 0$ only at $r = 0$; when $\mu = 0$, $\dot{r}$ has a single non-trivial zero at $r = 1$; and for $\mu > 0$, $\dot{r}$ has two non-trivial zeros. The Poincaré map P_μ therefore has no fixed points for $\mu < 0$; one for $\mu = 0$; and two for $\mu > 0$. Thus $\mu = \mu^* = 0$ is a bifurcation point for the family.

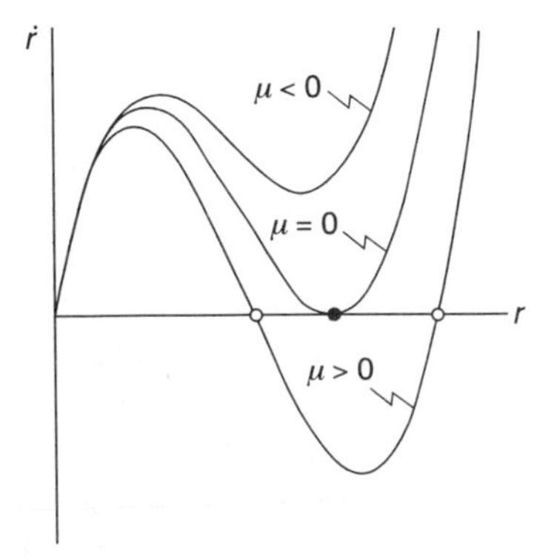

Fig. 6.24. Graphs of $\dot{r}$ given in (6.69) for: $\mu < 0$; $\mu = 0$; and $\mu > 0$.

We can approximate $P_0(x)$ near the single fixed point occurring at $x = x^* = 1$ as follows. Substitute $r = 1 + \rho$ into (6.69) with $\mu = 0$ to obtain

$$\dot{\rho} = \rho^2 + \rho^3. \tag{6.71}$$

If we confine attention to sufficiently small values of ρ, the cubic term in (6.71) can be neglected and the results of Example 1.5.2 used to obtain

$$\phi_t(1 + \rho) = 1 + \frac{\rho}{(1 - t\rho)}. \tag{6.72}$$

It follows that

$$\begin{aligned} P_0(x) &= \phi_{2\pi}(x) = \phi_{2\pi}(1 + \xi), \\ &= 1 + \frac{\xi}{(1 - 2\pi\xi)}, \\ &= 1 + \xi + 2\pi\xi^2 + \cdots, \end{aligned} \tag{6.73}$$

where $\xi = x - x^* = x - 1$ is a local coordinate at x^*.

Comparison of (6.73) with the Taylor expansion of $P_0(x)$ about $x = x^* = 1$ gives

$$\left.\frac{\mathrm{d}P_0}{\mathrm{d}x}\right|_{x^*} = 1, \tag{6.74}$$

$$\left.\frac{\mathrm{d}^2 P_0}{\mathrm{d}x^2}\right|_{x^*} = 4\pi. \tag{6.75}$$

Equation (6.74) shows that x^* is a non-hyperbolic fixed point with eigenvalue $+1$ while (6.75) shows that $P_0(x)$ is locally quadratic at x^*. Geometrically these equations mean that, at $\mu = 0$, the graphs $y = P_0(x)$ and $y = x$ meet tangentially at $x = x^*$ as shown in Fig. 6.25 curve (b). For $\mu < 0$ the

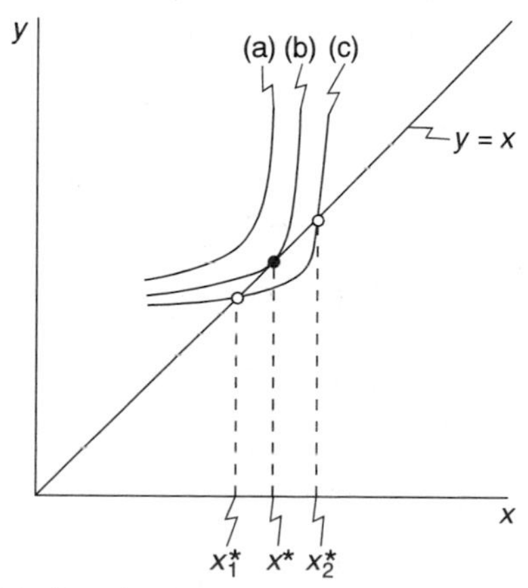

Fig. 6.25. Plots of $P_\mu(x)$ given by (6.70) for: μ negative (curve (a)); μ zero (curve (b)); and μ positive (curve (c)). Note that, for $\mu > 0$, $\mathrm{d}P_\mu/\mathrm{d}x$ is less than 1 at $x = x_1^*$ while it is greater than 1 at $x = x_2^*$. This is an example of a fold bifurcation.

corresponding graphs do not intersect and P_μ has no fixed points (Fig. 6.25 curve (a)). For $\mu > 0$ they intersect in two places and P_μ has two fixed points; one stable and the other unstable (Fig. 6.25 curve (c)). In other words, the change in the parameter μ moves a fold in the graph $y = P_\mu(x)$ through the graph $y = x$ of the identity. This behaviour characterizes the **fold bifurcation**.

It is often convenient to have a canonical family which exhibits a given bifurcation in a simple way. The family

$$f_v(y) = v + y + y^2, \tag{6.76}$$

with v and y near zero, fulfils this role for the fold bifurcation (as demonstrated in Exercise 6.26). Notice that, since both v and y remain close to zero, only the local behaviour of (6.76) at $(v, y) = (0, 0)$ is involved in the bifurcation. We say that the **local family** defined by (6.76) undergoes a fold bifurcation at $y = y^* = 0$ when $v = v^* = 0$. It can be shown that any local family that undergoes a fold bifurcation can be expressed in the form (6.76) by a suitable choice of local coordinate and parameter.

6.5.2 The flip bifurcation

A fixed point, x^*, of a one-dimensional map, P_0, is non-hyperbolic if the eigenvalue of the linearization of P_0 at x^* has modulus unity. The flip bifurcation arises from a fixed point with the eigenvalue -1.

Since $P_0(x^*) = x^*$ and $(\mathrm{d}P_0/\mathrm{d}x)|_{x^*} = -1$, Taylor expansion of P_0 about x^* gives

$$P_0(x) = P_0(x^* + \xi) = x^* - \xi + a\xi^2 + b\xi^3 + \cdots, \tag{6.77}$$

where $\xi = x - x^*$, $a = (\mathrm{d}^2 P_0/\mathrm{d}x^2)|_{x^*}/2!$ and $b = (\mathrm{d}^3 P_0/\mathrm{d}x^3)|_{x^*}/3!$. For sufficiently small ξ the linear term in (6.77) dominates and $P_0(x)$ for x near x^* takes the form shown in Fig. 6.26. The negative slope of $P_0(x)$ at $x = x^*$

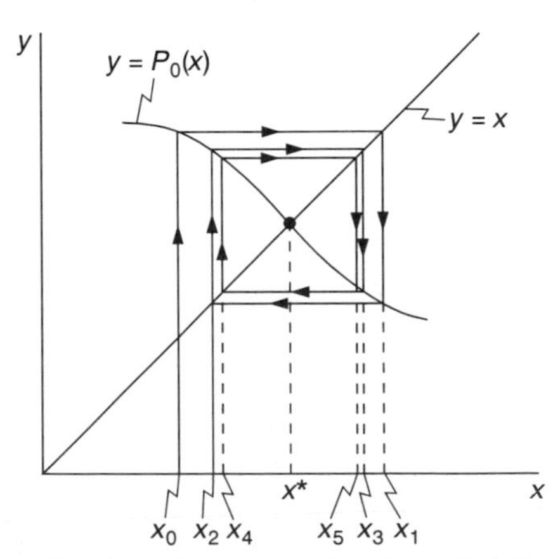

Fig. 6.26. Illustration of the iteration $x_{n+1} = P_0(x_n)$, $n = 0, 1, 2, \ldots$, for a typical choice of x_0 near the fixed point x^* of a map P_0 approximated by (6.77). The case $a = 0, b = 1$ is shown.

means that the orbit of a point, x_0, near to x^* switches or flips from one side of the fixed point to the other at each application of the map. Notice that such a map could not be the Poincaré map of a flow on an annulus containing $(x^*, 0)$ (as in section 6.5.1) but it could, for example, correspond to a flow on a Möbius strip (discussed in Exercise 4.17). The behaviour of those orbit points that lie on the same side of x^* is determined by P_0^2; indeed, they are simply the orbits of x_0 and $P_0(x_0)$ under P_0^2. Once again we can approximate $P_0^2(x)$ near x^* by its Taylor expansion. Differentiation of $P_0^2(x) = P_0(P_0(x))$, yields

$$\left.\frac{dP_0^2}{dx}\right|_{x^*} = \left.\frac{dP_0}{dx}\right|_{P_0(x^*)} \left.\frac{dP_0}{dx}\right|_{x^*} \tag{6.78}$$

and

$$\left.\frac{d^2P_0^2}{dx^2}\right|_{x^*} = \left.\frac{d^2P_0}{dx^2}\right|_{P_0(x^*)} \left[\left.\frac{dP_0}{dx}\right|_{x^*}\right]^2 + \left.\frac{dP_0}{dx}\right|_{P_0(x^*)} \left.\frac{d^2P_0}{dx^2}\right|_{x^*}. \tag{6.79}$$

Since $P_0(x^*) = x^*$ and $(dP_0/dx)|_{x^*} = -1$, (6.78) and (6.79) give

$$\left.\frac{dP_0^2}{dx}\right|_{x^*} = 1, \qquad \left.\frac{d^2P_0^2}{dx^2}\right|_{x^*} = 0. \tag{6.80}$$

Equation (6.80) shows that the fixed point of P_0^2 at $x = x^*$ has eigenvalue $+1$ but that, unlike the case considered in section 6.5.1, the leading non-linear terms are at least of order ξ^3. We can therefore write

$$P_0^2(x) = x^* + \xi + c\xi^3 + \cdots, \tag{6.81}$$

where $c = (d^3P_0^2/dx^3)|_{x^*}/3!$, for $|\xi|$ near zero. Figure 6.27(a) and (b) show graphs of $P_0^2(x)$ near $x = x^*$ for $c < 0$ and $c > 0$ in relation to the identity. Observe that the fold that characterized the discussion in section 6.51 is not present.

The family

$$P_\mu(x) = x^* - (1 + \mu)\xi + a\xi^2 + b\xi^3 + \cdots, \tag{6.82}$$

has the non-hyperbolic fixed point given in (6.77) at $\mu = 0$. As μ increases through zero, the eigenvalue of the linearization at x^* decreases through -1. Thus x^* is stable for $\mu < 0$ and unstable for $\mu > 0$. However, it is not only the stability of the fixed point that changes at $\mu = 0$. Given that both $|\xi|$ and $|\mu|$ are sufficiently small, then (6.82) gives

$$P_\mu^2(x) = x^* + (1 + 2\mu)\xi + \cdots. \tag{6.83}$$

There are two cases to consider.

1. $c < 0$. Equation (6.83) shows that, as μ *increases from zero*, the slope of $P_\mu^2(x)$ at $x = x^*$ becomes greater than 1 and P_μ^2 develops two *stable* fixed points for $\mu > 0$ as shown in Fig. 6.27(c).

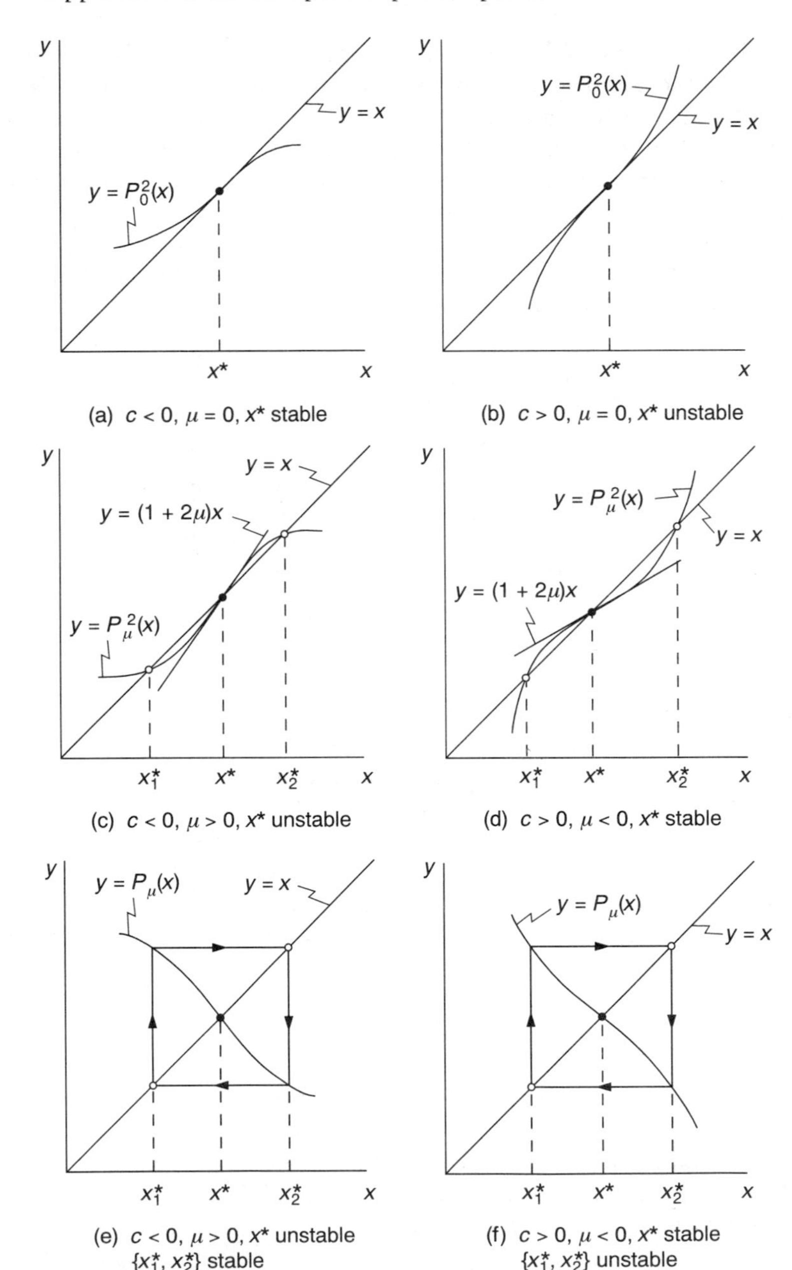

Fig. 6.27. Graphs showing how period-2 points can arise at flip bifurcations in the local family (6.82). In diagrams (a), (c), (e) the parameter c in (6.81) is assumed to be negative. The stable, non-hyperbolic fixed point x^* becomes unstable and a stable 2-cycle appears as μ increases from zero. The parameter c is positive in diagrams (b), (d), (f). The unstable, non-hyperbolic fixed point x^* becomes stable and an unstable 2-cycle is formed as μ decreases from zero.

2. $c > 0$. As μ *decreases from zero*, the slope of $P_\mu^2(x)$ at $x = x^*$ becomes less than one and P_μ^2 develops two *unstable* fixed points for $\mu < 0$ as shown in Fig. 6.27(d).

In both cases, the fixed points, x_1^* and x_2^*, of P_μ^2 are not fixed points of P_μ and (as shown in Fig. 6.27(e), (f)) $P_\mu(x_1^*) = x_2^*$, so that $\{x_1^*, x_2^*\}$ is a 2-cycle of P_μ. Thus, not unlike the Hopf bifurcation in section 6.3.2, as μ passes through zero the stability of the fixed point x^* of P_μ changes and a 2-cycle $\{x_1^*, x_2^*\}$ is created. This local bifurcation, associated with a non-hyperbolic fixed point with eigenvalue -1, is called a **flip bifurcation**. The canonical local family for the flip bifurcation takes the form

$$f_\nu(y) = (\nu - 1)y \pm y^3, \tag{6.84}$$

with ν and y near zero (as shown in Exercise 6.30).

6.5.3 The logistic map

A discrete analogue of the logistic equation (1.18) takes the form

$$x_{n+1} = F_\mu(x_n) = \mu x_n(1 - x_n) \tag{6.85}$$

and models populations (typically of insects) with distinct, non-overlapping generations. The variable x_n is the population in generation n measured in units of the carrying capacity of the environment. The parameter μ is usually restricted to the interval $(0, 4]$ to ensure that $x_n \geqslant 0$ for all n. Equation (6.85) then represents a one-parameter family of discrete dynamical systems defined on $[0, 1]$. The map F_μ is not the Poincaré map of a two-dimensional flow for any μ, indeed, it is not even a diffeomorphism on $[0, 1]$. However, the dynamics that it exhibits are of fundamental importance in their own right and they are relevant to Poincaré maps in higher dimensions.

Clearly, $F_\mu(0) = 0$ and the origin is a fixed point of F_μ for all $\mu \in (0, 4]$. Since $\mathrm{d}F_\mu/\mathrm{d}x|_{x=0} = \mu$, this point is stable for $0 < \mu < 1$ and the orbits of all points in $(0, 1]$ are attracted to it. For $\mu > 1$, the origin is unstable and F_μ has a non-trivial fixed point at $x^* = 1 - 1/\mu$. The stability of this fixed point is given by

$$\left.\frac{\mathrm{d}F_\mu}{\mathrm{d}x}\right|_{x=x^*} = \mu(1 - 2x^*) = 2 - \mu. \tag{6.86}$$

For $1 < \mu < 3$, x^* is stable with basin of attraction $(0, 1)$ but, at $\mu = 3$, F_μ undergoes a flip bifurcation as shown in Fig. 6.28. The fixed point x^* becomes unstable and a stable 2-cycle is created with basin of attraction $(0, 1)\backslash\mathscr{S}(x^*)$, where $\mathscr{S}(x^*) = \{x \,|\, F_\mu^p(x) = x^* \text{ for some } p \in \mathbb{Z}^+\}$ (see Exercise 6.33).

As μ increases above 3 the 2-cycle expands away from x^* and becomes less stable until, for $\mu = 1 + \sqrt{6} = 3.4494\ldots$, F_μ^2 undergoes a pair of flip bifurcations. The stable fixed points of F_μ^2 both become unstable and each one gives rise to a pair of stable fixed points of F_μ^4 corresponding to a stable

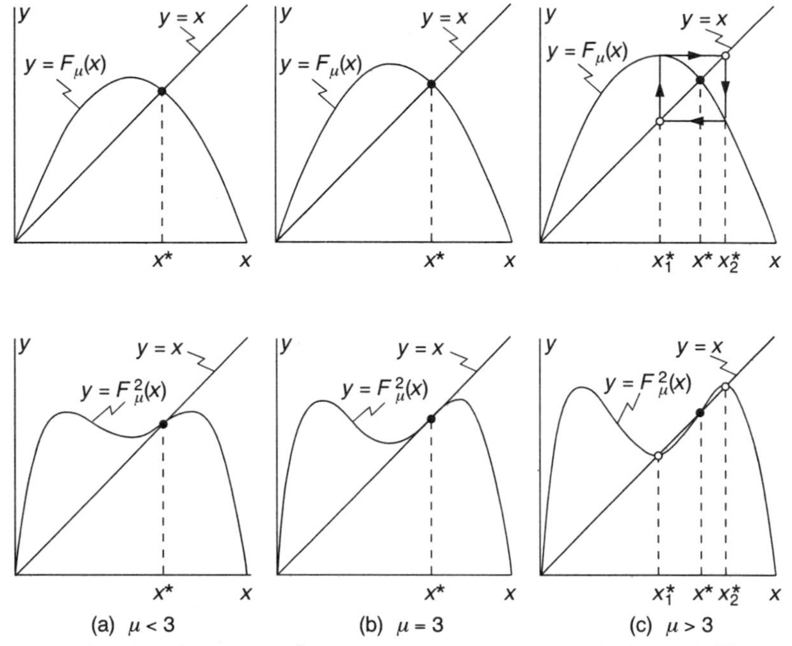

Fig. 6.28. Graphs of F_μ and F_μ^2 for: (a) $\mu < 3$; (b) $\mu = 3$; and (c) $\mu > 3$; illustrating the flip bifurcation undergone by the logistic map $F_\mu(x) = \mu x(1 - x)$ at $\mu = 3$.

4-cycle of F_μ (shown in Fig. 6.29). As μ increases further, the fixed points of F_μ^4 undergo flip bifurcations leading to an attracting 8-cycle of F_μ (Fig. 6.30) and so on.

The result of these repeated **period-doubling bifurcations** is a sequence or 'cascade' of stable 2^k-cycles. For each $k = 0, 1, 2, \ldots$, a 2^k-cycle appears at $\mu = \mu_k$, remains stable for an interval or 'window' in μ and then becomes unstable at $\mu = \mu_{k+1}$, giving way to a stable cycle of period 2^{k+1}. There is only one stable periodic orbit for each value of μ. The length of the window for which the 2^k-cycle is stable rapidly decreases as k increases. If we write $\Delta\mu_k = \mu_{k+1} - \mu_k$ then, for large k, $\Delta\mu_{k+1} \approx \delta^{-1}\Delta\mu_k$, where $\delta = 4.6692\ldots$ is known as the **Feigenbaum number**. For sufficiently large K, $\Sigma_{k=K}^{\infty}\Delta\mu_k \approx \Delta\mu_K\Sigma_{k=0}^{\infty}(\delta^{-1})^k = \Delta\mu_K/(1 - \delta^{-1})$ and the sequence $\{\mu_k\}_{k=0}^{\infty}$ accumulates at a finite value of $\mu = \mu_c = 3.5700\ldots$.

The fold bifurcation also plays an important role in the dynamics of F_μ. For example, Fig. 6.31 shows how two 3-cycles (one stable, one unstable) appear as a result of three simultaneous fold bifurcations in F_μ^3 at $\mu = \mu^* = 3.8284\ldots$. Moreover, as Fig. 6.32 indicates, the stable 3-cycle is the beginning of a cascade of stable $(3 \cdot 2^k)$-cycles that follow as a result of subsequent period doubling bifurcations. However, the existence of a period-3 orbit has more far-reaching consequences as the following result (due to Sarkovskii, 1964) shows.

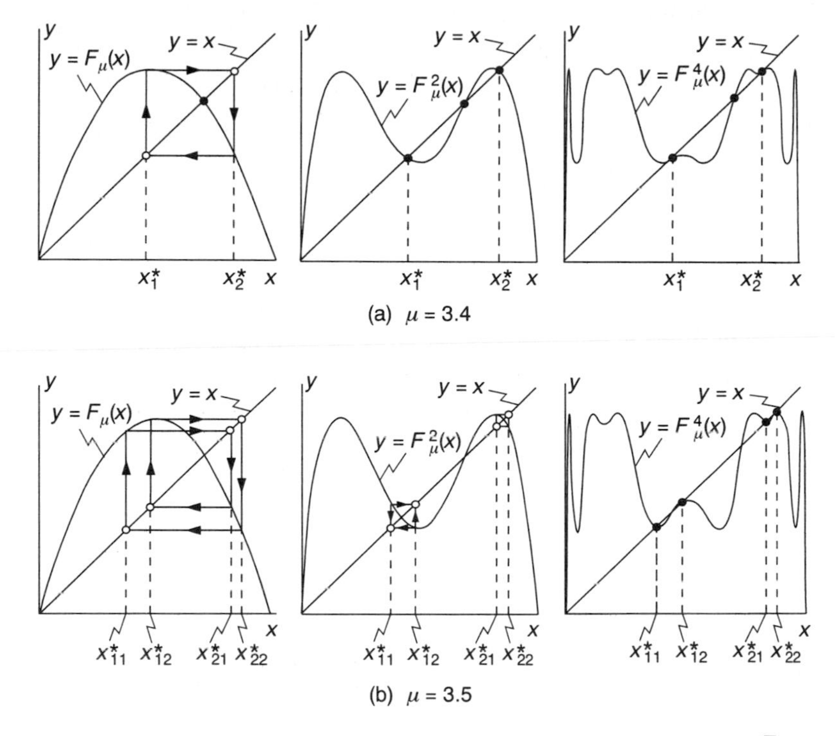

Fig. 6.29. Illustration of the flip bifurcations undergone by F_μ^2 at $\mu = 1 + \sqrt{6} \approx 3.45$. The stable fixed points x_1^* and x_2^* of F_μ^2 in (a) are unstable in (b) where a pair of stable 2-cycles of F_μ^2 occur. The four points involved are fixed points of F_μ^4 and form a 4-cycle of F_μ.

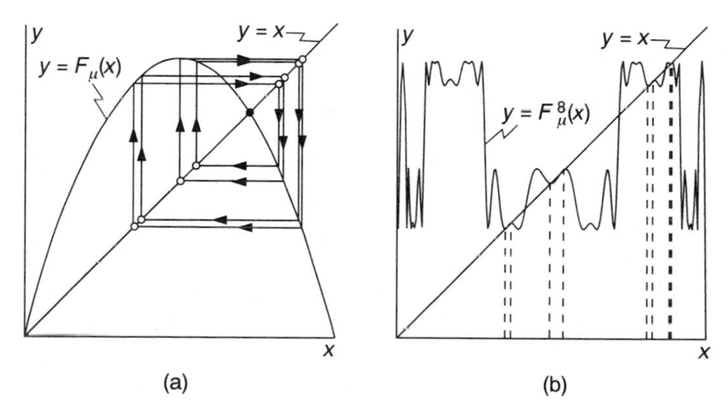

Fig. 6.30. Graphs of F_μ and F_μ^8 for $\mu = 3.555$. The four fixed points of F_μ^4 shown in Fig. 6.29 have 'period-doubled' and there are eight fixed points of F_μ^8 (see (b)) corresponding to an 8-cycle in F_μ (see (a)).

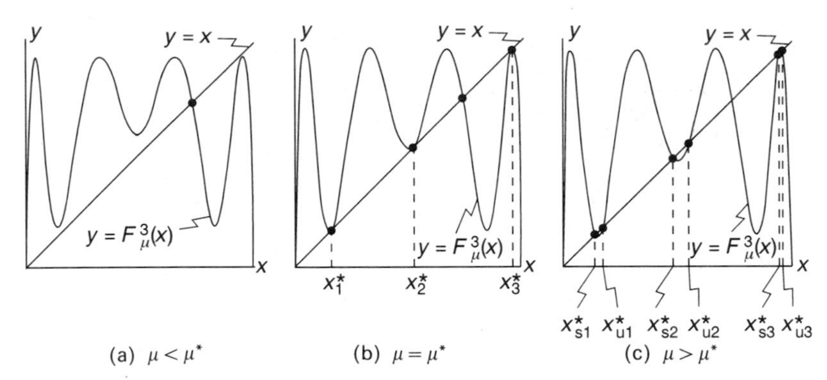

Fig. 6.31. Fixed points of F_μ^3 appear at three simultaneous fold bifurcations when $\mu = \mu^* = 3.8284\ldots$. For μ slightly greater than μ^* there are six fixed points: three stable and three unstable (see (c)) corresponding to two 3-cycles of F_μ.

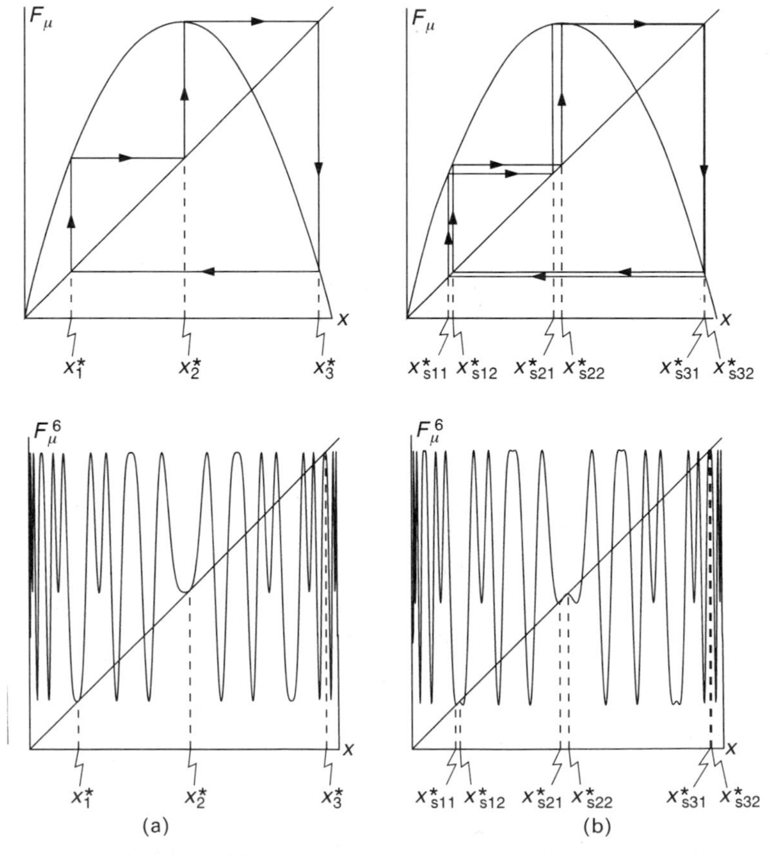

Fig. 6.32. Comparison of graphs of F_μ and F_μ^6 for: (a) $\mu = \mu^*$; (b) $\mu = 3.845$. Diagram (a) shows the 3-cycle of F_{μ^*} along with the corresponding fixed points of $F_{\mu^*}^6$. In (b) the 3-cycle has 'period-doubled' at a flip bifurcation of F_μ^3.

Theorem 6.5.1
Order the positive integers $\mathbb{Z}^+$ as follows

$$\begin{aligned} 3 \lhd 5 \lhd 7 \lhd \cdots \lhd 2\cdot3 \lhd 2\cdot5 \lhd 2\cdot7 \lhd \cdots \lhd 2^k\cdot3 \lhd 2^k\cdot5 \\ \lhd 2^k\cdot7 \lhd \cdots \lhd 2^k \lhd \cdots \lhd 2^3 \lhd 2^2 \lhd 2 \lhd 1. \end{aligned} \quad (6.87)$$

If $f:\mathbb{R}\to\mathbb{R}$ is a continuous map that has an orbit of period n then f has an orbit of period m for every $m \in \mathbb{Z}^+$ with $n \lhd m$.

If the domain of F_μ is extended to $\mathbb{R}$ then it is easy to verify that no periodic points occur in $\mathbb{R}\backslash[0, 1]$. Thus the recurrence predicted by Theorem 6.5.1 is confined to $[0, 1]$ and we conclude that the presence of a 3-cycle for F_μ implies the existence of m-cycles for all positive integers m.

Figure 6.33 illustrates how the stable periodic orbits of F_μ develop as μ increases from three to four. The scaling of the μ-axis is non-linear to emphasize the period-3 window but period-5 and period-6 windows are also visible for $\mu < \mu^*$. Observe that 6-cycles occur both above and below μ^*. This highlights the fact that a given periodicity occurs more than once. The Sarkovskii sequence only predicts the order of first occurrence of a particular period. Thus the 6-cycle that arises for $\mu > \mu^*$ does not contradict Theorem 6.5.1 because it is not the first occurrence of a period-6 orbit.

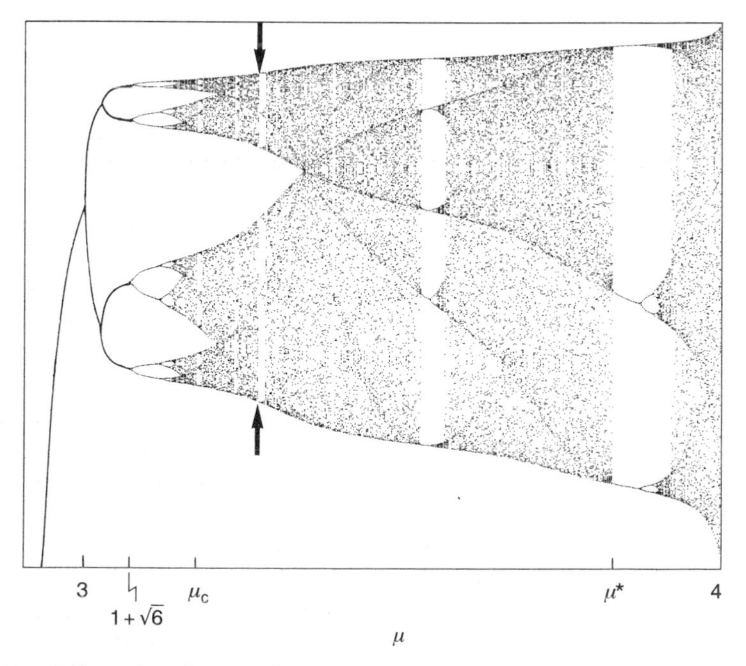

Fig. 6.33. Bifurcation diagram illustrating the cascades of stable periodic orbits of the logistic map F_μ. Observe the 6-cycle (arrowed) that occurs for $\mu < \mu^*$.

It can be shown (Guckenheimer *et al.*, 1977) that for each value of μ there is at most one stable periodic orbit. What is more, for each period m, there is a value of μ for which F_μ has a stable m-cycle. However, in their famous paper entitled 'Period three implies chaos', Li and Yorke (1975) showed that the presence of a 3-cycle is sufficient to ensure the existence of aperiodic orbits, i.e. orbits that do not ultimately approach any periodic orbit. For $\mu < 4$, the set of these aperiodic points has measure zero so that they are only rarely encountered. Since in practice it is impossible to distinguish an aperiodic orbit from a periodic orbit of long period, it might be tempting to argue that such orbits are not very significant. However, for $\mu = 4$ it can be proved that F_μ has no attracting periodic orbits and, consequently, no periodic orbit can be approached asymptotically. Under these circumstances aperiodic orbits are of central importance. Each such orbit forms a dense subset of $[0, 1]$ and its qualitative behaviour is said to be **chaotic** (see section 6.8).

The complicated behaviour exhibited by the logistic map is typical of a whole class of families of one-dimensional maps of a finite interval that have a single smooth maximum known as unimodal maps (Collet and Eckmann, 1980). All such families exhibit period-doubling similar to that described above for F_μ. In fact, not only do their cascades of stable 2^k-cycles all accumulate in an asymptotically geometric way, they all do so at the same rate given by the Feigenbaum number δ. As a consequence, δ is said to be a universal constant.

Chaotic behaviour in two-dimensional maps can arise from embedded unimodal maps. An example is the quadratic Hénon map, considered in Exercise 6.36, where an alternative form of the logistic map is combined with a linear contraction/expansion. However, a striking illustration of the fact that such maps are relevant to flows in more than two dimensions is provided by the system

$$\dot{x} = -y - z, \qquad \dot{y} = x + ay, \qquad \dot{z} = b + z(x - c), \tag{6.88}$$

suggested by Rössler (1976). The orbits of (6.88) circulate around the z-axis in a right-handed sense and repeatedly pass through the half-plane $y = 0$, $x < 0$. The Poincaré map, $\mathbf{P}(x, z)$, of (6.88) defined on this half-plane can be approximated numerically and the behaviour of its orbits investigated for various values of a, b, c. For example, for $a = 0.398$, $b = 2$, $c = 4$, $\mathbf{P}$ has an attracting set consisting of points with z-coordinates close to zero. On the other hand, Fig. 6.34(a) shows how successive x-coordinates, x_n, of $\mathbf{P}^n(x_0, z_0)$ are ultimately related. The quadratic curve shown is independent of (x_0, z_0) and, for these parameter values, the x-coordinates of the points of any orbit of $\mathbf{P}$ ultimately wander about within a finite interval of values in a chaotic way. Figure 6.34(b) shows the corresponding attracting set for the differential equation (6.88). It is referred to as a chaotic **folded band** or **Rössler attractor**. With $b = 2$, $c = 4$, values of a can be found for which $\mathbf{P}$ exhibits the same kind of period-doubling behaviour as the logistic map resulting in corres-

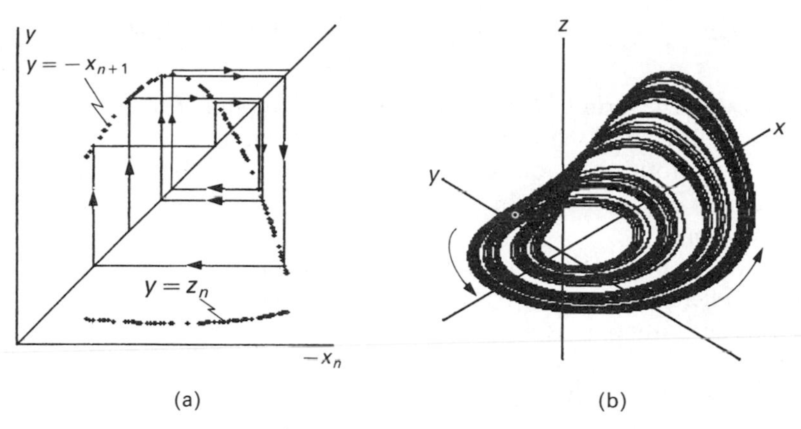

Fig. 6.34. (a) Plots of $(-x_n, -x_{n+1})$ and $(-x_n, z_n)$ obtained from the iteration $(x_{n+1}, z_{n+1}) = \mathbf{P}(x_n, z_n)$, where $\mathbf{P}$ is the Poincaré map of (6.88), with $a = 0.398$, $b = 2, c = 4$, defined on the half-plane $y = 0$, $x < 0$. The sample of successive iterates that is distinguished indicates that the x-coordinates wander through an interval of values in an erratic way. (b) Chaotic folded band attractor for the Rössler equations (6.88) with $a = 0.398$, $b = 2$, $c = 4$.

ponding 'doublings' of the closed trajectories in the flow (Thompson and Stewart, 1986, p. 242). Similar 'doublings' of closed orbits are found in the phase portraits of the Lorenz equations (Sparrow, 1982, pp. 57, 60, 61).

6.6 SOME BIFURCATIONS IN FAMILIES OF TWO-DIMENSIONAL MAPS

6.6.1 The child on a swing

The reader may remember the technique for building up the oscillations of a swing from their childhood. The best results are obtained by standing on the swing, crouching as it passes through the vertical and straightening up at the extremes of its oscillation. Judicious use of this method can build the amplitude of the oscillations to exhilarating proportions; over-enthusiasm can be disastrous.

Under the above circumstances the swing acts like a simple pendulum whose length varies periodically in time. Such an arrangement is known as a **parametrically forced** simple pendulum. The dynamical equations for a pendulum of average length l are

$$\dot{x}_1 = x_2, \qquad \dot{x}_2 = -\Omega(t)^2 \sin(x_1) - kx_2, \tag{6.89}$$

where

$$\Omega(t) = \omega_0(1 + \mu(t)), \tag{6.90}$$

with $\omega_0^2 = g/l$. The forcing term μ is periodic with period τ, i.e.

$$\mu(t+\tau) = \mu(t), \tag{6.91}$$

for all real t. The term kx_2 in (6.89) represents damping of the system. In the absence of forcing, (6.89) is an autonomous system and $(x_1, x_2) = (0, 0)$ is a stable (asymptotically stable) equilibrium for $k = 0$ ($k > 0$). The forcing term is capable of de-stabilizing this equilibrium and providing alternative stable periodic orbits.

Consider first the change in stability of the equilibrium at $(x_1, x_2) = (0, 0)$. The periodicity of $\mu(t)$ allows us to relate the non-autonomous planar system (6.89) to an autonomous system on the solid torus $\mathbb{R}^2 \times S^1$. For example, suppose there is no damping and that the oscillations are small. Then (6.89) can be approximated by

$$\dot{x}_1 = x_2, \qquad \dot{x}_2 = -\Omega(t)^2 x_1. \tag{6.92}$$

The change of variables

$$x_1' = \nu x_1, \qquad x_2' = x_2, \qquad t = \nu t, \tag{6.93}$$

in (6.92) gives

$$\frac{dx_1'}{dt'} = x_2', \qquad \frac{dx_2'}{dt'} = -\Omega_\nu(t')^2 x_1', \tag{6.94}$$

where $\Omega_\nu(t') = \nu^{-1}\Omega(t'/\nu)$ has period 2π if $\nu = 2\pi/\tau$. Dropping primes, we see that (6.94) can be written as the autonomous system

$$\dot{x}_1 = x_2, \qquad \dot{x}_2 = -\Omega_\nu(\theta)^2 x_1, \qquad \dot{\theta} = 1, \tag{6.95}$$

where

$$\Omega_\nu(\theta) = \frac{\omega_0}{\nu}(1 + \mu(\theta/\nu)). \tag{6.96}$$

The 2π-periodicity of $\Omega_\nu(\theta)$ means that we can identify θ and $\theta + 2n\pi$, n integer, so that the phase space of (6.95) is the solid torus. The equilibrium at $(x_1, x_2) = (0, 0)$ is represented by the null solution $(x_1(t) = x_2(t) \equiv 0)$ of (6.89) which in turn corresponds to a closed orbit in (6.95). The latter gives rise to a fixed point at the origin for the Poincaré map defined, for example, on the plane $\theta = 0$. The stability of the equilibrium at $(x_1, x_2) = (0, 0)$ is therefore given by the stability of this fixed point.

In general, we must resort to numerical approximations to the Poincaré map but the following choice of $\mu(t)$ (Arnold, 1973) allows an explicit expression to be obtained for it. Let

$$\mu(t) = \begin{cases} \nu\mu, & 0 \leqslant t < \tau/2 \\ -\nu\mu, & \tau/2 \leqslant t < \tau, \end{cases} \tag{6.97}$$

then (6.96) gives

$$\Omega_\nu(\theta) = \begin{cases} \omega + \mu = \omega_+, & 0 \leqslant \theta < \pi \\ \omega - \mu = \omega_-, & \pi \leqslant \theta < 2\pi, \end{cases} \tag{6.98}$$

where $\omega = \omega_0/\nu$. We can construct the Poincaré map on the $\theta = 0$ plane from the evolution operators of the linear systems

$$\dot{\boldsymbol{x}} = \boldsymbol{A}_{\pm}\boldsymbol{x} = \begin{bmatrix} 0 & 1 \\ -\omega_{\pm}^2 & 0 \end{bmatrix}\boldsymbol{x}. \tag{6.99}$$

For $0 \leqslant t < \pi$ the evolution operator is $\exp(t\boldsymbol{A}_+)$, while for $\pi \leqslant t < 2\pi$ it is $\exp(t\boldsymbol{A}_-)$. Consequently the Poincaré map is linear and represented by the matrix

$$\boldsymbol{P} = \exp(\pi\boldsymbol{A}_-)\exp(\pi\boldsymbol{A}_+). \tag{6.100}$$

It is not difficult to show (as in Exercise 6.38) that

$$\exp(\pi\boldsymbol{A}_{\pm}) = \begin{bmatrix} \cos(\pi\omega_{\pm}) & \omega_{\pm}^{-1}\sin(\pi\omega_{\pm}) \\ -\omega_{\pm}\sin(\pi\omega_{\pm}) & \cos(\pi\omega_{\pm}) \end{bmatrix}, \tag{6.101}$$

from which it follows that

$$\det(\boldsymbol{P}) = \det(\exp(\pi\boldsymbol{A}_-))\det(\exp(\pi\boldsymbol{A}_+)) = 1. \tag{6.102}$$

Hence, $\boldsymbol{P}$ is area-preserving and its eigenvalues are given by

$$\lambda_{1,2} = \tfrac{1}{2}\{\operatorname{tr}(\boldsymbol{P}) \pm \sqrt{\operatorname{tr}(\boldsymbol{P})^2 - 4}\}. \tag{6.103}$$

For $|\operatorname{tr}(\boldsymbol{P})| < 2$, the eigenvalues of $\boldsymbol{P}$ are complex conjugates of one another and have modulus unity, i.e.

$$\lambda_1 = \lambda_2^* = \cos\beta + \mathrm{i}\sin\beta, \tag{6.104}$$

where $\tan\beta = \mathrm{Im}\lambda_1/\mathrm{Re}\lambda_1$. Typically therefore the trajectories of $\boldsymbol{P}$ are confined to ellipses centred on the origin of the $\theta = 0$ plane. The fixed point of $\boldsymbol{P}$ at $(x_1, x_2) = (0, 0)$ is stable in the sense of Liapunov and is said to be of **elliptic** type. For $|\operatorname{tr}(\boldsymbol{P})| > 2$, the eigenvalues of $\boldsymbol{P}$ are real and $\lambda_1 = 1/\lambda_2$. If $|\lambda_1| > 1 (< 1)$ then $|\lambda_2| < 1 (> 1)$ and $(x_1, x_2) = (0, 0)$ is a fixed point of **saddle** type and therefore unstable in the sense of Liapunov.

We can think of μ and ω as parameters in a family of Poincaré maps corresponding to different forcing amplitudes and frequencies. Each value of (ω, μ) for which $\operatorname{tr}(\boldsymbol{P}) = \pm 2$ has a simple root represents a bifurcation point of the family where the stability of the equilibrium at $(x_1, x_2) = (0, 0)$ changes. Equation (6.100) allows $\operatorname{tr}(\boldsymbol{P})$ to be calculated from (6.101). Substitution of $\omega_{\pm} = \omega \pm \mu$ and a little rearrangement reduces $\operatorname{tr}(\boldsymbol{P}) = \pm 2$ to the form

$$\omega^2\cos(2\pi\omega) - \mu^2\cos(2\pi\mu) = \pm(\omega^2 - \mu^2). \tag{6.105}$$

Numerical approximations to the solutions of (6.105) for small μ are given in Fig. 6.35 where they form the boundaries between the regions of the ω, μ-plane where the equilibrium at $(x_1, x_2) = (0, 0)$ is stable and unstable. The zones of instability (shaded) are narrow 'tongues' emanating from the points $(\omega, \mu) = (n/2, 0)$, n integer, and it is in the neighbourhood of these parameter values that we might expect to encounter instability in practice. Indeed, the broadest zone of instability occurs for $\omega = \omega_0/\nu = \tfrac{1}{2}$ which suggests instability

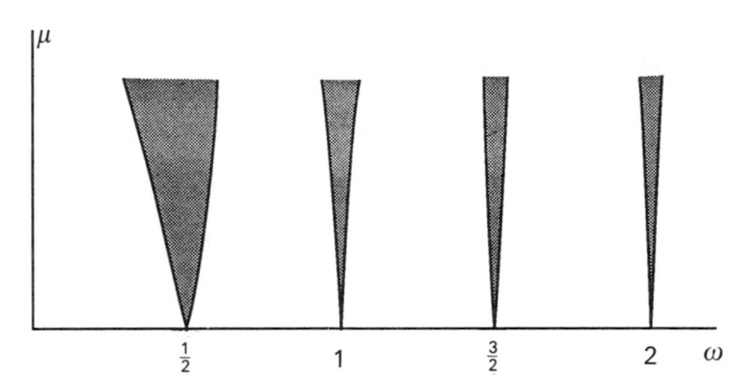

Fig. 6.35. Zones of instability for the null solution of the parametrically forced linear oscillator (6.92) when the forcing term takes the form (6.90) with $\mu(t)$ given by (6.97).

if the forcing frequency, ν, is twice the natural frequency, ω_0. This is certainly consistent with the movements required to pump up a swing.

While the linear approximation (6.92) exhibits the change of stability of the null solution of (6.89), it cannot provide information about the stable oscillatory motions that replace it, because the latter are determined by non-linear terms. Numerical solution of (6.89) with

$$\Omega(t)^2 = -\omega_0^2(1 + 2\mu\cos t) \tag{6.106}$$

(Eckmann, 1983) shows that, even in the presence of damping, a variety of stable motions are encountered as μ is increased from zero with $\omega_0 = \frac{1}{2}$. Damping has the effect of preventing the zones of instability from extending down to the ω-axis in Fig. 6.35. Consequently, the null solution remains stable until a threshold value of μ is exceeded. Subsequently, an approximately sinusoidal stable motion appears with period equal to twice the forcing period. This is the sustained oscillation of the judiciously pumped swing. For higher values of μ, periodic motions at the forcing period and multiples of it are detected but the stable motion becomes less sinusoidal and increasingly erratic, with the pendulum making complete circles around its pivot. Eventually, the motion of the pendulum fails to settle to any recognizable pattern of behaviour. It repeatedly switches between clockwise and anticlockwise rotation after apparently uncorrelated intervals of time.

6.6.2 The Duffing equation

Any planar non-autonomous differential equation that is periodic in time can be re-expressed in terms of an autonomous flow on the solid torus. The parametrically forced problem discussed in section 6.6.1 is one type of example but this approach can be applied to less subtly driven systems with equally spectacular results.

The **twin-well Duffing oscillator** has dynamical equation

$$\dot{x} = y, \qquad \dot{y} = x - x^3 - \delta y + \gamma\cos\omega t, \tag{6.107}$$

where δ, γ and ω are positive parameters. Equation (6.107) describes the displacement, x, of the free end of a long, slim, cantilevered steel beam in the following circumstances. The beam is mounted vertically with its upper end rigidly attached to a solid frame. Its free end is attracted by two magnets of equal strength, symmetrically placed on opposite sides of it. In the vertical position ($x = 0$) the net magnetic force on the beam is zero but slight displacement from this position results in attraction towards one or other of the magnets. Thus $x = 0$ is an unstable equilibrium for the beam and stable equilibria are symmetrically placed on either side of the vertical. The forcing term is provided by a transducer attached to the frame and driven by a signal generator (Guckenheimer and Holmes, 1983).

Equation (6.107) can be written as

$$\dot{x} = y, \qquad \dot{y} = x - x^3 - \delta y + \gamma \cos\theta, \qquad \dot{\theta} = \omega. \tag{6.108}$$

When t increases by $\tau = 2\pi/\omega$, θ increases by 2π and we can interpret θ as a cyclic variable so that $(x, y, \theta) \in \mathbb{R}^2 \times S^1$. We have already investigated (6.108) for $\gamma = 0$ in Example 4.3.3. The Poincaré map on the plane $\theta = \theta_0$ is then independent of θ_0 and given by ϕ_τ, where ϕ_t is the flow of the x, y-subsystem, (cf. equation (4.62)). The equilibria of the unforced beam, which correspond to the fixed points of ϕ_t, are evident in Fig. 4.12 and, for $\gamma = 0$, (6.108) has three closed orbits (two stable and one unstable) that are 'circular' in the sense that they coincide with the θ-coordinate lines passing through $(x, y, \theta) = (0, 0, 0)$, $(\pm 1, 0, 0)$.

When γ is greater than zero, (6.108) cannot be separated into two independent subsystems and the Poincaré map defined on the plane $\theta = \theta_0$ is no longer independent of θ_0. Observe that if ψ_t is the flow of (6.108) then (since $\theta = \omega t + \theta_0$ for all values of γ) the Poincaré map on the plane $\theta = \theta_0$ is given by the first two components of $\psi_\tau(x, y, \theta_0)$. However, it can be shown that the Poincaré maps defined on any two different planes of constant θ must be topologically conjugate. Consequently, all the Poincaré maps must have the same number of fixed points of the same topological type.

Example 6.6.1

Consider equation (6.108) with $\delta = 0.25$ and $\omega = 1$. Figure 6.36 shows a plot of the first two components of $\psi_t(1, 0, 0)$, $t > 0$, for $\gamma = 0.1$. The arrows indicate the sense of description of the projected orbit as t increases and the projections of $\psi_{n\tau}(1, 0, 0)$, $n \in \mathbb{Z}^+$, are distinguished. Interpret Fig. 6.36 in terms of the Poincaré map, $\mathbf{P}_{0.1}$, defined on the $\theta = 0$ plane and describe the asymptotic behaviour of $\psi_t(1, 0, 0)$ for large t.

Consider the Poincaré map defined on the $\theta = \theta_0$ plane. Explain, with the aid of a diagram, how the position of the fixed point of this map that lies near to $(1, 0)$, varies as θ_0 increases from zero to 2π.

Solution

The (x, y)-projections of $\psi_{n\tau}(1, 0, 0)$, $n \in \mathbb{Z}^+$, taken in order of increasing n, represent successive iterates of $(x, y) = (1, 0)$ under the Poincaré map defined

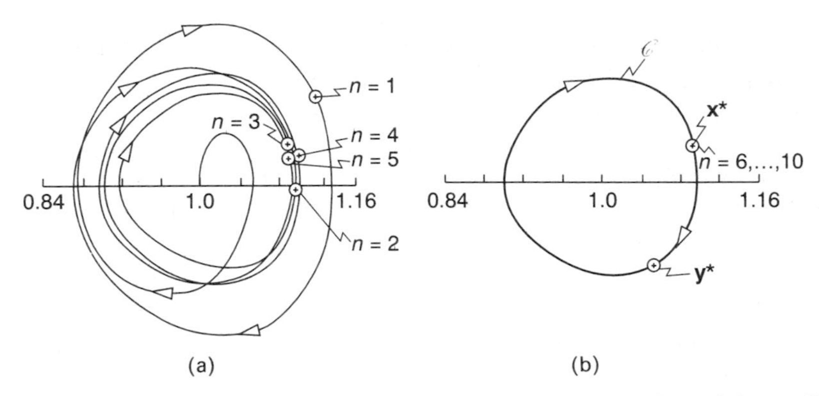

Fig. 6.36. Plots of the xy-projection of $\psi_t(1,0,0)$, where ψ_t is the flow of the Duffing equation (6.108), with $\omega = 1, \delta = 0.25, \gamma = 0.1$. Data are shown for: (a) $0 \leqslant t \leqslant 10\pi$; (b) $10\pi < t \leqslant 20\pi$ with the points with $t = 2\pi n, n = 1, 2, \ldots 10$, labelled. No change can be detected in either $\mathscr{C}$ or $\mathbf{x}^*$ at this scale.

on the $\theta = 0$ plane. Figure 6.36 shows that these iterates converge to a point, $\mathbf{x}^*$, near to $(x, y) = (1, 0)$. This is a fixed point of the $\theta = 0$ Poincaré map.

At the same time as the iterates converge to $\mathbf{x}^*$, the projections of the segments of trajectory between them settle down to a closed curve, $\mathscr{C}$, surrounding $(x, y) = (1, 0)$ and containing $\mathbf{x}^*$ (as shown in Fig. 6.36(b)). This closed curve represents the projection of the trajectory approached by $\psi_t(1,0,0)$ for large t. However, the third component of $\psi_t(1,0,0)$ is ωt and therefore, between successive visits to $\mathbf{x}^*$ the corresponding trajectory of (6.108) travels once around the solid torus in the θ-sense in such a way as to produce the projection $\mathscr{C}$ on the $\theta = 0$ plane. Thus $\{\psi_t(1,0,0) \mid t > 0\}$ tends to a closed, period-τ orbit that spirals about the θ-coordinate line passing through $(x, y, \theta) = (1, 0, 0)$.

The $\theta = \theta_0$ projection of the limiting orbit is the same closed curve $\mathscr{C}$ for all θ_0. Furthermore, the fixed points of the $\theta = \theta_0$ Poincaré map are given by the points of intersection of the limiting orbit with the plane $\theta = \theta_0$. Given that this orbit passes through the $\theta = 0$ plane at $\mathbf{x}^*$ when $t = 0$, it will intersect the plane $\theta = \theta_0, \theta_0 > 0$, when $t = \theta_0/\omega$. The point, $\mathbf{y}^*$, at which this intersection occurs will lie on the curve $\mathscr{C}$ at a distance (measured along $\mathscr{C}$ in the sense in which t increases) determined by $t = \theta_0/\omega$. Thus, the fixed point, $\mathbf{y}^*$, of the $\theta = \theta_0$ Poincaré map moves from $\mathbf{x}^*$ at $\theta_0 = 0$, around $\mathscr{C}$ in the direction of increasing t, and returns to $\mathbf{x}^*$ at $\theta_0 = 2\pi$ (as illustrated in Fig. 6.36(b)). □

$\mathbf{P}_{0.1}$ also has a stable fixed point near $(x, y) = (-1, 0)$ and a fixed point of saddle-type near the origin. As for $\gamma = 0$, every point not on the stable manifold of the saddle lies in the basin of attraction of one or other of the two stable fixed points. This situation persists as γ is increased further until, when

$\gamma \simeq 0.1598\ldots$, two new fixed points appear as shown in Fig. 6.37(a). Subsequently these new fixed points undergo period doubling bifurcations as shown in Figs. 6.37(b), (c) that culminate in a pair of chaotic, folded band attractors for $\gamma = 0.1923\ldots$ (Fig. 6.37(d)).

The shape of these attractors is not very clear in the $\theta = 0$ plane but the

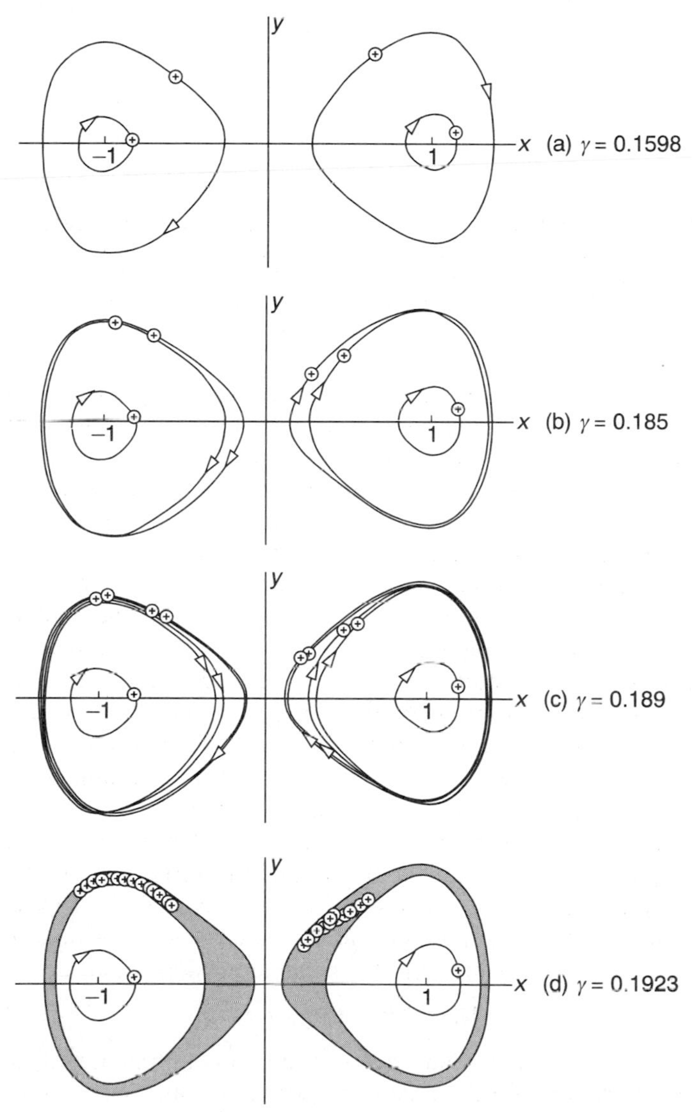

Fig. 6.37. Period-doubling of the two fixed points of the Poincaré map $\mathbf{P}_\gamma$ of the Duffing problem that appear at simultaneous saddle-node bifurcations when $\gamma = 0.1598$. In (d) 100 iterates of the Poincaré map $\mathbf{P}_\gamma$ are plotted; however, in the interest of clarity, the projected orbit is not shown. Its behaviour between iterates of $\mathbf{P}_\gamma$ is similar to that for earlier values of γ and it is confined to the shaded region.

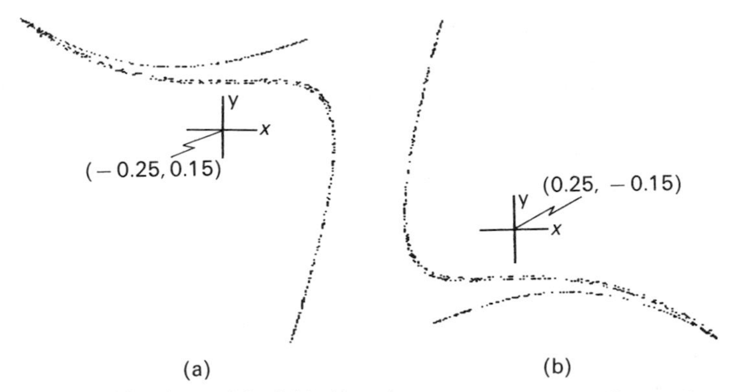

Fig. 6.38. Magnifications of the folded band attractors corresponding to those shown in Fig. 6.37 in planes of constant θ that clarify their structure. Data shown is for: (a) the left-hand attractor in the $\theta = \pi/2$ plane; and (b) the right-hand attractor in the plane $\theta = 3\pi/2$. The 'braided' nature of these attractors is just resolved at this magnification.

conjugacy of the Poincaré maps defined on different planes of constant θ means that they are topologically the same. This is illustrated in Fig. 6.38, where (a) shows the left-hand attractor for the Poincaré map on $\theta = \pi/2$, while (b) shows the right-hand attractor for that defined on the plane $\theta = 3\pi/2$. These particular planes are related by the transformation $(x, y) \to (-x, -y)$.

The magnification in Fig. 6.38 is just sufficient to resolve more than one 'strand' in some parts of the attractors shown. Closer examination reveals a **braided structure** with the braids or strands repeating on finer and finer scales as the attractor folds indefinitely within the image resolved at any given magnification. This structure is a common feature of chaotic attracting sets and it is best studied in more detail in maps that are more accessible than $\mathbf{P}_\gamma$. (Recall that each iteration of $\mathbf{P}_\gamma$ requires the numerical integration of (6.107) over a time interval of length $\tau = 2\pi/\omega$). The Hénon map given in Exercise 6.36 is a good example. Several thousands of iterations are readily obtainable with even the most modest microcomputer and the braided nature of its attractor is easily seen (Thompson and Stewart, 1986, pp. 181–2).

Typically we can expect the above fixed points of $\mathbf{P}_\gamma$ to appear at what amounts to a fold bifurcation embedded in the two-dimensional maps. The canonical local family exhibiting this behaviour takes the form

$$\mathbf{f}_\nu(x, y) = (\nu + x + x^2, \alpha y), \tag{6.109}$$

where x, y, ν are near zero and $|\alpha| \neq 1$. The first component of $\mathbf{f}_\nu$ undergoes a fold bifurcation while the second is just an expansion/contraction. The result is that a non-hyperbolic fixed point appears at $(x, y) = (0, 0)$ when $\nu = 0$ and this separates into two fixed points, one of node and one of saddle type, at $(x, y) = (\pm\sqrt{-\nu}, 0)$ for $\nu < 0$. Such behaviour is referred to as a **saddle–node bifurcation for maps**. A similar extension of the flip bifurcation can be responsible for period doubling in families of two-dimensional maps. This case is illustrated by

$$\mathbf{f}_\nu(x, y) = ((\nu - 1)x \pm x^3, \alpha y), \tag{6.110}$$

with x, y, ν near zero and $|\alpha| \neq 1$.

It must be emphasized that while the above discussion serves to illustrate the occurrence of period-doubling phenomena in families of two-dimensional maps, it fails to give any impression of how complicated the bifurcation diagram for (6.108) really is. Indeed, it does not even provide a complete description of the bifurcations that are known to occur as γ increases with $\delta = 0.25$ and $\omega = 1$. For example, the folded band attractors shown in Fig. 6.37(d) do not persist to $\gamma = 0.1924$. They disappear in what is colourfully referred to as a 'blue sky' bifurcation (the attractor disappears 'into the blue'). This bifurcation is similar in type to that occurring on the $B_2' B_1'$-boundary in Fig. 6.21. Further bifurcations of (6.108) for $\delta = 0.25$ and $\omega = 1$ are discussed in section 6.7.4.

6.7 AREA-PRESERVING MAPS, HOMOCLINIC TANGLES AND STRANGE ATTRACTORS

6.7.1 Introductory remarks

In Chapter 4 we noted that the Poincaré maps of recurrent Hamiltonian systems with two degrees of freedom can be area-preserving whether or not the system is integrable. Maps of this kind arise, for example, in celestial mechanics and, perhaps surprisingly, in the experimental study of elementary particles.

The problem of finding the motion of N point masses interacting via their mutual gravitational attractions is well known in astronomy. The special case of this N-body problem with $N = 2$ is integrable giving elliptic, parabolic and hyperbolic solutions (Goldstein, 1980). The three-body problem is a great deal more difficult. It is a Hamiltonian system with nine degrees of freedom which can at best be reduced to four degrees of freedom by using all known integrals. In order to gain some insight into this intractable problem, the so-called restricted three-body problem has been studied. The motion of a vanishingly small mass is considered to be restricted to the orbital plane of two large masses which describe circular orbits around their centre of mass. This point is taken as the origin of a rotating coordinate system whose x-axis lies along the line joining the two large masses. The small mass has two degrees of freedom and its equations of motion have a single integral. If its coordinates are (x, y) then the half-plane $y = 0, \dot{y} > 0$ defines a surface of section with coordinates $(x, \dot{x})$. The corresponding Poincaré map, which must be obtained by numerical integration, is area-preserving.

The connection with the study of elementary particles arises from the problem of containment of high-energy particles in beam–beam interaction experiments. Two intersecting beams of particles (typically protons) are accelerated to high energies in synchrotron and then allowed to interact over one or more small colliding regions. Collisions between the particles are relatively rare and the beams are required to make a number of trips around the storage rings of the synchrotron so they can pass through the colliding region(s) more than once. Even when collisions do not take place the beams experience a strong electromagnetic force when passing through a colliding region and it is important to know if they are stable to such a perturbation.

The particles do not follow circular paths around the storage rings but spiral about such circles according to

$$\frac{\mathrm{d}y}{\mathrm{d}\theta} = p, \qquad \frac{\mathrm{d}p}{\mathrm{d}\theta} = -Q^2 y. \tag{6.111}$$

Here y is the displacement of the particle perpendicular to the ring and θ is its angular position around the ring (which is proportional to the time). This simple harmonic motion is disrupted when the particle passes through a colliding region. For example, if there is only one such region, then the particle can be considered to receive an impulse from the electromagnetic force whenever θ passes through a multiple of 2π. The change in p generated by the impulse is a non-linear function, F, of the displacement y.

Consider a particle that exits from the colliding region in the state (y, p). The displacement of this particle perpendicular to the ring evolves according to (6.111) as θ increases by 2π. The colliding region is then encountered and p undergoes an additional change determined by F. Thus the values of y and p as the particle exits from the colliding region after one complete trip around the storage ring are given by

$$\begin{aligned} Y &= \cos(2\pi Q)y + Q^{-1}\sin(2\pi Q)p, \\ P &= -Q\sin(2\pi Q)y + \cos(2\pi Q)p + F(Y). \end{aligned} \tag{6.112}$$

It is not difficult to show that the determinant of the linearization of the mapping (6.112) is identically equal to unity. We conclude therefore that the stability of the beam is determined by the behaviour of an area-preserving map.

Let $\mathbf{x}^*$ be a fixed point of an area-preserving Poincaré map $\mathbf{P}$. Since $\det(D\boldsymbol{P}(\boldsymbol{x})) \equiv 1$, then, as in section 6.6.1, the fixed point of the linear map represented by $D\boldsymbol{P}(\boldsymbol{x}^*)$ is typically of either elliptic or saddle type. While the linearization theorem determines the nature of the non-linear fixed point in the latter case, it provides no information about the former. Area-preservation excludes the possibility of this non-linear fixed point being a focus (stable or unstable) so what type of qualitative behaviour should we expect near an elliptic fixed point in an area-preserving map?

If the linearized map has a complex conjugate pair of distinct eigenvalues with modulus unity, it is topologically conjugate to a rotation. For a rotation the origin is surrounded by a continuum of invariant circles and all points of the plane undergo the same angular displacement when the rotation is applied. The following example of an **area-preserving twist map** gives some insight into the effect that non-linear terms might have. The map is conveniently defined in terms of canonical polar coordinates, (τ, θ), by

$$\mathbf{T}:(\tau, \theta) \rightarrow (\tau, \theta + \alpha(\tau)), \tag{6.113}$$

where $\tau > 0$ and $\alpha'(\tau) = \mathrm{d}\alpha(\tau)/\mathrm{d}\tau \neq 0$ for all τ. The twist map $\mathbf{T}$ takes the circle of radius $\sqrt{2\tau}$ to itself but, unlike a rotation, the angular displacement depends on τ. If $\alpha(\tau_0)/2\pi$ is a rational number, say p/q in lowest terms, then each point of the circle, $\mathscr{C}$, of radius $\sqrt{2\tau_0}$ is a periodic point with period q. On the other hand, if $\alpha(\tau_0)/2\pi$ is an irrational number then no point of $\mathscr{C}$ is a periodic point and the orbit of each point of $\mathscr{C}$ is a dense subset of that circle. This leads to a subtly complex combination of topologically distinct types of behaviour. For example, between any two circles $\tau = a$ and $\tau = b, b > a > 0$, $\mathbf{T}$ has q-cycles, corresponding to every rational number p/q in the **rotation interval** $[\alpha(a)/2\pi, \alpha(b)/2\pi]$, intimately intermingled with the aperiodic orbits of points associated with every irrational number in that interval.

The twist map (6.113) does not represent the typical qualitative behaviour of area-preserving maps near to an elliptic fixed point. To obtain such behaviour it is necessary to consider general area-preserving perturbations of $\mathbf{T}$ that affect both radial and angular components.

Two theorems assure us that the resulting maps retain certain aspects of the behaviour of $\mathbf{T}$. The Poincaré–Birkhoff theorem gives, for each rational number p/q in a rotation interval of $\mathbf{T}$, the existence of at least two q-periodic orbits with average rotation per iteration (or **rotation number**) equal to p/q in units of 2π. The Kolmogorov–Arnold–Moser (KAM) theorem proves the existence, for sufficiently small perturbations, of invariant closed curves or 'KAM-circles' corresponding to some irrational rotation numbers in the rotation interval. A third theorem—the Aubry-Mather theorem—recognizes the existence of another class of invariant sets. These sets are associated with irrational rotation numbers but they are not invariant circles, having instead the structure of a Cantor set. They are interpreted as the result of break-up of irrational invariant circles as the perturbation increases. These theorems are considered in more detail in Arrowsmith and Place (1990).

6.7.2 Periodic orbits and island chains

Numerical experiments can help us to appreciate the orbit structure of a typical area-preserving map near an elliptic fixed point. Visually it is the Poincaré–Birkhoff periodic orbits that dominate the plots of successive iterates of the map. One of the two periodic orbits predicted by the Poincaré–Birkhoff theorem is elliptic and the other is of saddle type. Together they form what is called an **island chain** surrounding the elliptic fixed point. An example corresponding to $p/q = 1/5$ is shown in Fig. 6.39. What appear

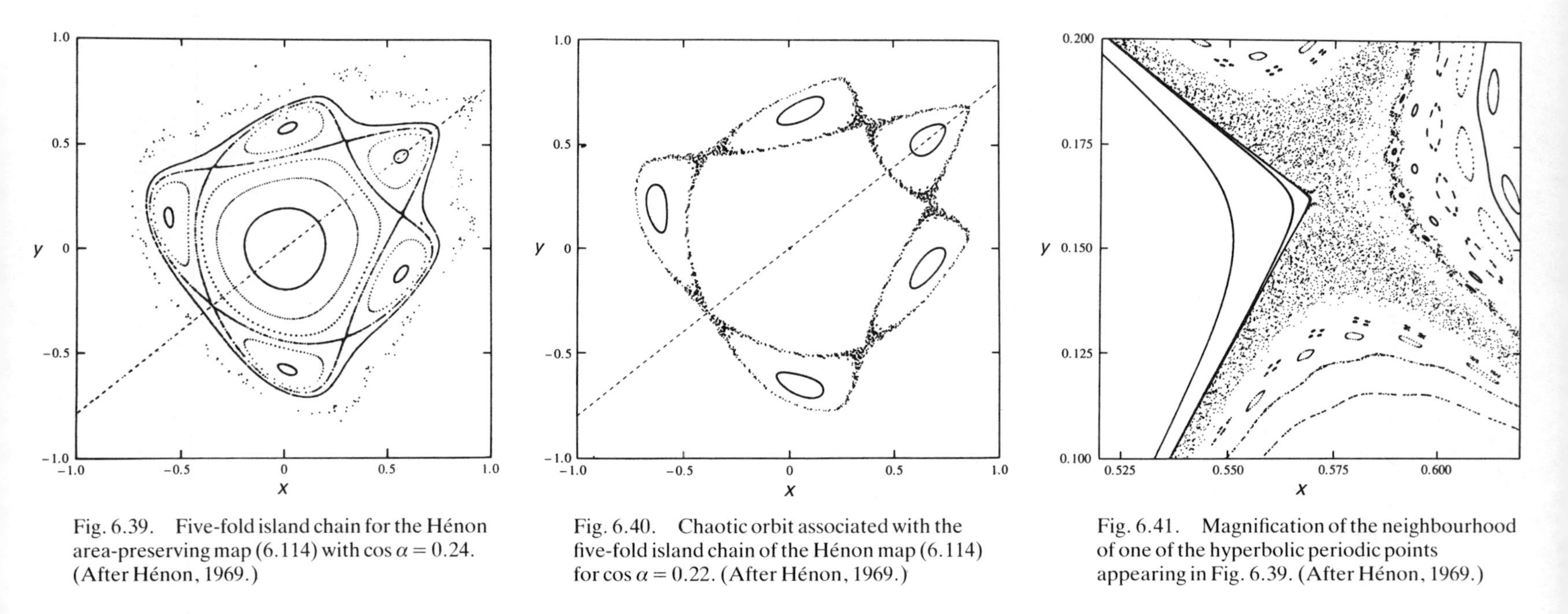

Fig. 6.39. Five-fold island chain for the Hénon area-preserving map (6.114) with $\cos\alpha = 0.24$. (After Hénon, 1969.)

Fig. 6.40. Chaotic orbit associated with the five-fold island chain of the Hénon map (6.114) for $\cos\alpha = 0.22$. (After Hénon, 1969.)

Fig. 6.41. Magnification of the neighbourhood of one of the hyperbolic periodic points appearing in Fig. 6.39. (After Hénon, 1969.)

to be separatrices of the points on the hyperbolic cycle form 'islands' that contain the elliptic periodic points. The map used in Fig. 6.39 is the quadratic, area-preserving map due to Hénon (1969). The iteration takes the form

$$\begin{aligned} x_{i+1} &= x_i \cos\alpha - y_i \sin\alpha + x_i^2 \sin\alpha, \\ y_{i+1} &= x_i \sin\alpha + y_i \cos\alpha - x_i^2 \cos\alpha, \end{aligned} \tag{6.114}$$

for $i = 0, 1, 2, \ldots$. This map is of particular interest because all quadratic, area-preserving maps with an elliptic fixed point are topologically conjugate to (6.114).

The 'shorelines' of the islands in Fig. 6.39 appear to be curves but Fig. 6.40 shows that this is not the case. Apart from five closed curves that serve to indicate the position of the elliptic periodic orbit, all the points on this figure are iterates of a single initial point. The orbit of this point winds around the island chain in an erratic manner, spreading out over a bounded, two-dimensional neighbourhood of the saddle-type periodic orbit and along the edges of the islands. Such orbits are often called **chaotic orbits** because of their irregular behaviour.

Fig. 6.41 is a close-up of one of the hyperbolic periodic points shown in Fig. 6.39. A chaotic orbit is apparent along with smaller adjacent island chains corresponding to rational numbers near to 1/5. What is more, further island chains are visible within the islands of the original chain. On reflection this is not surprising. The points of an elliptic q-periodic orbit of a map $\mathbf{P}$ are elliptic fixed points of $\mathbf{P}^q$. If $\mathbf{P}$ is area-preserving then so is $\mathbf{P}^q$ and therefore typical elliptic fixed points of $\mathbf{P}^q$ will be surrounded by island chains corresponding to q'-cycles of $\mathbf{P}^q$. Such points are qq'-periodic points of $\mathbf{P}$. This argument can be repeated for the elliptic points of the island chains of $\mathbf{P}^q$. If these islands are magnified their elliptic points will be surrounded by further island chains, and so on. We say that the island chain structure is **repeated on all scales**.

It should be noted that all the periodic points associated with a particular island chain have the same rotation number. For example, the qq'-periodic points described above have the same rotation number, p/q say, as the original period-q orbits predicted by the Poincaré–Birkhoff theorem. If the average rotation per iteration is p/q then an orbit of the elliptic q-cycle makes p trips around the elliptic fixed point at the origin for every q iterations of the map. The orbit of the qq'-cycle makes p trips around the elliptic fixed point and returns to the same island after q iterations but only to the next point of the q'-periodic orbit of $\mathbf{P}^q$ (Fig. 6.45). Hence after qq' iterations this orbit has made a total of pq' trips around the elliptic fixed point and the average rotation per iteration is $pq'/qq' = p/q$.

The above observation, together with the Poincaré–Birkhoff theorem, means that the effect of general area-preserving perturbations on $\mathbf{T}$ is to replace the rational invariant circles of (6.113) by island chains. Some of these chains, like those illustrated in Figs 6.39 and 6.40 are 'fat' and are easily detected numerically, others are so 'thin' as to be indistinguishable from closed curves at any practical plotting accuracy.

6.7.3 Chaotic orbits and homoclinic tangles

The occurrence of chaotic orbits is a result of a fundamental difference in the behaviour of the stable and unstable manifolds of saddle points in flows and maps. For example, Fig. 6.42(a) shows the stable and unstable separatrices of the saddle point in the flow of the differential equation

$$\dot{x} = y, \qquad \dot{y} = kx(x-1), \tag{6.115}$$

where k is a positive constant. Fig. 6.42(b) shows stable and unstable manifolds for the saddle point of the related area-preserving map

$$x_1 = x + y_1, \qquad y_1 = y + kx(x-1). \tag{6.116}$$

For a flow, if the stable and unstable separatrices of a saddle have one point in common then they must coincide, resulting in a **homoclinic saddle connection**. For a map, if the stable and unstable manifolds of a saddle meet in a point (called a **homoclinic point**) then they must meet at all forward and reverse iterates of that point. However, the manifolds do not have to coincide, so typically they intersect transversely. As the forward (reverse) iterates of a homoclinic point approach the saddle point they are compressed together on the stable (unstable) manifold and the segments of unstable (stable) manifold joining them are stretched out by the action of the saddle. The resulting configuration of the manifolds is appropriately called a **homoclinic tangle** (see Fig. 6.42(b)).

The map (6.116) is quadratic, area-preserving and has an elliptic fixed point at the origin. It is therefore conjugate to the Hénon map (6.114). Both maps have chaotic orbits associated with their saddle fixed points. In Fig. 6.39

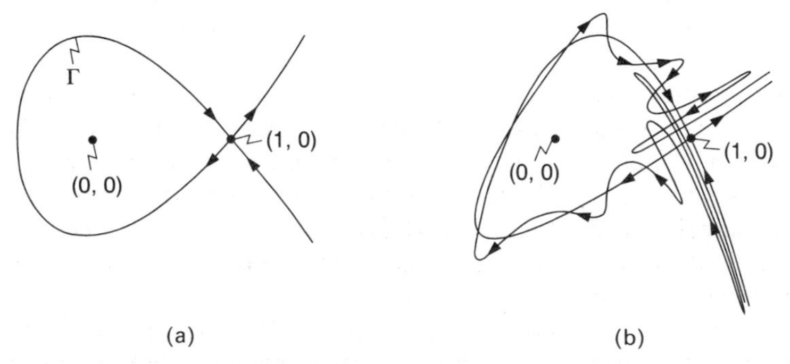

Fig. 6.42. (a) Homoclinic saddle connection, Γ, in the phase portrait of the differential equation (6.115). (b) Homoclinic tangle associated with the saddle fixed point of the area-preserving, planar diffeomorphism (6.116).

such an orbit is visible as a scattering of points outside the five-fold island chain. Typically, iterates of the map make a number of irregular revolutions round the elliptic point, always remaining outside the island chain, and eventually escape along the branch of the unstable manifold to the right of the saddle point. The number of revolutions achieved before escape depends sensitively on the starting point of the iteration.

We can use Fig. 6.43(a) to understand this behaviour. Consider a point P lying in the shaded tangle loop S_0. The forward images of this loop are $S_1, S_2, \ldots$ and so forward iterates of P escape along the unstable manifold of the saddle point. The inverse images of S_0 are $S_{-1}, S_{-2}, \ldots$. As i increases above 3, S_{-i} is stretched along the stable manifold, its width decreases and it wraps further and further around the elliptic point at (0, 0). By taking i large enough, we can find a reverse iterate, P', of P whose forward orbit will make any chosen number of revolutions around (0, 0) before arriving at P and subsequently escaping to infinity (as shown in Fig. 6.43(b)). What is more, note how S_{-i} narrows and contorts for large i and that it is pressed closely alongside S_{-i-1} and S_{-i+1}. It is not surprising that the orbit should depend sensitively on the choice of initial point.

Similar manifold tangles (called **heteroclinic tangles**) occur when the unstable manifold of one fixed point meets the stable manifold of another (as illustrated in Fig. 6.44). The periodic points of a q-cycle of a map $\mathbf{P}$ are fixed points of $\mathbf{P}^q$. If heteroclinic tangles form between adjacent fixed points of $\mathbf{P}^q$ as indicated in Fig. 6.45, then the resultant configuration of the manifolds is said to be a **homoclinic tangle for the periodic orbit**. It is tangles of this kind that are responsible for the chaotic orbits that appear in Figs 6.40 and 6.41. Figure 6.45 also highlights the fact that such tangles will be repeated on all scales.

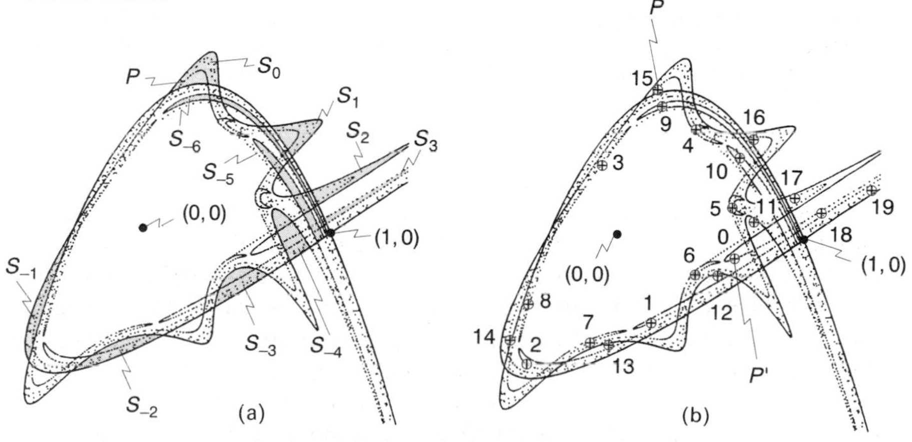

Fig. 6.43. (a) Images, S_i, and pre-images, S_{-i}, $i = 1, 2, \ldots$, of the tangle loop S_0 under the map (6.116). (b) The orbit under (6.116) of P' makes two trips around (0, 0) before entering S_0 at P on the 15th iteration and subsequently escaping along the unstable manifold of the saddle point at (1, 0) (after Arrowsmith and Place, 1990).

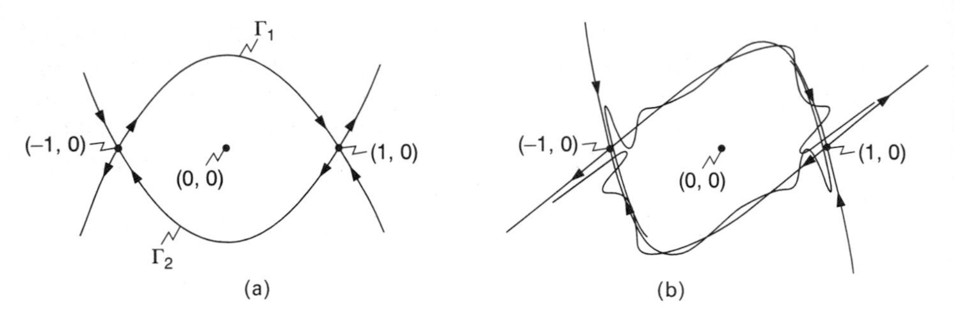

Fig. 6.44. (a) Heteroclinic saddle connections in the phase portrait of the differential equation $\dot{x} = y$, $\dot{y} = kx(x^2 - 1)$. (b) Heteroclinic tangle involving the stable and unstable manifolds of the two saddle points of the map $x_1 = x + y_1$, $y_1 = y + kx(x^2 - 1)$.

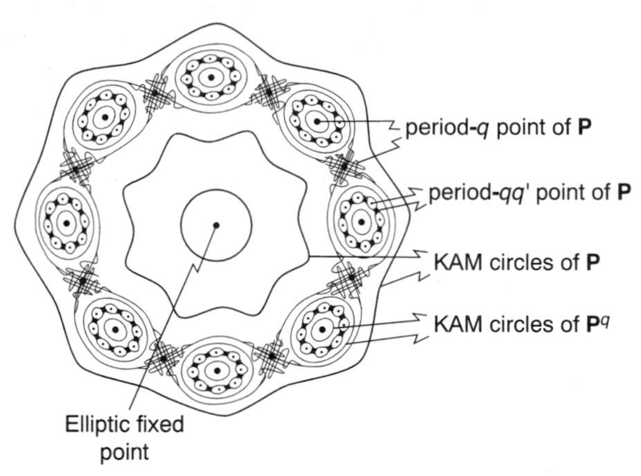

Fig. 6.45. Schematic diagram showing the typical orbit structure found in the neighbourhood of an elliptic fixed point in an area-preserving map **P**. Note: (i) the homoclinic tangles associated with the hyperbolic periodic orbits of the island chains confined by the KAM circles; (ii) the structure is repeated on all scales because the powers of **P** are also area-preserving.

Numerical experiment shows that the chaotic orbit associated with the island chain in Fig. 6.40 remains in the neighbourhood of the chain and shows no inclination to escape to infinity. This behaviour is attributed to the existence of irrational invariant circles (predicted by the KAM theorem) which act as a barrier to the orbit leaving the vicinity of the island chain. Without these restraining KAM circles, chaotic orbits can filter away from their island chain, join with other chaotic orbits to form larger chaotic regions and may ultimately escape from the elliptic fixed point. The problem of beam stability, outlined in section 6.7.1, is therefore related to the question of the break-up of KAM circles and the consequent merging of chaotic orbits (Thompson and Stewart, 1986; Percival, 1988; Arrowsmith and Place, 1990).

6.7.4 Strange attracting sets

Homoclinic tangles do not only occur in area-preserving maps. The transverse nature of the intersection of the manifolds means that the tangles persist even under non-area-preserving perturbations. This behaviour is in sharp contrast to the structural instability of a homoclinic saddle connection in a flow. For example, consider the effect of adding a damping term $-\delta y$ to the expressions for $\dot{y}$ in (6.115) and y_1 in (6.116). While the saddle connection cannot exist

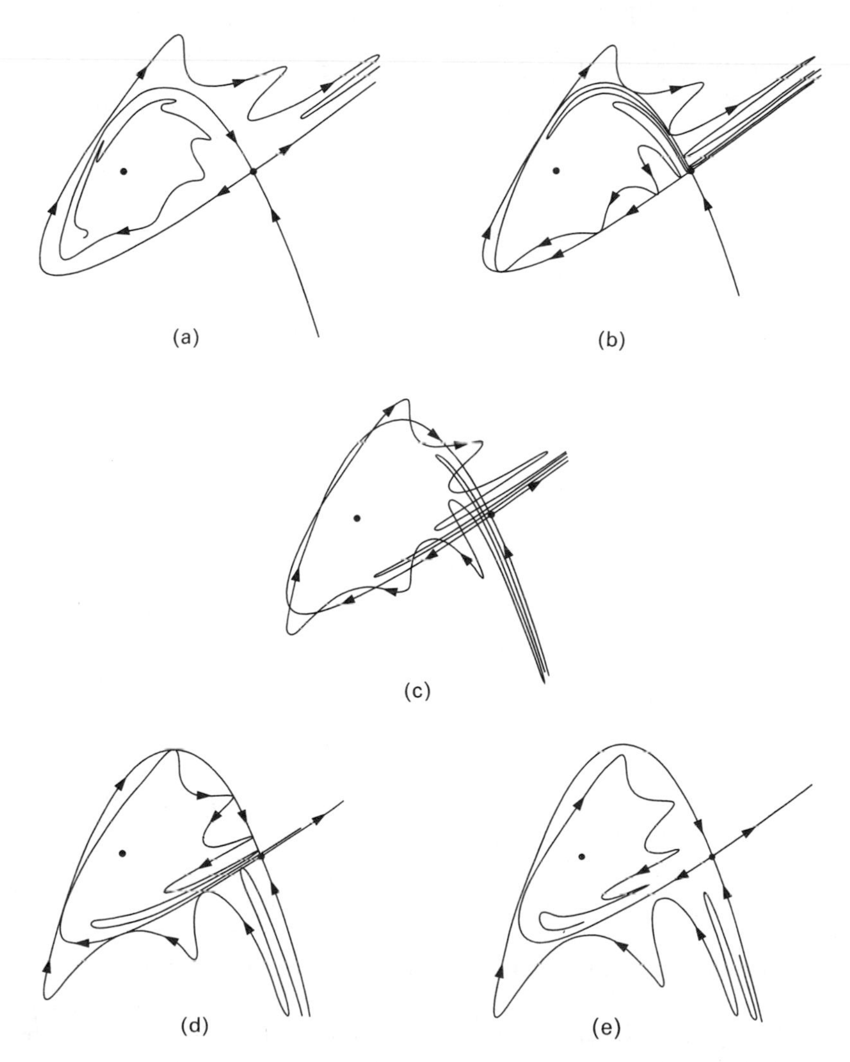

Fig. 6.46. Plots illustrating destruction of the homoclinic tangle in (6.116) by the addition of $-\delta y$ to the expression for y_1. Note that the stable and unstable manifolds are tangent at the limits of existence of the tangle in (b) and (d).

in the perturbed flow for any non-zero δ, Fig. 6.46 shows that the homoclinic tangle in (6.116) persists for an interval of δ-values.

If we increase δ from below this interval, then Fig. 6.46 illustrates a typical bifurcation in which a homoclinic tangle is formed. It is apparent that such bifurcations are preceded by extensive contortions of the manifolds. These contortions are necessary because infinitely many homoclinic tangencies must occur *simultaneously* when the tangle forms.

For $\delta > 0$, the perturbation, $\mathbf{P}$ say, of (6.116) considered above is an example of a **dissipative map**. It can be shown that $\det(D\boldsymbol{P}(\boldsymbol{x})) \equiv 1 - \delta$, so that $\mathbf{P}$ reduces the area of any region on which it acts. This observation raises an interesting prospect. If chaotic orbits are associated with homoclinic tangles and the tangles can exist in dissipative maps, then are not these circumstances conducive to the formation of chaotic or strange attracting sets? An attracting set involving these ingredients occurs in the Poincaré map, $\mathbf{P}_\gamma$, of the Duffing equation (6.108).

It follows from the definition of $\mathbf{P}_\gamma$ in terms of the flow, ψ_t, of (6.108) that

$$\det(D\boldsymbol{P}_\gamma(x, y)) = \det(D\psi_\tau(x, y, 0)). \tag{6.117}$$

Similar arguments to those used to justify the Liouville theorem in section 4.4, allow us to show that

$$\frac{\mathrm{d}}{\mathrm{d}t}[\det(D\psi_t(x, y, \theta))] = \operatorname{tr}(D\boldsymbol{X}(x, y, \theta)) \det(D\psi_t(x, y, \theta)), \tag{6.118}$$

where $\boldsymbol{X}(x, y, \theta)$ is the vector field of the flow ψ_t. Substitution of $\boldsymbol{X}$ from (6.108) gives

$$\operatorname{tr}(D\boldsymbol{X}(x, y, \theta)) = -\delta. \tag{6.119}$$

The differential equation (6.118) is then easily solved and we obtain

$$\det(D\boldsymbol{P}_\gamma(x, y)) = \exp(-\delta\tau). \tag{6.120}$$

Thus $\det(D\boldsymbol{P}_\gamma(x, y))$ is less than 1 for all (x, y) and we conclude that $\mathbf{P}_\gamma$ is a dissipative map.

Recall from section 6.6.2 that $\mathbf{P}_\gamma$ has a saddle fixed point (usually denoted by ${}^1D$) near the origin. As γ is increased from zero, a homoclinic tangle forms at this point. The critical value of γ is given by

$$\frac{\gamma}{\delta} = \frac{4\cosh(\pi\omega/2)}{3\sqrt{2}\pi\omega} \tag{6.121}$$

(Guckenheimer and Holmes, 1983, Arrowsmith and Place, 1990).

For $\omega = 1$ and $\delta = 0.25$, (6.121) gives $\gamma = \gamma_c = 0.1883\ldots$ Figure 6.47 shows segments of the stable and unstable manifolds near the saddle point for these parameter values. Simultaneous tangential intersections of the branches of the stable and unstable manifolds on the left- and right-hand sides of the saddle point are required by the symmetry of the vector field in (6.108)

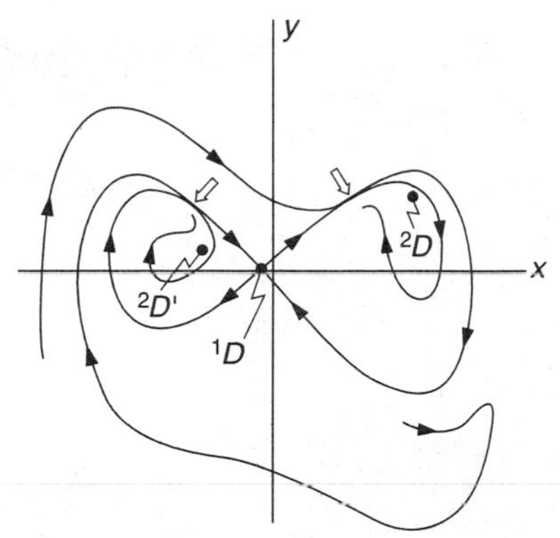

Fig. 6.47. Tangential intersections (arrowed) of the stable and unstable manifolds of the saddle point ${}^1D$ in the Poincaré map $\mathbf{P}_\gamma$ of the Duffing equation (6.108) with $\omega = 1$, $\delta = 0.25$ and $\gamma = 0.1883$. Note that for each of the intersections shown infinitely many other tangencies *must* occur. Approximate positions of the saddle points ${}^2D$ and ${}^2D'$ are also shown.

(cf. Exercise 6.41). Only one tangential intersection on each side is shown in Fig. 6.47 because the loops of the ensuing tangle are not easily resolved. However, there is no doubt that a homoclinic tangle forms at ${}^1D$ for γ greater than γ_c.

Note also that the two saddle–node bifurcations described in section 6.6.2 have already taken place when $\gamma = \gamma_c$. The saddle points (usually denoted by ${}^2D$ and ${}^2D'$) that develop from these bifurcations lie in the vicinity of the corresponding branches of the unstable manifold (Fig. 6.47). Both ${}^2D$ and ${}^2D'$ have stable and unstable manifolds of their own, which may be involved in further tangles.

Numerical experiments show that a chaotic attractor does not necessarily occur. For example, for $\gamma = 0.2$, the chaotic folded-band attractor discussed in section 6.6.2 has disappeared, the main homoclinic tangle described above is present but the only attractors that can be detected are the original foci near $(x, y) = (\pm 1, 0)$ (usually denoted by ${}^1S$ and ${}^1S'$).

Under such circumstances, orbits of $\mathbf{P}_\gamma$ starting in the vicinity of the tangle exhibit transient chaotic behaviour. Projected orbits of ψ_t that first wind around one focus, may subsequently cross the y-axis to wind around the other. Several crossovers of this kind may occur before the projected orbit approaches that of either ${}^1S$ or ${}^1S'$. The orbit, and even the limiting focus, depends sensitively on the choice of initial point.

Recall from section 4.3 that the stable manifold of the saddle point forms the boundary between the basins of attraction of the two foci. Numerical approximations to these basins of attraction for $\gamma = 0.2$ are shown in Fig. 6.48. Extensive contortions in the stable manifold are apparent but it is important to realize that the plots have a finite resolution. On magnifying what appear to be boundary 'curves' in this figure further layers of structure (corresponding

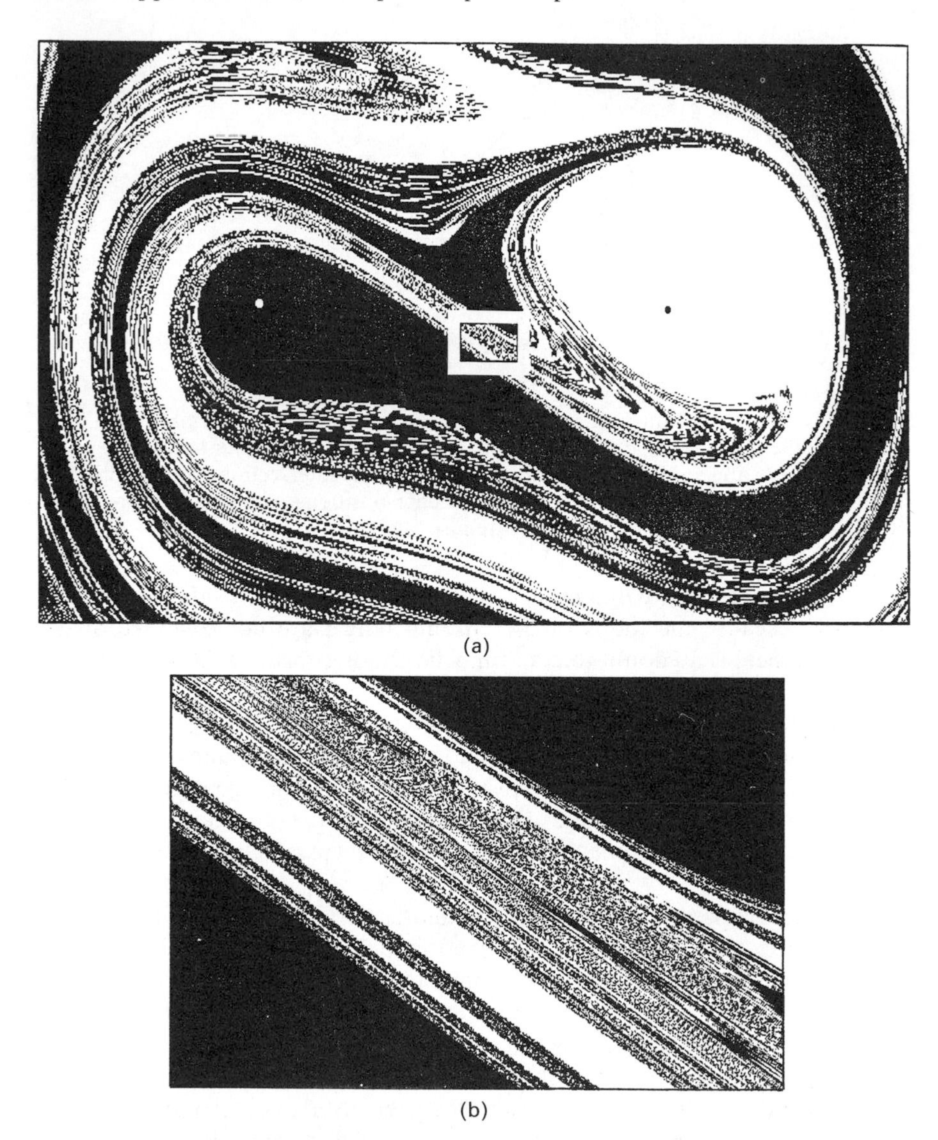

Fig. 6.48. Basins of attraction of the two foci ^1S and $^1S'$ in $\mathbf{P}_\gamma$ when the homoclinic tangle is present at ^1D. Data shown is for $\omega = 1$, $\delta = 0.25$ and $\gamma = 0.2$. The detail indicated in (a) is magnified to form the whole of (b).

to further folds in the manifold) are revealed. Since there are infinitely many folds this process can be continued indefinitely. The domains of attraction of the foci are said to have **fractal basin boundaries** when the homoclinic tangle is present.

The trapping region T_1, used in the discussion of (6.108) with $\gamma = 0$ in section 4.3, is still a trapping region for $\mathbf{P}_\gamma$ when $\gamma = 0.2$. We know that the attracting set within T_1 will contain: the foci $^1S, {}^1S'$; the saddles $^1D, {}^2D, {}^2D'$

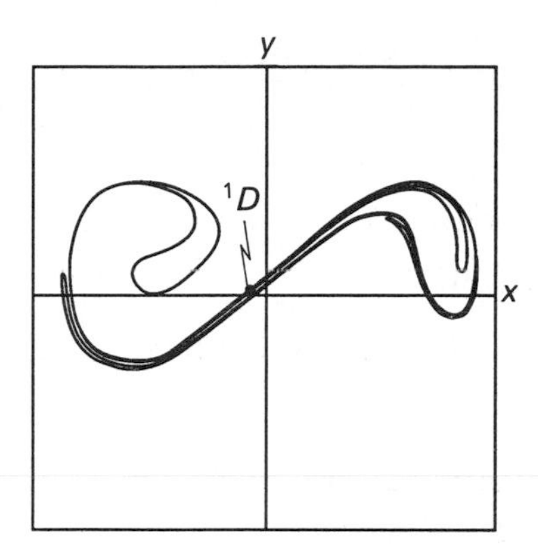

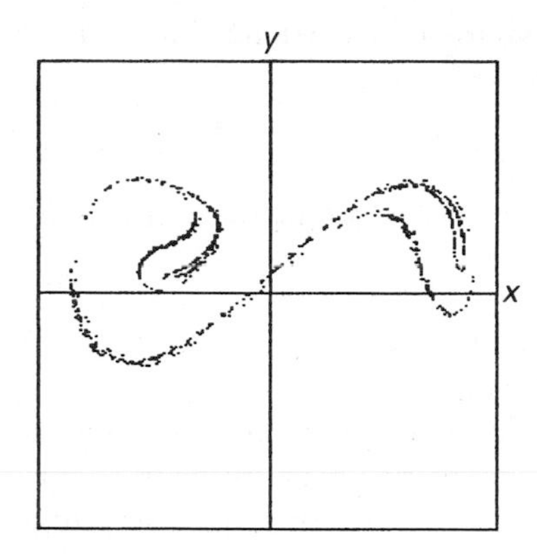

Fig. 6.49. Comparison of the unstable manifold of ^1D (in (a)) and the strange attractor (in (b)) for the Poincaré map $\mathbf{P}_\gamma$ of the Duffing equation (6.108) with $\omega = 1$, $\delta = 0.25$ and $\gamma = 0.28$. Diagram (b) consists of 1000 iterates of a single point near the origin.

and their unstable manifolds; and all the unstable periodic points left after the period-doubling of the nodes, ^2S and $^2S'$, described in section 6.6.2. However, this set is not an attractor. As we have noted, for this value of γ, the only attractors are ^1S and $^1S'$.

As γ is increased above 0.2, the situation described above persists but the saddle $^2D(^2D')$ moves towards the focus $^1S(^1S')$ until, when $\gamma \approx 0.266$, they coalesce and disappear in a reverse saddle–node bifurcation. The disappearance of the attractors ^1S and $^1S'$ is accompanied by the formation of what seems to be a chaotic or strange attractor that is numerically indistinguishable from the closure of the unstable manifold of ^1D. Figure 6.49) illustrates this by comparing a typical chaotic orbit with a numerical approximation to the unstable manifold for $\gamma = 0.28$.

The above discussion highlights the relationship between chaotic behaviour and homoclinic points in maps. However, the homoclinic orbit of a Poincaré map like $\mathbf{P}_\gamma$ corresponds to a complicated homoclinic saddle connection in the flow ψ_t of (6.108). This orbit repeatedly winds around the solid torus $\mathbb{R}^2 \times S^1$ meeting the xy-plane in successive points of the homoclinic orbit of $\mathbf{P}_\gamma$.

6.8 SYMBOLIC DYNAMICS

Thus far we have only encountered chaotic behaviour in the context of numerical experiments. Successive points of a 'chaotic' orbit appear in an erratic or unpredictable manner when plotted on a visual display unit. Symbolic dynamics provides us with a deeper understanding of the nature of such orbits. This approach 'codes' the dynamics in terms of symbol sequences and exhibits the random nature of chaotic orbits in an explicit way.

Consider the 'triangle' or 'tent' map

$$G(x) = \begin{cases} 2x, & 0 \leqslant x \leqslant \frac{1}{2} \\ 2(1-x), & \frac{1}{2} \leqslant x \leqslant 1. \end{cases} \tag{6.122}$$

This map is conjugate to the logistic map with $\mu = 4$; in fact

$$F_4(h(x)) = h(G(x)), \tag{6.123}$$

where $h(x) = \sin^2(\pi x/2), 0 \leqslant x \leqslant 1$. Therefore G and F_4 exhibit the same kind of dynamics.

The coding for (6.122) is obtained by using the positive integer powers of G to describe each point of $[0, 1]$ as a binary sequence. We define the kth element, σ_k, of the sequence, σ representing the point x by

$$\sigma_k = \begin{cases} 0, & \text{if } G^k(x) \leqslant \frac{1}{2} \\ 1, & \text{if } G^k(x) \geqslant \frac{1}{2}. \end{cases} \tag{6.124}$$

Figure 6.50 shows that the finite symbol blocks $\sigma^{(N)} = \{\sigma_0 \ldots \sigma_N\}$, N a positive integer, label the subintervals of a uniform dissection of $[0, 1]$. Each subinterval has length $1/2^{(N+1)}$ and therefore, in the limit when N tends to infinity, (6.124) ascribes an infinite, binary sequence or 'code' to each point x of $[0, 1]$. This sequence is unique for those points whose orbit does not contain $x = \frac{1}{2}$.

In terms of this coding, the map G takes a very simple form. If x is represented by $\sigma = \{\sigma_0 \sigma_1 \sigma_2 \ldots\}$ then, by the construction (6.124), $G(x)$ is represented by $\sigma' = \{\sigma_1 \sigma_2 \ldots\}$. Thus the code for $G(x)$ is obtained from that for x by shifting the elements of σ one place to the left and discarding the first element. We say that σ' is a **left-shift** of σ and write $\sigma' = \alpha(\sigma)$. It is

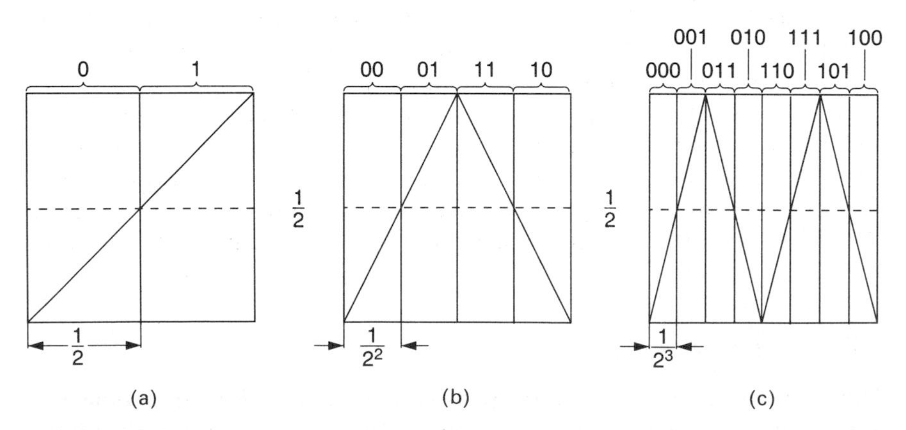

Fig. 6.50. Illustration of the coding of subintervals of $[0, 1]$ given by (6.124) with: (a) $k = 0$; (b) $k = 0$ and 1; (c) $k = 0, 1$ and 2. Each additional power of G separates the subintervals already labelled into two halves and assigns a zero to one half, and a one to the other. This new symbol is attached to the right-hand end of the current code.

important to note that the code σ is not the ordinary binary representation of the real number x. While this representation gives a binary coding of the points in $[0, 1]$, G is not a left-shift on this set of codes.

It can be shown that (with the exception of the pre-images of $x = \frac{1}{2}$) the procedure for obtaining σ from x defines a continuous bijection between $[0, 1]$ and the set, Σ, of all binary codes. It follows that the map G on $[0, 1]$ is conjugate to the left-shift map α on Σ. This conjugacy means that the qualitative behaviour of G is the same as that of α.

Since $\alpha(\sigma)$ is simply a left-shift of σ the orbit structure of α is easy to obtain. For example, the fixed points of α have the codes $\{000\ldots\}$ and $\{111\ldots\}$ that are invariant under a left-shift. These codes correspond, respectively, to the fixed points $x = 0$ and $x = \frac{2}{3}$ of G. It is clear from Fig. 6.50 that G has a period-2 orbit that spans $x = \frac{1}{2}$. The codes for the points of this 2-cycle are $\sigma_1 = \{010101\ldots\}$ and $\sigma_2 = \{101010\ldots\}$, corresponding respectively to the period-2 points of G at $x_1 < \frac{1}{2}$ and $x_2 > \frac{1}{2}$. Observe that $\sigma_2 = \alpha(\sigma_1)$ and $\sigma_1 = \alpha(\sigma_2)$ as required.

Any code that repeats periodically gives rise to a periodic orbit of α. Thus a code σ that consists of the symbol block $\sigma^{(q-1)} = \{\sigma_0\sigma_1\ldots\sigma_{q-1}\}$ repeated indefinitely belongs to a q-cycle of α and corresponds to a q-periodic point of G. It follows that both α and G have periodic orbits of all periods (note that all periodic orbits of G are unstable (cf. Fig. 6.50). However, it is possible to construct codes that never repeat themselves, e.g.

$$\{0\underbrace{10}_{2}\underbrace{010}_{3}001\ldots\underbrace{0\ldots01}_{n}\ldots\},$$

and therefore both α and G must also have aperiodic orbits.

If the codes σ and σ' both begin with the symbol block $\sigma^{(N-1)}$, then the corresponding points x and x' both belong to the subinterval of $[0, 1]$ specified by that symbol block and $|x - x'| < 2^{-N}$. Thus, for example, the metric

$$d(\sigma, \sigma') = 2^{-N}, \tag{6.125}$$

on Σ is compatible with the Euclidean metric on $[0, 1]$. When Σ is equipped with such a metric, we are able to recognize those orbits of G that exhibit chaotic behaviour.

Consider the code δ constructed by successively listing all symbol blocks of length n for each $n \in \mathbb{Z}^+$ and let σ be any code. Given any positive integer N, the sequence δ has within it the symbol block $\sigma^{(N-1)}$ consisting of the first N symbols of σ. Suppose this block begins at δ_k, then σ and $\alpha^k(\delta)$ both begin with the symbol block $\sigma^{(N-1)}$ and $d(\sigma, \alpha^k(\delta)) = 2^{-N}$. By allowing N to increase indefinitely, we conclude that the orbit of δ must come arbitrarily close to σ. Furthermore, since σ is arbitrary, the orbit of δ comes arbitrarily close to every element of Σ. Therefore the orbit of δ under α is dense in Σ. Correspondingly, if δ represents the point x_δ, then the orbit of x_δ under G is dense in $[0, 1]$.

If we plot the iterates $G^k(x_\delta)$, $k = 0, 1, 2, \ldots$, (as in Exercise 6.64) they exhibit 'chaotic' behaviour in the numerical sense of earlier sections. The iterates move between $[0, \frac{1}{2}]$ and $[\frac{1}{2}, 1]$ according as the leading symbol of the shifted code is zero or one, and ultimately appear to fill out $[0, 1]$ because the orbit of x_δ is dense in this interval.

While the code δ provides us with a chaotic orbit, it seems to be constructed in a very special way. The truly astonishing result is that almost all codes exhibit the same behaviour as δ. This follows from a classical result of Hardy (1979) which essentially says that the set of real numbers in $[0, 1]$ that have all possible finite sequences of zeros and ones in their binary expansions has measure one. This result means that iterates of a randomly chosen initial point will give rise to chaotic orbits with probability one. This is the case in spite of the fact that it can be shown (as in Exercise 6.63) that the periodic orbits of G are dense in $[0, 1]$. They are dense but they have measure zero.

The result of Hardy also allows us to appreciate the sense in which the behaviour of a chaotic orbit is as random as any coin tossing experiment. If we take a 'head' to be 'zero' and a 'tail' to be 'one', then repeated tossing of a fair coin will generate a symbol block. This block, however long, will be found within the code corresponding to a chaotic orbit. What is more, the zeros and ones of this code correspond to the appropriate iterate of G being to the 'left' or 'right' of $x = \frac{1}{2}$. Thus we are unable to distinguish the behaviour of this section of the orbit of G from the result of picking 'left' or 'right' by tossing a fair coin.

It is somewhat unsettling to find a deterministic system like (6.122) exhibiting the randomness associated with tossing a coin. However, are these misgivings well founded? After all, the equations of motion of the coin are themselves deterministic. The uncertainty arises from the sensitivity of the tumbling motion of the coin to the initial conditions provided by the 'flip'. Similarly, determinism in the orbit of G is only assured if the initial value of x is given precisely. If x' approximates x then symbolic dynamics allows us to show that the orbits of x and x' are ultimately independent of one another. This phenomenon is a well-documented feature of chaotic behaviour and is referred to as its **sensitive dependence on initial conditions**.

If x is near x', then the corresponding codes σ and σ' agree over a finite initial symbol block of length N, say. Since the application of G is given by a left-shift of σ and σ', the common symbol block is 'flushed out' in N iterations and subsequently the orbits are independent of one another. In terms of the orbits under G, this means that, for $k = 0, 1, \ldots, N-1$, $G^k(x)$ and $G^k(x')$ both lie on the same side of $x = \frac{1}{2}$ but they are no longer correlated in this way for $k = N, N+1, \ldots$.

What properties of G lead to the 'sensitive dependence on initial conditions' described above? Figure 6.50 shows that the effect of G is to stretch $[0, 1]$

uniformly to twice its original length, fold it in half and replace it on [0, 1] so that the fold coincides with unity. G is locally an expansion at every x except $x = \frac{1}{2}$. Typically the separation of two neighbouring points, x and x', doubles at each application of G and therefore their orbits diverge exponentially. However, the range of G is [0, 1] and the expansion is accommodated by the fold that occurs at $x = \frac{1}{2}$. The iterates of x and x' move apart but remain correlated (in the sense that they both move to the same side of $x = \frac{1}{2}$) until the interval between $G^k(x)$ and $G^k(x')$ spans $x = \frac{1}{2}$. For x and x' described by the codes of the previous paragraph this would occur when $k = N$.

The symbolic dynamics discussed above allows us to characterize the chaotic behaviour of maps on an interval like G and F_4 by saying that they have the same properties as a left-shift on the set of all infinite sequences of two symbols. The main features of this type of behaviour are: (1) the set of periodic points is dense in the interval; (2) there are orbits that are dense in the interval; and (3) there is sensitive dependence on initial conditions.

The same kind of dynamics can be found in other situations. Consider, for example,

$$G_v(x) = \begin{cases} 2vx, & -\infty < x \leqslant \frac{1}{2} \\ 2v(1-x), & \frac{1}{2} \leqslant x < \infty, \end{cases} \tag{6.126}$$

with $1 < v < \frac{3}{2}$. If $G_v^k(x)$ lies outside [0, 1] for some positive integer k then the orbit of x leaves [0, 1] and never returns. Figure 6.51 shows graphs of G_v, G_v^2 and G_v^3 restricted to [0, 1]. It can be seen that the interval is divided into two disjoint subintervals by the points that leave [0, 1] under G_v, four subintervals

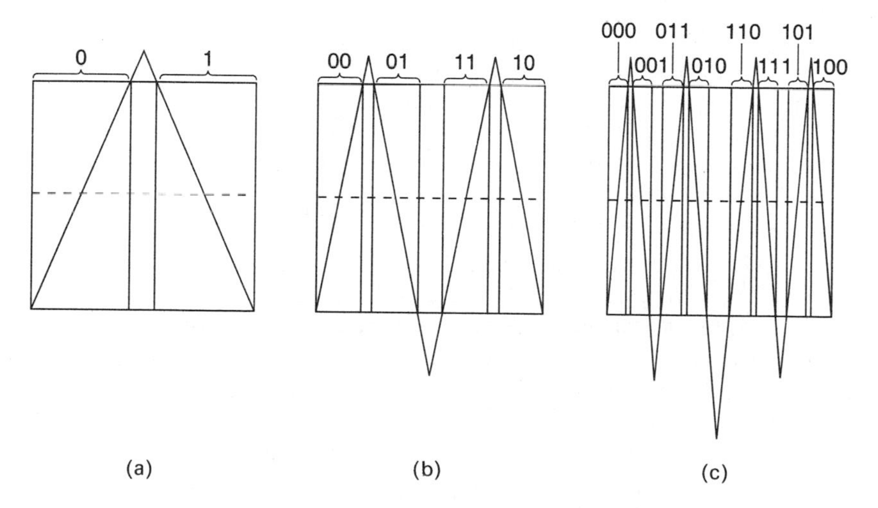

Fig. 6.51. Graphs of: (a) G_v; (b) G_v^2; (c) G_v^3; on [0, 1] for $v = 9/8$, together with coding obtained from (6.124) by omitting the final symbol.

by those that leave under G_v^2 and eight subintervals by those that leave under G_v^3. The sets of points that remain are of the same type as those appearing in the construction of the 'mid-third' Cantor set $\mathscr{C}$. The reader will recall that $\mathscr{C}$ is the intersection of a sequence of sets obtained by removing the middle third of $[0, 1]$, then removing the middle third of both $[0, \frac{1}{3}]$ and $[\frac{2}{3}, 1]$, and so on. We conclude therefore that G_v has an invariant Cantor set in $[0, 1]$ consisting of points that remain in $[0, 1]$ indefinitely. A coding of these points for which G_v corresponds to a left-shift can be achieved by using (6.124) and dropping the last symbol. The resulting subinterval labels are shown in Fig. 6.51. It follows that the restriction of G_v to its invariant Cantor set can exhibit chaotic behaviour.

The chaotic invariant set of the famous **Smale horseshoe map** can be treated

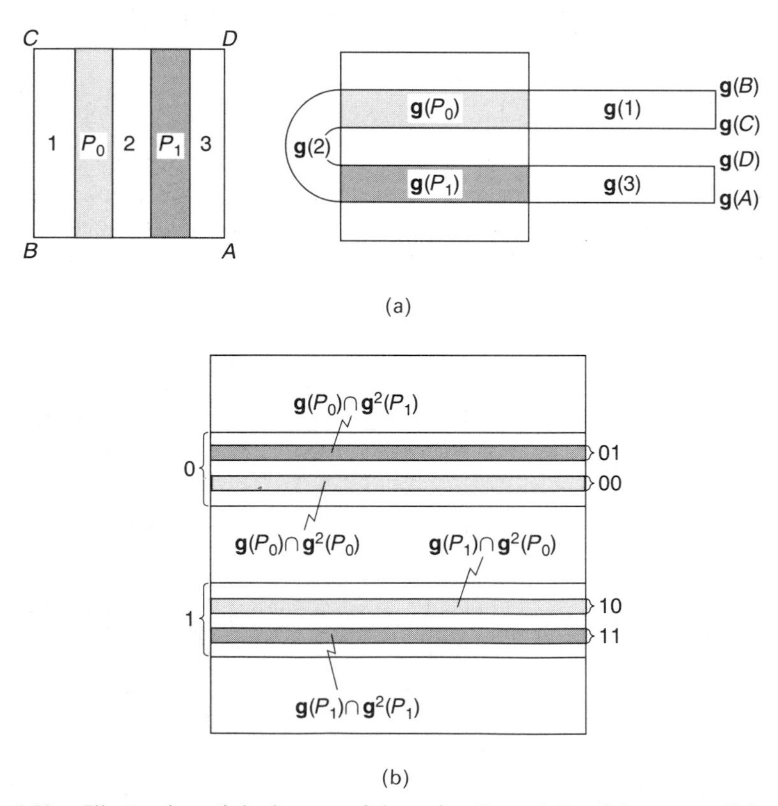

Fig. 6.52. Illustration of the images of the strips P_0 and P_1 of the square S (vertices ABCD) under $\mathbf{g}$ and $\mathbf{g}^2$ together with the codings of the horizontal strips containing points that remain in S after two applications of $\mathbf{g}$.

in a similar manner. The canonical form, g, of such a map is illustrated in Fig. 6.52(a). The square $S = I \times I$ is stretched by a factor of five horizontally, contracted by a factor of five vertically, bent into a horseshoe shape and placed on S as shown. Only points in the vertical strips P_0 and P_1 have images in S and these lie in the two horizontal strips $Q_0 = \mathbf{g}(P_0)$ and $Q_1 = \mathbf{g}(P_1)$, respectively. The images of P_0 and P_1 under $\mathbf{g}^2$ give rise to four horizontal

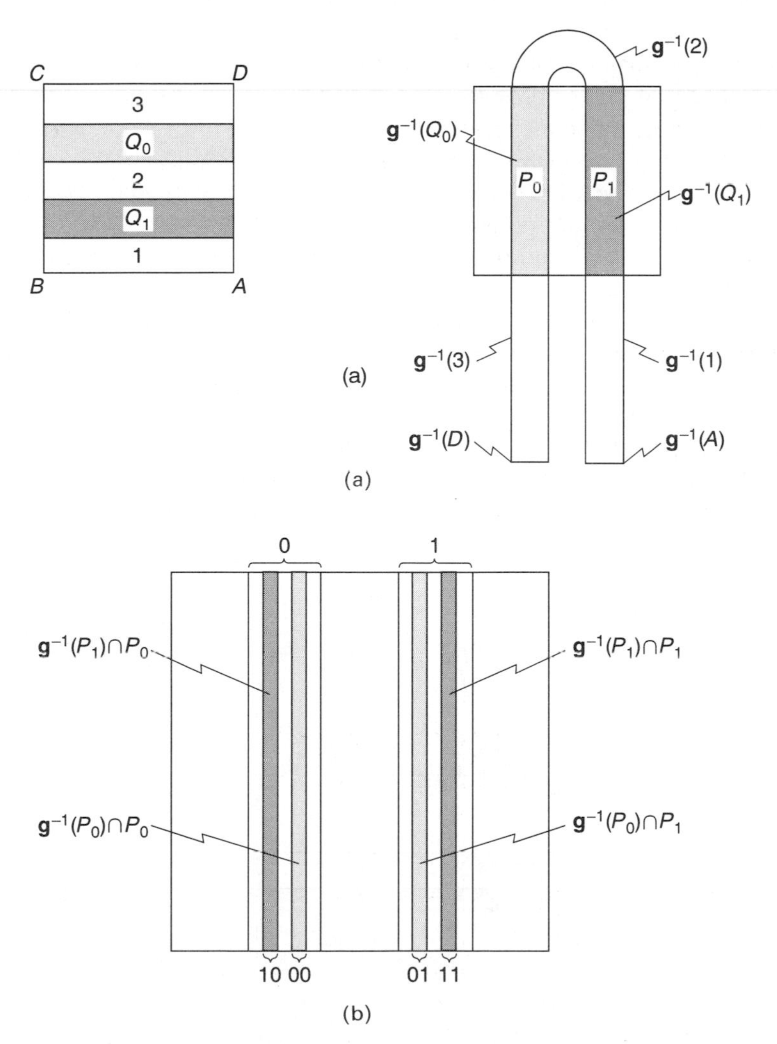

Fig. 6.53. Illustration of the images of the strips Q_0 and Q_1 under $\mathbf{g}^{-1}$ and $\mathbf{g}^{-2}$. Note that: (i) $\mathbf{g}^{-1}(Q_0) = P_0$ and $\mathbf{g}^{-1}(Q_1) = P_1$; (ii) new symbols are appended to the left-hand end of existing codes.

strips within Q_0 and Q_1. These strips can be assigned a unique symbol block according to whether they are the intersection of images under $\mathbf{g}$ and $\mathbf{g}^2$ of P_0 or P_1 (as in Fig. 6.52(b)). When restricted to S the images of P_0 and P_1 under $\mathbf{g}^k$ lie within their images under $\mathbf{g}^{k-1}$ for $k \geqslant 2$ and this system of coding can be extended indefinitely (Arrowsmith and Place, 1990). If we focus attention on vertical coordinates only, it is evident that the evolution of the horizontal strips corresponds to Cantor set formation and we find that the set of points of S that remain in S under indefinite forward iteration takes the form $I \times \mathscr{C}$, where $\mathscr{C}$ is a Cantor set.

The inverse horseshoe map, $\mathbf{g}^{-1}$, is illustrated in Fig. 6.53(a). A similar argument to that given above for $\mathbf{g}$ reveals that the set of points of S which remain in S under indefinite reverse iteration takes the form $\mathscr{C} \times I$. Moreover, $\mathbf{g}^{-1}$ can be used to assign a unique binary code to the horizontal coordinate of each of the vertical lines that make up this set. Notice that these codes are developed on the left rather than the right (Fig. 6.53(b)).

The codes from $\mathbf{g}^{-1}$ and $\mathbf{g}$ are combined to label points whose forward and reverse orbits under $\mathbf{g}$ remain in S. This is achieved by writing the code for a vertical line in $\mathscr{C} \times I$ followed by that of a horizontal line in $I \times \mathscr{C}$ and separating them by a binary point as in Fig. 6.54. The map $\mathbf{g}$ is conjugate to a left-shift on the resulting **bi-infinite** symbol sequence. These bi-infinite codes label points of the set $\Lambda = \mathscr{C} \times \mathscr{C}$ which is an invariant set for $\mathbf{g}$. The dynamics of $\boldsymbol{g}$ restricted to Λ therefore exhibit all the qualitative properties associated with the left-shift on the set Σ of all bi-infinite, binary codes.

The horseshoe map is of particular interest because it can be shown that

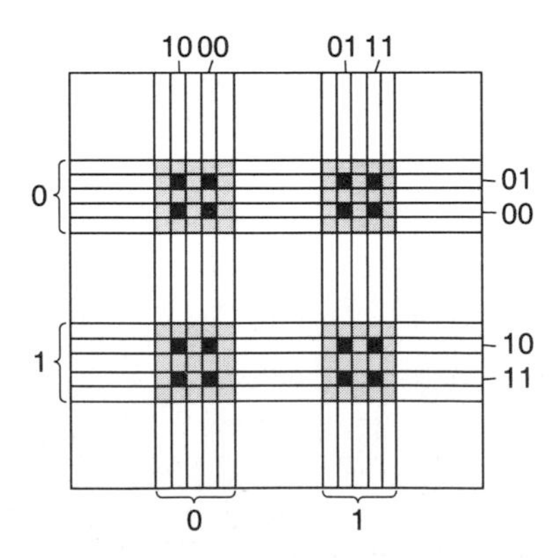

Fig. 6.54. Illustration of the coding of the 'boxes' formed by taking the intersection of the strips in Figs 6.52 and 6.53. Large light shaded boxes have codes (clockwise from top left): $\{0.0\}$, $\{1.0\}$, $\{1.1\}$, $\{0.1\}$, i.e. write down the code for the vertical strip followed by the code for the horizontal strip and insert a binary point between them. Within box $\{0.0\}$ the small dark shaded boxes have codes that are formed in a similar way: (clockwise from top left) $\{10.01\}$; $\{00.01\}$; $\{00.00\}$; $\{10.00\}$; and so on.

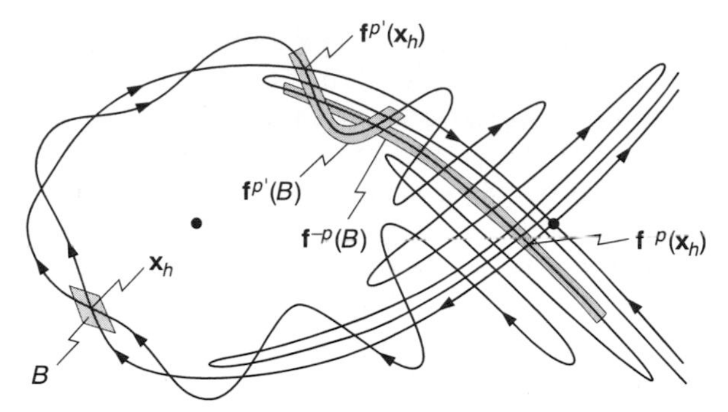

Fig. 6.55. Schematic diagram showing that a planar map $\mathbf{f}$ with homoclinic tangle can contain an embedded horseshoe map.

a homoclinic tangle can contain an embedded horseshoe map. Figure 6.55 indicates schematically how this can come about for a diffeomorphism $\mathbf{f}$ like (6.116). The 'box' B contains a homoclinic point $\mathbf{x}_h$ and therefore $\mathbf{f}^k(B)$ must contain $\mathbf{f}^k(\mathbf{x}_h)$ for all integers k. However, for $k > 0$ (< 0) $\mathbf{f}^k(B)$ is stretched along the unstable (stable) manifold and compressed along the stable (unstable) manifold. If $\mathbf{f}^{\;\,p}(B)$ meets $\mathbf{f}^{p'}(B)$ as shown in Fig. 6.55 then the map $\mathbf{f}^{p+p'}$ restricted to $\mathbf{f}^{-p}(B)$ is a horseshoe map.

Since $\mathbf{g}$ expands horizontally and contracts vertically, the set Λ is of saddle type and it is therefore not readily detected directly in numerical experiments. However, orbits passing close to Λ reflect the chaotic dynamics on it and behave in an erratic way. In the homoclinic tangles that we have encountered, orbits are repeatedly fed into the region that potentially contains an embedded horseshoe map, either by the contracting environment (section 6.7.4), or by the area-preserving constraints (section 6.7.3). Consequently, the chaotic behaviour on invariant sets like Λ may well play a significant role in the 'chaotic' orbits found in the corresponding numerical experiments.

6.9 NEW DIRECTIONS

6.9.1 Introductory remarks

In retrospect, it is noticeable how our desire to understand the qualitative behaviour of differential equations has forced us to broaden our concept of a dynamical system. For example, the need for Poincaré maps of flows led us to study diffeomorphisms, taking us from 'continuous' to 'discrete' systems. The great complexity of planar Poincaré maps and the difficulty of obtaining them by numerical integration has resulted in the study of simpler maps exhibiting specific types of behaviour. For example, to understand period-doubling in the context of three-dimensional flows is a daunting task (even if a Poincaré map is available). However, considerable progress has been made with this problem in terms of maps like the logistic map. Thus the class of maps whose dynamics we must consider has extended beyond diffeomorphisms to unimodal maps on an interval. The characterization of chaotic behaviour in the logistic and horseshoe maps led us to symbolic representations of their dynamics, so that maps on symbol sequences must also be included under the ever-growing umbrella of 'dynamical systems'.

However, it is not only our notion of a dynamical system that has developed; new features common to several types of dynamical system are apparent. For example, we have noted that the island chain structure in area-preserving maps and the 'braided' nature of strange attractors is repeated on all scales. By this we mean that there are small regions within these structures that, when magnified, are the same, or similar, to the original structure itself. This property of 'self-similarity' plays a central role in the theoretical understanding of period-doubling in unimodal maps. For example, recall that the flip bifurcation in F_μ is characterized by the evolution of F_μ^2 shown in Fig. 6.28. Now observe that the distinctive 'twin humps' also occur, on finer scales, in plots of F_μ^4 (Fig. 6.29) and F_μ^8 (Fig. 6.30).

To be more precise, it can be shown that there is an interval $I_{\mu,k} \subset [0, 1]$ such that the restriction $F_\mu^{2k} | I_{\mu,k}$ is conjugate to $\pm F_{\hat{\mu}(\mu,k)}$ (Exercise 6.66). For given k, $\hat{\mu}(\mu, k)$ is less than μ but increases with it and F_μ^{2k} period-doubles when $\hat{\mu}(\mu, k) = 3$.

The transformation that effects the above conjugacy for F_μ^2 is essentially a rescaling of the variable (as shown in Exercise 6.67). The process of transforming the variable and replacing μ by $\hat{\mu}(\mu, 1)$ is referred to as **renormalization** and we write $F_{\hat{\mu}(\mu,1)} = \mathscr{T}(F_\mu)$, where $\mathscr{T}$ is the **renormalization operator**. Investigation of the properties of $\mathscr{T}$ ultimately leads to an explanation of why the Feigenbaum number is a universal constant (Guckenheimer & Holmes, 1983; Manneville 1990).

In view of the vastness of the subject readers will have no difficulty in finding 'new directions' in which to expand their knowledge of dynamical systems. However, in this closing section we give the briefest of introductions of two topics that reflect the themes of self-similarity and symbolic dynamics emphasized above. We hope the reader will wish to find out more about them.

6.9.2 Iterated function schemes

The familiar 'mid-third' Cantor set, $\mathscr{C}$, is an example of a self-similar set where the repetition on all scales is exact. By construction, the magnification of either of the subsets of $\mathscr{C}$ in the intervals $[0, \frac{1}{3}]$ or $[\frac{2}{3}, 1]$ by a factor of three (combined with a shift of origin in the latter case) yields the Cantor set itself. A transverse section through the 'fractal basin boundary' mentioned in section 6.7.4 yields a set of points with an analogous self-similar property.

The term 'fractal' used above requires explanation. Some self-similar structures have the unusual property that they have a non-integer or **fractal dimension**. In order to recognize this we must use a precise definition of dimension. Let S be a set in $\mathbb{R}^n$ and let $N(S,\varepsilon)$ be the smallest number of closed balls of diameter ε required to cover the set S. Then the dimension D of S is given by

$$D = \lim_{\varepsilon\to 0}\left(-\frac{\ln(N(S,\varepsilon))}{\ln(\varepsilon)}\right). \tag{6.127}$$

For example, if S is the interval $[0,1]$, then $N(S,\varepsilon)=\varepsilon^{-1}$ and $D=1$, as expected. However, recall that the Cantor set $\mathscr{C}=\bigcap_{n=1}^{\infty} I_n$, with $I_1=[0,1]$, $I_2=[0,\frac{1}{3}]\bigcup[\frac{2}{3},1]$, $I_3=[0,\frac{1}{9}]\bigcup[\frac{2}{9},\frac{1}{3}]\bigcup[\frac{2}{3},\frac{7}{9}]\bigcup[\frac{8}{9},1],\ldots$. Therefore, $N(\mathscr{C},\frac{1}{3})=2$, $N(\mathscr{C},\frac{1}{9})=4,\ldots$ and, in general, $N(\mathscr{C},\frac{1}{3^n})=2^n$. Hence

$$\begin{aligned} D &= \lim_{\varepsilon\to 0}\left(-\frac{\ln(N(\mathscr{C},\varepsilon))}{\ln(\varepsilon)}\right), \\ &= \lim_{n\to\infty}\left(-\frac{\ln(2^n)}{\ln(3^{-n})}\right), \\ &= \frac{\ln 2}{\ln 3}. \end{aligned} \tag{6.128}$$

Thus $\mathscr{C}$ has a fractal dimension that is less than one.

In conventional dynamical systems, strange attracting sets with a self-similar structure occur in an uncontrolled way. A great deal of effort can be required to demonstrate that such sets are attractors or that the dynamics on them is really chaotic. Many outstanding problems in current research are of this type. For example, how does one prove that an attracting set such as the Hénon attractor contains a dense orbit. After all, numerical experiments could be detecting a periodic orbit of very long period.

An **iterated function scheme** is a new type of dynamical system that employs a collection of contracting maps to create chaotic attractors that have a fractal structure.

Example 6.9.1

Show that the contracting maps

$$f_1(x)=\tfrac{1}{3}x, \qquad f_2(x)=\tfrac{1}{3}x+\tfrac{2}{3} \tag{6.129}$$

leave the Cantor set $\mathscr{C}$ invariant.

Solution

By construction, each point x of $\mathscr{C}$ is uniquely determined by specifying a 'left' or 'right' subinterval in the set I_n for $n=2,3,\ldots$ as in Fig. 6.56. The length of the intervals in I_n tends to zero as n tends to infinity. Thus, if we calculate the position of the leading point of the interval containing x for each n, then x is given by the limit of this sequence as n tends to infinity. Hence

$$x=\sum_{n=1}^{\infty} a_n/3^n, \tag{6.130}$$

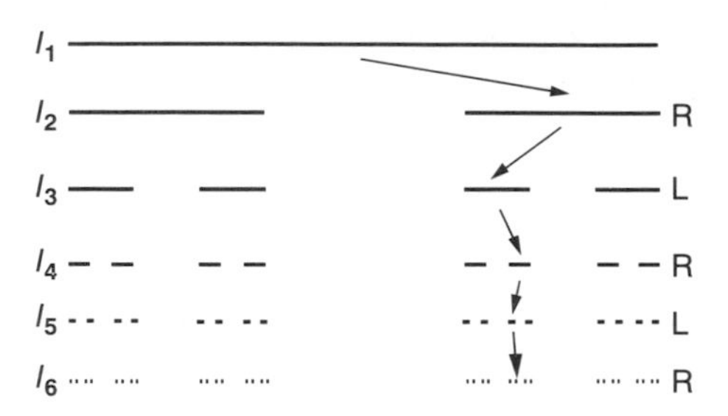

Fig. 6.56. Illustration of how each point of the mid-third Cantor set can be specified by choosing a 'left-' or 'right-hand' subinterval of I_n for every $n \in \mathbb{Z}^+$.

where a_n is 0 or 2 depending on whether a left or right subinterval is chosen at the nth stage. It follows that

$$f_1(x) = \sum_{n=1}^{\infty} a_n/3^{n+1} = \sum_{m=1}^{\infty} b_m/3^m, \tag{6.131}$$

where $b_1 = 0$, $b_m = a_{m-1}$ for $m > 1$. Thus $f_1(x)$ belongs to $\mathscr{C}$. An analogous result holds for f_2 with $b_1 = 2$, and therefore both f_1 and f_2 leave $\mathscr{C}$ invariant. □

The maps f_1 and f_2 in Example 6.9.1 are contracting, affine maps. Randomness is introduced in the manner in which they are used in the scheme. If $x_0 \in \mathbb{R}$, then its orbit might, for example, be given by $x_{n+1} = f(x_n)$, where

$$f = \begin{cases} f_1, & \text{with probability } \frac{1}{2} \\ f_2, & \text{otherwise.} \end{cases} \tag{6.132}$$

Thus the iterated function scheme is defined by specifying the contracting affine maps and the probabilities with which they are to be used.

The orbit of any point x_0 under the scheme (6.132) converges towards the Cantor set $\mathscr{C}$. More precisely, let $\mathscr{S}_N = \{x_n\}_{n=0}^{N}$ and define the distance, $d(\mathscr{S}_N, \mathscr{C})$, between $\mathscr{S}_N$ and $\mathscr{C}$ by

$$d(\mathscr{S}_N, \mathscr{C}) = \max_{x \in \mathscr{C}} \{ \min_{y \in \mathscr{S}_N} [|x - y|] \}. \tag{6.133}$$

If $d(\mathscr{S}_N, \mathscr{C})$ tends to zero as N tends to infinity then the 'tail' of the infinite sequence $\mathscr{S}_\infty$ approaches $\mathscr{C}$.

To summarize, since $\mathscr{C}$ is invariant under both f_1 and f_2, it is an invariant set for the iterated function scheme f. In fact, it is a minimal invariant set for f in that it has no proper subsets that are invariant. The maps f_1 and f_2 are both contractions and therefore $\mathscr{C}$ is an attracting set for f. Moreover, the random choice of f_1 or f_2 ensures that $f|\mathscr{C}$ behaves like a left-shift on two symbols. Hence $\mathscr{C}$ contains a dense orbit and is a **chaotic attractor** for f.

Example 6.9.2

Draw a diagram to illustrate the images T_i of the region $T = \{(x, y) | 0 \leqslant x \leqslant 1, 0 \leqslant y \leqslant 1, x + y \leqslant 1\}$ under the maps f_i, where

$$f_1(x, y) = (\tfrac{1}{2} x, \tfrac{1}{2} y), \tag{6.134}$$

$$f_2(x, y) = (\tfrac{1}{2}x, \tfrac{1}{2}y + \tfrac{1}{2}), \quad (6.135)$$

$$f_3(x, y) = (\tfrac{1}{2}x + \tfrac{1}{2}, \tfrac{1}{2}y). \quad (6.136)$$

Distinguish the points of T that do not lie in $T_1 \bigcup T_2 \bigcup T_3$. Find and illustrate the sets $T_{ij} = f_i(T_j), i, j = 1, 2, 3$. Explain how repeated applications of the maps $f_i, i = 1, 2, 3$, to T can be used to generate: $(0, 0)$; $(0, 1)$; $(1, 0)$; and $(\frac{1}{2}, \frac{1}{2})$.

Solution

The map f_1 simply halves the values of both coordinates, thus T_1 is the triangle with vertices $\{(0,0), (0, \frac{1}{2}), (\frac{1}{2}, 0)\}$. The maps f_2 and f_3 have the additional effect of translating this reduced triangle so that the 'right-angled' vertex occurs at $(0, \frac{1}{2})$ and $(\frac{1}{2}, 0)$, respectively. Hence the images $T_i, i = 1, 2, 3$, are as shown in Fig. 6.57(a). The interior points of the triangle with vertices $\{(0, \frac{1}{2}), (\frac{1}{2}, \frac{1}{2}), (\frac{1}{2}, 0)\}$ do not occur in $T_1 \bigcup T_2 \bigcup T_3$.

A further application of each map has the effect of halving the size of Fig. 6.57(a) and placing the result on T three times so that the right-angled vertices are at $(0, 0)$, $(0, \frac{1}{2})$ and $(\frac{1}{2}, 0)$. Hence the sets T_{ij} are as shown in Fig. 6.57(b). This diagram allows us to recognize the pattern associated with repeated application of a particular map. For example, $\bigcap_{n=0}^{\infty} f_1^n(T) = (0, 0)$, $\bigcap_{n=0}^{\infty} f_2^n(T) = (0, 1)$ and $\bigcap_{n=0}^{\infty} f_3^n(T) = (1, 0)$. To obtain the point $(\frac{1}{2}, \frac{1}{2})$, observe that it is the image under f_2 of $(1, 0)$ and therefore it can be obtained as $f_2(\bigcap_{n=0}^{\infty} f_3^n(T)) = (1, 0)$. □

The attractor associated with the iterated function scheme that uses the functions f_1, f_2 and f_3 with equal probabilities of one third is shown in Fig. 6.58. It is known as the **Sierpinski triangle** or '**gasket**'. Like the Cantor set, the Sierpinski gasket $\mathscr{S} = \bigcap_{n=0}^{\infty} \mathscr{S}_n$, where the closed sets $\mathscr{S}_n$ are given (Fig. 6.57) by $\mathscr{S}_0 = T$, $\mathscr{S}_1 = T \backslash T_0$, $\mathscr{S}_2 = (T_1 \backslash T_{01}) \bigcup (T_2 \backslash T_{02}) \bigcup (T_3 \backslash T_{03})$, and so on. It can be shown that $\mathscr{S}$ has fractal dimension $D = \ln 3 / \ln 2$.

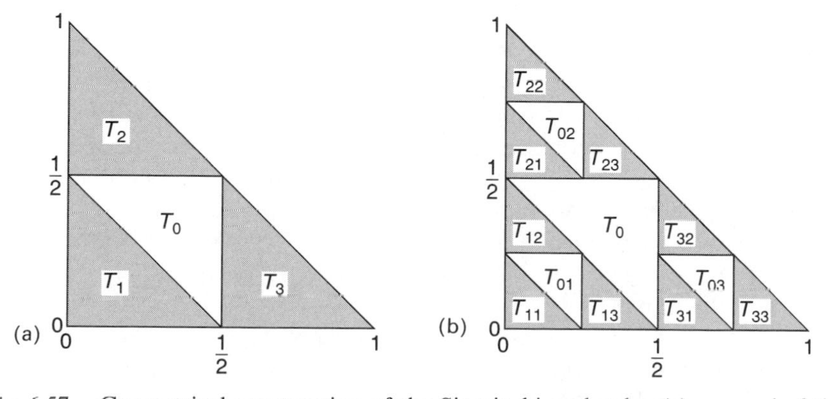

Fig. 6.57. Geometrical construction of the Sierpinski gasket by: (a) removal of the interior of T_0; (b) removal of interior of T_{0j}, $j = 1, 2, 3$. Repetition of this procedure (i.e. deleting centres from all triangles that remain at each stage) leads to a sparse set that is the Sierpinski gasket.

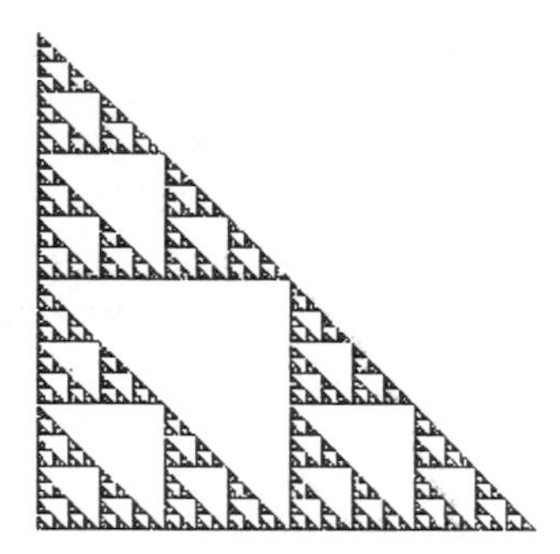

Fig. 6.58. Plot showing 20 000 applications of the iterated function scheme which uses the functions f_1, f_2, f_3 with equal probabilities of $\frac{1}{3}$. Similar results are obtained for all choices of initial point.

Quite apart from their interest as dynamical systems, iterated function schemes have potentially valuable applications in the field of image transmission. For example, in order to pass a picture of the Sierpinski gasket to a friend's computer, transfer of the maps (6.134)–(6.136) would involve less time and storage than the transfer of a large number of data points. Of course, for this idea to have practical value we must learn how to construct an iterated function scheme with an attractor that 'looks like' the picture we wish to transmit. To do this we must view our picture as a set of points that is 'close to' a union or **collage** of its images under a set of contracting affine maps that leave the set invariant (Exercise 6.72). The interested reader should consult the excellent book of Barnsley (1988) for more information about this intriguing topic.

6.9.3 Cellular automata

In section 6.8 we saw the advantages, in terms of precision and ease of operation, of representing a dynamical system in symbolic form. Cellular automata are discrete dynamical systems that are defined in symbolic form. They were conceived by von Neumann and Ulam (von Neumann, 1963, 1966) as models of self-reproducing biological systems but other physical and chemical systems can be represented in this way (Wolfram, 1986).

Elementary cellular automata consist of a one-dimensional array of sites each of which may be in one of two states: 'zero' or 'one'. The state of the system at time t is given by specifying the value, s_n^t, of the nth site for each $n \in \mathbb{Z}$. The evolution of the system is governed by a set of rules that determine s_n^{t+1} from $s_{n-1}^t, s_n^t, s_{n+1}^t$, i.e. the sites are deemed to interact via a short-range (near-neighbour) interaction only. For example, we might take the 'modulo-two' rule for which

$$s_n^{t+1} = (s_{n-1}^t + s_{n+1}^t) \bmod 2, \tag{6.137}$$

giving:

$$111 \to 0; \quad 110 \to 1; \quad 101 \to 0; \quad 100 \to 1;$$
$$011 \to 1; \quad 010 \to 0; \quad 001 \to 1; \quad 000 \to 0.$$

If we agree to always write the triplet codes in the same order, we can shorten the specification of the rule by only giving the right-hand sides of the above associations. Thus (6.137) would be given as '01011010'.

The state of the triplet of sites at time t must be one of eight possibilities and the state at time $t + 1$, associated with such a triple, can take one of two values. Thus there are $2^8 = 256$ possible rules for the time development of the state of the system. However, all but 32 of these are commonly considered illegal because either they fail to leave unchanged the 'quiescent' state consisting entirely of zeros or they fail to be reflection symmetric (e.g. 100 and 001, or 110 and 011 must yield the same values).

From the point of view of growth patterns, an interesting experiment is to consider the evolution of the system state in which all except one of the sites has value zero. Some rules erase the single non-zero site (e.g. 00000000 or 10100000), some fail to produce any growth (e.g. 00000100 or 00100100), others lead to uniform growth (e.g. 00110010 or 01111010), while others lead to fractal configurations of ones (as illustrated in Fig. 6.59). The pattern shown is somewhat similar to the Sierpinski gasket. However, the connection is closer than superficial scrutiny might suggest. It must be remembered that, because we have chosen to represent the time steps and site separations by finite distances, the growth pattern extends to infinity. If we wish to represent the pattern on a finite triangle then these distances have to tend to zero in an appropriate way and the points of Fig. 6.59 are squeezed together resulting in a pattern much more like Fig. 6.58.

Another line of investigation involves examining the evolution of an initial state consisting of an uncorrelated sequence of zeros and ones. Finite arrays of symbols satisfying periodic boundary conditions are commonly used in

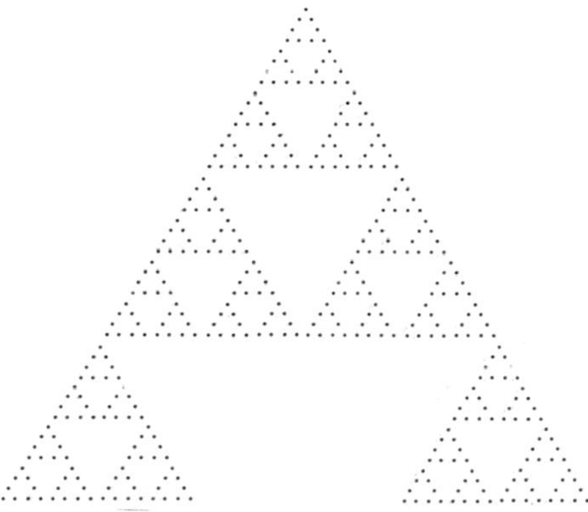

Fig. 6.59. Growth pattern of a single non-zero site generated by an elementary cellular automaton operating the modulo-two rule (6.137). Sites with value zero are unlabelled. Note the similarity to the Sierpinski gasket.

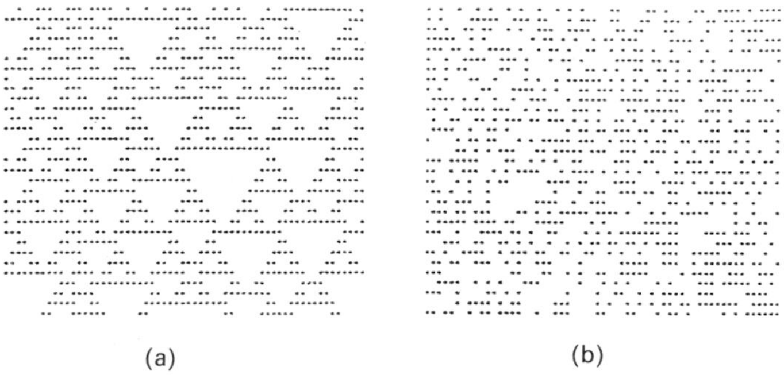

(a) (b)

Fig. 6.60. Illustration of self-organization in the elementary cellular automaton rule 01111110: (a) evolution of uncorrelated initial state; (b) corresponding sequence of uncorrelated states. Periodic boundary conditions have been used (after Wolfram, 1983).

computer experiments. This corresponds to the sites being arranged on a circle, so that the time development takes place on a cylinder. Fortunately, the choice of boundary conditions does not appear to introduce significant qualitative effects. Some elementary automata rules result in long-range correlations between sets of site values which show up as triangular patterns in the cumulative evolution of the uncorrelated initial state. Figure 6.60(a) shows the pattern produced by applying the rule 01111110 to a particular uncorrelated initial state in which the zeros and ones are chosen with equal probabilities. This should be compared to Fig. 6.60(b) in which each line is chosen independently in the same way as the initial state. The existence of correlations means that the rule exhibits a rudimentary form of self-organization.

Self-reproduction is another important property of elementary cellular automata. Observe that, in Fig. 6.61, the pattern '10111' is reproduced: twice at times 8, 16, 32,...; four times at times 24, 48, and so on. The rule used here

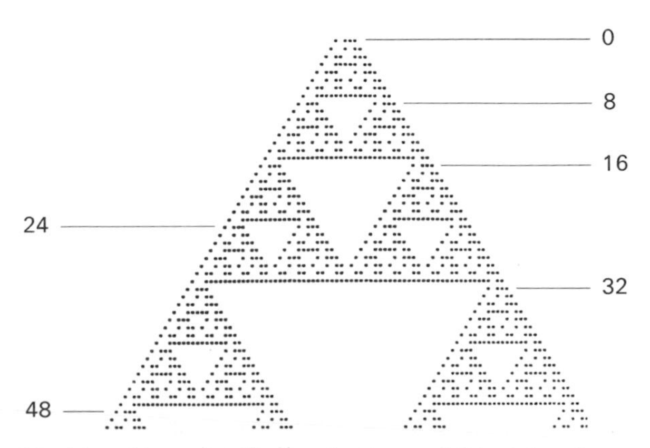

Fig. 6.61. Primitive form of self-reproduction exhibited by the modulo-two rule (6.137) (after Wolfram, 1983).

is the modulo-two rule (6.137) and it clearly shows a primitive form of self-reproduction.

Cellular automata can be defined in more than one dimension; for example, on square, cubic or hyper-cubic lattices. The number of legal rules increases rapidly with the dimension of the lattice. Two main types are distinguished: Type I which concern only orthogonally adjacent sites; and Type II where orthogonally and diagonally adjacent sites are involved. Orthogonally adjacent sites have a single Cartesian coordinate differing by one unit, whereas orthogonally and diagonally adjacent sites have no coordinates that differ by more than one unit. For a square lattice, for example, Type I automata rules involve five sites, while Type II rules involve nine.

The modulo-two rule has a natural extension to the square lattice in which the value of a site is given by the sum modulo two of the values of its four orthogonally adjacent neighbours. This is an example of a rule of Type I. Figure 6.62 shows the growth of a single non-zero site under this rule. Each time step is represented by a pattern on the square lattice and these planar patterns can be stacked to form a structure embedded in three-dimensional space that becomes self-similar (with fractal dimension $\log_2 5$) in the infinite time limit.

A widely studied example of a rule of Type II is Conway's game 'Life' (Gardner, 1971, 1972; Berlekamp, Conway and Guy, 1982). The game is defined on a square lattice and sites have value zero (dead) or one (alive). The value, σ, of a site at time $t + 1$ is determined by the sum, Σ, of the values of its eight neighbours at time t. If $\Sigma = 2$ then the value of the site remains unchanged; if $\Sigma = 3$ then σ is one; otherwise σ is zero. This rule gives rise to a fascinating family of isolated configurations of live sites; all colourfully named after the properties they exhibit. Some simple examples are shown in Fig. 6.63. The rather dull 'block' and 'beehive' simply remain invariant but the 'blinker' oscillates with period two and the 'glider' cycles with period two

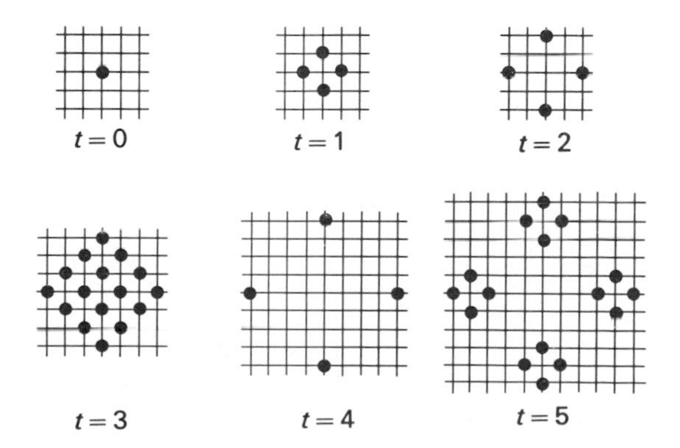

Fig. 6.62. Five time steps in the evolution of a single initial site according to the two-dimensional version of the modulo-two rule. Only sites with value 1 are distinguished.

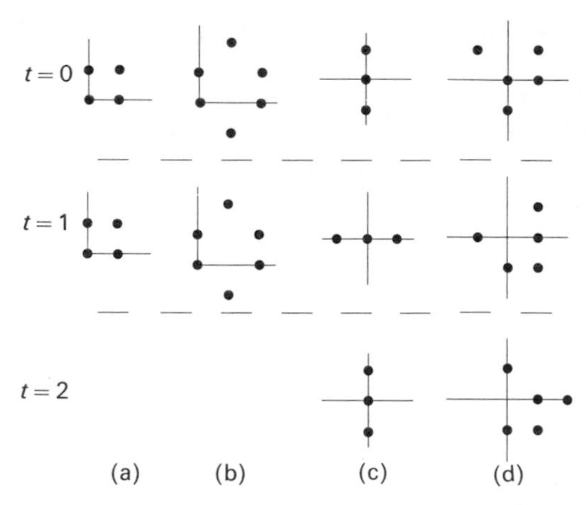

Fig. 6.63. Simple configurations from 'Life': (a) block; (b) beehive; (c) blinker; (d) glider.

and moves at the same time. This is only the beginning of a long list that includes 'glider guns' that emit gliders, 'eaters' that devour gliders and 'the harvester' that packages diagonal strings of live sites into blocks. There are even stable patterns of live sites that can repair themselves when they are disturbed by some isolated 'virus' sites but are completely eroded by others. 'Life' is a seriously obsessive game that should carry a health warning for your computer!

EXERCISES

Section 6.1

6.1 Show that the linearization of the system (6.2) at $(x, b) = (0, 0)$ has coefficient matrix

$$A = \begin{bmatrix} \varepsilon^{-1} & -\varepsilon^{-1} \\ 1 & 0 \end{bmatrix}.$$

Verify that the eigenvalues of A are given by

$$\lambda_1 = \varepsilon^{-1} - 1 + O(\varepsilon), \qquad \lambda_2 = 1 + O(\varepsilon).$$

Find corresponding eigenvectors $\boldsymbol{u}_1, \boldsymbol{u}_2$ and confirm that, to order ε

$$\boldsymbol{u}_1 = \begin{bmatrix} 1 \\ \varepsilon \end{bmatrix}, \qquad \boldsymbol{u}_2 = \begin{bmatrix} 1 \\ 1 - \varepsilon \end{bmatrix}.$$

6.2 Sketch $\dot{x} = 0$ and $\dot{b} = 0$ isoclines of (6.4) with $x_0 = \frac{1}{\sqrt{3}}, 1, \frac{2}{\sqrt{3}}$ in the xb-plane and find the unique fixed point $E = (x_0, b_0)$ of the system. Use the linearization theorem to show that E is: (a) a focus if $\frac{1}{\sqrt{3}} < x_0 < 1$; (b) a node if $1 < x_0 < \frac{2}{\sqrt{3}}$. Draw a diagram to indicate the changes that would take place if b_0 were greater than zero in Fig. 6.4.

6.3 Prove that the eigenvalues of the coefficient matrix in the linearized system (6.7) satisfy the cubic equation

$$f(\lambda) = \lambda^3 + 2(1 + \varepsilon^{-1})\lambda^2 + 2\varepsilon^{-1}\lambda + 2\varepsilon^{-1} = 0.$$

Assume that $0 < \varepsilon \ll 1$ and show that $f(\lambda)$ has zeros near $\lambda = \frac{1}{2}(-1 \pm i\sqrt{3})$. Deduce that the third zero is approximated by $2\varepsilon^{-1}$.

6.4 Show that the line OF in Fig. 6.7 has equation

$$b = -\frac{2a}{3}\sqrt{\frac{-a}{3}}, \qquad a \leqslant 0.$$

Hence confirm that the change in b alone induced by a triggering process must exceed $b_c = 2/(3\sqrt{3})$ if the 'fast action–slow return' dynamic of the system (6.6) is to occur.

Section 6.2

6.5 (a) Let the per capita growth rate of the proportion x_i of the population playing strategy i be given by $K[(\boldsymbol{Ax})_i - \boldsymbol{x}^{\mathrm{T}}\boldsymbol{Ax}]$, where K is a constant. Explain how the units of time can be chosen so that (6.13) holds.
(b) Let the pay-off matrix $\boldsymbol{B}$ be obtained by adding c to every entry in the kth column of $\boldsymbol{A}$. Show that:

$$\text{(i)}\ \ (\boldsymbol{Bx})_i = (\boldsymbol{Ax})_i + cx_k;$$

$$\text{(ii)}\ \ \boldsymbol{x}^{\mathrm{T}}\boldsymbol{Bx} = \boldsymbol{x}^{\mathrm{T}}\boldsymbol{Ax} + cx_k.$$

Deduce that the advantage of playing strategy i is the same for both pay-off matrices.

6.6 The pay-off matrix for the hawk-dove-bully-retaliator (HDBR) model of animal conflict is given by

$$\begin{bmatrix} -2 & 6 & 6 & -2 \\ 0 & 2 & 0 & 2 \\ 0 & 6 & 3 & 0 \\ -2 & 2 & 6 & 2 \end{bmatrix}.$$

Let x_1, x_2, x_3, respectively, be the proportions of doves, bullies and retaliators in a population that contains no hawks and obtain the dynamical equations for this subgame. Show that:
(a) DR is a line of fixed points;
(b) $\dot{x}_1 \leqslant 0$ and $\dot{x}_3 \geqslant 0$ for all points in the phase space;
(c) $\dot{x}_2 = 0$ on the curve

$$(x_1, x_2, x_3) = (\tfrac{1}{6}(2 + x_2)(1 - x_2), x_2, \tfrac{1}{6}(4 - x_2)(1 - x_2)),$$

where $0 \leqslant x_2 \leqslant 1$.

Hence sketch the phase portrait for this subsystem.

Section 6.3

6.7 Consider the following one-parameter families of differential equations defined on $\mathbb{R}$:
(a) $\dot{x} = \mu - x^2$;
(b) $\dot{x} = \mu x - x^2$;
(c) $\dot{x} = -(1 + \mu^2)x^2$.
Verify that all three families have the same non-hyperbolic fixed point at the origin when $\mu = 0$. Draw diagrams in the μx-plane to illustrate the local bifurcations, if any, that these families undergo at $\mu = 0$.

6.8 Consider the following one-parameter families of differential equations on the plane:
(a) $\dot{x}_1 = \mu x_1, \qquad \dot{x}_2 = -x_2$;
(b) $\dot{x}_1 = \mu x_1 - x_2, \qquad \dot{x}_2 = x_1 + \mu x_2$;
(c) $\dot{x}_1 = -x_1 - \mu x_2, \qquad \dot{x}_2 = \mu x_1 - x_2$.
Verify that each system has a non-hyperbolic fixed point at the origin when $\mu = 0$. Sketch phase portraits for $\mu < 0$, $\mu = 0$, $\mu > 0$ and describe the bifurcation, if any, which takes place at $\mu = 0$.

6.9 Sketch $\dot{x}_1 = 0$ and $\dot{x}_2 = 0$ isoclines for the planar system

$$\dot{x}_1 = x_1, \qquad \dot{x}_2 = \mu - x_2^2,$$

when (a) $\mu < 0$; (b) $\mu = 0$; (c) $\mu > 0$. Confirm that there are no fixed points in (a), one in (b) and two in (c). Verify that the single fixed point occurring in (b) is non-hyperbolic and show that in (c) the fixed point at $(0, +\sqrt{\mu})$ is a saddle point while that at $(0, -\sqrt{\mu})$ is a node. Hence obtain Fig. 6.13.

6.10 Sketch bifurcation diagrams of the type illustrated in Fig. 6.14 for each of the following one-parameter families of planar differential equations
(a) $\dot{r} = r(\mu - r^2), \qquad \dot{\theta} = 1$;
(b) $\dot{r} = r(-\mu - r^2), \qquad \dot{\theta} = 1$;
(c) $\dot{r} = r(\mu + r^2), \qquad \dot{\theta} = 1$;
(d) $\dot{r} = r(-\mu + r^2), \qquad \dot{\theta} = 1$.
Explain why it is reasonable to say that all four families exhibit Hopf bifurcations while it can be argued that the bifurcation exhibited by $\{(a), (b)\}$ is qualitatively different from that in $\{(c), (d)\}$.

6.11 Consider the one-parameter family of planar differential equations

$$\dot{x}_1 = x_1, \qquad \dot{x}_2 = \mu - x_2^4.$$

Sketch phase portraits for members of this family with: (a) $\mu < 0$; (b) $\mu = 0$; (c) $\mu > 0$ and compare your results with Fig. 6.13.

6.12 Sketch $\dot{x}_2$ as a function of x_2 for members of the family

$$\dot{x}_1 = x_1, \qquad \dot{x}_2 = -(x_2^2 - \mu)(x_2^2 - \mu/2),$$

with: (a) $\mu < 0$; (b) $\mu = 0$; (c) $\mu > 0$. Confirm that the non-hyperbolic fixed point occurring at $\mu = 0$ is the same as that in Exercise 6.11 and describe the fixed point bifurcation that takes place at $\mu = 0$.

6.13 Show that the family

$$\dot{x}_1 = x_1, \qquad \dot{x}_2 = \mu x_2 - x_2^3,$$

has:

(a) one hyperbolic fixed point for $\mu < 0$;

(b) a non-hyperbolic fixed point with zero determinant and non-zero trace for $\mu = 0$;

(c) three hyperbolic points (two saddles and a node) for $\mu > 0$.

The bifurcation that takes place in this family is known as a **symmetric saddle–node bifurcation.** What is the fundamental difference between this bifurcation and the saddle–node bifurcation in (6.21)?

6.14 A special case of a prey–predator model (due to Tanner, 1975) takes the form

$$\dot{H} = \frac{11}{5}H\left(1 - \frac{H}{11}\right) - \frac{10HP}{(2+3H)},$$

$$\dot{P} = P\left(1 - \frac{P(\beta + H)}{\gamma H}\right),$$

where H and P are herbivore (prey) and predator populations respectively. The quantities β and γ are positive constants. Sketch $\dot{H} = 0$ and $\dot{P} = 0$ isoclines and suggest how β and γ can be chosen so that the system has (a) zero; (b) one; and (c) two non-trivial fixed points. What can be deduced about the single fixed point in (b) from the configuration of the $\dot{H} = 0$ and $\dot{P} = 0$ isoclines?

6.15 State the distinct qualitative types of phase portraits obtained as the parameter μ varies from $-\infty$ to $+\infty$ in the systems:

(a) $\dot{r} = -r^2(r + \mu), \quad \dot{\theta} = 1$; (b) $\dot{r} = \mu r(r + \mu)^2, \quad \dot{\theta} = 1$;

(c) $\dot{r} = r(\mu - r)(\mu - 2r), \quad \dot{\theta} = 1$; (d) $\dot{r} = r(\mu - r^2), \quad \dot{\theta} = 1$;

(e) $\dot{r} = r^2\mu, \quad \dot{\theta} = \mu$; (f) $\dot{r} = r^2, \quad \dot{\theta} = 1 - \mu^2$.

6.16 Consider the family of planar differential equations

$$\begin{aligned} \dot{x}_1 &= \mu x_1 - 2x_2 - 2x_1(x_1^2 + x_2^2)^2, \\ \dot{x}_2 &= 2x_1 - \mu x_2 - x_2(x_1^2 + x_2^2)^2. \end{aligned} \tag{1}$$

Show that $V(x_1, x_2) = x_1^2 + x_2^2$ is a strong Liapunov function for (1) with $\mu = 0$ and hence or otherwise show that the family undergoes a Hopf bifurcation at the origin when $\mu = 0$.

6.17 Use the stability index to deduce that the following systems are asymptotically stable at the origin:

(a) $\dot{x}_1 = x_2 - x_1^3 + x_1 x_2^2, \quad \dot{x}_2 = -x_1 - x_1 x_2^2$;

(b) $\dot{x}_1 = x_2 - x_1^2 \sin x_1, \quad \dot{x}_2 = -x_1 + x_1 x_2 + 2x_1^2$;

(c) $\dot{x}_1 = x_2 - x_1^2 + 2x_1 x_2 + x_2^2, \quad \dot{x}_2 = -x_1 + x_1 x_2 + x_2^2$.

6.18 Prove that all the following one-parameter systems undergo Hopf bifurcations at $\mu = 0$ such that a stable limit cycle surrounds the origin for $\mu > 0$:
(a) $\dot{x}_1 = x_2 - x_1^3, \qquad \dot{x}_2 = -x_1 + \mu x_2 - x_1^2 x_2;$
(b) $\dot{x}_1 = \mu x_1 + x_2 - x_1^3 \cos x_1, \qquad \dot{x}_2 = -x_1 + \mu x_2;$
(c) $\dot{x}_1 = \mu x_1 + x_2 + \mu x_1^2 - x_1^2 - x_1 x_2^2, \qquad \dot{x}_2 = -x_1 + x_2^2.$

6.19 Show that Rayleigh's equation

$$\ddot{x} + \dot{x}^3 - \mu \dot{x} + x = 0$$

undergoes a Hopf bifurcation at $\mu = 0$. Describe the phase portraits near and at $\mu = 0$.

6.20 Prove that the fixed point at the origin of the system

$$\dot{x}_1 = (\mu - 3)x_1 + (5 + 2\mu)x_2 - 2(x_1 - x_2)^3$$
$$\dot{x}_2 = -2x_1 + (3 + \mu)x_2 - (x_1 - x_2)^3$$

has purely imaginary eigenvalues when $\mu = 0$. Find new coordinates y_1, y_2 so that when $\mu = 0$ the linearized part of the system has the correct form for checking stability. Hence, or otherwise, show that the system undergoes a Hopf bifurcation to stable limit cycles as μ increases through 0.

6.21 Prove that the one-parameter system

$$\dot{x}_1 = -\mu x_1 - x_2, \qquad \dot{x}_2 = x_1 + x_2^3$$

undergoes a Hopf bifurcation at $\mu = 0$ to unstable limit cycles surrounding a stable focus for $\mu > 0$.

Section 6.4

6.22 Show that the number of fixed points of (6.68) is determined by the value of $\bar{\mu}_1$ as follows: (a) none for $\bar{\mu}_1 > 0$; one for $\bar{\mu}_1 = 0$; and (c) two for $\bar{\mu}_1 < 0$. Show that the linearized system at the fixed point occurring in (b) has non-zero trace and zero determinant for all $\bar{\mu}_2 \neq 0$. What bifurcation can be expected to take place as $\bar{\mu}_1$ decreases through zero at fixed non-zero $\bar{\mu}_2$?

6.23 Show that the fixed points of the system (6.68) with $\bar{\mu}_1 < 0$ are given by $\bar{P}_1 = (\varepsilon\bar{\mu}_2 - \varepsilon^2, -\varepsilon)$ and $\bar{P}_2 = (-\varepsilon\bar{\mu}_2 - \varepsilon^2, \varepsilon)$, where $\varepsilon = \sqrt{-\bar{\mu}_1}$. Show that $\bar{P}_1$ is a node/focus, while $\bar{P}_2$ is a saddle point.

Verify that $\bar{P}_1$ is: (a) asymptotically stable for $\bar{\mu}_2 > 2\sqrt{-\bar{\mu}_1}$; and (b) unstable for $\bar{\mu}_2 < 2\sqrt{-\bar{\mu}_1}$. What bifurcation will take place generically when $\bar{\mu}_2$ decreases through $\bar{\mu}_2 = 2\sqrt{-\bar{\mu}_1}$ with $\bar{\mu}_1$ fixed and negative?

6.24 Sketch phase portraits of (6.68) to illustrate the bifurcation that occurs on the $B_2'B_1'$-boundary in Fig. 6.21. Your answer should include a diagram corresponding to a parameter point that lies *on* the boundary.

Section 6.5

6.25 Show that the Poincaré map P_0 given in (6.73) has inverse given by

$$P_0^{-1}(1+\xi) = 1 + \xi - 2\pi\xi^2 + \cdots.$$

Sketch the graphs $y = P_0^{-1}(x)$ and $y = x$ in the neighbourhood of $x = 1$. If P_μ is given by (6.70), explain what happens to the non-trivial fixed points of P_μ^{-1} as μ increases through zero.

6.26 Consider the local family defined by

$$f_v(y) = v + y + y^2$$

with v and y near zero. Show that:
(a) f_v has no fixed point for $v > 0$;
(b) f_0 has one fixed point at $y = 0$ and $(\mathrm{d}f_0/\mathrm{d}y)|_{y=0} = +1$;
(c) for $v < 0, f_v$ has a stable fixed point at $y = -\sqrt{-v}$ and an unstable fixed point at $y = +\sqrt{-v}$.

Draw a diagram illustrating how the fixed point structure of the family changes as v increases through zero. Indicate how this diagram changes if v is replaced by $-v$ in the definition of f_v.

6.27 Consider the local family defined by

$$f_\mu(x) = \mu x + x - x^3$$

with μ and x near zero. Draw, on the same diagram, sketches of $f_\mu(x)$ for (a) $\mu = -\varepsilon$; (b) $\mu = 0$; (c) $\mu = \varepsilon$, $0 < \varepsilon \ll 1$. Verify that $x = 0$ is a fixed point of f_μ for all μ: stable for $\mu \leqslant 0$ and unstable for $\mu > 0$. Confirm that, for $\mu > 0, f_\mu$ also has two stable fixed points at $x = \pm\sqrt{\mu}$. Draw a diagram in the μx-plane, showing the positions and stabilities of the fixed points of f_μ as μ increases through zero. The bifurcation taking place at $\mu = 0$ in this family is called the **pitchfork** bifurcation. What is the fundamental difference between the pitchfork and fold bifurcations?

6.28 Consider the local family defined by

$$f_v(y) = (v-1)y + \delta y^3$$

with v and y near zero. Assume that $\delta = +1$ and sketch $f_v(y)$ in the neighbourhood of the origin for $v = -\varepsilon$, 0, ε where $0 < \varepsilon \ll 1$. Verify that $f_v(y)$ has a fixed point at $y = 0$ for all v near zero and describe how its stability changes as v increases through zero.

6.29 Show that, with f_v defined as in Exercise 6.28,

$$f_v^2(y) = (v-1)^2 y + \delta(v-1)[1 + (v-1)^2]y^3 + \cdots.$$

Take $\delta = +1$, sketch $f_v^2(y)$ in the neighbourhood of the origin for $v = -\varepsilon$, 0, ε, where $0 < \varepsilon \ll 1$. Confirm that, for $v < 0$, f_v has a 2-cycle that collapses towards the origin as v increases to zero.

6.30 Draw a bifurcation diagram illustrating how the 1- and 2-cycle structure of f_ν, considered in Exercises 6.28 and 6.29, changes as ν increases through zero. Consider how the results obtained in Exercises 6.28 and 6.29 are affected if $\delta = -1$ in f_ν and draw the corresponding bifurcation diagram.

6.31 Use the Taylor expansion (6.77) of $P_0(x)$ to verify the form (6.81) for $P_0^2(x)$ and show that $c = -2(a^2 + b)$. Draw diagrams to illustrate the local behaviour of $P_0^2(x)$ at $x = x^*$ if: (a) $a \neq 0$, $b = 0$; (b) $a = 0$, $b \neq 0$. Why do these diagrams suggest that the canonical family (6.84) corresponds to (b) but not to (a)?

6.32 Verify that the logistic map $F_\mu(x)$ has a fixed point at $x^* = 1 - \mu^{-1}$ for $\mu > 1$. Show that

$$F_\mu(x) = x^* + (2 - \mu)(x - x^*) - \mu(x - x^*)^2.$$

Sketch $F_3^2(x)$ for x near x^* (cf. Exercise 6.31) and deduce that F_μ undergoes a flip bifurcation at x^* for $\mu = 3$. Confirm that a stable 2-cycle is created as μ increases through 3.

6.33 Prove that the logistic map F_μ satisfies

$$F_\mu(x) = F_\mu(1 - x).$$

Show that the fixed point $x^* = 1 - \mu^{-1}$ of $F_\mu(x)$ is unstable for $\mu > 3$. Use graphs of the identity and $F_\mu(x)$, $\mu > 3$, to show that there is a countable infinity of points in $(0, 1)$ whose orbits reach x^* after a finite number of iterations. Such points are said to be **eventually periodic** with period 1. What property of $F_\mu(x)$ is responsible for this behaviour?

6.34 Prove that the non-trivial fixed points of $F_\mu^2(x)$ are given by

$$\mu^3x^3 - 2\mu^3x^2 + \mu^2(1 + \mu)x - (\mu^2 - 1) = 0$$

and deduce that for $\mu > 3$ the period-2 points of F_μ satisfy

$$\mu^2x^2 - \mu(\mu + 1)x + (\mu + 1) = 0.$$

Hence or otherwise show that F_μ^2 period-doubles at $\mu = 1 + \sqrt{6}$.

6.35 Assume that $F_{\mu^*}^3(x_1^*) = x_1^*$ and $F_{\mu^*}(x_1^*) \neq x_1^*$. Explain how two further fixed points, x_2^* and x_3^*, of $F_{\mu^*}^3$ can be obtained.

Let $G(x) = F_{\mu^*}^3(x)$, show that

$$G'(x) = F_\mu'(F_\mu^2(x))F_\mu'(F_\mu(x))F_\mu'(x),$$

where $'$ denotes $\mathrm{d}/\mathrm{d}x$. Given that $G'(x_1^*) = 1$, show that $G'(x_i^*) = 1$, for $i = 2, 3$. Comment on how this result relates to Fig. 6.31.

6.36 (a) Find the affine change of coordinate, $x \mapsto \xi$, that transforms the family of maps $F_\mu(x) = \mu x(1 - x)$ into $f_a(\xi) = 1 - a\xi^2$ and obtain a in terms of μ.
(b) Write a computer program to plot successive points of the iteration

$$x_{n+1} = 1 - ax_n^2 + y_n,$$
$$y_{n+1} = bx_n,$$

for $a = 1.4$ and $b = 0.3$ with $(x_0, y_0) = (0, 0)$. Show that the iteration rapidly approaches a 'chaotic' attracting set. The limit set of this iteration is known as the **Hénon attractor** (Thompson and Stewart, 1986, p. 177).

6.37 Let ϕ_t be the flow of the autonomous differential equation $\dot{\mathbf{x}} = \mathbf{X}(\mathbf{x})$. The Euler method approximates $\phi_{(k+1)h}(\mathbf{x})$ by $\phi_{kh}(\mathbf{x}) + h\mathbf{X}(\phi_{kh}(\mathbf{x}))$, for $k = 0, 1, 2, \ldots$, where h is a suitably chosen step length.

Write a computer program to iterate the Euler approximation to the return map of the Rössler equations (6.88) defined on the half-plane $y = 0, x < 0$ and plot the x-coordinate pairs $(-x_n, -x_{n+1})$, $n = 0, 1, 2 \ldots$, as in Fig. 6.34(a). Take $h = 0.005$, $b = 2$, $c = 4$ and construct plots for: (a) $a = 0.3$; (b) $a = 0.35$; (c) $a = 0.375$. Interpret your results.

Section 6.6

6.38 Consider $\boldsymbol{A}_{\pm}$ given in (6.99). Show that $\boldsymbol{A}^2_{\pm} = -\omega^2_{\pm}\boldsymbol{I}$, where $\boldsymbol{I}$ is the 2×2 unit matrix. Deduce that

$$\boldsymbol{A}^{2j}_{\pm} = (-1)^j \omega^{2j}_{\pm} \boldsymbol{I}$$

and

$$\boldsymbol{A}^{2j+1}_{\pm} = (-1)^j \omega^{2j+1}_{\pm} \begin{bmatrix} 0 & \mathrm{u}^{-1}_{\pm} \\ \omega_{\pm} & 0 \end{bmatrix}.$$

Usc thc dcfinition of thc cxponential matrix given in (2.63) to obtain $\exp(\pi \boldsymbol{A}_{\pm})$.

6.39 Let $\mathbf{P}$ be an area-preserving linear planar diffeomorphism represented by $\boldsymbol{P}$. Write down the real Jordan form, $\boldsymbol{J}$, of $\boldsymbol{P}$ when $\boldsymbol{P}$ has: (a) real eigenvalues; (b) complex eigenvalues. Verify that $y_1 y_2 = \text{constant}$ and $y_1^2 + y_2^2 = \text{constant}$ are invariant curves for $\boldsymbol{J}\boldsymbol{y}$ in (a) and (b), respectively. Given that $\boldsymbol{J} = \boldsymbol{M}^{-1}\boldsymbol{P}\boldsymbol{M}$, where $\boldsymbol{M}$ is a non-singular matrix, show that $\mathbf{P}$ always has invariant curves of the form

$$Ax_1^2 + 2Bx_1x_2 + Cx_2^2 = \text{constant}.$$

Obtain A, B and C in terms of the elements of $\boldsymbol{M}$ and explain how cases (a) and (b) are distinguished. Illustrate your answer by sketching typical invariant curves for $\mathbf{P}$ in both cases.

6.40 Use the expression (6.101) for $\exp(\pi \boldsymbol{A}_{\pm})$ to show that

$$\operatorname{tr}(\boldsymbol{P}) = 2\cos(\pi\omega_-)\cos(\pi\omega_+) - \sin(\pi\omega_-)\sin(\pi\omega_+)\left(\frac{\omega_-}{\omega_+} + \frac{\omega_+}{\omega_-}\right),$$

where $\boldsymbol{P}$ is defined in (6.100). Given that $\omega_{\pm} = \omega \pm \mu$, deduce that $\operatorname{tr}(\boldsymbol{P}) = \pm 2$ can be written in the form

$$\omega^2 \cos(2\pi\omega) - \mu^2 \cos(2\pi\mu) = \pm(\omega^2 - \mu^2).$$

6.41 Let $\psi_t(x, y, \theta)$ be a flow defined on $\mathbb{R}^2 \times S^1$, which, like that of (6.108), circulates around the torus in the positive θ sense. Suppose that **P, Q** are Poincaré maps of ψ_t defined on the $\theta = \theta_0$ and $\theta = \theta_1$ planes, respectively. Express **P** and **Q** in terms of ψ_t and prove that **P** and **Q** are topologically conjugate to one another.

Assume that ψ_t is the flow (6.108) and $\theta_1 = \theta_0 + \pi$. Show that $\mathbf{P} = \mathbf{R}^{-1}\mathbf{Q}\mathbf{R}$, where $\mathbf{R}(x, y) = (-x, -y)$. Use a copy of Fig. 6.38 to interpret this result geometrically.

6.42 Show that the Hénon map

$$\mathbf{H}_{a,b}(x, y) = (1 - ax^2 + y, bx),$$

where a, b are real parameters, has two fixed points if and only if $4a + (1 - b)^2 > 0$. Assume $0 < b < 1$, $4a = 3(1 - b)^2$ and obtain the eigenvalues of $D\mathbf{H}_{a,b}$ at both fixed points. Verify that one fixed point is non-hyperbolic while the other is of saddle type. What bifurcation do you expect to be associated with such parameter values?

6.43 Use the computer program constructed in Exercise 6.36 for iteration of the Hénon map $\mathbf{H}_{a,b}(x, y)$ to investigate: (a) period-doubling as a is varied with $b = 0.3$; (b) the 'braided nature' of the attracting set that occurs for $a = 1.4$, $b = 0.3$.

6.44 Consider the local family of planar diffeomorphisms given in (6.109). Show that $\mathbf{f}_0(x, y)$ has a non-hyperbolic fixed point $(x, y) = (0, 0)$ at which only one eigenvalue of $D\mathbf{f}_0$ is equal to $+1$. Find the fixed points of $\mathbf{f}_\nu$ when $\nu < 0$ and use the linearization theorem to determine their topological type. Compare the bifurcation that takes place at $\nu = 0$ in (6.109) with the saddle–node bifurcation for flows discussed in section 6.3.2.

6.45 Consider a planar diffeomorphism $\mathbf{f}(x, y) = (f_1(x), f_2(y))$. Given that $f_1(x)$ has a 2-cycle $\{x_1^*, x_2^*\}$ and $f_2(y)$ has a fixed point y^*, show that **f** has a 2-cycle $\{(x_1^*, y^*), (x_2^*, y^*)\}$. If the 2-cycle of f_1 is asymptotically stable, how does the stability of y^* affect the stability of the 2-cycle in **f**?

Use the results of Exercise 6.30 to decide which of the local families given in (6.110) gives behaviour consistent with that shown in Fig. 6.37.

Section 6.7

6.46 Let **P** be the mapping defined in (6.112). Show that

$$\det(D\mathbf{P}(y, p)) \equiv 1$$

for any differentiable function F.

Find the transformation matrix $\boldsymbol{M}$ relating the yp- and yY-planes and explain why the orbits of **P** can be exhibited equally well in either plane. Draw a schematic diagram illustrating how points in the yY-plane are mapped using $\boldsymbol{M}$ and **P**.

6.47 Let $(y_{i+1}, p_{i+1}) = \mathbf{P}(y_i, p_i)$, $i = 0, 1, 2, \ldots$, where **P** is defined by (6.112).

Show that

$$y_{i+1} + y_{i-1} = 2y_i \cos 2\pi Q + Q^{-1} F(y_i) \sin 2\pi Q.$$

Write a computer program that uses this expression to calculate y_{i+1} given y_i and y_{i-1} and plot the iterates (y_i, y_{i+1}) of any given point (y_0, y_1). Use this program to investigate the nature of the orbits of **P** for

$$F(y) = 2B[1 - \exp(-y^2/2)]/y$$

when $Q \approx \frac{1}{3}$, $\frac{1}{2} \leqslant B \leqslant 1$ and $-10 \leqslant y_0, y_1 \leqslant 10$.

6.48 The phase portrait of the Hamiltonian system

$$\dot{x} = y, \qquad \dot{y} = kx(x-1)$$

is shown below.

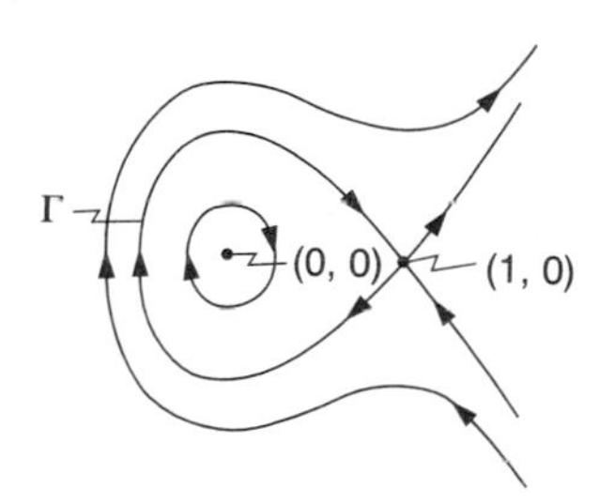

If ϕ_t is the flow of this system explain why its time-1 map, ϕ_1, is an area-preserving map of the xy-plane.

Obtain a linear approximation to ϕ_1 for points (x, y) that are sufficiently close to the elliptic fixed point at the origin. Hence show that the rotation numbers of the invariant circles of ϕ_1 of small enough radius are approximately, equal to $\sqrt{k}/2\pi$.

Consider the orbit of a point on the saddle connection Γ under ϕ_1. Show that the cumulative angular displacement after N iterations must be less that 2π. What does this imply about the rotation number of an invariant circle of ϕ_1 that is close to Γ.

Describe the dynamics of ϕ_1 restricted to the inside of Γ.

6.49 Sketch the phase portrait of the Hamiltonian system

$$\dot{x} = y \qquad \dot{y} = kx(x^2 - 1).$$

Let ϕ_1 be the time-1 map of the flow of this differential equation. Find the largest neighbourhood of the origin on which ϕ_1 behaves like an area-preserving twist map and find its rotation interval.

6.50 Confirm that the Hénon map defined by (6.114) is an area-preserving map. Show that for $\alpha \neq 0$ the fixed points of (6.114) lie on the line

$y = x\tan\frac{1}{2}\alpha$ and, hence or otherwise, deduce that (6.114) has a non-trivial fixed point at $(x, y) = (2\tan\frac{1}{2}\alpha, 2\tan^2\frac{1}{2}\alpha)$. Verify that this fixed point is of saddle type.

6.51 Write a computer program to plot points of the iteration (6.114). Investigate the occurrence of: (a) a 4-fold island chain for $\cos\alpha = -0.01$; (b) a five-fold island chain for $\cos\alpha = 0.24$; (c) a six-fold island chain for $\cos\alpha = 0.4$. Demonstrate the existence of a chaotic orbit that escapes to infinity in all three cases.

6.52 The stable (unstable) manifold $W^s_\phi(W^u_\phi)$ of a saddle point $\mathbf{x}^*$ of a flow ϕ_t is $\{\mathbf{x} \mid \phi_t(\mathbf{x}) \to \mathbf{x}^* \text{ as } t \to \infty(-\infty)\}$. Let $\mathbf{x}_1 \in W^u_\phi$, $\mathbf{x}_2 \in W^s_\phi$ and suppose that the orbits of $\mathbf{x}_1$ and $\mathbf{x}_2$ meet at $\mathbf{y}$. Use the flow property (1.56) to prove that the trajectory of $\mathbf{y}$ is a subset of $W^s_\phi \cap W^u_\phi$.

Verify that the corresponding result holds for the stable and unstable manifolds, $W^s_\mathbf{P}$ and $W^u_\mathbf{P}$, of a saddle fixed point $\mathbf{x}^*$ of a diffeomorphism $\mathbf{P}$. Comment on the difference between Fig. 6.42(a) and Fig. 6.42(b).

6.53 The planar map $\mathbf{P}$: $(x, y) \mapsto (x_1, y_1)$ takes the form

$$x_1 = x + y_1, \qquad y_1 = y + f(x),$$

where $f(x)$ is a differentiable function of x. Show that:

(a) $\mathbf{P}$ is an area-preserving diffeomorphism of the plane;

(b) the fixed points of $\mathbf{P}$ are the same as those of the differential equation $\dot{x} = y$, $\dot{y} = f(x)$.

6.54 Consider the map $\mathbf{P}$ defined by

$$x_1 = x + y_1, \qquad y_1 = y + kx(x-1).$$

Verify that $\mathbf{P}$ has a fixed point of saddle type at $(x, y) = (1, 0)$ and find the eigenvalues and eigenvectors of $D\mathbf{P}(1, 0)$.

Write a computer program that plots N iterates under $\mathbf{P}$ of each of the points $(x_j, y_j) = (1, 0) + j\varepsilon(u, v)$, $j = 1, 2\ldots, M$, where $(u, v)^T$ is an eigenvector of $D\mathbf{P}(1, 0)$ and ε is a constant. Take $k = 1.5$ and use this program to approximate the branch of the unstable manifold of $(1, 0)$ that is involved in the homoclinic tangle shown in Fig. 6.42(b). Given that the stable manifold of $\mathbf{P}$ at $(1, 0)$ is the unstable manifold of $\mathbf{P}^{-1}$ at that point, extend the program to produce Fig. 6.42(b).

6.55 Let $\mathbf{P}$: $(x, y) \mapsto (x_1, y_1)$ where

$$x_1 = x + y_1, \qquad y_1 = y + kx(x^2 - 1).$$

Verify that $\mathbf{P}$ has fixed points of saddle type at $(x, y) = (\pm 1, 0)$. Modify the computer program of Exercise 6.54 to obtain the branches of stable and unstable manifolds involved in Fig. 6.43(b).

6.56 Show that the map $\mathbf{P}$: $(x, y) \mapsto (x_1, y_1)$ with

$$x_1 = x + y_1, \qquad y_1 = y + kx(x-1) - \delta y,$$

$0 < \delta < 1$, is a dissipative diffeomorphism of the plane. Let $A(D)$ be the area of a compact region, D, in the xy-plane. Prove that $A(\mathbf{P}(D)) = (1-\delta)A(D)$.

6.57 Let ψ_t be the flow of (6.108), $\mathbf{P}_\gamma$ be the Poincaré map defined on the $\theta = 0$ plane and assume that ψ_τ: $(x, y, \theta) \mapsto (X, Y, \Theta)$, $\tau = 2\pi/\omega$. Write down an expression for Θ in terms of x, y, θ, explain its consequences for $D\psi_\tau(x, y, \theta)$ and deduce (6.117).

6.58 Let ϕ_t: $\mathbb{R}^n \mapsto \mathbb{R}^n$ be the flow of the differential equation $\dot{\mathbf{x}} = \mathbf{X}(\mathbf{x})$. Show that

$$\det(D\phi_{t+h}(\boldsymbol{x})) = \det(D\phi_t(\phi_h(\boldsymbol{x})))\det(D\phi_h(\boldsymbol{x})).$$

Use the 'small h' approximation for $\det(D(\phi_h(\mathbf{x}))$ developed in section 4.4 to deduce (6.118).

6.59 Fourth-order Runge–Kutta methods are a popular choice for the numerical approximation of $\mathbf{P}_\gamma$. Algorithms for the solution of a system of two first-order differential equations by this method can be found in many elementary texts on numerical analysis (e.g. Flannery *et al.*, 1987, p. 553).

Consider (6.107) with $\omega = 1$, $\delta = 0.25$, $\gamma = 0.2$. Use a fourth-order Runge–Kutta method with a step length $h = 2\pi/100$ to plot the first two components of $\{\psi_t(x, y, 0) \mid 0 \leqslant t \leqslant 100\pi\}$, where ψ_t is the flow of (6.108). Arrange for the points corresponding to iterates of $\mathbf{P}_\gamma$ to be distinguished from other projected orbit points. Take $(x, y) = (0, y_0)$ and find values of y_0 in the interval $(-0.142, -0.141)$ that illustrate the sensitive dependence of transient chaotic orbits on the choice of initial point. Confirm that the chaotic behaviour persists if γ is increased to 0.28.

Section 6.8

6.60 Sketch $h(x) = \sin^2(\pi x/2)$ for $0 \leqslant x \leqslant 1$. Confirm that h: $[0, 1] \to [0, 1]$ is a continuous bijection and show that

$$F_4(h(x)) = h(G(x))$$

where F_μ and G are defined by (6.85) and (6.122) respectively.

6.61 Draw a diagram illustrating the uniform dissection of $[0, 1]$ obtained by using the third power of the tent map G and obtain the finite symbol blocks $\{\sigma_0, \sigma_1, \sigma_2, \sigma_3\}$ labelling each subinterval. Write down the ordinary binary representation $\cdot s_1 s_2 s_3 s_4$ of the initial point of each subinterval and confirm that the two representations are related by

$$\sigma_0 = s_1, \qquad \sigma_k = (s_k + s_{k+1}) \bmod 2, \qquad k = 1, 2, \ldots$$
$$s_1 = \sigma_0, \qquad s_{k+1} = (\sigma_k + s_k) \bmod 2, \qquad k = 1, 2, \ldots.$$

Verify that $\sigma = \{11111\ldots\}$ represents the point $x = 2/3$. Let $\{x_1, x_2\}$ be the 2-cycle of G corresponding to the codes $\sigma_1 = \{010101\ldots\}$ and $\sigma_2 = \{101010\ldots\}$. Find x_1 and x_2.

6.62 Show that the **doubling map** $f:[0,1)\to[0,1)$ defined by

$$f(x) = 2x \bmod 1$$

can be represented as a left-shift α on the 'code' given by the ordinary binary representation of the number x. Describe the main features of the dynamics of f that follow from this observation.

6.63 (a) Write down the codes for two period-3 points of G that belong to distinct periodic orbits. How many distinct 3-cycles are there? Explain your answer.
(b) Let $\sigma=\{\sigma_0,\sigma_1,\sigma_2\ldots\}$ be the code corresponding to an arbitrary point x of $[0,1]$. Write down the code for a periodic point x' of G that satisfies $|x-x'|<2^{-N}$. What does this imply about the periodic points of G?

6.64 Write a computer program to print out the iteration of a point $x\in[0,1]$ under G. Explain why the obvious approach to this problem is unlikely to yield a true representation of the orbit of x. Describe how you would modify your program to obtain a realistic representation of one hundred iterates of the point x_δ defined in section 6.8.

6.65 Let $\mathbf{g}$ be the horseshoe map described in the text. Draw a schematic diagram illustrating:
(a) $\mathbf{g}(P_i)$, $\mathbf{g}^2(P_i)$, $\mathbf{g}^3(P_i)$;
(b) P_i, $\mathbf{g}^{-1}(P_i)$, $\mathbf{g}^{-2}(P_i)$;
for $i=0,1$. Use this diagram to extend the block codes given on Fig. 6.52(b) and Fig. 6.53(b) to length three.

How many disjoint 'boxes' within $\mathscr{S}=I\times I$ are represented by the combination of the above symbol blocks and, given that $I=[-1,1]$, what are their dimensions. Draw a diagram indicating where the boxes labelled by

$$\{000.000\}, \{111.111\}, \{111.000\}, \{000.111\}$$

are to be found.

Section 6.9

6.66 Consider the logistic map F_μ defined in (6.85). Suppose $2<\mu<4$ and let $x^*=1-\mu^{-1}$. For $x\in[1-x^*,x^*]$, sketch the curves in the xy-plane given by the parametric forms: (a) $(x,F_\mu^2(x))$; (b) $(x,F_\mu^2(x)-x^*)$; (c) $(\alpha(x-x^*),\alpha(F_\mu^2(x)-x^*))$, where $\alpha=(1-2x^*)^{-1}$. Explain how curve (c) is related to the graph of F_μ^2.

Plot curve (c) for $\mu=3$ and compare it with the graph of $F_{5/4}$. What does this comparison suggest about $F_3^2|[\frac{1}{3},\frac{2}{3}]$ and $F_{5/4}$?

6.67 Consider the symmetric form of the logistic map $f_a(x)=1-ax^2$. Show that

$$f_a^2(x) = 1-a+2a^2x^2+O(x^4).$$

Evaluate $h^{-1}(f_a^2(h(y)))$ correct to order y^2, where $y = h^{-1}(x) = kx$, k real. Find $k(a)$ and $\hat{a}(a)$ such that f_a^2 is conjugate to $f_{\hat{a}(a)}$ to this order of approximation.

Show that the above result means that $f_a^{2^i}$ is conjugate to $f_{\hat{a}^i(a)}$, $i = 2, 3, \ldots$. Comment on the significance of a stable fixed point of $\hat{a}$.

6.68 Let $I = [0, 1]$. Use (6.127) to verify that the dimension of: (a) the square $I \times I$ is two; (b) the cube $I \times I \times I$ is three.

6.69 Let x be a point of the 'mid-third' cantor set $\mathscr{C}$. Use the representation of x given in (6.130) to verify that

$$f_2(x) = \tfrac{1}{3}x + \tfrac{2}{3}$$

lies in $\mathscr{C}$.

Consider the orbit of a given point x_0 under an iterated function scheme like (6.132) that uses f_1 with probability $\frac{3}{4}$ and f_2 with probability $\frac{1}{4}$. How would you expect the overall appearance of a computer plot of the iterates of x_0 with this scheme to differ from that of x_0 under (6.132).

6.70 Verify that the sides of the triangle T in Fig. 6.57 are subsets of the Sierpinski gasket $\mathscr{S}$. Explain why this means that the fractal dimension of $\mathscr{S}$ is greater than unity. Use (6.127) to show that $D = \log_2(3)$. Find the fractal dimension of $T \backslash \mathscr{S}$.

6.71 The **Sierpinski carpet** is obtained by partitioning the unit square into nine equal squares and removing the interior of the middle one. This process is then repeated in each of the remaining eight squares. The subset remaining after three such steps is illustrated below. Show that the fractal dimension of the carpet is $\log_3(8)$.

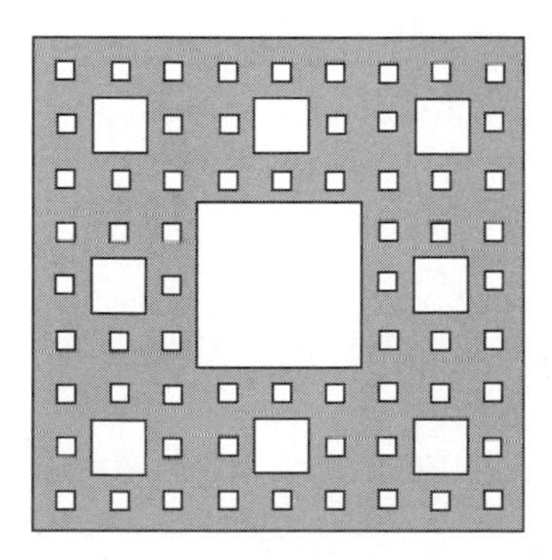

6.72 Construct a computer program to apply the iterated function scheme that uses the following maps $\mathbf{f}_1, \mathbf{f}_2, \ldots, \mathbf{f}_9$ with equal probability where: $\mathbf{f}_i(x, y) = (0.2y, 0.25x) + \mathbf{v}_i$, $i = 1, \ldots, 6$ and $\mathbf{v}_1 = \mathbf{0}$, $\mathbf{v}_2 = (0, 2.5)$, $\mathbf{v}_3 = (3.5, 0)$, $\mathbf{v}_4 = (3.5, 2.5)$, $\mathbf{v}_5 = (5.5, 0)$ and $\mathbf{v}_6 = (5.5, 2.5)$, $\mathbf{f}_7(x, y) = (0.2y, 0.3x) + (9, 2)$, $\mathbf{f}_8(x, y) = (0.1x, 0.2y) + (9, 0)$ and $\mathbf{f}_9(x, y) = (0.25x, 0.2y) + (1, 2)$.

6.73 Let $\alpha_1\alpha_2\alpha_3 \cdots \alpha_8, \alpha_i = 0, 1, i = 1, \ldots, 8$, be an elementary cellular automata rule. Explain why this rule:
(a) is illegal if $\alpha_8 = 1$;
(b) erases a single non-zero site if it is legal and $\alpha_4 = 0, \alpha_6 = 0$ and $\alpha_7 = 0$;
(c) maintains a single non-zero site if it is legal and $\alpha_4 = 0$, $\alpha_6 = 1$ and $\alpha_7 = 0$.
Show that there are only two legal rules that give maximal uniform growth from a single non-zero site.

6.74 Examine how the 'modulo-two' rule $\alpha = 01011010$ acts on an initial state with a single non-zero site. Show that any rule of the form $\alpha_1\alpha_2 01\alpha_5 010$ will have the same growth pattern as α. How many legal elementary cellular automata rules generate this growth pattern?

6.75 Consider the following Type I cellular automaton rule (Gardner, 1971*a*) defined on a square lattice. If a site has 0, 2, 4 live (orthogonally adjacent) neighbours at time t, then its value is *zero* at $t + 1$; if it has 1, 3 live neighbours at time t, then its value is *one* at $t + 1$. Show that the 'tromino' consisting of three live sites at the points $(0, -1)$, $(0, 0)$, $(1, 0)$ in an otherwise dead lattice, self-replicates four-fold after two time steps.

6.76 In the game 'Life', what is the minimum number of time steps before the glider orientation shown at $t = 0$ in Fig. 6.63(d) is restored? Obtain the coordinates of the central site of the glider at this time relative to its initial position.

Bibliography

BOOKS

Andronov, A. A. and Chaikin, C. E. (1949) *Theory of Oscillations*, Princeton University Press, Princeton, NJ.
Arnold, V. I. (1973) *Ordinary Differential Equations*, MIT Press, Cambridge, MA.
Arnold, V. I. (1978) *Mathematical Methods of Classical Mechanics*, Graduate Texts in Mathematics, Vol. 60, Springer-Verlag, New York, NY.
Arnold, V. I. and Avez, A. (1968) *Ergodic Problems of Classical Mechanics*, W. A. Benjamin, New York, NY.
Arrowsmith, D. K. and Place, C. M. (1990) *An Introduction to Dynamical Systems*, CUP, Cambridge.
Barnett, S. (1975) *Introduction to Mathematical Control Theory*, OUP, Oxford.
Barnsley, M. (1988) *Fractals Everywhere*, Academic Press, London.
Berlekamp, E. R., Conway, J. H. and Guy, R. K. (1982) *Winning Ways for Your Mathematical Plays*, Academic Press, New York, NY.
Braun, M. (1975) *Differential Equations and Their Applications: An Introduction to Applied Mathematics*, Applied Mathematical Sciences, Vol. 15, Springer-Verlag, New York, NY.
Goldstein, H. (1980) *Classical Mechanics*, Addison Wesley, Reading, MA.
Guckenheimer, J. and Holmes, P. J. (1983) *Non-linear Oscillations, Dynamical Systems and Bifurcations of Vector Fields*, Applied Mathematical Sciences, Vol. 42, Springer-Verlag, New York, NY.
Haberman, R. (1977) *Mathematical Models*, Prentice-Hall, Englewood Cliffs, NJ.
Hardy, G. H. (1979) *An Introduction to the Theory of Numbers*, OUP, Oxford.
Hartman, P. (1964) *Ordinary Differential Equations*, Wiley, New York, NY.
Hartley, B. and Hawkes, T. O. (1970) *Rings, Modules and Linear Algebra*, Chapman & Hall, London.
Hirsch, M. W. and Smale, S. (1974) *Differential Equations, Dynamical Systems and Linear Algebra*, Academic Press, London.
Hayes, P. (1975) *Mathematical Methods in the Social and Managerial Sciences*, Wiley, New York, NY.
Jordan, D. W. and Smith, P. (1987) *Non-linear Ordinary Differential Equations*, Clarendon Press, Oxford.
Manneville, P. (1990) *Dissipative Structures and Weak Turbulence*, Academic Press, London.

Marsden, J. E. and McCracken, M. (1976) *The Hopf Bifurcation and its Applications*, Applied Mathematical Sciences, Vol. 19, Springer-Verlag, New York, NY.

May, R. M. (1974) *Stability and Complexity in Model Ecosystems*, Princeton University Press, Princeton, NJ.

Neumann, J. von. (1966) *Theory of self-reproducing automata*, (edited and completed by A. W. Burks), University of Illinois Press, Urbana, IL.

Petrovski, I. G. (1966) *Ordinary Differential Equations*, Prentice-Hall, Englewood Cliffs, NJ.

Pielou, E. C. (1977) *Mathematical Ecology*, Wiley, New York, NY.

Press, W. H., Flannery, B. P., Teukolsky, W. T. and Vetterling, W. T. (1987) *Numerical Recipes*, CUP, Cambridge.

Smith, J. Maynard (1974) *Models in Ecology*, CUP, London.

Sparrow, C. (1982) *The Lorenz Equations: Bifurcations, Chaos and Strange Attractors*, Applied Mathematical Sciences, Vol. 41, Springer-Verlag, New York, NY.

Swan, G. W. (1977) *Some Current Mathematical Topics in Cancer Research*, Xerox University Microfilms, Ann Arbor, Michigan, MI.

Takens, F. (1974) *Applications of Global Analysis I*, Comm. Math. Inst. Rijksuniversitiet Utrecht, 3.

Thompson, J. M. T. and Stewart, H. B. (1986) *Non-linear Dynamics and Chaos*, Wiley, New York, NY.

Wolfram, S. (1986) *Theory and Applications of Cellular Automata*, World Scientific, Singapore.

PAPERS

Collet, P. and Eckmann, J.-P. (1980) Iterated Maps on the Interval as Dynamical Systems. *Progress in Physics*, **1**, Birkhauser-Boston, Boston.

Eckmann, J.-P. (1983) Les Houches, Session XXXVI, 1981. *Chaotic Behaviour of Deterministic Systems*, (eds G. Iooss, R. H. G. Helleman and R. Stora), pp. 455–510, North-Holland.

Gardner, M. (1971) Mathematical Games. *Sci. Amer.*, **224**, (a) February, (b) 112–17, March, (c) 106–9, April, 114–17.

Gardner, M. (1972) Mathematical Games. *Sci. Amer.*, **226**, January, 104–7.

Guckenheimer, J., Oster, G. and Ipaktchi, A. (1977) The dynamics of density dependent population models. *J. Math. Biol.*, **4**, 101–47.

Helleman, R. H. G. (1980) Self-generated chaotic behaviour in non-linear mechanics. *Fundamental Problems in Statistical Mechanics*, (ed. E. G. D. Cohen), **5**, pp. 165–233, North-Holland.

Hénon, M. (1969) Numerical study of quadratic area-preserving mappings. *Quart. Appl. Math.*, **27**, 291–311.

Lefever, R. and Nicolis, G. (1971) Chemical instabilities and sustained oscillations. *J. Theor. Biol.*, **30**, 267–84.

Li, T. Y. and Yorke, J. A. (1975) Period three implies chaos. *Amer. Math. Monthly*, **82**, 985–92.

Neumann, J. von. (1963) The general and logical theory of automata. *Collected Works*, (ed. A. H. Taub), Vol. 5, 288–328, Macmillan, New York, NY.

Percival, I. C. (1988) Order and Chaos in Nonlinear Physical Systems, (eds S. Lundqvist, N.H. March and M. P. Tosi), pp. 361–86, Plenum Press, New York, NY.

Rescigno, A. and De Lisi, C. (1977) Immune surveillance and neoplasia II. *Bull. Math. Biol.*, **39**, 487–97.

Rössler, O. E. (1976) An equation for continuous chaos. *Phys. Lett.*, **57A**, 397–8.

Sarkovskii, A. N. (1964) Coexistence of cycles of a continuous map of a line into itself. *Ukrainian Math. J.*, **16**, 61–71.

Soberon, J. M. and Martinez del Rio, C. (1981) The dynamics of plant–pollinator interaction. *J. Theor. Biol.*, **91**, 363–78.

Swan, G. W. (1979) Immunological surveillance and neoplastic development. *Rocky Mountain J. Math.*, **9**, 143–8.

Tanner, J. T. (1975) The stability and intrinsic growth rate of prey and predator populations. *Ecol.*, **56**, 855–67.

Wolfram, S. (1983) Statistical Mechanics of Cellular Automata. *Rev. Mod. Phys.*, **55**, 601–44.

Zeeman, E. C. (1973) *Differential Equations for the Heartbeat and Nerve Impulse.* Salvador Symposium on Dynamical Systems, Academic Press, 683–741.

Zeeman, E. C. (1979) *Population Dynamics from Game Theory.* Int. Conf. Global Theory of Dynamical Systems, Northwestern University, Evanston, IL.

Hints to exercises

CHAPTER 1

1. (a) $x = 1 + t + Ce^t$.
   (b) $x = te^t + Ce^t$.
   (c) $x = -5e^{\cos t}\operatorname{cosec} t + C \operatorname{cosec} t$, $(n-1)\pi < t < n\pi$, n integer.
2. (a) $x = Ce^{t^2/2}, x > 0$.
   $x = C'\ e^{t^2/2}, x < 0$.
   $x \equiv 0$.
   (b) $x = C/t, t < 0$.
   $x = C'/t, t > 0$.
   (c) $x = \sqrt{(C - t^2)}, -\sqrt{C} < t < \sqrt{C}$.
   $x = -\sqrt{(C' - t^2)}, -\sqrt{C'} < t < \sqrt{C'}$.
   (d) $x = \dfrac{C}{\sinh t}, t < 0$.
   $x = \dfrac{C'}{\sinh t}, t > 0$.
3. $F(t, x) = x \ln(tx)$.
4. (a) $x = \dfrac{1}{C - t}, t < C$.
   $x = \dfrac{1}{C' - t}, t > C'$.
   $x \equiv 0$.
   (b) $x = (C - t)^{-1/2}, t < C$. $x = -(C' - t)^{-1/2}, t < C'$.
   $x \equiv 0$.
   (c) $x = (1 + Ce^t)/(1 - Ce^t), x < -1, t > -\ln C, C \in \mathbb{R}^+$.
   $x = (1 - C'e^t)/(1 + C'e^t), -1 < x < 1, -\infty < t < \infty, C' \in \mathbb{R}^+$.
   $x = (1 + \bar{C}e^t)/(1 - \bar{C}e^t), x > 1, t < -\ln \bar{C}, \bar{C} \in \mathbb{R}^+$.
   $x \equiv -1. x \equiv 1$.
   (d) $x = (t - C)^3, t < C;\ x \equiv 0, C \leqslant t \leqslant C'$;
   $x = (t - C')^3, C' < t$.

$x = (t - C)^3, t < C;\ x \equiv 0, C \leqslant t.$
$x \equiv 0, t < C;\ x = (t - C)^3, C \leqslant t.$
$x \equiv 0.$

(e) $x \equiv -1, t < C - \pi/2;$
$x = \sin(t - C), C - \pi/2 \leqslant t \leqslant C + \pi/2;$
$x \equiv 1, t > C + \pi/2.$
$x \equiv 1,\ x \equiv -1.$
$x \equiv 0, t \leqslant C;\ x = (t - C)^2, t > C.$
$x \equiv 0.$
Yes for both (e) and (f).

5. 4(d) $x = (t - C)^3, t \leqslant C;$
$x \equiv 0, t > C,$
satisfies $x(0) = 0$ for all negative C.
4(f) $x \equiv 0, t < C;$
$x = (t - C)^2, t > C,$
satisfies $x(0) = 0$ for all positive C.

4(e) $$x = \begin{cases} -1, & t < C - \pi/2 \\ \sin(t - C), & C - \pi/2 \leqslant t \leqslant C + \pi/2 \\ 1, & t > C + \pi/2 \end{cases}$$ satisfies $x(0) = +1$
for $C < -\pi/2$ and $x(0) = -1$ for $C > \pi/2$.

6. (a) $\xi(-t)$ is a solution.
(b) $-\xi(t)$ is a solution.
7. If $C = h(t_0, x_0)$ then $C = h(\alpha t_0, \alpha x_0), \forall \alpha \neq 0.$
Substituting $u = x/t$ in $\dot{x} = e^{x/t}$ results in $\displaystyle\int^x \frac{du}{e^u - u} = \log t + C;$
the integral cannot be evaluated using elementary calculus.
8. $\ddot{x} = \dot{x}\dfrac{d}{dx}(\dot{x}) = (a - 2bx)x(a - bx)$ gives the regions of concavity and convexity in the tx-plane.
9. $\dot{y} = -ay + b;\ y = \dfrac{b}{a} + Ce^{-at}.$
11. (b), (d) and (e) are autonomous equations; the isoclines are $x =$ constant lines
12. (a) $1(R)$; (b) $-1(A)$, $0(R)$, $1(A)$; (c) $0(S)$; (d) $-1(A)$, $0(S)$, $2(R)$; (e) none [$(A) \equiv$ attractor, $(S) \equiv$ shunt, $(R) \equiv$ repellor.]
13. $\{(a), (b), (f)\}$, $\{(c), (e)\}$, $\{(d)\}$.
14. $\lambda \leqslant 0 \leftarrow\bullet\rightarrow$; $\lambda > 0\,(\lambda \neq 1) \leftarrow\bullet\rightarrow\bullet\leftarrow\bullet\rightarrow$; $\lambda = 1 \leftarrow\bullet\rightarrow\bullet\rightarrow$.
15. 16; 2^{n+1}.
16. $\dot{y} = y(y - c)$ transforms to $k\dot{x} = (kx + l)(kx + (l - c)); k = 1$ and $l - c = -a, l = -b$ or $l - c = -b, l = -a.$
17. Let $y = x^3 + ax - b$. The numbers of real roots of $y = 0$ are determined by:
$a > 0 \Rightarrow \dfrac{dy}{dx} > 0 \Rightarrow$ one root;

$$a<0 \Rightarrow y_{\max} = -\frac{2a}{3}\sqrt{\left(-\frac{a}{3}\right)} - b,\ y_{\min} = +\frac{2a}{3}\sqrt{\left(-\frac{a}{3}\right)} - b;$$

(i) $y_{\max} < 0 \Rightarrow 1$ root;
(ii) $y_{\min} > 0 \Rightarrow 1$ root;
(iii) $y_{\max} > 0 > y_{\min} \Rightarrow 3$ roots.

18. b/q.
19. (a) $(0,0)$, $(c/d, a/b)$; (b) $(n\pi, 0)$, n integer; (c) $(0,0)$; (d) $(0,0)$, $(2,0)$, $(0,2)$, $(2/3, 2/3)$; (e) $(n\pi, (2m+1)\pi/2)$, n, m integers.
20. (a) $\dot{x}_1 = -x_2, \dot{x}_2 = x_1$; (b) $\dot{x}_1 = -\frac{1}{2}x_2, \dot{x}_2 = 2x_1$; (c) $\dot{x}_1 = x_1, \dot{x}_2 = -2x_2$; (d) $\dot{x}_1 = x_2, \dot{x}_2 = x_1$; (e) $\dot{x}_1 = x_1 + x_2, \dot{x}_2 = 2x_2$.
21. $\{(a), (b)\}$, $\{(c), (d)\}$, $\{(e)\}$
22. Both systems satisfy the same differential equation
$$\frac{dx_2}{dx_1} = \frac{X_2(x_1, x_2)}{X_1(x_1, x_2)}.$$
23. $\dot{y}_1 = y_1, \dot{y}_2 = -y_2$.
24. Assume another solution $(x_1'(t), x_2'(t))$ and consider the differences $x_1 - x_1'$, $x_2 - x_2'$.
25. See Exercise 1.9.
26. Invariance of the system equations under $\mathbf{x} \to -\mathbf{x}$ means that $-\dot{\mathbf{x}} = \mathbf{X}(-\mathbf{x})$. Thus $\mathbf{X}(\mathbf{x}) = -\mathbf{X}(-\mathbf{x})$ (see Exercise 1.6).
30. (a) The isoclines are all straight lines passing through the point $(2, 1)$. Moreover the slope is always perpendicular to the isocline (closed orbits exist).
    (b) Locate the regions of constant $\dot{x}_1$, $\dot{x}_2$ sign and consider trajectory directions (no closed orbits).
    (c) The isoclines are radial and the tangents to the trajectories make acute angles with outward direction (no closed orbits).
31. Slope k-isocline satisfies $kx_2 + x_2^2 = -x_1$. Solutions:
$$e^{2x_1}(x_2^2 + x_1 - \tfrac{1}{2}) = C.$$
32. $$\frac{x^2}{1-x^2} = Ce^{2t}$$
33. (a) $\phi_t(x) = \sinh^{-1}(e^t \sinh x)$; (b) $\phi_t(x) = x^{(e^t)}$.
34. (a) $x_1(t) = e^t x_1(0)$, $x_2(t) = \dfrac{e^t x_1(0)}{2} + e^{-t}\left[x_2(0) - \dfrac{x_1(0)}{2}\right]$, all t.
    (b) $x_2(t) = \dfrac{1}{[1/x_2(0)] - t}$; substitute in $\dot{x}_1 = x_1 x_2$, $t < \dfrac{1}{x_2(0)}$, $t > \dfrac{1}{x_2(0)}$.
36. Solve $\dot{r} = r(1-r)$ by separation of variables and partial fractions. Yes.
37. Let $\phi_t(\mathbf{x}) = \mathbf{u}(t) = (u_1(t), u_2(t))^T$, then $\dot{\mathbf{u}}(t) = d\phi_t(x)/dt = \beta(-u_2(t), u_1(t))^T = \mathbf{X}(\mathbf{u}(t))$. Since $\phi_0(\mathbf{x}) = \mathbf{x}$, take $t = 0$ to obtain $\mathbf{X}(\mathbf{x}) = (-x_2, x_1)^T$. Trajectories are concentric circles centred on the origin to which $(-x_2, x_1)$ is tangent at every point (x_1, x_2).

38. For $x_0 > 0$, $\ln\left(\dfrac{x}{x_0}\right) = \dfrac{t^2}{2} - \dfrac{t_0^2}{2}$ using initial conditions. For $x_0 = 0$, $x \equiv 0$.

CHAPTER 2

2. $\{(a)\}$, $\{(b), (f)\}$, $\{(c), (d)\}$, $\{e\}$.
3. $\boldsymbol{J}, \boldsymbol{M} = \begin{bmatrix} 3 & 1 \\ 0 & 3 \end{bmatrix}, \begin{bmatrix} 1 & -1 \\ -1 & 2 \end{bmatrix}; \begin{bmatrix} 2 & -2 \\ 2 & 2 \end{bmatrix}, \begin{bmatrix} -1 & 1 \\ 3 & -2 \end{bmatrix};$
$\begin{bmatrix} 3 & 0 \\ 0 & 2 \end{bmatrix}, \begin{bmatrix} -1 & 2 \\ 4 & -7 \end{bmatrix}.$
4. $\boldsymbol{J}, \boldsymbol{M} = \begin{bmatrix} 2 & 0 \\ 0 & 1 \end{bmatrix}, \begin{bmatrix} 1 & 1 \\ 2 & 1 \end{bmatrix}; \begin{bmatrix} 0 & -1 \\ 1 & 0 \end{bmatrix}, \begin{bmatrix} 41 & -1 \\ 58 & 0 \end{bmatrix}; \begin{bmatrix} 3 & 1 \\ 0 & 3 \end{bmatrix}, \begin{bmatrix} -2 & -1/3 \\ 3 & 0 \end{bmatrix}.$
5. $\dot{y}_1 = -y_1, \dot{y}_2 = -2y_2 - 3y_3, \dot{y}_3 = -2y_2 - 4y_3$.
6. Similar matrices have the same eigenvalues. Check that the Jordan matrices in Proposition 2.2.1 are determined uniquely by their eigenvalues for types (a) and (b). Are $\begin{bmatrix} \lambda & 1 \\ 0 & \lambda \end{bmatrix}$ and $\begin{bmatrix} \lambda & 0 \\ 0 & \lambda \end{bmatrix}$ similar?
8. Use Exercise 2.4 to obtain canonical systems, solve these and use $\boldsymbol{x} = \boldsymbol{M}\boldsymbol{y}$.
11. $\dot{x}_1 = -\frac{3}{2}x_1 + \frac{5}{2}x_2, \qquad \dot{x}_2 = -\frac{1}{2}x_1 - \frac{1}{2}x_2$.
12. $x_1(t) = (2e^{-2t} - e^{-3t})x_1(0) + (e^{-3t} - e^{-2t})x_2(0)$;
$x_2(t) = (2e^{-2t} - 2e^{-3t})x_1(0) + (-e^{-2t} + 2e^{-3t})x_2(0)$.
Put $y_1 = 4e^{-2t} = x_1 + 3x_2$. Compare coefficients of e^{-2t} and e^{-3t} and obtain simultaneous equations for $x_1(0)$ and $x_2(0)$ $[x_1(0) = x_2(0) = 1]$.
13. (a) unstable node; (b) saddle; (c) centre; (d) stable focus; (e) unstable improper node.
14. $x_1 = y_1 + 3y_2, x_2 = 2y_1 + 4y_2;\ \dot{y}_1 = -y_1, \dot{y}_2 = -3y_2$.
15. Linear transformations are continuous and so

$$\lim_{t\to\infty} (\boldsymbol{N}\boldsymbol{x}(t)) = \boldsymbol{N}\left(\lim_{t\to\infty} \boldsymbol{x}(t)\right).$$

$$\lim_{t\to\infty} \boldsymbol{y}(t) = \lim_{t\to\infty} \boldsymbol{N}\boldsymbol{x}(t) = \boldsymbol{N}\left(\lim_{t\to\infty} \boldsymbol{x}(t)\right) = \boldsymbol{N}\boldsymbol{0} = \boldsymbol{0}.$$

If $\dot{\boldsymbol{x}} = \boldsymbol{A}\boldsymbol{x}$ has a stable fixed point at $\boldsymbol{0}$, then $\lim_{t\to\infty} \boldsymbol{x}(t) = \boldsymbol{0}$ for all trajectories $\boldsymbol{x}(t)$ and so $\lim_{t\to\infty} \boldsymbol{y}(t) = 0$ for all trajectories of $\dot{\boldsymbol{y}} = \boldsymbol{N}\boldsymbol{A}\boldsymbol{N}^{-1}\boldsymbol{y}$.

16. If $\boldsymbol{y}(t) = \boldsymbol{N}\boldsymbol{x}(t)$, then $\lim_{t\to-\infty} \boldsymbol{y}(t) = \lim_{t\to-\infty} \boldsymbol{N}\boldsymbol{x}(t) = \boldsymbol{0}$. If 0 is a saddle point then there exist two trajectories $\mathbf{x}(t)$ and $\mathbf{x}'(t)$ such that $\lim_{t\to+\infty} = \lim_{t\to-\infty} \mathbf{x}'(t) = \mathbf{0}$. The corresponding trajectories $\boldsymbol{y}(t) = \boldsymbol{N}\boldsymbol{x}(t)$, $\boldsymbol{y}'(t) = \boldsymbol{N}\boldsymbol{x}'(t)$ satisfy $\lim_{t\to\infty} \boldsymbol{y}(t) = \lim_{t\to-\infty} \boldsymbol{y}'(t) = \mathbf{0}$. The saddle is the only fixed point of linear systems with this property.

17. Mathematical merit is that number of qualitatively distinct types of phase portrait is reduced. Practical disadvantage is that increasing and decreasing behaviour are taken as qualitatively the same.
18. The linear transformation must preserve straight lines and so either $y_1 = ax_1, y_2 = dx_2$ or $y_1 = bx_2$, $y_2 = cx_1$. Then consider
$$y_2 = C'y_1^{\mu'} \Rightarrow x_2 = \frac{C'a^{\mu'}}{d} x_1^{\mu'} \Rightarrow \mu = \mu';$$
$$y_2 = C'y_1^{\mu'} \Rightarrow x_1 = \frac{C'b^{\mu'}}{c} x_2^{\mu'} \Rightarrow \mu = \frac{1}{\mu'}.$$
Consider $\mu = \lambda_2/\lambda_1, \mu' = v_2/v_1$.
19. If $\lambda_2 k = \lambda_1$, then $\dot{y}_1 = \lambda_1 y_1, \dot{y}_2 = \lambda_1 y_2$ which is qualitatively equivalent to $\dot{y}_1 = \varepsilon y_1, \dot{y}_2 = \varepsilon y_2$ with $\varepsilon = \operatorname{sgn}(\lambda_1)$.

21. (a) $\begin{bmatrix} 1 & 0 \\ 0 & -1 \end{bmatrix}$; (b) $\begin{bmatrix} -1 & 0 \\ 0 & -1 \end{bmatrix}$; (c) $\begin{bmatrix} 0 & -1 \\ 1 & 0 \end{bmatrix}$; (d) $\begin{bmatrix} 0 & 1 \\ 1 & 0 \end{bmatrix}$.

22. (a) $\begin{bmatrix} e^{2t} & 0 \\ 0 & e^{3t} \end{bmatrix}$; (b) $\begin{bmatrix} e^t & 2e^t(e^t - 1) \\ 0 & e^{2t} \end{bmatrix}$; (c) $\frac{e^{6t}}{7}\begin{bmatrix} 3 & 4 \\ 3 & 4 \end{bmatrix} + \frac{e^{-t}}{7}\begin{bmatrix} 4 & -4 \\ -3 & 3 \end{bmatrix}$;

(d) $\frac{e^{-2t}}{\sqrt{2}}\begin{bmatrix} \sqrt{2}\cos\beta t & \sin\beta t \\ -2\sin\beta t & \sqrt{2}\cos\beta t \end{bmatrix}$, $\beta = 2\sqrt{2}$; (e) $e^{-3t}\begin{bmatrix} 1-t & t \\ -t & 1+t \end{bmatrix}$.

23. (a) $\frac{e^{-t}}{4}\begin{bmatrix} 7 & -7 \\ 3 & -3 \end{bmatrix} + \frac{e^{-5t}}{4}\begin{bmatrix} -3 & 7 \\ -3 & 7 \end{bmatrix}$;

(b) $e^{4t}\begin{bmatrix} \cos 2t - \sin 2t & 2\sin 2t \\ -\sin 2t & \cos 2t + \sin 2t \end{bmatrix}$.

24. Let $y_1 = x_1, y_2 = x_3, y_3 = x_2, y_4 = x_4$.

25. $\frac{(P+Q)^k}{k!} = \sum_{l=0}^{k} \frac{1}{l!(k-l)!} P^l Q^{k-l}$.

26. (a) $\boldsymbol{C}^2 = \boldsymbol{0}$; $\quad e^{\boldsymbol{A}t} = e^{\lambda_0 t}(\boldsymbol{I} + t\boldsymbol{C})$.
(b) $e^{\alpha t\boldsymbol{I}} = e^{\alpha t}\boldsymbol{I}$;

$$e^{\beta t\boldsymbol{D}} = \sum_{n=0}^{\infty} \frac{(-1)^n}{(2n)!}(\beta t)^{2n}\boldsymbol{I} + \sum_{n=0}^{\infty} \frac{(-1)^n}{(2n+1)!}(\beta t)^{2n+2}\boldsymbol{D},$$

$$\boldsymbol{D} = \begin{bmatrix} 0 & -1 \\ 1 & 0 \end{bmatrix}.$$

27. The characteristic equation of $\boldsymbol{A}$ is

$$\lambda^2 - (\lambda_1 + \lambda_2)\lambda + \lambda_1\lambda_2 = 0$$

where λ_1, λ_2 are the eigenvalues. The Cayley–Hamilton Theorem states

$$\boldsymbol{A}^2 - (\lambda_1 + \lambda_2)\boldsymbol{A} + \lambda_1\lambda_2\boldsymbol{I} = \boldsymbol{0} \qquad (1)$$

and so

$$(A - \lambda_1 I)(A - \lambda_2 I) = 0.$$

Calculate $(A - \lambda_1 I)^2$ by expanding and then substituting for A^2 from (1).

28. Show $e^{At} = e^{\lambda_0 t}.e^{Qt}$ and $Q^3 = 0$. Use (2.63).

$$e^{At} = e^{\lambda_0 t}\left[I + tQ + \cdots + \frac{t^{n-1}Q^{n-1}}{(n-1)!} \right].$$

29. (a) $y_1 = x_1 + 1, y_2 = x_2 + 1$; (d) $y_1 = x_1 + \frac{1}{4}, y_2 = x_2 - \frac{1}{2}$; (e) $y_1 = x_1 - 1/3, y_2 = x_2 + 5/3, y_3 = x_3 + 1/3$.

30. $x(t) = \{\frac{1}{2}(a-b) + \frac{1}{2}t\}\begin{bmatrix} 1 \\ -1 \end{bmatrix} + \{\frac{1}{2}e^{2t}(a+b) - \frac{1}{4}(1 - e^{2t})\}\begin{bmatrix} 1 \\ 1 \end{bmatrix}.$

31. If $x = My$, then $\dot{x} = Ax + h$ transforms to

$$\dot{y}(t) = M^{-1}AMy(t) + M^{-1}h(t).$$

If A has real distinct eigenvalues λ_1, λ_2 choose matrix M such that

$$M^{-1}AM = \begin{bmatrix} \lambda_1 & 0 \\ 0 & \lambda_2 \end{bmatrix}.$$

No; if $h(t) \not\equiv 0$ then $M^{-1}h(t) \not\equiv 0$.

32. (a) Convert to linear system; (b) use isoclines.
33. For example,

$$M = \begin{bmatrix} 1 & -2 & 0 \\ 0 & -1 & 0 \\ 0 & -1 & 1 \end{bmatrix} \text{ and } a = 1, b = 2, c = 3.$$

Unstable focus in the y_1y_2-plane. Repellor on the y_3-axis.

34. $x_1 = 5e^{2t} - (4t+5)e^t, x_2 = 2e^{2t} - 2e^t, x_3 = e^{2t}$.
35. $e^{At} =$:

$$\text{(a) } e^{\lambda t}\begin{bmatrix} 1 & t & \frac{t^2}{2} & \frac{t^3}{6} \\ 0 & 1 & t & \frac{t^2}{2} \\ 0 & 0 & 1 & t \\ 0 & 0 & 0 & 1 \end{bmatrix};$$

$$\text{(b) } \left[\begin{array}{c|c} e^{\alpha t}\begin{bmatrix} \cos\beta t & -\sin\beta t \\ \sin\beta t & \cos\beta t \end{bmatrix} & 0 \\ \hline 0 & e^{\lambda t}\begin{bmatrix} 1 & t \\ 0 & 1 \end{bmatrix} \end{array}\right];$$

(c)
$$\left[\begin{array}{c|c} e^{\alpha t}\begin{bmatrix} \cos\beta t & -\sin\beta t \\ \sin\beta t & \cos\beta t \end{bmatrix} & \mathit{0} \\ \hline \mathit{0} & e^{\gamma t}\begin{bmatrix} \cos\delta t & -\sin\delta t \\ \sin\delta t & \cos\delta t \end{bmatrix} \end{array}\right];$$

(d)
$$\left[\begin{array}{c|c} e^{\alpha t}\begin{bmatrix} \cos\beta t & -\sin\beta t \\ \sin\beta t & \cos\beta t \end{bmatrix} & te^{\alpha t}\begin{bmatrix} \cos\beta t & -\sin\beta t \\ \sin\beta t & \cos\beta t \end{bmatrix} \\ \hline \mathit{0} & e^{\alpha t}\begin{bmatrix} \cos\beta t & -\sin\beta t \\ \sin\beta t & \cos\beta t \end{bmatrix} \end{array}\right].$$

36. Subsystems are formed from the following subsets of coordinates: $\{x_1\}$, $\{x_2, x_3\}$, $\{x_4\}$, $\{x_5, x_6\}$ and are respectively a repellor, centre, repellor and unstable improper node.

CHAPTER 3

2. Let $y_1 = f(r)\cos\theta$, $y_2 = f(r)\sin\theta$. Show the circle $x_1^2 + x_2^2 = r^2$ maps onto the circle $y_1^2 + y_2^2 = (f(r))^2$. The result illustrates the qualitative equivalence of the global phase portrait and the local phase portrait at the origin.
4. (i) $\dot{y}_1 = y_2, \dot{y}_2 = 2y_1 - 3y_2$;
   (ii) $\dot{y}_1 = 2y_1 + y_2, \dot{y}_2 = y_1$;
   (iii) $\dot{y}_1 = 2e^{-1/2}y_1 - e^{-1}y_2, \dot{y}_2 = -y_2$.
5. (a) $(1, -1)$, saddle; $(1, 1)$, stable node; $(2, -2)$, unstable focus; $(2, 2)$, saddle.
   (b) $(\pm 1, 0)$ saddle; $(0, 0)$ centre, no classification;
   (c) $(m\pi, 0)$, unstable improper node (m even), saddle (m odd);
   (d) $(0, -1)$, stable focus;
   (e) $(0, 0)$, non-simple; $(2, -2)$, saddle;
   (f) $(0, 0)$, stable focus;
   (g) $(-1, -1)$, stable focus; $(4, 4)$, unstable focus.
6. Solve $\dfrac{dx_2}{dx_1} = \dfrac{x_1 - x_1^5}{-x_2}$. The function $f(x_1, x_2) = 3(x_1^2 + x_2^2) - x_1^6$ has a minimum at the origin and so locally the level curves of f are closed. The linearization theorem is not applicable because the linerization is a centre.
7. (a) $(1, 0)$; (b) $(1, 2), (1, 0)$; (c) $(\sqrt{2}, 1), (-\sqrt{2}, 1)$.
8. $x_1 \dfrac{du}{dx_1} = u - x_1^3$; $u = Cx_1 - \dfrac{x_1^3}{2}$.
9. Yes; the identity map suffices for qualitative equivalence. Both systems satisfy $\dfrac{dx_2}{dx_1} = \dfrac{x_2}{x_1}$ and both have a unique fixed point at the origin.
10. If a trajectory lies on the line $x_2 = kx_1$, then $\dfrac{dx_2}{dx_1} = k$.

11. The line of fixed point is $x_1^2 = x_2^3$. The linearization at the fixed point (k^3, k^2), k real, is

$$\begin{bmatrix} 2k^3 & -3k^4 \\ 2k^9 & -3k^{10} \end{bmatrix}.$$

Yes, assume the fixed point is simple and use linearization theorem to obtain a contradiction; Yes.

12. $dx_2/dx_1|_{x_1=0} \to \infty$ as $|x_2| \to 0$ suggests the existence of saddle connections in the form of figures of eight lying in the first and third quadrants.
13. (d) $\dot{V}(x_1, x_2) = -2x_1^2(\sin x_1)^2 - 2x_2^2 - 2x_2^6$ is negative definite when $x_1^2 + x_2^2 < \pi^2$.

    (e) $\dot{V}(x_1, x_2) = -2x_1^2(1 - x_2) - 2x_2^2(1 - x_1)$ is negative definite when $x_1^2 + x_2^2 < 1$.
14. The domain of stability is $\mathbb{R}^2$ for (a), (b) and (c) and $\{(x_1, x_2) | x_1^2 + x_2^2 < r^2\}$ where $r = \pi$ for (d) and $r = 1$ for (e).
15. Asymptotically stable: (a) and (b).
    Neutrally stable: (c) and (d).
16. The system $\dot{\mathbf{x}} = -\mathbf{X}(\mathbf{x})$ has an asymptotically stable fixed point at the origin. Let $\mathbf{x}_0$ be such that $\lim_{t\to\infty} \phi_t(\mathbf{x}_0) = \mathbf{0}$. Choose a neighbourhood N of $\mathbf{0}$ not containing $\mathbf{x}_0$. The trajectory through $\mathbf{x}_0$ of the system $\dot{\mathbf{x}} = \mathbf{X}(\mathbf{x})$ satisfies $\lim_{t\to-\infty} \phi_t(\mathbf{x}_0) = \mathbf{0}$. Use this property to show that the origin is unstable. Use the function $V(x_1, x_2) = x_1^2 + x_2^2$ in (a) to (c).
17. Use the function $V(x_1, x_2) = x_1^2 + x_2^2$ in (a) and (b) and $V(x_1, x_2) = x_1^4 + 2x_2^2$ in (c) and (d).
18. If V is positive definite then $V(1, 0)$ is positive and so a is positive; also

$$V(x_1, x_2) = a\left(x_1 + \frac{b}{a}x_2\right)^2 + \left(c - \frac{b^2}{a}\right)x_2^2$$

and thus a and $c - b^2/a$ are positive. Try $a = 5, b = 1, c = 2$; then

$$V(x_1, x_2) = 5\left(x_1 + \frac{x_2}{5}\right)^2 + \frac{9}{5}x_2^2.$$

For $V(x_1, x_2) < 9/5, x_2^2 < 1$ and so there is a domain of stability defined by $25x_1^2 + 10x_1x_2 + 10x_2^2 < 9$.

19. (a) $V(x_1, x_2) = x_1^2 + x_2^2$, $\dot{V}(x_1, x_2) = -2r^2(1 - r^2)(1 + r^2)$; $x_1^2 + x_2^2 < 1$.

    (b) $V(x_1, x_2) = x_1^6 + 3x_2^2$, $\dot{V} = -x_2^2(1 - x_2^2)$; $x_1^6 + 3x_2^2 < 3$.
20. $\dot{V}(x_1, x_2) = \dfrac{2x_1^2}{a^2}(x_1 - a) + \dfrac{2x_2^2}{b^2}(x_2 - b)$ is negative definite for $\dfrac{x_1^2}{a^2} + \dfrac{x_2^2}{b^2} < 1$.
21. $V(x_1, x_2) = \frac{1}{2}x_1^4 + \frac{1}{2}x_1^2 - x_1x_2 + x_2^2$ satisfies the hypotheses of Theorem 3.5.3.

22. $\dot{V}(x_1, x_2) = 3(x_1^2 + x_2^2)^2$.
23. $V(x_1, x_2) = x_1 - x_2$.
One separatrix is $x_2 = 0$ where $\dot{x}_1 = x_1^4$. Thus the phase portrait on the x_1-axis is a shunt and so the fixed point is unstable.
24. $\dot{y}_1 = \dot{x}_1 + 3x_2^2\dot{x}_2 = 0, \dot{y}_2 = \dot{x}_2 + 2x_2\dot{x}_2 = 1$. y_1, y_2 are differentiable functions of x_1 and x_2; inverse transformations

$$x_1 = y_1 - x_2^3 \text{ and } x_2 = \frac{-1 + \sqrt{(1 + 4y_2)}}{2}$$

are differentiable functions of y_1 and y_2 providing $y_2 > -\frac{1}{4}$ (this condition is necessarily satisfied if $y_2 = x_2 + x_2^2$).
27. $(-1, 0)$ saddle; $(0, 0)$ unstable node; $(1, 0)$ saddle. Observe that the lines $x_1 = 0, \pm 1$ and $x_2 = 0$ are all unions of trajectories.
28. (a) $\dfrac{x_1^3}{3} + x_1 - \dfrac{x_2^2}{2}$, $D = \mathbb{R}^2$;
(b) $x_1 + \ln|x_1| + x_2 + \ln|x_2|$, $D = \{(x_1, x_2) | x_1 x_2 \neq 0\}$;
(c) $\dfrac{1}{x_2} - \sin x_1$, $D = \{(x_1, x_2) | |x_1| < \pi/2, x_2 \neq 0\}$;
(d) $x_1^2 e^{x_2} - x_1 \sin_{x_2}$, $D = \mathbb{R}^2$.
29. $\dfrac{du}{dx_1} = \dfrac{-(u^2 - 1)}{x_1(3 - u)}$; $(x_1 - x_2) = C(x_1 + x_2)^2$. Introduce new coordinates $y_1 = x_1 + x_2$ and $y_2 = x_1 - x_2$ and plot curves for $C = 0, \pm 1, \pm 2, \pm 3$.
30. The trajectories of both systems lie on the solutions of the differential equation $\dfrac{dx_2}{dx_1} = -\dfrac{x_2}{x_1}$. The first system has a line of fixed points $(x_1 = x_2^2)$ whereas the second system has a saddle point at the origin.
31. $x_2^2 = (\ln x_1)^2 + C$.
32. $\boldsymbol{J} = \begin{bmatrix} \lambda_1 & 0 \\ 0 & \lambda_2 \end{bmatrix}$, $(x_2/x_1)^{\lambda_2/\lambda_1}$, $D = \mathbb{R}^2 \backslash \{x_2\text{-axis}\}$

$\boldsymbol{J} = \begin{bmatrix} 0 & -\beta \\ \beta & 0 \end{bmatrix}$, $x_1^2 + x_2^2$, $D = \mathbb{R}^2$.

$\boldsymbol{J} = \begin{bmatrix} \alpha & -\beta \\ \beta & \alpha \end{bmatrix}$, $re^{-\alpha\theta/\beta}$, $\beta \neq 0$, $D = \mathbb{R}^2 \backslash \{0\}$.

Let the trajectory $\mathbf{x}(t)$ tend to a fixed point $\mathbf{x}_0$ as $t \to \infty$. A first integral f satisfies $f(\mathbf{x}(t)) = f(\mathbf{x}_0)$, by continuity of f. If $\mathbf{x}_0$ is asymptotically stable then $f(\mathbf{x}) = f(\mathbf{x}_0)$ *for all* $\mathbf{x}$ in some neighbourhood of $\mathbf{x}_0$.
33. $H = \frac{1}{2}x_2^2 + \dfrac{x_1^2}{2} - \dfrac{\alpha x_1^3}{3}$, centre at $(0, 0)$, saddle at $(\alpha^{-1}, 0)$.
34. Show that the stationary points of $\tilde{H}$ have angular coordinates given by $\cos(5\theta) = 0$ and radial coordinates satisfying $\pm 5r^3 + 4r^2 - 2\mu = 0$.

Approximate latter for r and μ small to obtain fixed points near $r = \sqrt{\mu/2}$. Obtain topological types from nature of extreme points of $\tilde{H}$ and phase curves from sketches of r-dependence of $\tilde{H}(r, \theta)$ as θ is changed.

35. Let $r = |\mathbf{x}|$ and C_r be the circle of radius r centred on the origin.

    (*a*) $0 < r < 1$: $L_\alpha(\mathbf{x}) = C_1$, $L_\omega(\mathbf{x}) = \{\mathbf{0}\}$;
    $r = 1$: $L_\alpha(\mathbf{x}) = C_1$, $L_\omega(\mathbf{x}) = C_1$;
    $r > 1$: $L_\alpha(\mathbf{x}) = \varnothing$, $L_\omega(\mathbf{x}) = C_1$.

    (*b*) $0 < r < 1$: $L_\alpha(\mathbf{x}) = \varnothing$, $L_\omega(\mathbf{x}) = C_1$;
    $r \geqslant 1$: $L_\alpha(\mathbf{x}) = C_r$, $L_\omega(\mathbf{x}) = C_r$.

    (*c*) $0 < r \leqslant 1$: $L_\alpha(\mathbf{x}) = C_r$, $L_\omega(\mathbf{x}) = C_r$;
    $r > 1$: $L_\alpha(\mathbf{x}) = C_1$, $L_\omega(\mathbf{x}) = \varnothing$.

36. $P_1(r) = r/[r + (1 - r)\exp(-2\pi a)]$.
37. Graph of $P(x)$ is reflected in the line $y = x$ and the direction of the iterations is reversed. Original system has limit cycle that is stable for $r < 1$ and unstable to $r > 1$. Limit cycle of time-reversal is stable for $r > 1$ and unstable for $r < 1$.
38. Circular limit cycle of radius $r = x^*$: $\varepsilon > 0$, unstable; $\varepsilon < 0$, stable.
41. (i) implies trajectories enter D at every point of ∂D so, for $t > 0$, $\phi_t(D) \subset D$ but (ii) allows possibility that ∂D is a trajectory when $\phi_t(D) = D$. Example illustrates that trapping region can occur in case (ii).
42. Two cases: (c), $\{\mathbf{0}\}$; (e), $\{(x_1, x_2) | (x_1^2 + x_2^2)^{1/2} = 1\}$.
43. If r is the radial distance, $\dot{r} = r(1 - r^2) - F\sin\theta$. When $F = 0$, the limit cycle is given by $r \equiv 1$.
44. The system has no fixed points. If there exists at least one limit cycle choose a region R bounded by such a trajectory which contains no other limit cycles. Use Poincaré–Bendixson theory to obtain a contradiction by showing a fixed point should exist in R.
45. $\dfrac{\partial X_1}{\partial x_1} + \dfrac{\partial X_2}{\partial x_2} = 2\text{–}4(x_1^2 + x_2^2)$. A limit cycle either intersects $x_1^2 + x_2^2 = \frac{1}{2}$ or contains this curve in its bounded region.
30. Locate the subregions of R of constant sign in $\dot{x}_1$ and $\dot{x}_2$. Consider the behaviour of the trajectories through the points on the parabola $x_2 = -x_1^2 + 3x_1 + 1$, and show that they spiral in an anticlockwise sense about the fixed point.

CHAPTER 4

1. Eigenvalues (a) 2, $-2 \pm \mathrm{i}\sqrt{2}$; $\dot{x}_1 = x_1$, $\dot{x}_2 = -x_2$, $\dot{x}_3 = -x_3$; (b) 2,4,8; $\dot{x}_1 = x_1$, $\dot{x}_2 = x_2$, $\dot{x}_3 = x_3$.
2. The four quadrants of the $\alpha\lambda$-plane give distinct hyperbolic behaviour: $\alpha, \lambda > 0$-unstable; $\alpha < 0$, $\lambda > 0$-saddle ($n_u = 1$, $n_s = 2$); $\alpha, \lambda < 0$-stable; $\alpha > 0$, $\lambda < 0$-saddle($n_u = 2$, $n_s = 1$).

3. Eigenvalues are $2, 1, -1$, therefore saddle point. Eigenvectors can be chosen to be $(1, -4, -3)$, $(-1, 1, 1)$ and $(1, -1, 0)$ respectively.
4. Fixed points: $(0,0,0)$-unstable; $(1,0,0)$-saddle; $(0,1,1)$-linearization, theorem not applicable.
5. Since $f(\lambda)$ has real coefficients, complex roots λ_2, λ_3 satisfy $\lambda_3 = \lambda_2^*$. Thus sum of the roots of $f(\lambda) = 0$ is $\lambda_1 + 2\mathrm{Re}(\lambda_2) = 1 + b + \sigma$.
6. (i) x-axis; (ii) and (iv) xy-plane; (iii) y-axis.
8. A 2-torus. C_1 and C_2 are the radii of the 2-torus and the qualitative behaviour of the flow is independent of these dimensions. The topological type of the flow is determined by β_1, β_2.
9. (a) $n_u = 2$, $n_s = 0$, OP; (b) $n_u = 1$, $n_s = 1$, OP; (c) $n_u = 2$, $n_s = 0$, OR; (d) $n_u = 1$, $n_s = 1$, OR; (e) $n_u = 0$, $n_s = 2$, OP; (f) $n_u = 2$, $n_s = 0$; OP; (g) $n_u = 0$, $n_s = 2$, OR; (h) $n_u = 1$, $n_s = 1$, OR. Topologically conjugate groups: $\{(a), (f)\}$; $\{(d), (h)\}$.
10. Eigenvalues: $\frac{1}{2}$; $2 \pm i$. E^s is the eigenspace of the eigenvalue $\frac{1}{2}$ which has basis $\{(2, 0, 1)\}$. The eigenvalue $2 + i$ has a (complex) eigenspace with basis $\{(1, i, -1)\}$. Taking real and imaginary parts gives the real two dimensional eigenspace E^u with basis $\{(1, 0, -1), (0, 1, 0)\}$. $\mathbf{A}|E^s$ is a contraction; $\mathbf{A}|E^u$ is a spiralling expansion; $\mathbf{A}$ is of saddle type.
11. (i) and (ii) together with $\boldsymbol{P} \sim \boldsymbol{P}$ implies $\sim$ is an equivalence relation on $\mathscr{S}$. Thus $\mathscr{S}$ can be partitioned into equivalence classes of topologically conjugate diffeomorphisms.
12. (a) 4; (b) 8; (c) $4m$.
13. Assume $\boldsymbol{A}\boldsymbol{u}_j = \lambda_j \boldsymbol{u}_j$ and substitute (2.63) into $\boldsymbol{B}\boldsymbol{u}_j$. Observe: (i) $\prod_j \exp(\lambda_j t) > 0$; (ii) if $\mathrm{Re}(\lambda_j) \neq 0$, then $|\exp(\lambda_j t)| \neq 1, t \neq 0$. When $t = 0$, $\boldsymbol{B} = \boldsymbol{I}$.
14. If eigenvalues of $\boldsymbol{A}$ are $\alpha \pm i\beta$, then $\det(\boldsymbol{A}) = \alpha^2 + \beta^2$.
15. $\alpha = \ln \lambda_2 / \ln \lambda_1$. For (4.41) $\alpha = -\lambda$, examine y and dy/dx as $|x| \to 0$.
16. $\mathbf{P}^q(\mathbf{P}^l(\mathbf{x}^*)) = \mathbf{P}^{q+l}(\mathbf{x}^*) = \mathbf{P}^l(\mathbf{P}^q(\mathbf{x}^*)) = \mathbf{P}^l(\mathbf{x}^*)$, $l = 1, \ldots, q$. If N is a neighbourhood of $\mathbf{x}^*$, then $N' = \mathbf{P}^l(N)$ is a neigbourhood of $\mathbf{P}^l(\mathbf{x}^*)$. Since $\mathbf{P}^q\mathbf{P}^l = \mathbf{P}^l\mathbf{P}^q$, then $\mathbf{P}^q|N$ is topologically conjugate to $\mathbf{P}^q|N'$ so that $\mathbf{x}^*$ and $\mathbf{P}^l(\mathbf{x}^*)$ have the same topological type. Observe that the symmetry of the phase portrait means that, if the four stable (saddle) fixed points are labelled $\mathbf{x}_0, \mathbf{x}_1, \mathbf{x}_2, \mathbf{x}_3$ in a counterclockwise sense, then $\mathbf{x}^*_{(i+1)\mathrm{mod}4} = \mathbf{Q}(\mathbf{x}_i^*)$. Hence $\mathbf{Q}^4(\mathbf{x}_i^*) = \mathbf{x}_i^*$, $i = 0, \ldots, 3$, and $\{\mathbf{x}_0, \mathbf{x}_1, \mathbf{x}_2, \mathbf{x}_3\}$ is a 4-cycle of stable (saddle) type. The union of the stable and unstable manifolds of the fixed and periodic points of $\mathbf{Q}$ covers the whole plane.
17. $P_1(r) = r$; $P_2(r) = -r$. No, because it is orientation reversing (OR). Map (i). Note map (ii) is OR and therefore corresponds to a flow on a Klein bottle rather than a solid torus.
18. Show that $x = 1$ is a fixed point of P for which $0 < P'(1) < 1$.
19. Interiors given by strict inequalities. (a) is a trapping region when $\lambda < 1$. Let $D_r = \{(x, y) | x^2 + y^2 \leq r\}$, then $\mathbf{P}^n(T) = D_{\lambda^n}$ and D_{λ^n} is a subset of the interior of $D_{\lambda^{n-1}}$. $A_{\mathbf{P}} = \{\mathbf{0}\}$.

20. The transformation $\mathbf{y}=\mathbf{P}(\mathbf{x})$, where $\mathbf{P}$ is a diffeomorphism, is simply a differentiable change of coordinates. Thus $\mathbf{P}$ maps a neighbourhood, N, of $\mathbf{x}$ to a neighbourhood $N'=\mathbf{P}(N)$ of $\mathbf{y}=\mathbf{P}(\mathbf{x})$ in a bijective way. Moreover, near $\mathbf{x}, \mathbf{P}$ is approximated by the non-singular linear transformation $D\mathbf{P}(\mathbf{x})$. With this in mind, can conclude: (i) true, boundary points are preserved under diffeomorphism; (ii) false, $\mathbf{P}(T) \subseteq \operatorname{int}(T)$ and therefore $\mathbf{P}(T) \cap \operatorname{bd}(T)=\varnothing$; (iii) true, interior points are preserved under diffeomorphism; (iv) false, simple closed curves are preserved under diffeomorphism; (v) true, $\operatorname{int}(T) \neq \varnothing$ and $\operatorname{int}(\mathbf{P}(T))=\mathbf{P}(\operatorname{int}(T))$.
21. $H(x, y)=C, -\frac{1}{4}<C<0$, forms the boundary of a set that is not connected and a trapping region is a connected set. Each connected component of $\{(x, y) \mid H(x, y) \leqslant C\}$ is a trapping region for a fixed point attractor.
23. The local behaviour of $\mathbf{P}$ at $\mathbf{x} \in \mathbb{R}^2$ is given by the non-singular, linear transformation $D\mathbf{P}(\mathbf{x})$. Thus $\mathbf{P}$ preserves interior and boundary points of both S and T. Recall (4.62) and note any phase curve is of the form $S=\{\phi_t(\mathbf{x}) \mid t \in \mathbb{R}\}$. Observe $\phi_\tau(S)=\{\phi_{t+\tau}(\mathbf{x}) \mid t \in \mathbb{R}\}=S$. The fixed points and the unstable manifold lie entirely within T_C, $C>0$.
24. $\dot{f}_1=r_1, \dot{r}_1=0$. The set $f_1=C_1>0$ is confined to a circle of radius r_1 in the x_1x_2-plane and is unrestricted in the x_3x_4-plane. Thus it is a solid torus $\mathbb{R}^2 \times S^1$ in $\mathbb{R}^4$. The phase portrait for the x_3x_4-subsystem is that shown in Exercise 3.34 where the phase curves are the level sets of f_2. The trajectories of the system on $\mathbb{R}^4$ are constrained to the surfaces generated as these phase curves are swept around the solid torus by angular variable θ_1. In particular the flow on the solid torus has eleven closed orbits: six of centre and five of saddle type. The stable and unstable manifolds of the latter are the surfaces generated by the saddle connections in the phase portrait in Exercise 3.34. All other orbits lie on invariant tori.
25. $\dot{q}=\partial H/\partial p, \dot{p}=-\partial H/\partial q$ transforms to

$$\dot{\theta}=-\frac{1}{r}\frac{\partial \tilde{H}}{\partial r}, \quad \dot{r}=\frac{1}{r}\frac{\partial \tilde{H}}{\partial \theta},$$

where $\tilde{H}(\theta, r)=H(q(\theta, r), p(\theta, r))$.
26. $\dot{q}=\partial H/\partial p, \dot{p}=-\partial H/\partial q$ transforms to

$$\dot{\theta}=-\frac{\partial \tilde{H}}{\partial \tau}, \quad \dot{\tau}=\frac{\partial \tilde{H}}{\partial \theta},$$

where $\tilde{H}(\theta, \tau)=H(q(\theta, \tau), p(\theta, \tau))$.
27. (a) Observe $\det(\boldsymbol{S})=(-1)^n \det\begin{bmatrix} -\boldsymbol{I}_n & \boldsymbol{0} \\ \boldsymbol{0} & \boldsymbol{I}_n \end{bmatrix}$.

(b) Let $\mathbf{y}=\mathbf{h}(\mathbf{x})$:
(i) note that for domain $\mathscr{D}$

$$\int_{\mathbf{h}(\mathscr{D})} dy_1 \cdots dy_{2n}=\int_{\mathscr{D}} |\det(D\boldsymbol{h}(\mathbf{x}))| \, dx_1 \cdots dx_{2n},$$

take determinants on both sides of (4.92) to show that

$$|\det(D\boldsymbol{h}(\boldsymbol{x}))| = 1;$$

(ii) differentiate $\boldsymbol{h}(\boldsymbol{h}^{-1}(\boldsymbol{y})) = \boldsymbol{y}$ with respect to $\boldsymbol{y}$ and $\boldsymbol{h}^{-1}(\boldsymbol{h}(\boldsymbol{x})) = \boldsymbol{x}$ with respect to $\boldsymbol{x}$ to show that $D\boldsymbol{h}^{-1}(\boldsymbol{y}) = [D\boldsymbol{h}(\boldsymbol{x})]^{-1}$, use (4.92) to calculate $[D\boldsymbol{h}^{-1}(\boldsymbol{y})]^{\mathrm{T}}\boldsymbol{S}D\boldsymbol{h}^{-1}(\boldsymbol{y})$.
(iii) Operate on both sides of (4.92) from the left with $\boldsymbol{S}^{\mathrm{T}}$ and from the right with $[D\boldsymbol{h}(\boldsymbol{x})]^{-1}$.

28. Substitute

$$D\boldsymbol{h}(\boldsymbol{x}) = \begin{bmatrix} \boldsymbol{A} & \boldsymbol{B} \\ \boldsymbol{C} & \boldsymbol{D} \end{bmatrix}$$

into left hand side of (4.92) with $\boldsymbol{S}$ given by (4.90) to obtain the alternative form.

29. $\mathbf{P}(\theta_1, \tau_1) = (\theta_1 + 2\pi\mu, \tau_1)$. Invariant curves are lines of constant τ_1 but, since θ_1 is an angular coordinate, θ_1 and $\theta_1 + 2\pi$ are to be identified. Orbits of $\mathbf{P}$ are periodic (quasi-periodic) if ω_1 and ω_2 are rationally dependent (independent).

30. If $m \times m$ matrix $\boldsymbol{A}$ has eigenvalues α_i, $i = 1, \ldots, m$, then $\boldsymbol{I} + \boldsymbol{A}$ has eigenvalues $1 + \alpha_i$. Equation (4.105) implies $\det(D\phi_h(\boldsymbol{x})) = \det(\boldsymbol{I} + hD\boldsymbol{X}(\boldsymbol{x}) + O(h^2))$. (ii) gives $\det(D\phi_h(\boldsymbol{x})) = \prod_{i=1}^{m}(1 + h\alpha_i + O(h^2)) = 1 + h\sum_{i=1}^{m}\alpha_i + O(h^2)$ and (i) gives result.

31. $\mathbf{P}$ $[\mathbf{P}^{-1}]$ stretches S by λ in the x- $[y$-$]$ direction, contracts it by λ in the y-$[x$-$]$ direction and translates it so that the corner at $(1, 1)$ moves to $(\lambda, 1/\lambda)$ $[(1/\lambda, \lambda)]$. If $\mathbf{P}$ preserves area then $\text{area}(\mathbf{P}(U)) = \text{area}(U)$ for all planar sets U. Hence area $(\mathbf{P}^{k+1}(S)) = \text{area}(\mathbf{P}(\mathbf{P}^k(S)) = \text{area}(\mathbf{P}^k(S))$, $k = 0, \ldots n$. As $n \to \infty$ $(-\infty)$, $\mathbf{P}^n(S)$ stretches indefinitely along and moves arbitrarily close to the x-(y-) axis.

32. Use Exercise 4.28 with $n = 1$ so that $\boldsymbol{A}, \boldsymbol{B}, \boldsymbol{C}, \boldsymbol{D}$ are scalars a, b, c, d. Note from form of $\boldsymbol{S}$ that $ad - bc = 1$ which implies $\mathbf{h}$ is both area- and orientation-preserving. Conversely, area- and orientation-preserving $\mathbf{h}$ implies $ad - bc = 1$ which is sufficient to prove that $\mathbf{h}$ satisfies (4.92) with $n = 1$.

CHAPTER 5

3. The principal directions are $(1, \lambda_1), (1, \lambda_2)$ where $\lambda_1, \lambda_2 (\lambda_1 > \lambda_2)$ are the eigenvalues. Show that as $k \to \infty, \lambda_1 \to 0$ and $\lambda_2 \to -\infty$.

4. $\dfrac{\mathrm{d}^2 i}{\mathrm{d}t^2} + \dfrac{1}{CR}\dfrac{\mathrm{d}i}{\mathrm{d}t} + \dfrac{1}{CL}i = 0.$

5. $L\dfrac{\mathrm{d}j_R}{\mathrm{d}t} + Rj_R = E_0.$

6. $C\dfrac{dv_C}{dt}=\dfrac{v_C}{R}$.

8. $\dfrac{D}{U+D}=\dfrac{kD_0}{(k+\mu)D_0+(kU_0-\mu D_0)e^{-kt}}$, $k=\lambda-v-\mu$.

   $\dfrac{D}{U+D}\to\dfrac{k}{k+\mu}$ as $t\to\infty$.

9. $x=\dfrac{1}{\omega}\sin\omega t$, $y=\cos t$.

10. $M=\begin{bmatrix}1 & 1\\ 1 & -1\end{bmatrix}$, $D=\begin{bmatrix}-k/m & 0\\ 0 & -3k/m\end{bmatrix}$.

    Normal modes: $y_1\equiv 0$, the particles undergo oscillations that are symmetric relative to the mid-point of AB; $y_2\equiv 0$, oscillations occur with particles fixed at a distance l apart.

11. (a) The equilibrium state is $(\alpha G_0/[\alpha-1], G_0/[\alpha-1])$; introduce local coordinates and use trace-det classification.
    (b) The equilibrium state is $(\alpha G_0/[\alpha(1-k)-1], G_0/[\alpha(1-k)-1])$ with $k<(\alpha-1)/\alpha(=A)$. The stable equilibrium point moves off to infinity as $k\to A$.

13. $A=\begin{bmatrix}0 & 1\\ -1 & -5/2\end{bmatrix}$, $e^{A\tau}=\dfrac{1}{3}\begin{bmatrix}4e^{-\tau/2}-e^{-2\tau} & 2e^{-\tau/2}-2e^{-2\tau}\\ -2e^{-\tau/2}+2e^{-2\tau} & -e^{-\tau/2}+4e^{-2\tau}\end{bmatrix}$.

14. The steady state solutions are:

    $$i_1=\frac{E_0}{\sqrt{(R^2+\omega^2L^2)}}\cos(\omega t-\phi),\ \text{where } \tan\phi=\omega L/R;$$

    $$i_2=-E_0C\omega\sin\omega t.$$

    The current i_1+i_2 has amplitude

    $$\frac{E_0C}{L}\left[R^2+\left(\omega L-\frac{1}{\omega C}\right)^2\right]^{1/2}$$

15. The fixed points are at: (a) $(0,0)$, unstable node; (b) $(2,0)$, stable improper node; (c) $(0,2)$, stable improper node; (d) $(2/3,2/3)$, saddle. The principal directions are: (a) $(1,0)$ and $(0,1)$; (b) $(1,0)$; (c) $(0,1)$; (d) $(1,1)$ and $(1,-1)$.

16. The fixed points occur at $O=(0,0)$, $A=(0,v)$, $B=(1,0)$ and $C=((1-v)/(1-4v^2), v(1-4v)/(1-4v^2))$ and depend on v as follows:

    $$A:\begin{cases}\text{saddle } 0<v<1\\ \text{stable node } v>1\end{cases}$$

    $$B:\begin{cases}\text{saddle } v<\frac{1}{4}\\ \text{stable node } v>\frac{1}{4}\end{cases};$$

$C:\begin{cases}\text{stable node } \nu < \frac{1}{4} \\ \text{saddle } \nu > 1\end{cases}$

(C is not in the first quadrant for $\frac{1}{4} < \nu < 1$)

O: unstable node $\nu > 0$.

17. Fixed points at $O = (0, 0)$, $A = (0, -1/\alpha)$. $B = (1/\alpha, 0)$, $C = ([1+\alpha]/[1+\alpha^2], [1-\alpha]/[1+\alpha^2])$. Linearization at C has trace $-2\alpha/(1+\alpha^2)$, determinant $(1-\alpha^2)/(1+\alpha^2)$ has hence $\Delta < 0$.

18. $$\bar{x}_1 = \frac{1}{T}\int_0^T x_1 \mathrm{d}t = \frac{1}{T}\int_0^T \left(\frac{c}{d} + \frac{\dot{x}_2}{x_2}\right)\mathrm{d}t = \frac{c}{d}.$$

$$\int_{x_2(0)}^{x_2(T)} \frac{\mathrm{d}x_2}{x_2} = 0, \text{ since } x_2(0) = x_2(T).\ \bar{x}_2 = a/b.$$

With harvesting $\dot{x}_1 = x_1[(a-\varepsilon) - bx_2]$, $\dot{x}_2 = -x_2[(c+\varepsilon) - dx_1]$,

$$\bar{x}_1 = \frac{c+\varepsilon}{d}, \quad \bar{x}_2 = \frac{a-\varepsilon}{b}.$$

19. The peak of the parabola occurs at $y_1 = (k-d)/2$ which is greater than 1. Use this to deduce that $\mathrm{tr}(\boldsymbol{W})$ (see (5.94)) is negative. A phase portrait with just one (stable) limit cycle must contain an unstable fixed point.
20. $\dot{B} = \gamma\dot{P}$ when $B = \gamma P$.
    For $\mu(P) = b + cP$, fixed points are: $O = (0, 0)$, unstable node; $S = (-b/c, 0)$, saddle; $T = ((\gamma - b)/c, \gamma(\gamma - b)/c)$, stable node or focus. The fixed point T has the positive quadrant as a domain of stability.
21. $y = -x + \frac{1}{2}\ln x - c_0$ (orientation is given by x decreasing for $x, y > 0$). The number of susceptibles (x) decreases and the infectives (y) increases to a maximum before decreasing to zero.
22. The fixed point $(\frac{1}{2}, 1)$ is a stable focus. The epidemic sustains a non-zero number of infectives.
23. $I = 1 - S + (1/\sigma)\log(S/S_0)$ (at $t = 0$, $I_0 + S_0 = 1$, $R_0 = 0$).
    (a) $\sigma S_0 \leqslant 1$, then $\sigma S(t) < 1$ for all positive $t(\dot{S} < 0)$. Hence $\dot{I} = \gamma I\ (\sigma S - 1)$ is negative and $\dot{I} = 0$ if and only if $I = 0$.
    (b) $\sigma S_0 \geqslant 1$, S decreases and so let $t = t_0$ satisfy $\sigma S(t_0) = 1$. Then $\dot{I}$ is positive for $t < t_0$ and negative for $t > t_0$.
    Note $S = S_L$ when $I = 0$. To show that S_L is unique, prove that I is an increasing function of S in $(0, 1/\sigma)$ where $I(0) < 0$ and $I(1/\sigma) > 0$.
24. A first integral is $f(X, Y) = h(X) + g(Y)$ where $h(X) = \frac{1}{2}\alpha X^2 - \beta X$ and $g(Y) = bY - (a/k)\ln(A + kY)$. The functions $h(X)$ and $g(Y)$ both have global minima for positive values of X and Y respectively.
25. A first integral is $f(x_1, x_2) = x_2^2/2 + \omega_0^2(1 - \cos x_1)$. The linearized system at the fixed points $(2n\pi, 0)$, n integer, is a centre and at the fixed points $((2n+1)\pi, 0)$, n integer, is a saddle.

26. At the fixed point $(0, \nu^{-1})$, eigenvalues are $-\mu + 1/\nu$, $-\nu$; at the fixed point $(\pm\sqrt{(1-\mu\nu)}, \mu)$, the trace is $-\nu$ and the determinant is $2(1-\mu\nu)$.
27. $r\dot{r} = x_1\dot{x}_1 + x_2\dot{x}_2$; $r^2\dot{\theta} = x_1\dot{x}_2 - x_2\dot{x}_1$; $\dot{r} = \varepsilon r_0 \sin^2\theta(1 - r_0^2\cos^2\theta) + 0(\varepsilon^2)$; $\dot{\theta} = -1 + \varepsilon(r_0^2\cos^2\theta - 1) + 0(\varepsilon^2)$. $\Delta r = \varepsilon\pi r_0(1 - r_0^2/4)$.
   $\Delta r > 0$, $r_0 < 2$, $\Delta r < 0$, $r_0 > 2$, to first order in ε and so $\Delta r = 0$ for $r_0 = 2 + O(\varepsilon)$.
28. (a) Obtain the second order equations in x and y given by the system $\dot{x} = y - \varepsilon(\frac{1}{3}x^3 - x)$, $\dot{y} = -x$.
   (b) If $x_2 = \varepsilon\omega$, then $\dot{x}_1 = \varepsilon(\omega - \frac{1}{3}x_1^3 + x_1)$, $\dot{\omega} = -x_1/\varepsilon$. As $\varepsilon \to \infty$, $\dot{x}_1 \to \infty$ except near $\omega = \frac{1}{3}x_1^3 - x_1$.
29. Introduce a capacitor C in parallel, the circuit equations become $C(dv_C/dt) = j_L - v_C^3 + v_C$, $L(dj/dt) = -v_C$. The system equations are of Liénard type and hence the circuit oscillates.
30. $$T_1 = \int_q^Q \frac{dp}{(ap - bp^2)}, \quad T_2 = \int_Q^q \frac{dp}{(Ap - Bp^2)}.$$
   The population cycle has period $T_1 + T_2$.
31. Obtain the system equations
   $$\ddot{Y} + k\dot{Y} + Y = L \quad (\dot{Y} > 0),$$
   $$\ddot{Y} + k\dot{Y} + Y = -M \quad (\dot{Y} < 0).$$
   Use the substitution $Y' = Y - L$ in both equations to draw an analogy with the system considered in section 4.5.2.
32. $Y_1(t) = G_0 t e^{1-t}$, $0 \leqslant t < 1$, $Y_1(t) = G_0$, $t \geqslant 1$. The general solution of the differential equation for $t > 1$ is $Y(t) = (At + B)e^{-t} + G_0$, where A and B are constants. Regardless of the choice of A and B, $Y(t) \to G_0$ as $t \to \infty$.

CHAPTER 6

4. The line OF is given by $x^3 + ax + b = 0$ and $3x^2 + a = 0$.
7. (a) $\mu < 0$ no fixed point; $\mu = 0$ one fixed point; $\mu > 0$ two fixed points.
   (b) $\mu \neq 0$ two fixed points; $\mu = 0$ one fixed point (Note that $\mu < 0$ qualitatively distinct from $\mu > 0$ if Definition 1.2.1 is used). (c) same type of non-hyperbolic fixed point for all μ, therefore no bifurcation.
8. (a) $\mu < 0$ stable node; $\mu = 0$ x_1-axis line of fixed points; $\mu > 0$ saddle.
   (b) $\mu < 0$ stable focus; $\mu = 0$ centre; $\mu > 0$ unstable focus. (c) $\mu < 0$ clockwise stable focus; $\mu = 0$ stable star node; $\mu > 0$ anti-clockwise stable focus; therefore no bifurcation (see Fig. 2.11).
10. All four families exhibit a change in stability of a fixed point accompanied by the appearance/disappearance of a limit cycle. The fixed point at $\mu = 0$ is stable in $\{(a), (b)\}$ and unstable in $\{(c), (d)\}$.
11. (a) no fixed points; (b) one fixed point; (c) saddle and node fixed points.

Results take the same form as Fig. 6.13 but the fixed point in (b) is not a (generic) saddle node.

12. A pair of saddle points and a pair of nodes grow from the origin as μ increases from zero.
13. Note that $\mu = 0$ gives $\dot{x}_2 = -x_2^2$ in (6.21) and $\dot{x}_2 = -x_2^3$ for system in the question. Hence non-hyperbolic fixed point at $\mu = 0$ is different in the two systems.
14. The determinant of the linearization at the single fixed point is zero because isoclines are tangential.
15. (a) $\mu < 0$: unstable focus, stable limit cycle at $r = -\mu$;
    $\mu \geqslant 0$: stable focus.
    (b) $\mu < 0$: stable focus, semi-stable limit cycle at $r = -\mu$;
    $\mu = 0$: centre;
    $\mu > 0$: unstable focus,
    (c) $\mu \leqslant 0$: unstable focus;
    $\mu > 0$: unstable focus, stable limit cycle at $r = \mu/2$ and unstable limit cycle at $r = \mu$.
    (d) $\mu \leqslant 0$: stable focus;
    $\mu > 0$: unstable focus surrounded by a stable limit cycle at $r = \sqrt{\mu}$.
    (e) $\mu < 0$: stable focus (clockwise);
    $\mu = 0$: plane of fixed points;
    $\mu > 0$: unstable focus (anti-clockwise).
    (f) $|\mu| < 1$: unstable focus (anti-clockwise);
    $|\mu| = 1$: unstable star node;
    $|\mu| > 1$: unstable focus (clockwise).
17. (a) $I = -4$; (b) $I = -2$; (c) $I = -2$.
19. Let $\dot{x}_1 = x_2$, $\dot{x}_2 = -x_1 + \mu x_2 - x_2^3$. The system bifurcates from a stable fixed point to a stable limit cycle surrounding an unstable fixed point as μ passes through 0.
20. Use $\boldsymbol{x} = \begin{bmatrix} 1 & -2 \\ 1 & -1 \end{bmatrix} \boldsymbol{y}$ to obtain $\dot{y}_1 = \mu y_1 + y_2$, $\dot{y}_2 = -y_1 + \mu y_2 - y_2^3$.
21. Consider the system which has trajectories with reverse orientation. Show that this system undergoes a Hopf bifurcation to stable limit cycles at $\mu = 0$.
22. Saddle-node bifurcation.
23. Hopf bifurcation.
25. Observe (6.70) implies $P_0^{-1}(x) = \phi_{2\pi}^{-1}(x) = \phi_{-2\pi}(x)$, then use (6.72). P_μ^{-1} has the same fixed points as P_μ but of opposite stability because local quadratic curvature occurs in the opposite sense.
27. The leading non-linear term in f_0 is $-x^3$ instead of $-x^2$.
31. Calculate $P_0(P_0(x^* + \xi))$. Behaviour corresponding to both choices of sign in (6.84) can be achieved in (b) but not in (a).
32. Use Taylor's Theorem with $F'_\mu(x^*) = 2 - \mu$, $F''_\mu(x^*) = -2\mu$.
33. $F'_\mu(x^*) < -1$ for $\mu > 3$. F_μ is non-bijective.

34. Recall that the fixed point $x^* = 1 - 1/\mu$ is a solution of $F_\mu^2(x) = x$. If $x_\pm^*$ are period-2 points of F_μ, then F_μ^2 period-doubles when $DF_\mu^2(x_+^*) = DF_\mu^2(x_-^*) = DF_\mu(x_+^*)DF_\mu(x_-^*) = -1$. Note that $x_+^* + x_-^*$ and $x_+^* x_-^*$ are given directly by coefficients in the quadratic equation in the question.
35. Let $x_2^* = F_{\mu*}(x_1^*)$, $x_3^* = F_\mu^2(x_1^*)$ and verify that $F_\mu^3(x_i^*) = x_i^*$, $i = 2, 3$. Use chain rule to calculate G' and show that $G'(x_j^*) = \prod_{i=1}^3 F_\mu'(x_i^*)$, $j = 1, 2, 3$. Tangency of $y = F_{\mu*}^3$ and $y = x$ occurs simultaneously at points $x = x_1^*$, x_2^*, x_3^*.
36. Let $\xi = h(x) = \alpha x + \beta$ and consider $F_\mu(x) = h^{-1}(f_a(h(x)))$. Show that $\alpha = \mu/a$, $\beta = -\mu/2a$ with $a = \mu(\mu - 2)/4$.
37. Results illustrate period-doubling approach to chaotic band attractor shown in Fig. 6.34(b).
38. Recognize series expansions of sine and cosine in (2.63) to obtain (6.101).
39. (a) $\boldsymbol{J} = \begin{bmatrix} \lambda & 0 \\ 0 & 1/\lambda \end{bmatrix}$; (b) $\boldsymbol{J} = \begin{bmatrix} \cos\beta & -\sin\beta \\ \sin\beta & \cos\beta \end{bmatrix}$.

    If $\boldsymbol{M} = \begin{bmatrix} a & b \\ c & d \end{bmatrix}$ and $\Delta = ad - bc$ then:

    (a) $A = -cd\Delta^{-2}$, $2B = (ad + bc)\Delta^{-2}$, $C = -ab\Delta^{-2}$;

    (b) $A = (c^2 + d^2)\Delta^{-2}$, $2B = -2(ac + bd)\Delta^{-2}$, $C = (a^2 + b^2)\Delta^{-2}$.

    (a) $AC - B^2 < 0$; (b) $AC - B^2 > 0$.
41. $\mathbf{P}(x, y)$ ($\mathbf{Q}(x, y)$) is given by the first two components of $\psi_\tau(x, y, \theta_0)$ ($\psi_\tau(x, y, \theta_1)$), where $\tau = 2\pi/\omega$. Let $T = (\theta_1 - \theta_0)/\omega$ and interpret the first two components of

    $$\psi_T(\psi_\tau(x, y, \theta_0)) = \psi_\tau(\psi_T(x, y, \theta_0)).$$

    Verify that the vector field (and hence flow) of (6.108) is invariant under $(x, y, \theta) \mapsto (-x, -y, \theta + \pi)$.
42. Consider discriminant of fixed point equation. Fixed points satisfy $2a\mathbf{x}_\pm^* = (1 - b)[-1 \pm 2]$. Eigenvalues: -1, b at $\mathbf{x}_+^*$; solutions of $-3(1 - b) = b\lambda^{-1} - \lambda = f(\lambda)$ at $\mathbf{x}_-^*$. For latter, graph of $f(\lambda)$ gives one solution < 1 and other > 1.
44. Two fixed points for $v < 0$. Topological types depend on α ($|\alpha| \neq 1$) but one is always a contraction/expansion while the other is of saddle type.
45. y^* asymptotically stable implies 2-cycle of $\mathbf{f}$ is asymptotically stable. y^* unstable implies 2-cycle of $\mathbf{f}$ is of saddle type (i.e. unstable). Numerical convergence found in Fig. 6.37 is consistent with behaviour of local family (6.110) with positive sign and $|\alpha| < 1$.
46. $\begin{bmatrix} y \\ Y \end{bmatrix} = \begin{bmatrix} 1 & 0 \\ \cos(2\pi Q) & Q^{-1}\sin(2\pi Q) \end{bmatrix} \begin{bmatrix} y \\ p \end{bmatrix}$. Diagram: transform to y, p-plane; apply $\mathbf{P}$; transform back.
47. Use $p_i = -Q\sin(2\pi Q)y_{i-1} + \cos(2\pi Q)p_{i-1} + F(y_i)$ and express p_i and p_{i-1} in terms of y's using Exercise 6.46. For numerical results see Helleman (1980), pp. 175–7.

48. Liouville Theorem. Linear approximation of system at origin is $\dot{x} = y$, $\dot{y} = -kx$; obtain ϕ_1 as in Exercise 6.38. ϕ_1 is conjugate to a rotation through angle $\sqrt{k}$. Rotation number (i.e. rotation per iteration in units of 2π) is $\sqrt{k}/2\pi$. Rotation number tends to zero as invariant circle approaches Γ. Inside Γ, ϕ_1 behaves like a twist map with rotation interval $[0, \sqrt{k}/2\pi]$.
49. Region enclosed by the saddle connection. Rotation interval $[0, \sqrt{k}/2\pi]$.
50. Verify that determinant of derivative map is unity. To obtain line manipulate the fixed point equations $x_{i+1} = x_i = x^*$, $y_{i+1} = y_i = x^*$. To get fixed point use $y = tx$, $t = \tan\frac{1}{2}\alpha$, and the t-formulae in one fixed point equation. For saddle, show trace of linearization is greater than two and note that this means that the eigenvalues are real and not equal to unity.
52. Show that: (i) the trajectories through $\mathbf{x}_1$ and $\mathbf{x}_2$ are subsets of W^u_ϕ and W^s_ϕ, respectively; (ii) the orbits of $\mathbf{x}_1$, $\mathbf{x}_2$ and $\mathbf{y}$ coincide. For diffeomorphism $\mathbf{P}$ use analogous arguments with 'discrete' time parameter. The continuous nature of the trajectories of flows together with (ii) above means that branches of W^u_ϕ and W^s_ϕ coincide as in Fig. 6.42(a). The discrete nature of the orbits of diffeomorphisms means that (ii) only restricts a discrete subset of points in $W^u_\mathbf{P}$ and $W^s_\mathbf{P}$. In general, this allows $W^u_\mathbf{P}$ and $W^s_\mathbf{P}$ to intersect as in Fig. 6.42(b). Note that not all diffeomorphisms exhibit such (typical) behaviour, e.g. the time-one map, ϕ_1, of the flow in Exercise 6.48.
53. (a) Show that $\det(D\boldsymbol{P}(x, y)) \equiv 1$; (b) consider $x_1 = x$, $y_1 = y$.
54. Eigenvalues are $\lambda_\pm = \frac{1}{2}(2 + k \pm \sqrt{4k + k^2})$ with eigenvectors $\mathbf{u}_\pm = (1, \frac{1}{2}[-k \pm \sqrt{4k + k^2}])$.
55. Replace k by $2k$ in hint 54.
56. Show that $\det(D\boldsymbol{P}(x, y)) \equiv 1 - \delta$ and note that $A(\mathbf{P}(D)) = \int_{\mathbf{P}(D)} dxdy = \int_D |\det(D\boldsymbol{P}(x, y))| dxdy$.
57. $\Theta = \theta + \omega\tau$ which is independent of (x, y). Observe

$$\psi_\tau(x, y, 0) = \begin{bmatrix} \boldsymbol{P}_\gamma(x, y) \\ \omega\tau \end{bmatrix} \Rightarrow D\psi_\tau(x, y, 0) = \begin{bmatrix} D\boldsymbol{P}_\gamma(x, y) & \cdot \\ \mathbf{0} & 1 \end{bmatrix}.$$

58. The flow property (1.56) gives $\phi_{t+h}(\mathbf{x}) = \phi_t(\phi_h(\mathbf{x}))$. The chain rule implies $D\phi_{t+h}(\boldsymbol{x}) = D\phi_t(\phi_h(\boldsymbol{x}))D\phi_h(x)$. Use $\det(\boldsymbol{AB}) = \det(\boldsymbol{A})\det(\boldsymbol{B})$.
60. Show that $F_4(h(x)) = \sin^2 \pi x = h(G(x))$.
61. $\sigma = \{11111\ldots\}$ gives $s = \{1010101\ldots\}$ and so $x = 1/2 - 1/8 + 1/32 - \cdots = 2/3$. $x_1 = 2/5$, $x_2 = 4/5$.
62. Write $x = \sum_{i=1}^\infty a_i/2^i$. Show that $f(x) = \sum_{i=1}^\infty a_{i+1}/2^i$. Main features of dynamics are: the set of periodic points is dense in $[0, 1]$; orbits exist that are dense in $[0, 1]$; there is sensitive dependence on initial conditions.
63. (a) Codes for period-3 points in distinct orbits are $\{001001001\ldots\}$ and $\{110110110\ldots\}$. There are two distinct 3-cycles. Of eight possible symbol

blocks of length three, two give fixed points and the remainder form two period-3 orbits generated by 001 and 110. (b) x' can be given by any periodic code with the first N-symbols the same as σ. The implication is that the set of periodic points is dense in the interval.

64. Computer truncates binary representation of x and repeated multiplication by two flushes this out after a finite number of iterations. To obtain realistic representation use symbolic dynamics. Store initial symbol block from δ, apply left shift and plot x corresponding to the initial point of the interval defined by a symbol block chosen to match resolution of your graphics.
65. $(2^3)^2$ boxes. Each box has length 2/125.
66. Curve (c) is $-1 \times$ a uniform magnification of the part of the graph of F_μ^2 lying between $1 - x^*$ and x^*. Suggests topological conjugacy.
67. $k(a) = (a-1)^{-1}$, $\hat{a}(a) = 2a^2(a-1)$. If $\hat{a}^i(a) \to a^*$ as $i \to \infty$ then period-doubling sequence accumulates at a^*.
68. Given the positive integer n, choose (a) squares (b) cubes of side $1/n$ and apply (6.127) with $\varepsilon = 1/n$.
69. Given x is represented by $\{a_n\}$, $f_2(x)$ is represented by $\{b_n\}$, where $b_1 = 2$ and $b_n = a_{n+1}$ for $n > 1$. Initially intensity of image near zero will be greater than that near one. However, intensity would even out after a sufficiently large number of iterations because of finite resolution of the graphics.
70. Observe that when the open triangles are removed, the edges of the original triangle T remain intact. Hence $\mathscr{S}$ contains line segments and is therefore at least one dimensional. Show that $N(\mathscr{S}, \varepsilon) = 3^n$ for $\varepsilon = 2^{-n}\sqrt{2}$, $n = 0, 1, 2, \ldots$. $\dim(T \backslash \mathscr{S}) = \dim(T) - \dim(\mathscr{S})$.
71. Let S be the Sierpinski carpet. Show that $N(S, \varepsilon) = (3^2 - 1)^n$ for $\varepsilon = 3^{-n}\sqrt{2}$ by considering covering squares at each stage of the construction of S.
73. (a) quiescent state not invariant;
    (b) $\alpha_1\alpha_2\alpha_3 0 \alpha_5 000$ acts on 100, 010, 001, 000 to give 0;
    (c) $\alpha_1\alpha_2\alpha_3 0 \alpha_5 100$ acts on 100, 001 to give 0 and 010 to give 1.
    For maximal growth from legal rule need $\alpha_1 = \alpha_2 = \alpha_4 = \alpha_5 = \alpha_6 = \alpha_7 = 1$, $\alpha_8 = 0$ and $\alpha_3 = 0, 1$.
74. Observe 111, 110, 011, are not involved in the development of a single site. Eight.
76. 4, (1, − 1).

Index

Action-angle variables 149
Affine system 55, 175
Algebraic type 52
Animal conflict model 218
Area-preserving map
  Hénon **263**
  Poincaré 155, 259
  twist 261
Asymptotic stability 84, 89, 229
Attracting set 103, 140, 141
Attractor 9, 141
Aubry–Mather theorem 261
Autonomous equations 6, 12

Basin of attraction 91
Beats 175
Bifurcation 212
  cusp 237
  flip 242, 245, 258
  fold 240, 242
  Hopf 225, 228
  period-doubling 246, 259
  pitchfork 293
  saddle-node
    flows 224, 226
    maps 258
    symmetric 291
Bifurcation point 224
Braided structure 258

Canonical local family
  flip 245, 293
  fold 242, 293
Canonical system 42
Cantor set **276**, 280
  fractal dimension 281
  horseshoe map 277, 278
  iterated function scheme 281
  tent map 276
Cellular automata 284
Centre 46
Chaotic (strange) attractor
  Duffing oscillator **271**
  folded band (Rössler) 251, 257
  Hénon 295
  iterated function scheme 282
Chaotic orbit
  area preserving map 263
  homoclinic tangles **264**
  horseshoe map **278**
  logistic map 250
Chaotic region 266
Characteristic
  cubic 189
  folded 194
  neon 194
  resistor 189
  surface 216
  triode 201
Chemical oscillator 231
Closed orbit 46, 102, 129
  hyperbolic 132
Collage 284
Competing species 180
Competitive exclusion 182
Complementary function 56

Conservative system 97
Coupled pendula 172
Critical damping 165
Cusp bifurcation 237

Decoupled system 17, 58
Derivative
  along a curve 88
  directional 96
Diffeomorphism 132
Differentiable manifold 128
Dissipative map 268
Domain of stability 91
Doubling map 300
Duffing oscillator
  twin well 254
Dynamical equations 162, 223
  capacitor 169
  inductor 168
  mutual-inductance 201
  Newton's law of cooling 11
  Newton's second law of motion 163

Economic model 170
Eigenspace
  centre 156
  stable/unstable 121
Electrical circuit theory 167
Elliptic fixed point 253, 260, 261, 266
Evolution operator 23, 52
Exact differential equation 27, 28
Exponential matrix 52

Family of differential equations 212, 223
Feigenbaum number 246, 250, 280
First integral 96, 147
Fixed point
  phase portrait/flow 9, 14
    hyperbolic 80, 120, 122
    isolated 11, 14, 16
    non-hyperbolic 125
    neutrally stable 86
    non-simple 46, 81
    simple 43, 77
    stable 84, 89
    stable/unstable manifold 123, 298
    unstable 87, 92
  Poincaré map/diffeomorphism 104, 131
  attracting/repelling 104
  elliptic 253, 260, 261, 266
  hyperbolic 132
  stable/unstable manifold 132
Flip bifurcation 242, 245, 258
Flow 23
Flow box theorem 94
Focus
  linear 46
  non-linear 80
Fold bifurcation 240, 242
Folded band (Rössler) attractor 251, 257
Forcing terms 176
Fractal basin boundary 270
Fractal dimension 280
  Cantor set 281
  Sierpinski carpet 301
  Sierpinski gasket 283, 301

Generic property 226
Global phase portrait 71, 237
Green's theorem 109

Hamiltonian 98, 148
  flow 154
  system 148
    generalized coordinates/momenta 148
    integrable 149
    normal form 149
Hamilton's equations 98, 148
Harmonic oscillator
  free 165
  forced 176
  overdamped 166
  second order form 163, 170, 171
  underdamped 165
Heartbeat model 212
Hénon
  attractor 295
  map
    area-preserving 263, 297, 298
    quadratic 258, 294, 296

Heteroclinic tangle 265
Holling–Tanner model 185
Homoclinic point 264
Homoclinic tangle
 periodic orbit 265
 saddle point 264
Homogeneous differential equation 28
Hopf bifurcation 225, 228, 239
 theorem 229

Improper node
 linear 45
 non-linear 80
Integrating factor 27
Invariant circle 142, 260, 261
Invariant set 107
Invariant torus 152
Island chain 261
Isocline 5, 20
Iterated function scheme 281
Iteration 104

Jordan form
 2×2 38
 3×3 57
 4×4 61
 $n \times n$ 62
Jump assumption 193

Kirchhoff Laws 168
Kolmogorov–Arnold–Moser (KAM) theorem 261
 KAM-circle 261

Left shift 272
Level curve 88, 96
Level surface 220
Liapunov
 function 88
 strong 90
 weak 90
 stability
  theorem 89
Liénard equation 191
Liénard plane 191
Life game 287
Limit cycle 103
 criterion for non-existence 109
 in modelling 185
 semi-stable 104
 stable 104
 unstable 104
Limit point 101
Limit set 101
Linear change of variable 35
Linear diffeomorphism
 classification 134
 hyperbolic 133
 orientation-preserving 134
 orientation-reversing 134
Linearization
 Poincaré map/diffeomorphism 132
 theorem
  differential equations 77, 123
  Poincaré maps/diffeomorphism 133, 135
 vector field 74
Linearized system 74
Linear mapping 35, 48
Linear part 74
Linear system 11, 35
 algebraic type 52
 classification of 52
 coefficient matrix of 35
 homogeneous 55
 non-homogeneous 55
 non-simple 46
 qualitative (topological) type 52
 simple 43
Liouville's theorem 154
Local bifurcation 225
Local coordinates 75
Local family 242
Local phase portrait 71
Logistic law 12
Logistic map 245
 chaotic orbits 250
 eventually periodic orbits 294
 period-doubling cascade 246, 249

Maximal solution 1
Momentum 163
  generalized 148

Natural frequency 165
Negative dcfinite 88
Negative semi-definite 88
Neighbourhood 71
Node
  linear 43
  non-linear 80
Non-autonomous
    equation 32, 34
Normal modes 174
Normal coordinates 174

Ohm's law 168
Orbit
  in phase portrait 13
  of Poincaré map/
    diffeomorphism 131
  in terms of flow 26
Ordinary point 93

Parametrically forced
    pendulum 251, 254
Partially decoupled
    system 18
Particular integral 56
Partitioned matrices 40,
    57, 61, 62
  application of 173
Period-doubling bifurcation
  one-dimension 246
  two-dimensions 257, 259,
    296
Periodic orbit 136
  Poincaré–Birkhoff 261
  stable/unstable manifold
    137
Periodic point
  phase portrait/flow 46, 127
  Poincaré map/
    diffeomorphism 136
Phase line 11
Phase plane 13
Phase point 11, 15, 27
Phase portrait
  construction of 17
  one-dimension 8
  qualitative type of 52
  restriction of 71
  two-dimensions 13
Piecewise modelling 195
Pitchfork bifurcation 293
Plant-pollinator model 226
Poincaré (first return)
    map 104, 129
  area-preserving 155, 259
  for Hamiltonian
    flow 153, 155
Poincaré–Bendixson
    theorem 106
Poincaré–Birkhoff
    theorem 261
Polar coordinates 17
Positive definite 88
Positive semi-definite 88
Positively invariant set 107
Prey–predator
    problem 183, 185
Principal directions 48, 81

q-cycle 136
Qualitative behaviour 4, 8,
    14, 16, 120, 125, 128, 132
Qualitative equivalence 4,
    9, 10, 16, 51
  of families of differential
    equations 239
Quasi-periodic motion 128, 138

Rationally independent
    numbers 127
Rayleigh equation 209, 292
Regularization 193
Relaxation oscillations 190
Renormalization 280, 300,
    301
Repellor 9
Resonance 177
Resonant frequency 179
Robust system 185
Rössler attractor 251, 257
Rotation interval 261, 297
Rotation number 261, 297

Saddle connection 95, 99, 105, 107
  bifurcation 239
  heteroclinic 265
  homoclinic 264
Saddle-node bifurcation
  flows 224, 226
  maps 258
  symmetric 291
Saddle point
  linear 44, 121
  non-linear 80, 123
  spiral 122, 156
Sawtooth oscillations 197
Sensitive dependence on initial conditions 274
Separable differential equation 27
Separation of variables 6, 27
Separatrix 44, 80
Shunt 9
Sierpinski carpet 301
Sierpinski gasket 283, 285
Similar matrices 36
Similarity classes 37
Similarity types 38
Simple pendulum 208
Simply connected region 109
Smale horseshoe map 276
Solution curve 3
Solution of a differential equation 1, 12
  existence and uniqueness 2, 20, 21
  fixed point 8, 14
  periodic 46, 127
  steady state 177
  transient 177
Spiral attracting/repelling 46
Spiral saddle point 122
Stability
  asymptotic 84
  neutral 86
  in sense of Liapunov 84
  structural 185, 207, 224
Star node
  linear 44
  non-linear 80
State of dynamical system 11, 162
Strange (chaotic) attractor
  Duffing oscillator 271
  Hénon 295
  iterated function scheme 282
  Rössler (folded band) 251, 257
Structural stability 185, 207, 224
Suspension
  of a diffeomorphism 133
Symbol sequence 271, 278, 300
Symbolic dynamics 271
  doubling map 300
  horseshoe map 278
  tent map 272
Symmetry 5, 28, 32
Symplectic (canonical) transformation 151

Taylor expansion 76
Tent map 272
Topological conjugacy 133
Trajectory
  phase portrait/flow 13, 26
  Poincaré map/diffeomorphism 131
Transients 177
Trapping region
  flow 106
  Poincaré map 141
Tumour growth model 232
Twist map 261

Unimodal map 250

Van der Pol equation 104, 191
Vector field 20
  linear part of 74
Volterra–Lotka equations 183
  structural instability 207